Lecture Notes in Mathematics 1880

S. Attal · A. Joye · C.-A. Pillet (Eds.)

Open Quantum Systems I

The Hamiltonian Approach

Springer

Editors

Stéphane Attal
Institut Camille Jordan
Université Claude Bernard Lyon 1
21 av. Claude Bernard
69622 Villeurbanne Cedex
France
e-mail: attal@math.univ-lyon1.fr

Alain Joye
Institut Fourier
Université de Grenoble 1
BP 74
38402 Saint-Martin d'Hères Cedex
France
e-mail: alain.joye@ujf-grenoble.fr

Claude-Alain Pillet
CPT-CNRS, UMR 6207
Université du Sud Toulon-Var
BP 20132
83957 La Garde Cedex
France
e-mail: pillet@univ-tln.fr

Library of Congress Control Number: 2006923432

Mathematics Subject Classification (2000): 37A60, 37A30, 47A05, 47D06, 47L30, 47L90, 60H10, 60J25, 81Q10, 81S25, 82C10, 82C70

ISSN print edition: 0075-8434
ISSN electronic edition: 1617-9692
ISBN 3-540-30991-8 Springer Berlin Heidelberg New York
ISBN 978-3-540-30991-8 Springer Berlin Heidelberg New York

DOI 10.1007/b128449

Springer is a part of Springer Science+Business Media
springer.com

Typesetting: by the authors and SPI Publisher Services using a Springer LaTeX package
Cover design: *design & production* GmbH, Heidelberg

Printed on acid-free paper SPIN: 11602606 VA41/3100/SPI 5 4 3 2 1 0

Preface

This is the first in a series of three volumes dedicated to the lecture notes of the Summer School "Open Quantum Systems" which took place at the Institut Fourier in Grenoble from June 16th to July 4th 2003. The contributions presented in these volumes are revised and expanded versions of the notes provided to the students during the School.

Closed vs. Open Systems

By definition, the time evolution of a *closed* physical system $\mathcal{S}$ is deterministic. It is usually described by a differential equation $\dot{x}_t = X(x_t)$ on the manifold M of possible configurations of the system. If the initial configuration $x_0 \in M$ is known then the solution of the corresponding initial value problem yields the configuration x_t at any future time t. This applies to classical as well as to quantum systems. In the classical case M is the phase space of the system and x_t describes the positions and velocities of the various components (or degrees of freedom) of $\mathcal{S}$ at time t. In the quantum case, according to the orthodox interpretation of quantum mechanics, M is a Hilbert space and x_t a unit vector – the wave function – describing the quantum state of the system at time t. In both cases the knowledge of the state x_t allows to predict the result of any measurement made on $\mathcal{S}$ at time t. Of course, what we mean by the result of a measurement depends on whether the system is classical or quantum, but we should not be concerned with this distinction here. The only relevant point is that x_t carries the maximal amount of information on the system $\mathcal{S}$ at time t which is compatible with the laws of physics.

In principle any physical system $\mathcal{S}$ that is not closed can be considered as part of a larger but closed system. It suffices to consider with $\mathcal{S}$ the set $\mathcal{R}$ of all systems which interact, in a way or another, with $\mathcal{S}$. The joint system $\mathcal{S} \vee \mathcal{R}$ is closed and from the knowledge of its state x_t at time t we can retrieve all the information on its subsystem $\mathcal{S}$. In this case we say that the system $\mathcal{S}$ is *open* and that $\mathcal{R}$ is its *environment.* There are however some practical problems with this simple picture. Since the joint system $\mathcal{S} \vee \mathcal{R}$ can be really big (*e.g.,* the entire universe) it may be difficult, if not impossible, to write down its evolution equation. There is no solution to this problem. The pragmatic way to bypass it is to neglect parts of the environment $\mathcal{R}$ which, a priori, are supposed to be of negligible effect on the evolution of the subsystem $\mathcal{S}$. For example, when dealing with the motion of a charged particle it is often reasonable to neglect all but the electromagnetic interactions and suppose that the environment consists merely in the electromagnetic field. Moreover, if the particle moves in a very sparse environment like intergalactic space then we can consider that it is the only source in the Maxwell equations which governs the evolution of $\mathcal{R}$. Assuming that we can write down and solve the evolution equation of the joint system $\mathcal{S} \vee \mathcal{R}$ we nevertheless hit a second problem: how to choose the initial configuration of the environment ? If $\mathcal{R}$ has a very large (*e.g.*, infinite) number of degrees of freedom then it is *practically* impossible to determine its configuration at some initial time $t = 0$. Moreover, the dynamics of the joint system is very likely to be chaotic, *i.e.*, to display some sort of instability or sensitive dependence on the initial condition. The slightest error in the initial configuration will be rapidly amplified and ruin our hope to predict the state of the system at some later time. Thus, instead of specifying a single initial configuration of $\mathcal{R}$ we should provide a statistical ensemble of typical configurations. Accordingly, the best we can hope for is a statistical information on the state of our open system $\mathcal{S}$ at some later time t. The resulting theory of open systems is intrinsically probabilistic. It can be considered as a part of statistical mechanics at the interface with the ergodic theory of stochastic processes and dynamical systems.

The paradigm of this statistical approach to open systems is the theory of Brownian motion initiated by Einstein in one of his celebrated 1905 papers [3] (see also [4] for further developments). An account on this theory can be found in almost any textbook on statistical mechanics (see for example [9]). Brownian motion had a deep impact not only on physics but also on mathematics, leading to the development of the theory of stochastic processes (see for example [12]).

Open systems appeared quite early in the development of quantum mechanics. Indeed, to explain the finite lifetime of the excited states of an atom and to compute the width of the corresponding spectral lines it is necessary to take into account the interaction of the electrons with the electromagnetic field. Einstein's seminal paper [5] on atomic radiation theory can be considered as the first attempt to use a Markov process – or more precisely a master equation – to describe the dynamics of a quantum open system. The theory of master equations and its application to radiation theory and quantum statistical mechanics was subsequently developed by Pauli [8], Wigner and Weisskopf [13], and van Hove [11]. The mathematical theory of the quantum Markov semigroups associated with these master equations started to develop more than 30 years later, after the works of Davies [2] and Lindblad [7]. It further led to the development of quantum stochastic processes.

To illustrate the philosophy of the modern approach to open systems let us consider a simple, classical, microscopic model of Brownian motion. Even though this model is not realistic from a physical point of view it has the advantage of being exactly solvable. In fact such models are often used in the physics literature (see [10, 6, 1]).

Brownian Motion: A Simple Microscopic Model

In a cubic crystal denote by q_x the deviation of an atom from its equilibrium position $x \in \Lambda_N = \{-N, \ldots, N\}^3 \subset \mathbb{Z}^3$ and by p_x the corresponding momentum. Suppose that the inter-atomic forces are harmonic and only acts between nearest neighbors of the crystal lattice. In appropriate units the Hamiltonian of the crystal is then

$$\sum_{x \in \Lambda_N} \frac{p_x^2}{2} + \sum_{x,y \in \mathbb{Z}^3} \frac{\kappa_{xy}}{4}(q_x - q_y)^2,$$

where

$$\kappa_{xy} = \begin{cases} 1 \text{ if } |x-y| = 1; \\ 0 \text{ otherwise;} \end{cases}$$

and Dirichlet boundary conditions are imposed by setting $q_x = 0$ for $x \in \mathbb{Z}^3 \setminus \Lambda_N$. If the atom at site $x = 0$ is replaced by a heavy impurity of mass $M \gg 1$ then the Hamiltonian becomes

$$H \equiv \sum_{x \in \Lambda_N} \frac{p_x^2}{2m_x} + \sum_{x,y \in \mathbb{Z}^3} \frac{\kappa_{xy}}{4}(q_x - q_y)^2,$$

where

$$m_x = \begin{cases} M \text{ if } x = 0; \\ 1 \text{ otherwise.} \end{cases}$$

We shall consider the heavy impurity at $x = 0$ as an open system $\mathcal{S}$ whose environment $\mathcal{R}$ is made of the $(2N+1)^3 - 1$ remaining atoms of the crystal. To write down the equation of motion in a convenient form let us introduce some notation. We set $\Lambda_N^* = \Lambda_N \setminus \{0\}$, $q = (q_x)_{x \in \Lambda_N^*}$, $p = (p_x)_{x \in \Lambda_N^*}$, $Q = q_0$, $P = p_0$. For $x \in \mathbb{Z}^3$ we denote by δ_x the Kronecker delta function at x and by $|x|$ the Euclidean norm of x. We also set $\chi = \sum_{|x|=1} \delta_x$. The motion of the joint system $\mathcal{S} \vee \mathcal{R}$ is governed by the following linear system

$$\begin{aligned} \dot{q} &= p, & \dot{p} &= -\Omega_0^2 q + Q\chi, \\ M\dot{Q} &= P, & \dot{P} &= -\omega_0^2 Q + (\chi, q), \end{aligned} \tag{1}$$

where $-\Omega_0^2$ is the discrete Dirichlet Laplacian on Λ_N^* and $\omega_0^2 = 6$. According to the open system philosophy described in the previous paragraph we should supply some appropriate statistical ensemble of initial states of the environment. To motivate the choice of this ensemble suppose that in the remote past the impurity was pinned at some fixed position, say $Q = P = 0$, and that at time $t = 0$ the resulting system has reached thermal equilibrium at some temperature $T > 0$. The positions and momenta in the crystal will be distributed according to the Gibbs-Boltzmann canonical ensemble corresponding to the pinned Hamiltonian $H_0 = H|_{Q=P=0}$,

$$H_0 = \frac{1}{2}\left((p,p) + (q, \Omega_0^2 q)\right).$$

This ensemble is given by the Gaussian measure

$$d\mu = Z^{-1} \mathrm{e}^{-\beta \mathrm{H}_0(q,p)} \mathrm{d}q\mathrm{d}p,$$

where Z is a normalization factor and $\beta = 1/k_B T$ with k_B the Boltzmann constant.

At time $t = 0$ we release the impurity. The subsequent evolution of the system is determined by the Cauchy problem for Equ. (1). The evolution of the environment can be expressed by means of the Duhamel formula

$$q(t) = \cos(\Omega_0 t)q(0) + \frac{\sin(\Omega_0 t)}{\Omega_0} p(0) + \int_0^t \frac{\sin(\Omega_0(t-s))}{\Omega_0} \chi Q(s)\, ds.$$

Inserting this relation into the equation of motion for Q leads to

$$M\ddot{Q} = -\omega_0^2 Q + \int_0^t K(t-s)Q(s)\, ds + \xi(t), \tag{2}$$

where the integral kernel K is given by

$$K(t) = (\chi, \frac{\sin(\Omega_0 t)}{\Omega_0} \chi), \tag{3}$$

and

$$\xi(t) = \left(\chi, \cos(\Omega_0 t)q(0) + \frac{\sin(\Omega_0 t)}{\Omega_0} p(0) \right).$$

Since $q(0), p(0)$ are jointly Gaussian random variables, $\xi(t)$ is a Gaussian stochastic process. It is a simple exercise to compute its mean and covariance

$$\mathbb{E}(\xi(t)) = 0, \quad \mathbb{E}(\xi(t)\xi(s)) = C(t-s) = \frac{1}{\beta}(\chi, \frac{\cos(\Omega_0(t-s))}{\Omega_0^2} \chi). \tag{4}$$

We note in particular that this process is stationary. The term $\xi(t)$ in Equ. (2) is the noise generated by the fluctuations of the environment. It vanishes if the environment is initially at rest. The integral in Equ. (2) is the force exerted by the environment on the impurity in reaction to its motion. Note that this dissipative term is independent of the state of the environment. The dissipative and the fluctuating forces are related by the so called *fluctuation-dissipation theorem*

$$K(t) = -\beta \partial_t C(t). \tag{5}$$

The solution $z^t = (Q(t), P(t))$ of the random integro-differential equation (2) defines a family of stochastic processes indexed by the initial condition z^0. These processes provide a statistical description of the motion of our open system. An invariant measure ρ for the process z^t is a measure on $\mathbb{R}^3 \times \mathbb{R}^3$ such that

$$\int f(z^t)\, d\rho(z^0) = \int f(z)\, d\rho(z),$$

holds for all reasonable functions f and all $t \in \mathbb{R}$. Such a measure describes a steady state of the system. If one can show that for any initial distribution ρ_0 which is absolutely continuous with respect to Lebesgue measure one has

$$\lim_{t\to\infty} \int f(z^t)\, d\rho_0(z^0) = \int f(z)\, d\rho(z), \tag{6}$$

then the steady state ρ provides a good statistical description of the dynamics on large time scales. One of the main problem in the theory of open systems is to show that such a natural steady state exists and to study its properties.

The Hamiltonian Approach

Remark that in our example, such a steady state fails to exist since the motion of the joint system is clearly quasi-periodic. However, in a real situation the number of atoms in the crystal is very large, of the order of Avogadro's number $N_A \simeq 6 \cdot 10^{23}$. In this case the recurrence time of the system becomes so large that it makes sense to take the limit $N \to \infty$. In this limit $-\Omega_0^2$ becomes the discrete Dirichlet Laplacian on the infinite lattice $\mathbb{Z}^3 \setminus \{0\}$. This is a well defined, bounded, negative operator on the Hilbert space $\ell^2(\mathbb{Z}^3)$. Thus, Equ. (2),(3), (4) and (5) still make sense in this limit. In the sequel we only consider the resulting infinite system.

We distinguish two main approaches to the study of open systems. The first one, the Hamiltonian approach, deals directly with the dynamics of the joint system $\mathcal{S} \vee \mathcal{R}$. We briefly discuss the second one, the Markovian approach, in the next paragraph.

In the Hamiltonian approach we rewrite the equation of motion (1) as

$$\dot{Z} = -i\tilde{\Omega}Z,$$

where $\tilde{\Omega}^2 = m^{-1/2}\Omega^2 m^{-1/2}$ with $m = I + (M-1)\delta_0(\delta_0, \cdot\,)$ the operator of multiplication by m_x and $-\Omega^2$ is the discrete Laplacian on $\mathbb{Z}^3$. The complex variable Z is given by $Z = \tilde{\Omega}^{1/2}m^{1/2}\tilde{q} + i\tilde{\Omega}^{-1/2}m^{-1/2}\tilde{p}$ and $\tilde{q} = (q_x)_{x\in\mathbb{Z}^3}$, $\tilde{p} = (p_x)_{x\in\mathbb{Z}^3}$. It follows from elementary spectral analysis that for $M > 1$ the operator $\tilde{\Omega}$ is self-adjoint with purely absolutely continuous spectrum $\sigma(\tilde{\Omega}) = \sigma_{\mathrm{ac}}(\tilde{\Omega}) = [0, \sqrt{2}\omega_0]$ on $\ell^2(\mathbb{Z}^3)$. A simple argument involving the scattering theory for the pair $\Omega_0^2 \oplus \omega_0^2/M, \tilde{\Omega}^2$ shows that the system $\mathcal{S}$ has a unique steady state ρ such that (6) holds for all ρ_0 which are absolutely continuous with respect to Lebesgue measure. Moreover, ρ is the marginal on $\mathcal{S}$ of the infinite dimensional Gaussian measure $Z^{-1}e^{-\beta H}dpdqdPdQ$ which describes the thermal equilibrium state of the joint system at temperature $T = 1/k_B\beta$. This is easily computed to be the Gaussian measure

$$\rho(dP, dQ) = \mathcal{N}^{-1}e^{-\beta(P^2/2M + \omega^2 Q^2/2)}dPdQ,$$

where $\mathcal{N}$ is a normalization factor and

$$\omega^2 = \frac{1}{(\delta_0, \Omega^{-2}\delta_0)}.$$

The Markovian Approach

A remarkable feature of Equ. (2) is the memory effect induced by the kernel K. As a result the process z^t is non-Markovian, *i.e.*, for $s > 0$, z^{t+s} does not only depend on z^t and $\{\xi(u) \mid u \in [t, t+s]\}$ but also on the full history $\{z^u \mid u \in [0, t]\}$. The only way to avoid this effect is to have K proportional to the derivative of a delta function. By Relation (5) this means that ξ should be a white noise. This is certainly not the case with our choice of initial conditions. However, as we shall see, it is possible to obtain a Markov process in some particular scaling limits. This is not a uniquely defined procedure: different scaling limits correspond to different physical regimes and lead to distinct Markov processes.

As a simple illustration let us consider the particular scaling limit

$$Q_M(t) \equiv M^{1/4} Q(M^{1/2} t), \quad M \to \infty.$$

of our model. For finite M the equation of motion for Q_M reads

$$\ddot{Q}_M(t) = -\omega_0^2 Q_M(t) + \int_0^t K_M(t-s) Q_M(s)\, ds + \xi_M(t),$$

where

$$K_M(t) \equiv M^{1/2} K(M^{1/2} t),$$

and the scaled process $\xi_M(t) \equiv M^{1/4} \xi(M^{1/2} t)$ has covariance

$$C_M(t) \equiv M^{1/2} C(M^{1/2} t).$$

One can show that $C(t)$ is in $L^1(\mathbb{R})$ and that $\sigma = \int C(t)\, dt > 0$. It follows that, in distributional sense,

$$\lim_{M\to\infty} C_M(t) = \sigma\delta(t), \qquad \lim_{M\to\infty} K_M(t) = 0.$$

We conclude that the limiting equation for Q is

$$\ddot{Q}(t) = -\omega_0^2 Q(t) + \sigma^{1/2} \eta(t),$$

where η is white noise, *i.e.*, $\mathbb{E}(\eta(t)\eta(s)) = \delta(t-s)$. The solution $(Q(t), \dot{Q}(t))$ is a Markov process on $\mathbb{R}^3 \times \mathbb{R}^3$ with generator

$$L = -\frac{\sigma}{2} \Delta_P^2 - P \cdot \nabla_Q + \omega_0^2 Q \cdot \nabla_P.$$

It is a simple exerçise to show that the unique invariant measure of this process is the Lebesgue measure. Moreover, one can show that for any initial condition (Q_0, P_0) and any function $f \in L^1(\mathbb{R}^3 \times \mathbb{R}^3, dQdP)$ one has

$$\lim_{t\to\infty} \mathbb{E}(f(Q(t), \dot{Q}(t))) = \int f(Q, P)\, dQdP,$$

a scaled version of return to equilibrium.

It is worth pointing out that in many instances of classical or quantum open systems the dynamics of the joint system $\mathcal{S} \vee \mathcal{R}$ is too complicated to be controlled analytically or even numerically. Thus, the Hamiltonian approach is inefficient and the Markovian approximation becomes the only available option. The study of the Markovian dynamics of open systems is the main subject of the second volume in this series. The third volume is devoted to applications of the techniques introduced in the first two volumes. It aims at leading the reader to the front of the current research on open quantum systems.

Organization of this Volume

This first volume is devoted to the Hamiltonian approach. Its purpose is to develop the mathematical framework necessary to define and study the dynamics and thermodynamics of quantum systems with infinitely many degrees of freedom.

The first two lectures by A. Joye provide a minimal background in operator theory and statistical mechanics. The third lecture by S. Attal is an introduction to the theory of operator algebras which is the natural framework for quantum mechanics of many degrees of freedom. Quantum dynamical systems and their ergodic theory are the main object of the fourth lecture by C.-A. Pillet. The fifth lecture by M. Merkli deals with the most common instances of environments in quantum physics, the ideal Bose and Fermi gases. Finally the last lecture by V. Jakšić introduces one of the main tool in the study of quantum dynamical systems: spectral analysis.

Lyon, Grenoble and Toulon,
September 2005

Stéphane Attal
Alain Joye
Claude-Alain Pillet

References

1. Caldeira, A.O., Leggett, A.J.: Path integral approach to quantum Brownian motion. Physica A **121** (1983), 587.
2. Davies, E.B.: Markovian master equations. Commun. Math. Phys. **39** (1974), 91.
3. Einstein, A: Uber die von der molekularkinetischen Theorie der Wärme geforderte Bewegung von in ruhenden Flüssigkeiten suspendierten Teilchen. Ann. Phys. **17** (1905), 549.
4. Einstein, A: *Investigations on the Theory of Brownian Movement.* Dover, New York 1956.
5. Einstein, A: Zur Quantentheorie der Strahlung. Physik. Zeitschr. **18** (1917), 121.
6. Ford, G.W., Kac, M., Mazur, P.: Statistical mechanics of assemblies of coupled oscillators. J. Math. Phys. **6** (1965), 504.
7. Lindblad, G.: Completely positive maps and entropy inequalities. Commun. Math. Phys. **40** (1975), 147.
8. Pauli, W.: Festschrift zum 60. Gebürtstage A. Sommerfeld. S. Hirzel, Leipzig 1928.
9. Reif, F.: *Fundamentals of Statistical and Thermal Physics.* McGraw-Hill, New York 1965.
10. Schwinger, J.: Brownian motion of a quantum oscillator. J. Math. Phys. **2** (1961), 407.
11. Van Hove, L.: Master equation and approach to equilibrium for quantum systems. In *Fundamental Problems in Statistical Mechanics.* E.G.D. Cohen ed., North Holland, Amsterdam 1962.
12. Wax, N. (Editor):*Noise and Stochastic Processes.* Dover, New York 1954.
13. Weisskopf, V., Wigner, E.: Berechnung der natürlichen Linienbreite auf Grund der Diracschen Lichttheorie. Zeitschr. für Physik **63** (1930), 54.

Contents

List of Contributors

Stéphane Attal
Institut Camille Jordan
Université Claude Bernard Lyon1
21 av. Claude Bernard
69622 Villeurbanne Cedex
France
email: attal@math.univ-lyon1.fr

Vojkan Jakšić
Department of Mathematics and Statistics
McGill University
805 Sherbrooke Street West
Montreal, QC, H3A 2K6
Canada
e-mail: jaksic@math.mcgill.ca

Alain Joye
Institut Fourier
Université de Grenoble 1
BP 74
38402 Saint-Martin d'Hères Cedex
France
email: alain.joye@ujf-grenoble.fr

Marco Merkli
Department of Mathematics and Statistics
McGill University
805 Sherbrooke Street West
Montreal, QC, H3A 2K6
Canada
email: merkli@math.mcgill.ca

Claude-Alain Pillet
CPT-CNRS (UMR 6207)
Université du Sud Toulon-Var
BP 20132
83957 La Garde Cedex
France
email: pillet@univ-tln.fr

Introduction to the Theory of Linear Operators

Alain Joye

Institut Fourier, Université de Grenoble 1,
BP 74, 38402 Saint-Martin-d'Hères Cedex, France
e-mail: alain.joye@ujf-grenoble.fr

1 Introduction

The purpose of this first set of lectures about Linear Operator Theory is to provide the basics regarding the mathematical key features of unbounded operators to readers that are not familiar with such technical aspects. It is a necessity to deal with such operators if one wishes to study Quantum Mechanics since such objects appear as soon as one wishes to consider, say, a free quantum particle in $\mathbb{R}$. The topics covered by these lectures are quite basic and can be found in numerous classical textbooks, some of which are listed at the end of these notes. They have been selected in order to provide the reader with the minimal background allowing to proceed to the more advanced subjects that will be treated in subsequent lectures, and also according to their relevance regarding the main subject of this school on Open Quantum Systems. Obviously, there is no claim about originality in the presented material. The reader is assumed to be familiar with the theory of bounded operators on Banach spaces and with some of the classical abstract Theorems in Functional Analysis.

2 Generalities about Unbounded Operators

Let us start by setting the stage, introducing the basic notions necessary to study linear operators. While we will mainly work in Hilbert spaces, we state the general definitions in Banach spaces.

If $\mathcal{B}$ is a Banach space over $\mathbb{C}$ with norm $\|\cdot\|$ and T is a bounded linear operator on $\mathcal{B}$, i.e. $T : \mathcal{B} \to \mathcal{B}$, its norm is given by

$$\|T\| = \sup_{\varphi \neq 0} \frac{\|T\varphi\|}{\|\varphi\|} < \infty.$$

Now, consider the position operator of Quantum Mechanics $q = \text{mult}\, x$ on $L^2(\mathbb{R})$, acting as $(q\varphi)(x) = x\varphi(x)$. It is readily seen to be unbounded since one can find a sequence of normalized functions $\varphi_n \in L^2(\mathbb{R})$, $n \in \mathbb{N}$, such that $\|q\varphi_n\| \to \infty$ as $n \to \infty$, and, there are functions of $L^2(\mathbb{R})$ which are no longer $L^2(\mathbb{R})$ when multiplied by the independent variable. We shall adopt the following definition of (possibly unbounded) operators.

Definition 2.1. *A linear operator on $\mathcal{B}$ is a pair (A, D) where $D \subset \mathcal{B}$ is a dense linear subspace of $\mathcal{B}$ and $A : D \to \mathcal{B}$ is linear.*

We will nevertheless often talk about the operator A and call the subspace D the domain of A. It will sometimes be denoted by $\text{Dom}(A)$.

Definition 2.2. *If $(\tilde{A}, \tilde{D})$ is another linear operator such that $\tilde{D} \supset D$ and $\tilde{A}\varphi = A\varphi$ for all $\varphi \in D$, the operator $\tilde{A}$ defines an* extension *of A and one denotes this fact by $A \subset \tilde{A}$.*

That the precise definition of the domain of a linear operator is important for the study of its properties is shown by the following

Example 2.3. : Let H be defined on $L^2[a,b]$, $a < b$ finite, as the differential operator $H\varphi(x) = -\varphi''(x)$, where the prime denotes differentiation. Introduce the dense sets D_D and D_N in $L^2[a,b]$ by

$$D_D = \{\varphi \in C^2[a,b] \,|\, \varphi(a) = \varphi(b) = 0\} \tag{1}$$

$$D_N = \{\varphi \in C^2[a,b] \,|\, \varphi'(a) = \varphi'(b) = 0\}\,. \tag{2}$$

It is easily checked that 0 is an eigenvalue of (H, D_N) but not of (H, D_D). The boundary conditions appearing in (1), (2) respectively are called Dirichlet and Neumann boundary conditions respectively.

The notion of continuity naturally associated with bounded linear operators is replaced for unbounded operators by that of closedness.

Definition 2.4. *Let (A, D) be an operator on $\mathcal{B}$. It is said to be* closed *if for any sequence $\varphi_n \in D$ such that*

$$\varphi_n \to \varphi \in \mathcal{B} \quad and \quad A\varphi_n \to \psi \in \mathcal{B},$$

it follows that $\varphi \in D$ and $A\varphi = \psi$.

Remark 2.5. *i.* In terms of the *graph* of the operator A, denoted by $\Gamma(A)$ and given by

$$\Gamma(A) = \{(\varphi, \psi) \in \mathcal{B} \times \mathcal{B} \,|\, \varphi \in D,\ \psi = A\varphi\},$$

we have the equivalence

$$A \text{ closed} \Longleftrightarrow \Gamma(A) \text{ closed (for the norm } \|(\varphi,\psi)\|^2 = \|\varphi\|^2 + \|\psi\|^2).$$

ii. If $D = \mathcal{B}$, then A is closed if and only if A is bounded, by the Closed Graph Theorem[a].

iii. If A is bounded and closed, then $\bar{D} = \mathcal{B}$ so that it is possible to extend A to the whole of $\mathcal{B}$ as a bounded operator.

iv. If $A : D \to D' \subset \mathcal{B}$ is one to one and onto, then A is closed is equivalent to $A^{-1} : D' \to D$ is closed. This last property can be seen by introducing the *inverse graph* of A, $\Gamma'(A) = \{(x, y) \in \mathcal{B} \times \mathcal{B} \,|\, y \in D, x = Ay\}$ and noticing that A closed iff $\Gamma'(A)$ is closed and $\Gamma(A) = \Gamma'(A^{-1})$.

The notion of spectrum of operators is a key issue for applications in Quantum Mechanics. Here are the relevant definitions.

Definition 2.6. *The* spectrum $\sigma(A)$ *of an operator* (A, D) *on* $\mathcal{B}$ *is defined by its complement* $\sigma(A)^C = \rho(A)$, *where the* resolvent set *of* A *is given by*

$$\rho(A) = \{z \in \mathbb{C} \,|\, (A - z) : D \to \mathcal{B} \text{ is one to one and onto, and } (A - z)^{-1} : \mathcal{B} \to D \text{ is a bounded operator.}\}$$

The operator $R(z) = (A - z)^{-1}$ *is called the* resolvent *of* A.

Actually, $A - z$ is to be understood as $A - z\mathbb{1}$, where $\mathbb{1}$ denotes the identity operator.

Here are a few of the basic properties related to these notions.

Proposition 2.7. *With the notations above,*

i. *If* $\sigma(A) \neq \mathbb{C}$, *then* A *is closed.*

ii. *If* $z \in \rho(A)$ *and* $u \in \mathbb{C}$ *is such that* $|u| < \|R(z)\|^{-1}$, *then* $z + u \in \rho(A)$. *Thus,* $\rho(A)$ *is open and* $\sigma(A)$ *is closed.*

iii. *The resolvent is an analytic map from* $\rho(A)$ *to* $\mathcal{L}(\mathcal{B})$, *the set of bounded linear operators on* $\mathcal{B}$, *and the following identities hold for any* $z, w \in \rho(A)$,

$$R(z) - R(w) = (z - w)R(z)R(w) \tag{3}$$

$$\frac{d^n}{dz^n} R(z) = n!\, R^{n+1}(z).$$

[a] If $T : \mathcal{X} \to \mathcal{Y}$, where $\mathcal{X}$ and $\mathcal{Y}$ are two Banach spaces, then T is bounded iff the graph of T is closed.

Remark 2.8. Identity (3) is called the first resolvent identity. As a consequence, we get that the resolvents at two different points of the resolvent set commute, i.e.

$$[R(z), R(w)] = 0, \ \forall z, w \in \rho(A).$$

Proof. i) If $z \in \rho(A)$, then $R(z)$ is one to one and bounded thus closed and remark iv) above applies.
ii) We need to show that $R(z+u)$ exists and is bounded from $\mathcal{B}$ to D. We have on D

$$(A - z - u)\varphi = (\mathbb{1} - u(A-z)^{-1})(A-z)\varphi = (\mathbb{1} - uR(z))(A-z)\varphi,$$

where $|u|\|R(z)\| < 1$ by assumption. Hence, the Neumann series

$$\sum_{n\geq 0} T^n = (\mathbb{1} - T)^{-1} \text{ where } T : \mathcal{B} \to \mathcal{B} \text{ is such that } \|T\| < 1, \tag{4}$$

shows that the natural candidate for $(A - z - u)^{-1}$ is $R(z)(\mathbb{1} - uR(z))^{-1} : \mathcal{B} \to D$. Then one checks that on $\mathcal{B}$

$$(A - z - u)R(z)(\mathbb{1} - uR(z))^{-1} = (\mathbb{1} - uR(z))(\mathbb{1} - uR(z))^{-1} = \mathbb{1}$$

and that on D

$$\begin{aligned} R(z)(\mathbb{1} - uR(z))^{-1}(A - z - u) &= (\mathbb{1} - uR(z))^{-1}R(z)(A - z - u) \\ &= (\mathbb{1} - uR(z))^{-1}(\mathbb{1} - uR(z)) = \mathbb{1}_D, \end{aligned}$$

where $\mathbb{1}_D$ denotes the identity of D.
iii) By (4) we can write

$$R(z+u) = \sum_{n\geq 0} u^n R^{n+1}(z)$$

so that we get the analyticity of the resolvent and the expression for its derivatives. Finally for $\varphi \in D$

$$((A - z) - (A - w))\varphi = (w - z)\varphi$$

so that, for any $\psi \in \mathcal{B}$,

$$R(z)((A - z) - (A - w))R(w)\psi = R(w)\psi - R(z)\psi = R(z)R(w)(w - z)\psi,$$

where $R(w)\psi \in D$. □

Note that in the bounded case, the spectrum of an operator is never empty nor equal to $\mathbb{C}$, whereas there exist closed unbounded operators with empty spectrum or empty resolvent set. Consider for example, $T = i\frac{d}{dx}$ on $L^2[0,1]$ on the following dense sets. If $AC^2[0,1]$ denotes the set of absolutely continuous functions on $[0,1]$ whose derivatives are in $L^2[0,1]$, (hence in $L^1[0,1]$), set

$$D_1 = \{\varphi \,|\, \varphi \in AC^2[0,1]\}, \quad D_0 = \{\varphi \,|\, \varphi \in AC^2[0,1] \text{ and } \varphi(0) = 0\}.$$

Then, one checks that (T, D_1) and (T, D_0) are closed and such that $\sigma_1(T) = \mathbb{C}$ and $\sigma_0(T) = \emptyset$ (with the obvious notations).

To avoid potential problems related to the fact that certain operators can be *a priori* defined on dense sets on which they may not be closed, one introduces the following notions.

Definition 2.9. *An operator* (A, D) *is* closable *if it possesses a closed extension* $(\tilde{A}, \tilde{D})$.

Lemma 2.10. *If* (A, D) *is closable, then there exists a unique extension* $(\bar{A}, \bar{D})$ *called the* closure *of* (A, D) *characterized by the fact that* $\bar{A} \subseteq \tilde{A}$ *for any closed extension* $(\tilde{A}, \tilde{D})$ *of* (A, D).

Proof. Let

$$\bar{D} = \{\varphi \in \mathcal{B} \,|\, \exists \varphi_n \in D \text{ and } \psi \in \mathcal{B} \text{ with } \varphi_n \to \varphi \text{ and } A\varphi_n \to \psi\}. \quad (5)$$

For any closed extension $\tilde{A}$ of A and any $\varphi \in \bar{D}$, we have $\varphi \in \tilde{D}$ and $\tilde{A}\varphi = \psi$ is uniquely determined by φ. Let us define $(\bar{A}, \bar{D})$ by $\bar{A}\varphi = \psi$, for all $\varphi \in \bar{D}$. Then $\bar{A}$ is an extension of A and any closed extension $A \subseteq \tilde{A}$ is such that $\bar{A} \subseteq \tilde{A}$. The graph $\Gamma(\bar{A})$ of $\bar{A}$ satisfies $\Gamma(\bar{A}) = \overline{\Gamma(A)}$, so that $\bar{A}$ is closed. □

Note also that the closure of a closed operator coincide with the operator itself. Also, before ending this section, note that there exist non closable operators. Fortunately enough, such operators do not play an essential role in Quantum Mechanics, as we will shortly see.

3 Adjoint, Symmetric and Self-adjoint Operators

The arena of Quantum Mechanics is a complex Hilbert space $\mathcal{H}$ where the notion of scalar product $\langle \cdot | \cdot \rangle$ gives rise to a norm denoted by $\| \cdot \|$. Operators that are self-adjoint with respect to that product play a particularly important role, as they correspond to the observables of the theory. We shall assume the following convention regarding the positive definite sesquilinear form $\langle \cdot | \cdot \rangle$ on $\mathcal{H} \times \mathcal{H}$: it is linear in the right variable and thus anti-linear in the left variable. We shall also always assume that our Hilbert space is separable, i.e. it admits a countable basis, and we shall identify the dual $\mathcal{H}'$ of $\mathcal{H}$ with $\mathcal{H}$, since $\forall l \in \mathcal{H}', \exists! \psi \in \mathcal{H}$ such that $l(\cdot) = \langle \psi | \cdot \rangle$.

Let us make the first steps towards self-adjunction.

Definition 3.1. *An operator* (H, D) *in* $\mathcal{H}$ *is said to be* symmetric *if* $\forall \varphi, \psi \in D \subseteq \mathcal{H}$

$$\langle \varphi | H\psi \rangle = \langle H\varphi | \psi \rangle.$$

For example, the operators $(-\frac{d^2}{dx^2}, D_D)$ and $(-\frac{d^2}{dx^2}, D_N)$ introduced above are symmetric, as shown by integration by parts.

Remark 3.2. If H is symmetric, its eigenvalues are real.

The next property is related to an earlier remark concerning the role of non closable operators in Quantum Mechanics.

Proposition 3.3. *Any symmetric operator (H, D) is closable and its closure is symmetric.*

This Proposition will allow us to consider that any symmetric operator is closed from now on.

Proof. Let $\bar{D} \supseteq D$ as in (5) and $\chi \in D$, $\varphi \in \bar{D}$. We compute for any such χ,

$$\langle\varphi|H\chi\rangle = \lim_n \langle\varphi_n|H\chi\rangle = \lim_n \langle H\varphi_n|\chi\rangle = \langle\psi|\chi\rangle. \tag{6}$$

As D is dense by assumption, the vector ψ is uniquely determined by the linear, bounded form $l_\psi : D \to \mathbb{C}$ such that $l_\psi(\chi) = \langle\psi|\chi\rangle$. In other words, ψ is characterized by φ uniquely. One then defines $\bar{H}$ on $\bar{D}$ by $\bar{H}\varphi = \psi$ and linearity is easily checked. As, by construction, $\Gamma(\bar{H}) = \overline{\Gamma(H)}$ is closed, $\bar{H}$ is a closed extension of H. Let us finally check the symmetry property. If $\chi_n \in D$ is such that $\chi_n \to \chi \in \bar{D}$, with $H\chi_n \to \eta$ and $\varphi \in \bar{D}$, (6) says

$$\langle\varphi|H\chi_n\rangle = \langle\bar{H}\varphi|\chi_n\rangle.$$

Taking the limit $n \to \infty$, we get from the above

$$\lim_n \langle\varphi|H\chi_n\rangle = \langle\varphi|\eta\rangle = \langle\varphi|\bar{H}\chi\rangle = \lim_n \langle\bar{H}\varphi|\chi_n\rangle = \langle\bar{H}\varphi|\chi\rangle.$$

□

When dealing with bounded operators, symmetric and self-adjoint operators are identical. It is not necessarily true in the unbounded case. As one of the most powerful tools in linear operator theory, namely the Spectral Theorem, applies only to self-adjoint operators, we will develop some criteria to distinguish symmetric and self-adjoint operators.

Definition 3.4. *Let (A, D) be an operator on $\mathcal{H}$. The* adjoint *of A, denoted by (A^*, D^*), is determined as follows: D^* is the set of $\psi \in \mathcal{H}$ such that there exists a $\chi \in \mathcal{H}$ so that*

$$\langle\psi|A\varphi\rangle = \langle\chi|\varphi\rangle, \ \forall\varphi \in D.$$

As D is dense, χ is unique, so that one sets $A^*\psi = \chi$ and checks easily the linearity. Therefore,

$$\langle\psi|A\varphi\rangle = \langle A^*\psi|\varphi\rangle, \ \forall\varphi \in D, \psi \in D^*.$$

In other words, $\psi \in D^*$ iff the linear form $l(\cdot) = \langle\psi|A\cdot\rangle : D \to \mathbb{C}$ is bounded. In that case, Riesz Lemma implies the existence of a unique χ such that $\langle\psi|A\cdot\rangle = \langle\chi|\cdot\rangle$. Note also that D^* is not necessarily dense.

Let us proceed with some properties of the adjoint.

Proposition 3.5. *Let (A, D) be an operator on $\mathcal{H}$.*

i. *The adjoint (A^*, D^*) of (A, D) is closed. If, moreover, A is closable, then D^* is dense.*
ii. *If A is closable, $\bar{A} = A^{**}$.*
iii. *If $A \subseteq B$, then $B^* \subseteq A^*$.*

Proof. i) Let $(\psi, \chi) \in D^* \times \mathcal{H}$ belong to $\Gamma(A^*)$. This is equivalent to saying

$$\langle \psi | A\varphi \rangle = \langle \chi | \varphi \rangle, \quad \forall \varphi \in D,$$

which is equivalent to $(\psi, \chi) \in M^\perp$, where

$$M = \{(A\varphi, -\varphi) \in \mathcal{H} \times \mathcal{H}, \mid \varphi \in D\},$$

with the scalar product $\langle\langle(\varphi_1, \varphi_2)|(\psi_1, \psi_2)\rangle\rangle = \langle \varphi_1 | \psi_1 \rangle + \langle \varphi_2 | \psi_2 \rangle$. As $M^\perp$ is closed, $\Gamma(A^*)$ is closed too. Assume now A is closable and suppose there exists $\eta \in \mathcal{H}$ such that $\langle \psi | \eta \rangle = 0$, for all $\psi \in D^*$. This implies that $(\eta, 0)$ is orthogonal to $\Gamma(A^*)$. But,

$$\Gamma(A^*)^\perp = M^{\perp\perp} = \overline{M}.$$

Therefore, there exists $\varphi_n \in D$, such that $\varphi_n \to 0$ and $A\varphi_n \to \eta$. As A is closable, $\eta = \bar{A}0 = 0$, i.e. $(D^*)^\perp = 0$ and $\overline{D^*} = (D^*)^{\perp\perp} = \mathcal{H}$.
ii) Define a unitary operator V on $\mathcal{H} \times \mathcal{H}$ by

$$V(\varphi, \psi) = (\psi, -\varphi).$$

It has the property $V(E^\perp) = (V(E))^\perp$, for any linear subspace $E \subseteq \mathcal{H} \times \mathcal{H}$. In particular, we have just seen

$$\Gamma(A^*) = (V(\Gamma(A)))^\perp$$

so that

$$\begin{aligned}\overline{\Gamma(A)} &= (\Gamma(A)^\perp)^\perp = ((V^2\Gamma(A))^\perp)^\perp \\ &= (V(V(\Gamma(A))^\perp))^\perp = (V(\Gamma(A^*)))^\perp = \Gamma(A^{**}),\end{aligned}$$

i.e. $\bar{A} = A^{**}$.
iii) Follows readily from the definition. □

When H is symmetric, we get from the definition and properties above that H^* is a closed extension of H. This motivates the

Definition 3.6. *An operator (H, D) is* self-adjoint *whenever it coincides with its adjoint (H^*, D^*). It is therefore closed.*
An operator (H, D) is essentially self-adjoint *if it is symmetric and its closure $(\bar{H}, \bar{D})$ is self-adjoint.*

Therefore, we have in general for a symmetric operator,

$$H \subseteq \bar{H} = H^{**} \subseteq H^*, \quad \text{and} \quad H^* = \overline{H^*} = H^{***} = \bar{H}^*.$$

In case H is essentially self-adjoint,

$$H \subseteq \bar{H} = H^{**} = H^*.$$

We now head towards our general criterion for (essential) self-adjointness. We need a few more

Definition 3.7. *For* (H, D) *symmetric and denoting its adjoint by* (H^*, D^*), *the* deficiency subspaces $L^{\pm}$ *are defined by*

$$\begin{aligned} L^{\pm} &= \{\varphi \in D^* \mid H^*\varphi = \pm i\varphi\} = \{\varphi \in \mathcal{H} \mid \langle H\psi|\varphi\rangle = \pm i\langle\psi|\varphi\rangle \ \forall\psi \in D\} \\ &= Ran(H \pm i)^{\perp} = Ker(H^* \mp i). \end{aligned}$$

The deficiency indices *are the dimensions of* $L^{\pm}$*, which can be finite or infinite.*

To get an understanding of these names, recall that one can always write

$$\mathcal{H} = \text{Ker}(H^* \mp i) \oplus \overline{\text{Ran}(H \pm i)} \equiv L^{\pm} \oplus \overline{\text{Ran}(H \pm i)}. \tag{7}$$

Note that the definitions of $L^{\pm}$ is invariant if one replaces H by its closure $\bar{H}$.

For (H, D) symmetric and any $\varphi \in D$ observe that

$$\|(H + i)\varphi\|^2 = \|H\varphi\|^2 + \|\varphi\|^2 = \|(H - i)\varphi\|^2 \neq 0.$$

This calls for the next

Definition 3.8. *Let* (H, D) *be symmetric. The* Cayley transform *of* H *is the isometric operator*

$$U = (H - i)(H + i)^{-1} : Ran(H + i) \to Ran(H - i).$$

It enjoys the following property.

Lemma 3.9. *The symmetric extensions of* H *are in one to one correspondence with the isometric extensions of* U.

Proof. Let $(\tilde{H}, \tilde{D})$ be a symmetric extension of (H, D) and $\tilde{U}$ be its Cayley transform. We have

$$\varphi \in \text{Ran}(H \pm i) \iff \exists\psi \in D \subseteq \tilde{D} \ \text{ such that } \ \varphi = (H \pm i)\psi = (\tilde{H} \pm i)\psi,$$

hence $\text{Ran}(H \pm i) \subset \text{Ran}(\tilde{H} \pm i)$, and

$$\tilde{U}\varphi = (\tilde{H} - i)(\tilde{H} + i)^{-1}\varphi = U\varphi, \quad \forall\varphi \in \text{Ran}(H \pm i). \tag{8}$$

Conversely, let $\tilde{U} : M^+ \to M^-$, be a isometric extension of U, where $\text{Ran}(H \pm i) \subseteq M^{\pm}$. We need to construct a symmetric extension of H whose Cayley transform is $\tilde{U}$. Algebraically this means, see (8),

$$\tilde{H} = (\tilde{U} - \mathbb{1})^{-1}\frac{2}{i} - i. \tag{9}$$

Let us show that 1 is not an eigenvalue of $\tilde{U}$. If $\varphi \in M^+$ is a corresponding eigenvector, and $\psi = (H+i)\chi$, where $\chi \in D$, then

$$2i\langle\varphi|\chi\rangle = \langle\varphi|(H+i)\chi - (H-i)\chi\rangle = \langle\varphi|\psi - U\psi\rangle$$
$$= \langle\varphi|\psi\rangle - \langle\tilde{U}\varphi|\tilde{U}\psi\rangle = 0.$$

By density of D, $\varphi = 0$, so that we can define $\tilde{H}$ by (9) on $\tilde{D} = (\tilde{U} - \mathbb{1})M^+$. It is not difficult to check that $\tilde{H}$ is a symmetric extension of H. □

We can now state the

Theorem 3.10. *If (H, D) is symmetric on $\mathcal{H}$, there exist self-adjoint extensions of H if and only if the deficiency indices are equal. Moreover, the following statements are equivalent:*

1. H is essentially self-adjoint.
2. The deficiency indices are both zero.
3. H possesses exactly one self-adjoint extension.

Proof. 1)⇒ 3): Let J be a self-adjoint extension of H. Then $H \subseteq J = J^*$ and $J \supseteq \bar{H}$. Hence $J = J^* \subseteq \bar{H}^* = \bar{H}$, so that $J = \bar{H}$.
1)⇒ 2): We can assume that H is closed so $H = \bar{H} = H^*$. For any $\varphi \in L^{\pm} = \mathrm{Ker}(H^* \mp i)$,

$$0 = \|(H^* \mp i)\varphi\|^2 = \|(H \mp i)\varphi\|^2 = \|H\varphi\|^2 + \|\varphi\|^2 \geq \|\varphi\|^2, \quad L^{\pm} = \{0\}. \tag{10}$$

2) ⇒ 1): Consider $(H+i) : D \to \mathrm{Ran}(H+i)$. By (10) above, this operator is one to one, and we can define $(H+i)^{-1} : \mathrm{Ran}(H+i) \to D$. By the same estimate it satisfies

$$\|(H+i)^{-1}\psi\|^2 \leq \|(H+i)(H+i)^{-1}\psi\|^2 = \|\psi\|^2.$$

As H can be assumed to be closed (i.e. $H = \bar{H}$) and $L^+ = \{0\}$, we get that $\mathrm{Ran}(H+i)$ is closed so that $\mathcal{H} = \mathrm{Ran}(H+i)$, due to (7). Therefore, for any $\varphi \in D^*$, there exists a $\psi \in D$ such that $(H^* + i)\varphi = (H+i)\psi$. As $H \subseteq H^*$,

$$(H^* + i)(\varphi - \psi) = 0, \text{ i.e. } \varphi - \psi \in Ker(H^* + i) = \{0\},$$

we get that $\varphi \in D$ and $H = H^*$, which is what we set out to prove.
3) ⇒ 2): if K is a self-adjoint extension of H, its deficiency indices are zero (by 2)). Therefore, (see (7)), its Cayley transform V is a unitary extension of U, the Cayley transform of H. In particular, $V|_{L^+} : L^+ \to L^-$ is one to one and onto, so that the deficiency indices of H are equal. That yields the first part of the Theorem. Now assume these indices are not zero. By the preceding Lemma, there exist an infinite number of symmetric extensions of H, parametrized by all isometries from L^+ to L^-. In particular, there exist extensions with zero deficiency indices, which by 2) and 1) are self-adjoint, contradicting the fact that K is the unique self-adjoint extension of H. □

Remark 3.11. It is a good exercise to prove that in case (H,D) is symmetric and $H \geq 0$, i.e. $\langle\varphi|H\varphi\rangle \geq 0$ for any $\varphi \in D$, then H is essentially self-adjoint iff $\text{Ker}(H^* + 1) = \{0\}$.

As a first application, we give a key property of self-adjoint operators for the role they play in the Quantum dogma concerning measure of observables. It is the following fact concerning their spectrum.

Theorem 3.12. *Let $H = H^*$. Then, $\sigma(H) \subseteq \mathbb{R}$ and,*

$$\|(H-z)^{-1}\| \leq \frac{1}{|\Im z|}, \text{ if } z \notin \mathbb{R}. \tag{11}$$

Moreover, for any z in the resolvent set of H,

$$(H-\bar{z})^{-1} = ((H-z)^{-1})^*. \tag{12}$$

Proof. Let $\varphi \in D$, D being the domain of H and $z = x + iy$, with $y \neq 0$. Then

$$\|(H-x-iy)\varphi\|^2 = \|(H-x)\varphi\|^2 + y^2\|\varphi\|^2 \geq y^2\|\varphi\|^2. \tag{13}$$

This implies

$$\text{Ker}(H-z) = \text{Ker}(H^*-z) = \{0\} \text{ i.e. } \overline{\text{Ran}(H-z)} = \mathcal{H},$$

and $H - z$ is invertible on $\text{Ran}(H-z)$. (13) shows that $(H-z)$ is bounded with the required bound, and as the resolvent is closed, it can be extended on $\mathcal{H}$ with the same bound. Equality (12) is readily checked. □

As an application of the first part of Theorem 3.10, consider a symmetric operator (H,D) which commutes with a *conjugation* C. More precisely:

C is anti-linear, $C^2 = \mathbb{1}$ and $\|C\varphi\| = \|\varphi\|$. Hence $\langle\varphi|\psi\rangle = \langle C\psi|C\varphi\rangle$. Moreover, $C : D \to D$ and $CH = HC$ on D.

Under such circumstances, the deficiency indices of H are equal and there exist self-adjoint extensions of H.

Indeed, one first deduces that $C(D) = D$. Then, for any $\varphi^+ \in L^+ = \text{Ker}(H^* - i)$ and $\psi \in D$, we compute

$$0 = \overline{\langle\varphi^+|(H+i)\psi\rangle} = \langle C\varphi^+|C(H+i)\psi\rangle = \langle C\varphi^+|(H-i)C\psi\rangle,$$

so that $C\varphi^+ \in \text{Ran}(H-i)^\perp = \text{Ker}(H^*+i) = L^-$. In other words, one has $C : L^+ \to L^-$, and one shows similarly that $C : L^- \to L^+$. As C is isometric, the dimensions of L^+ and L^- are the same.

A particular case where this happens is that of the complex conjugation and a differential operator on $\mathbb{R}^n$, with real valued coefficients.

An example of direct application of this criterion is the following. Consider the symmetric operator $H\varphi = i\varphi'$ on the domain $C_0^\infty(0,\infty) \subset L^2(0,\infty)$. A vector $\psi \in D^*$ iff there exists $\chi \in L^2(0,\infty)$ such that $\langle\psi|H\varphi\rangle = \langle\chi|\varphi\rangle$, for all $\varphi \in C_0^\infty(0,\infty)$. Expressing the scalar products this means

$$\int \chi(x)\bar{\varphi}(x)dx = -i\int \psi(x)\bar{\varphi}'(x)dx = iD_xT_\psi(\bar{\varphi}),$$

where T_ψ denotes the distribution associated with ψ. In other words, we have $\psi \in W^{1,2}(0,\infty) = D^*$ and $H^*\psi = i\psi$ in the weak sense. Elements of $\mathrm{Ker}(H^* \mp i)$ satisfy

$$H^*\psi = \pm i\psi \iff \psi' = \pm\psi \iff \psi(x) = ce^{\pm x}\begin{cases}\notin L^2(0,\infty)\\ \in L^2(0,\infty)\end{cases}$$

Hence there is no self-adjoint extension of that operator. If it is considered on $C_0^\infty(0,1) \subset L^2(0,1)$, the above shows that the deficiency indices are both 1 and there exist infinitely many self-adjoint extensions of it.

Specializing a little, we get a criterion for operators whose spectrum consists of eigenvalues only.

Corollary 3.13. *Let (H,D) symmetric on $\mathcal{H}$ such that there exists an orthonormal basis $\{\varphi_n\}_{n\in\mathbb{N}}$ of $\mathcal{H}$ of eigenvectors of H satisfying for any $n \in \mathbb{N}$, $\varphi_n \in D$ and $H\varphi_n = \lambda_n\varphi_n$, with $\lambda_n \in \mathbb{R}$. Then H is essentially self-adjoint and $\sigma(\bar{H}) = \{\lambda_n \,|\, n \in \mathbb{N}\}$.*

Proof. Just note that any vector φ in $L^\pm$ satisfies in particular

$$\langle H\varphi_n|\varphi\rangle = \pm i\langle\varphi_n|\varphi\rangle = \lambda_n\langle\varphi_n|\varphi\rangle,$$

so that $\langle\varphi_n|\varphi\rangle = 0$ for any n. This means that $L^+ = \{0\}$, hence that H is essentially self-adjoint.

Then $\bar{H}$, as an extension of H admits the φ_n's as eigenvectors with the same eigenvalues and as the spectrum is a closed set, we get $\sigma(\bar{H}) \supset \overline{\{\lambda_n \,|\, n \in \mathbb{N}\}}$. If λ does not belong to the latter set, we define R_λ by $R_\lambda\varphi_n = \frac{1}{\lambda_n - \lambda}\varphi_n$, for all $n \in \mathbb{N}$. Using the fact that $\bar{H}$ is closed, it is not difficult to see that R_λ is the resolvent of $\bar{H}$ at λ, which yields the result. □

As a first example of application we get that $-\frac{d^2}{dx^2}$ on $C^2(a,b)$ (or $C^\infty(a,b)$) with Dirichlet boundary conditions is essentially self-adjoint with spectrum

$$\left\{\frac{n^2\pi^2}{(b-a)^2}\,\middle|\, n \in \mathbb{N}^*\right\},$$

as the corresponding eigenvectors

$$\varphi_n(x) = \left(\frac{2}{b-a}\right)^{1/2}\sin\left(\frac{n\pi(x-a)}{b-a}\right), \quad n \in \mathbb{N}^*,$$

are known to form a basis of $L^2[a,b]$ by the theory of Fourier series.

Another standard operator is the harmonic oscillator defined on $L^2(\mathbb{R})$ by the differential operator

$$H_{osc} = -\frac{1}{2}\frac{d^2}{dx^2} + \frac{x^2}{2}$$

with dense domain $\mathcal{S}$ the Schwartz functions. This operator is symmetric by integration by parts, and it is a standard exercise, using creation and annihilation operators $b^\dagger = (x - \partial_x)/\sqrt{2}, b = (x + \partial_x)/\sqrt{2}$ to show that the solutions of

$$H_{osc}\varphi_n(x) = \lambda_n \varphi_n(x), \quad n \in \mathbb{N},$$

are given by $\lambda_n = n + 1/2$ with eigenvector

$$\varphi_n(x) = c_n H_n(x) e^{-x^2/2}, \quad \text{with } H_n(x) = (-1)^n e^{x^2} \frac{d^n}{dx^n} e^{-x^2},$$

and $c_n = (2^n n! \sqrt{\pi})^{-1/2}$. These eigenvectors also form a basis of $L^2(\mathbb{R})$, so that this operator is essentially self-adjoint with spectrum $\mathbb{N} + 1/2$. Note that we cannot work on C_0^∞ to apply this criterion here.

Another popular way to prove that an operator is self-adjoint is to compare it to another operator known to be self-adjoint and use a perturbative argument to get self-adjointness of the former.

Let (H, D) be a self-adjoint operator on $\mathcal{H}$ and let $(A, D(A))$ be symmetric with domain $D(A) \supseteq D$.

Definition 3.14. *The operator A has a* relative bound $\alpha \geq 0$ *with respect to H if there exists $c < \infty$ such that*

$$\|A\varphi\| \leq \alpha \|H\varphi\| + c\|\varphi\|, \quad \forall \varphi \in D. \tag{14}$$

The infimum over such relative bounds is the relative bound of A w.r.t. H.

Remark 3.15. The definition of the relative bound is unchanged if we replace (14) by the slightly stronger condition

$$\|A\varphi\|^2 \leq \alpha^2 \|H\varphi\|^2 + c^2\|\varphi\|^2, \quad \forall \varphi \in D.$$

Lemma 3.16. *Let $K : D \to \mathcal{H}$ be such that $K\varphi = H\varphi + A\varphi$. If $0 \leq \alpha < 1$, is the relative bound of A w.r.t. H, K is closed and symmetric. Moreover, $\|A(H + i\lambda)^{-1}\| < 1$, if $\lambda \in \mathbb{R}$ has large enough modulus.*

Proof. The symmetry of K is clear. Let us consider $\varphi_n \in D$ such that $\varphi_n \to \varphi$ and $K\varphi_n \to \psi$. Then, by assumption,

$$\begin{aligned}\|H\varphi_n - H\varphi_m\| &\leq \|K\varphi_n - K\varphi_m\| + \|A\varphi_n - A\varphi_m\| \\ &\leq \|K\varphi_n - K\varphi_m\| + \alpha\|H\varphi_n - H\varphi_m\| + c\|\varphi_n - \varphi_m\|,\end{aligned}$$

so that

$$\|H\varphi_n - H\varphi_m\| \leq \frac{1}{1-\alpha}\|K\varphi_n - K\varphi_m\| + \frac{c}{1-\alpha}\|\varphi_n - \varphi_m\| \to 0 \text{ as } n, m \to \infty.$$

H being closed, we deduce from the above that $\varphi \in D$ and $H\varphi_n \to H\varphi$. Then, from (14), we get $A\varphi_n - A\varphi \to 0$ from which follows $K\varphi_n \to K\varphi = \psi$.

The proof of the statement concerning the resolvent reads as follows. Let $\psi \in \mathcal{H}$, $\varphi = (H + i\lambda)^{-1}\psi$ and $0 \leq \alpha < \beta < 1$. Then, for $|\lambda| > 0$ large enough

$$\begin{aligned}\|A\varphi\|^2 &\leq (\alpha\|H\varphi\| + c\|\varphi\|)^2 \leq \beta^2\left(\|H\varphi\|^2 + \lambda^2\|\varphi\|^2\right) \\ &= \beta^2\|(H + i\lambda)\varphi\|^2 = \beta^2\|\psi\|^2.\end{aligned}$$

Hence $\|A(H + i\lambda)^{-1}\psi\| \leq \beta\|\psi\|$. □

This leads to the

Theorem 3.17. *If H is self-adjoint and A is symmetric with relative bound $\alpha < 1$ w.r.t. H, then $K = H + A$ is self-adjoint on the same domain as that of H.*

Proof. Let $|\lambda|$ be large enough. From the formal expressions

$$(H + A + i\lambda)^{-1} - (H + i\lambda)^{-1} = -(H + A + i\lambda)^{-1}(A)(H + i\lambda)^{-1} \iff$$

$$(H + A + i\lambda)^{-1} = (H + i\lambda)^{-1}(\mathbb{1} + A(H + i\lambda)^{-1})^{-1}$$

we see that the natural candidate for the resolvent of K is

$$R_\lambda = (H + i\lambda)^{-1}\sum_{n\in\mathbb{N}}(-A(H + i\lambda)^{-1})^n.$$

By assumption on $|\lambda|$, this sum converges in norm and $\mathrm{Ran}(R_\lambda) = D$. Routine manipulations show that $(H + A + i\lambda)R_\lambda = R_\lambda(H + A + i\lambda) = \mathbb{1}$ so that $\mathrm{Ran}(H + A + i\lambda) = \mathcal{H}$. This implies that the deficiency indices of $K = H + A$ are both zero, and since it is closed, K is self-adjoint. Note that one uses the fact that $\dim\ker(K^* - i\lambda)$ is constant for $\lambda > 0$ and $\lambda < 0$. □

4 Spectral Theorem

Let us start this section by the presentation of another example of self-adjoint operator, which will play a key role in the Spectral Theorem, we set out to prove here. Before getting to work, let us specify right away that we shall not provide here a full proof of the version of the Spectral Theorem we chose. Some parts of it, of a purely analytical character, will be presented as facts whose detailed full proofs can be found in Davies's book [1]. But we hope to convey the main ideas of the proof in these notes.

Consider $E \subseteq \mathbb{R}^N$ a Borel set and μ a Borel non-negative measure on E. Let $\mathcal{H} = L^2(E, d\mu)$ be the usual set of measurable functions $f : E \to \mathbb{C}$ such that

$\|f\|^2 = \int_E |f(x)|^2 d\mu < \infty$, with identification of functions that coincide almost everywhere.

Let $a : E \to \mathbb{R}$ be measurable and such that the restriction of a to any bounded set of E is bounded. We set

$$D = \{f \in \mathcal{H} | \int_E (1 + a^2(x))|f(x)|^2 d\mu < \infty\},$$

which is dense, and we define the *multiplication* operator (A, D) by

$$(Af)(x) = a(x)f(x), \quad \forall f \in D.$$

Lemma 4.1. *(A, D) is self-adjoint and if L^2_c denotes the set of functions of $\mathcal{H}$ which are zero outside a compact subset of E, then A is essentially self-adjoint on L^2_c.*

Proof. A is clearly symmetric. If $z \notin \mathbb{R}$, the bounded operator $R(z)$ given by

$$(R(z)f)(x) = (a(x) - z)^{-1} f(x)$$

is easily seen to be the inverse of $(A - z)$. Hence, $\sigma(A) \neq \mathbb{C}$, so that A is closed. Moreover, the deficiency indices of A are both seen to be zero:

$$\begin{aligned} A^* f = if &\Leftrightarrow \forall h \in D \int \overline{Ah} f d\mu = i \int \overline{h} f d\mu \\ &\Leftrightarrow \int (a(x) - i)\overline{h} f d\mu = 0, \\ &\Leftrightarrow f = 0 \;\; \mu \text{ a.e.} \end{aligned}$$

So that A is closed and essentially self-adjoint, hence self-adjoint.

Concerning the last statement, we need to show that A is the closure of its restriction to L^2_c. If $f \in D$ and $n \in \mathbb{N}$, we define

$$f_n(x) = \begin{cases} f(x) \text{ if } x \in E, |x| \leq n, \\ 0 \quad \text{otherwise.} \end{cases}$$

Hence $|f_n(x)| \leq |f(x)|$ and $f_n \in L^2_c \in D$. By Lebesgue dominated convergence Theorem, one checks that $f_n \to f$ and $Af_n \to Af$ as $n \to \infty$. □

Lemma 4.2. *The spectrum and resolvent of A are such that*

$$\sigma(A) = \textit{essential range of } a = \{\lambda \in \mathbb{R} \,|\, \mu(\{x \,|\, |a(x) - \lambda| < \epsilon\}) > 0, \forall \epsilon > 0\}.$$

If $\lambda \notin \sigma(A)$, then

$$\|(A - \lambda)^{-1}\| = \frac{1}{\textit{dist}\,(\lambda, \sigma(A))}.$$

Proof. If λ is not in the essential range of a, it is readily checked that the multiplication operator by $(a(x)-\lambda)^{-1}$ is bounded (outside of a set of zero μ measure). Also one sees that this operator yields the inverse of $a-\lambda$ for such λ's, which, consequently, belong to $\rho(A)$. Conversely, let us take λ in the essential range of a and show that $\lambda \in \sigma(A)$. We define sets of positive μ measures by

$$S_m = \{x \in E \,|\, |\lambda - a(x)| < 2^{-m}\}.$$

Let χ_m be the characteristic function of S_m, which is a non zero element of $L^2(E, d\mu)$. Then

$$\|(A-\lambda)\chi_m\|^2 = \int_{S_m} |\chi_m|^2 |a(x)-\lambda|^2 d\mu \leq 2^{-2m} \int_{S_m} |\chi_m|^2 d\mu = 2^{-2m}\|\chi_m\|^2,$$

which shows that $(A-\lambda)^{-1}$ cannot be bounded. Finally, if λ is not in the essential range of a, we set

$$\|(a(\cdot)-\lambda)^{-1}\|_\infty = \text{essential supremum of } (a(\cdot)-\lambda)^{-1},$$

where we recall that for a measurable function f

$$\|f\|_\infty = \inf\{K > 0 \,|\, |f(x)| \leq K \ \mu \text{ a.e.}\}.$$

We immediately get that $\|a(\cdot)-\lambda\|_\infty$ is an upper bound for the norm of the resolvent, as, for any $\epsilon > 0$ there exists a set $S \subset E$ of positive measure such that $|(a(x)-\lambda)^{-1}| \geq K - \epsilon, \forall x \in S$. Considering the characteristic function of this set, one sees that the upper bound is actually reached and corresponds with the distance of λ to the spectrum of A. □

4.1 Functional Calculus

Let us now come to the steps leading to the Spectral Theorem. The general setting is as follows. One has a self-adjoint operator (H, D), D dense in a separable Hilbert space $\mathcal{H}$. We first want to define a functional calculus, allowing us to take functions of self-adjoint operators. If H is a multiplication operator by a real valued function h, as in the above example, then $f(H)$, for a reasonable function $f : \mathbb{R} \to \mathbb{C}$, is easily conceivable as the multiplication by $f \circ h$. We are going to define a function of an operator H in a quite general setting by means of an explicit formula due to Helffer and Sjöstrand and we will check that this formula has the properties we expect of such an operation. Finally, we will also see that any operator can be seen as a multiplication operator on some $L^2(d\mu)$ space.

Let us introduce the notation $< z >= (1+|z|^2)^{1/2}$ and the set of functions we will work with. Let $\beta \in \mathbb{R}$ and S^β be the set of complex valued $C^\infty(\mathbb{R})$ functions such that there exists a c_n so that

$$|f^{(n)}(x)| = \left|\frac{d^n}{dx^n} f(x)\right| \leq c_n < x >^{\beta - n}, \ \forall x \in \mathbb{R} \ \forall n \in \mathbb{N}.$$

We set $\mathcal{A} = \cup_{\beta<0} S^\beta$ and we define norms $\|\cdot\|_n$ on $\mathcal{A}$, for any $n \geq 1$, by

$$\|f\|_n = \sum_{r=0}^{n} \int_{-\infty}^{\infty} |f^{(r)}(x)| < x >^{r-1} \, dx.$$

This set of functions enjoys the following properties:
$\mathcal{A}$ is an algebra for the multiplication of functions, it contains the rational functions which decay to zero at ∞ and have non-vanishing denominator on the real axis. Moreover, it is not difficult to see that

$$\|f\|_n < \infty \;\Rightarrow\; f' \in L^1(\mathbb{R}), \text{ and } f(x) \to 0 \text{ as } |x| \to \infty$$

and that

$$\|f - f_k\|_n \to 0, \text{ as } k \to \infty \;\Rightarrow\; \sup_{x\in\mathbb{R}} |f(x) - f_k(x)| \to 0, \text{ as } k \to \infty.$$

Definition 4.3. *A map which to any $f \in \mathcal{E} \subset L^\infty(\mathbb{R})$ associates $f(H) \in \mathcal{L}(\mathcal{H})$ is a* functional calculus *if the following properties are true.*

1. *$f \mapsto f(H)$ is linear and multiplicative, (i.e. $fg \mapsto f(H)g(H)$)*
2. *$\bar{f}(H) = (f(H))^*$, $\forall f \in \mathcal{E}$*
3. *$\|f(H)\| \leq \|f\|_\infty$, $\forall f \in \mathcal{E}$*
4. *If $w \notin \mathbb{R}$ and $r_w(x) = (x-w)^{-1}$, then $r_w(H) = (H-w)^{-1}$*
5. *If $f \in C_0^\infty(\mathbb{R})$ such that supp$(f) \cap \sigma(H) = \emptyset$ then $f(H) = 0$.*

For $f \in C^\infty$, we define its quasi-analytic extension $\tilde{f} : \mathbb{C} \to \mathbb{C}$ by

$$\tilde{f}(z) = \left(\sum_{r=0}^{n} f^{(r)}(x) \frac{(iy)^r}{r!} \right) \sigma(x, y)$$

with $z = x + iy$, $n \geq 1$, $\sigma(x, y) = \tau(y/ < x >)$, where $\tau \in C_0^\infty$ is equal to one on $[-1, 1]$, has support in $[-2, 2]$. We are naturally abusing notations as $\tilde{f}$ is not analytic in general, but it is C^∞. Its support is confined to the set $|y| \leq 2 < x >$ due to the presence of τ. Also, the projection on the x axis of the support of $\tilde{f}$ is equal to the support of f. The choice of τ and n will turn out to have no importance for us.

Explicit computations yield

$$\frac{\partial}{\partial \bar{z}} \tilde{f}(z) = \frac{1}{2} \left(\frac{\partial}{\partial x} + i \frac{\partial}{\partial y} \right) \tilde{f}(z) = \tag{15}$$

$$\left(\sum_{r=0}^{n} f^{(r)}(x) \frac{(iy)^r}{r!} \right) \frac{(\sigma_x(x, y) + i\sigma_y(x, y))}{2} + f^{(n+1)}(x) \frac{(iy)^n}{n!} \frac{\sigma(x, y)}{2}.$$

As supp$(\sigma_x(x, y))$ and supp$(\sigma_y(x, y))$ are included in supp$(\tau'(y/ < x >)$, i.e. in the set $< x > \leq |y| \leq 2 < x >$, if x is fixed and $y \to 0$,

$$\left| \frac{\partial}{\partial \bar{z}} \tilde{f}(z) \right| = O(|y|^n),$$

which justifies the name quasi-analytic extension as y goes to zero.

Definition 4.4. *For any* $f \in \mathcal{A}$ *and any self-adjoint operator* H *on* $\mathcal{H}$ *the* Helffer-Sjöstrand formula *for* $f(H)$ *reads*

$$f(H) = \frac{1}{\pi} \int_{\mathbb{C}} \frac{\partial}{\partial \bar{z}} \tilde{f}(z)(H - z)^{-1} dxdy \in \mathcal{L}(\mathcal{H}). \tag{16}$$

Remark 4.5. This formula allows to compute functions of operators by means of their resolvent only. Therefore it holds for bounded as well as unbounded operators. Moreover, being explicit, it can yield useful bounds in concrete cases. Note also that it is linear in f.

We need to describe a little bit more in what sense this integral holds.

Lemma 4.6. *The expression (16) converges in norm and the following bound holds*

$$\|f(H)\| \leq c_n \|f\|_{n+1}, \ \forall f \in \mathcal{A} \text{ and } n \geq 1, \tag{17}$$

Proof. The integrand is bounded and C^∞ on $\mathbb{C} \setminus \mathbb{R}$, therefore (16) converges in norm as a limit of Riemann sums on any compact of $\mathbb{C} \setminus \mathbb{R}$. It remains to deal with the limit when these sets tend to the whole of $\mathbb{C}$. Let us introduce the sets

$$U = \{(x, y) \mid <x> \leq |y| \leq 2 <x>\} \supseteq \operatorname{supp} \tau'(y/<x>)$$
$$V = \{(x, y) \mid 0 \leq |y| \leq 2 <x>\} \supseteq \operatorname{supp} \tau(y/<x>).$$

We easily get by explicit computations that

$$|\sigma_x(x, y) + i\sigma_y(x, y)| \leq \frac{c\chi_U(x, y)}{<x>},$$

where χ_U is the characteristic function of the set U. Using the bound (11) on the resolvent, (15), and the fact that $|y| \simeq <x>$ on U, we can bound the integrand of (16) by a constant times

$$\sum_{r=0}^{n} |f^{(r)}(x)| <x>^{r-2} \chi_U(x, y) + |f^{(n+1)}(x)||y|^{n-1} \chi_V(x, y).$$

After integration on y at fixed x, the integrand of the remaining integral in x is bounded by a constant times

$$\sum_{r=0}^{n} |f^{(r)}(x)| <x>^{r-1} + |f^{(n+1)}(x)| <x>^n,$$

hence the announced bound. □

We need a few more properties regarding formula (16) before we can show it defines a functional calculus.

It is sometimes easier to deal with C_0^∞ functions rather then with functions of $\mathcal{A}$. The following Lemma shows this is harmless.

Lemma 4.7. *$C_0^\infty(\mathbb{R})$ is dense in $\mathcal{A}$ for the norms $\|\cdot\|_n$.*

Proof. We use the classical technique of mollifiers. Let $\Phi \geq 0$ be smooth with the same conditions of support as τ. Set $\Phi_m(x) = \Phi(x/m)$ for all $x \in \mathbb{R}$ and $f_m = \Phi_m f$. Hence, $f_m \in \mathcal{A}$ and support considerations yield

$$\|f - f_m\|_{n+1} = \sum_{r=0}^{n+1} \int_{\mathbb{R}} \left| \frac{d^r}{dx^r}(f(x)(1 - \Phi_m(x))) \right| < x >^{r-1} dx$$

$$\leq c_n \sum_{r=0}^{n+1} \int_{|x|>m} |f^{(r)}(x)| < x >^{r-1} dx \to 0, \text{ as } m \to \infty.$$

□

The next Lemma will be useful several times in the sequel.

Lemma 4.8. *If $F \in C_0^\infty(\mathbb{C})$ and $F(z) = O(y^2)$ as $y \to 0$ at fixed real x, then*

$$\frac{1}{\pi} \int_{\mathbb{C}} \frac{\partial}{\partial \bar{z}} F(z)(H - z)^{-1} dx dy = 0. \tag{18}$$

Proof. Suppose supp$F \subset \{|x| < N, |y| < N\}$ and let Ω_δ, $\delta > 0$ small such that $\Omega_\delta \subset \{|x| < N, \delta < |y| < N\}$. We want to apply Stokes Theorem to the above integral. Recall that

$$\frac{\partial}{\partial \bar{z}} = \frac{1}{2}\left(\frac{\partial}{\partial x} + i \frac{\partial}{\partial y}\right), \quad \frac{\partial}{\partial z} = \frac{1}{2}\left(\frac{\partial}{\partial x} - i \frac{\partial}{\partial y}\right)$$

$$\iff d\bar{z} = dx - idy, \; dz = dx + idy$$

so that $d\bar{z} \wedge dz = 2idx \wedge dy = 2idxdy$. Moreover, since $\frac{\partial}{\partial \bar{z}}(H - z)^{-1} = 0$ by analyticity,

$$\begin{aligned} d(F(z)(H - z)^{-1} dz) &= \frac{\partial}{\partial z}(F(z)(H - z)^{-1}) dz \wedge dz \\ &\quad + \frac{\partial}{\partial \bar{z}}(F(z)(H - z)^{-1}) d\bar{z} \wedge dz \\ &= \frac{\partial F}{\partial \bar{z}}(z)(H - z)^{-1} d\bar{z} \wedge dz. \end{aligned}$$

Therefore, if I denotes (18), we get by Stokes Theorem

$$\begin{aligned} I &= \lim_{\delta \to 0} \frac{1}{2\pi i} \int_{\Omega_\delta} d(F(z)(H - z)^{-1} dz) = \lim_{\delta \to 0} \frac{1}{2\pi i} \int_{\partial \Omega_\delta} F(z)(H - z)^{-1} dz \\ &= \lim_{\delta \to 0} \frac{1}{2\pi i} \int_{\substack{y=\delta \\ y=-\delta \\ |x|<N}} F(z)(H - z)^{-1} dz. \end{aligned}$$

Hence the bound

$$|I| \leq \lim_{\delta\to 0} \frac{1}{2\pi}\int_{-N}^{N} (|F(x+i\delta| + |F(x-i\delta|)\frac{1}{\delta}dx = \lim_{\delta\to 0} O(\delta) = 0,$$

where we used Taylor's formula $F(x,y) = \frac{y^2}{2}F_{yy}(x,\theta(y,x)y)$ with $\theta(x,y) \in (0,1)$, so that $|F(x,y)| \leq c(x)y^2$. □

Remark 4.9. It follows from the above proof that if f has compact support, we can write

$$f(H) = \lim_{\delta\to 0}\frac{1}{2\pi i}\int_{\partial\Omega_\delta} \tilde{f}(z)(H-z)^{-1}dz.$$

Neglecting support considerations, if $\tilde{f}$ was analytic, this is the way we would naturally define $f(H)$.

We can now show a comforting fact about our definition (16) of $f(H)$.

Lemma 4.10. *If $f \in \mathcal{A}$ and $n \geq 1$, then $f(H)$ is independent of σ and n.*

Proof. By density of C_0^∞ in $\mathcal{A}$ for the norms $\|\cdot\|_n$ and Lemma 4.6, we can assume $f \in C_0^\infty$. Let $\tilde{f}_{\sigma_1,n}$ and $\tilde{f}_{\sigma_2,n}$ be associated with σ_1 and σ_2. Then

$$\tilde{f}_{\sigma_1,n} - \tilde{f}_{\sigma_2,n} = \left(\sum_{r=0}^{n} f^{(r)}(x)\frac{(iy)^r}{r!}\right)(\tau_1(y/<x>) - \tau_2(y/<x>)),$$

is identically zero for y small enough, so Lemma 4.8 applies. Similarly, if $m > n \geq 1$, with similar notations,

$$\tilde{f}_{\sigma,m} - \tilde{f}_{\sigma,n} = \sum_{r=n+1}^{m} f^{(r)}(x)\frac{(iy)^r}{r!}\sigma(x,y) = O(y^2), \text{ as } y\to 0, \ x \text{ fixed},$$

and Lemma 4.8 applies again. □

We are now in a position to show that formula (16) possesses the properties of a functional calculus.

Proposition 4.11. *With the notations above,*

a. If $f \in C_0^\infty$ and supp$(f) \cap \sigma(H) = \emptyset$, then $f(H) = 0$.
b. $(fg)(H) = f(H)g(H)$, for all $f, g \in \mathcal{A}$.
c. $\bar{f}(H) = f(H)^*$ and $\|f(H)\| \leq \|f\|_\infty$.
d. $r_w(H) = (H-w)^{-1}$, $w \notin \mathbb{R}$.

Proof. a) In that case, since the compact set supp(f) and the closed set $\sigma(H)$ are disjoint, we can consider a finite number of contours $\gamma_1, \cdots, \gamma_r$ surrounding a region W disjoint from $\sigma(H)$ containing the support of $\tilde{f}$. By Stokes Theorem again

$$f(H) = \frac{1}{\pi}\int_{\mathbb{C}} \frac{\partial}{\partial\bar{z}}\tilde{f}(z)(H-z)^{-1}dxdy = \frac{1}{2\pi i}\sum_{j=1}^{r}\int_{\gamma_j} \tilde{f}(z)(H-z)^{-1}dz \equiv 0,$$

by our choice of γ_j.
b) Assume first $f, g \in C_0^\infty$, so that $K = \text{supp}(\tilde{f})$ and $L = \text{supp}(\tilde{g})$ are compact.

$$f(H)g(H) = \frac{1}{\pi^2}\int_{K\times L} \frac{\partial}{\partial \bar{z}}\tilde{f}(z)\frac{\partial}{\partial \bar{w}}\tilde{g}(w)(H-z)^{-1}(H-w)^{-1}dxdydudv$$
$$= \frac{1}{\pi^2}\int_{K\times L} \frac{\partial}{\partial \bar{z}}\tilde{f}(z)\frac{\partial}{\partial \bar{w}}\tilde{g}(w)\frac{(H-w)^{-1}-(H-z)^{-1}}{w-z}dxdydudv.$$

Then one uses the formula (easily proven using Stokes again)

$$\frac{1}{\pi}\int_K \frac{\partial}{\partial \bar{z}}\tilde{f}(z)\frac{dxdy}{w-z} = \tilde{f}(w),$$

the equivalent one for $\tilde{g}$ and one gets, changing variables to z,

$$f(H)g(H) = \frac{1}{\pi}\int_{K\cap L}\left(\tilde{g}(z)\frac{\partial}{\partial \bar{z}}\tilde{f}(z) + \tilde{f}(z)\frac{\partial}{\partial \bar{z}}\tilde{g}(z)\right)(H-z)^{-1}dxdy$$
$$= \frac{1}{\pi}\int_{K\cap L}\frac{\partial}{\partial \bar{z}}(\tilde{f}(z)\tilde{g}(z))(H-z)^{-1}dxdy.$$

It remains to see that if $k(z) = \tilde{f}(z)\tilde{g}(z) - \widetilde{fg}(z)$, the integral of $\frac{\partial}{\partial \bar{z}}k(z)$ against the resolvent on $\mathbb{C}$ is zero. But this is again a consequence of Lemma 4.8, since k has compact support and explicit computations yield $k(z) = O(y^2)$ as $y \to 0$ with x fixed. The generalization to functions of $\mathcal{A}$ is proven along the same lines as Lemma 4.7 with Lemma 4.6.
c) The first point follows from $(H-z)^{-1*} = (H-\bar{z})^{-1}$, the convergence in norm of (16) and the fact that $\overline{\tilde{f}(z)} = \tilde{\bar{f}}(\bar{z})$ if τ is even, which we can always assume. For the second point, take $f \in \mathcal{A}$ and $c > 0$ such that $\|f\|_\infty \leq c$. Defining $g(x) = c - (c^2 - |f(x)|^2)^{1/2}$, one checks that $g \in \mathcal{A}$ as well. The identity $f\bar{f} - 2cg + g^2 = 0$ in the algebra $\mathcal{A}$ implies with the above

$$f(H)f(H)^* - 2cg(H) + g(H)g(H)^* = 0$$
$$\Leftrightarrow f(H)^*f(H) + (c-g(H))^*(c-g(H)) = c^2.$$

Thus, for any $\psi \in \mathcal{H}$, it follows

$$\|f(H)\psi\|^2 \leq \|f(H)\psi\|^2 + \|(c-g(H))\psi\|^2 \leq c^2\|\psi\|^2,$$

where $c \geq \|f\|_\infty$ is arbitrary.
d) Let us take $n = 1$ and assume $\Im w > 0$. We further choose

$$\sigma(x,y) = \tau(\lambda y / < x >),$$

where $\lambda \geq 1$ will be chosen large enough so that w does not belong to the support of σ and then kept fixed in the rest of the argument. The sole effect of this manipulation is to change the support of τ, but everything we have done so far remains true for

$\lambda > 1$ and fixed. Let us define, for $m > 0$ large,

$$\Omega_m = \{(x,y) \,|\, |x| < m \text{ and } <x>/m < |y| < 2m\}.$$

Then, by definition and Stokes,

$$\begin{aligned} r_w(H) &= \lim_{m\to\infty} \frac{1}{\pi} \int_{\Omega_m} \frac{\partial}{\partial \bar{z}} \tilde{r}_w(z)(H-z)^{-1} dx dy \\ &= \lim_{m\to\infty} \frac{1}{2\pi i} \int_{\partial\Omega_m} \tilde{r}_w(z)(H-z)^{-1} dz, \end{aligned} \tag{19}$$

where, since $n = 1$,

$$\tilde{r}_w(z) = (r_w(x) + r'_w(x) iy)\sigma(x,y).$$

At this point, we want to replace $\tilde{r}_w(z)$ by $r_w(z)$ in (19). Indeed, it can be shown using the above explicit formula that

$$\lim_{m\to\infty} \left\| \int_{\partial\Omega_m} (r_w(z) - \tilde{r}_w(z))(H-z)^{-1} dz \right\| = 0,$$

by support considerations and elementary estimates on the different pieces of $\partial\Omega_m$. Admitting this fact we have

$$\begin{aligned} r_w(H) &= \lim_{m\to\infty} \frac{1}{2\pi i} \int_{\partial\Omega_m} r_w(z)(H-z)^{-1} dz \\ &= \text{res}(r_w(z)(H-z)^{-1})|_{z=w} = (H-w)^{-1}, \end{aligned}$$

due to the analyticity of the resolvent inside Ω_m. □

We can now state the first Spectral Theorem for the set $C_\infty(\mathbb{R})$ of continuous functions that vanish at infinity

$$C_\infty(\mathbb{R}) = \{f \in C(\mathbb{R}) \,|\, \forall \epsilon > 0, \exists K \text{ compact with } |f(x)| < \epsilon \text{ if } x \notin K\}.$$

Theorem 4.12. *There exists a unique linear map $f \mapsto f(H)$ from C_∞ to $\mathcal{L}(\mathcal{H})$ which is a functional calculus.*

Proof. Replacing C_∞ by $\mathcal{A}$ we have existence. Now, $C_0^\infty \subset \mathcal{A} \subset C_\infty$ and it is a classical fact that $\overline{C_0^\infty}^{\|\cdot\|_\infty} = C_\infty$, [4]. Hence $\mathcal{A}$ is dense in C_∞ in the sup norm. As $\|f(H)\| \leq \|f\|_\infty \, \forall f \in \mathcal{A}$, a density argument yields an extension of the map to C_∞ with convergence in norm. It is routine to check that all properties listed in Proposition 4.11 remain true for $f \in C_\infty$. The uniqueness property is shown as follows. If there exists another functional calculus, then, by hypothesis, it must agree with ours on the set of functions R

$$R = \{\sum_{i=1}^{n} \lambda_i r_{w_i}, \text{ where } \lambda_i \in \mathbb{C},\ w_i \notin \mathbb{R}\}.$$

But, it is a classical result also that the set R satisfies the hypothesis of the Stone-Weierstrass Theorem[b] and $R = C_\infty$, so that the two functional calculus must coincide everywhere. □

We shall pursue in two directions. We first want to show that any self-adjoint operator can be represented as a multiplication operator on some L^2 space. Then we shall extend the functional calculus to bounded measurable functions.

4.2 L^2 Spectral Representation

Let (H, D) be self-adjoint on $\mathcal{H}$.

Definition 4.13. *A closed linear subspace L of $\mathcal{H}$ is said* invariant *under H if $(H-z)^{-1}L \subseteq L$ for any $z \notin \mathbb{R}$.*

Remark 4.14. It is an exercise to show that if $\varphi \in L \cap D$, then $H\varphi \in L$, as expected.

Also, for any $\lambda \in \mathbb{R}$, $\mathrm{Ker}(H - \lambda)$ is invariant. If it is positive, the dimension of this subspace is called the multiplicity of the eigenvalue λ.

Lemma 4.15. *If L is invariant under $H = H^*$, then $L^\perp$ is invariant also. Moreover, $f(H)L \subseteq L$, for all $f \in C_\infty(\mathbb{R})$.*

Proof. The first point is straightforward and the second follows from the approximation of the integral representation (16) of $f(H)$ for $f \in \mathcal{A}$ by a norm convergent limit of Riemann sums and by a density argument for $f \in C_\infty(\mathbb{R})$. □

Definition 4.16. *For (H, D) self-adjoint on $\mathcal{H}$, the* cyclic subspace generated by the vector $v \in \mathcal{H}$ *is the subspace*

$$L = \overline{span\,\{(H - z)^{-1}v,\ z \notin \mathbb{R}\}}.$$

Remark 4.17. *i.* Cyclic subspaces are invariant under H, as easily checked.

ii. If the vector v chosen to generate the cyclic subspace is an eigenvector, then, this subspace is $\mathbb{C}v$.

iii. If the cyclic subspace corresponding to some vector v coincides with $\mathcal{H}$, we say that v is a cyclic vector for H.

iv. In the finite dimensional case, the matrix H has a cyclic vector v iff the spectrum of H is simple, i.e. all eigenvalues have multiplicity one.

These subspaces allow to structure the Hilbert space with respect to the action of H.

Lemma 4.18. *For (H, D) self-adjoint on $\mathcal{H}$, there exists a sequence of orthogonal cyclic subspaces $L_n \subset \mathcal{H}$ with cyclic vector v_n such that $\mathcal{H} = \overline{\oplus_{n=1}^N L_n}$, with N finite or not.*

[b] Let X be locally compact and consider $C_\infty(X)$. If B is a subalgebra of $C_\infty(X)$ that separates points and satisfies $f \in B \Rightarrow \bar{f} \in B$, then B is dense in $C_\infty(X)$ for $\|\cdot\|_\infty$

Proof. As $\mathcal{H}$ is assumed to be separable, there exists an orthonormal basis $\{f_j\}_{j\in\mathbb{N}}$ of $\mathcal{H}$. Let L_1 be the subspace corresponding to f_1. By induction, let us assume orthogonal cyclic subspaces $L_1, L_2, \cdots, L_n$ are given. Let $m(n)$ be the smallest integer such that $f_{m(n)} \notin L_1 \oplus \cdots \oplus L_n$ and let $g_{m(n)}$ be the component of $f_{m(n)}$ orthogonal to that subspace. We let L_{n+1} be the cyclic subspace generated by the vector $g_{m(n)}$. Then we have $L_{n+1} \perp L_r$, for all $r \le n$ and $f_{m(n)} \in L_1 \oplus \cdots \oplus L_n \oplus L_{n+1}$. Then either the induction continues indefinitely and $N = \infty$, or at some point, such a $m(n)$ does not exist and the sum is finite. □

The above allows us to consider each $H|_{L_n}$, $n = 1, 2 \cdots, N$ separately. Note, however, that the decomposition is not canonical.

Theorem 4.19. *Let (H, D) be self-adjoint on $\mathcal{H}$, separable. Let $S = \sigma(H) \subset \mathbb{R}$. Then there exists a finite positive measure μ on $S \times \mathbb{N}$ and a unitary operator $U : \mathcal{H} \to L^2 \equiv L^2(S \times \mathbb{N}, d\mu)$ such that if*

$$\begin{aligned} h : S \times \mathbb{N} &\to \mathbb{R} \\ (x, n) &\mapsto x, \end{aligned}$$

then $\xi \in \mathcal{H}$ belongs to D if and only if $hU\xi \in L^2$. Moreover,

$$UHU^{-1}\psi = h\psi, \ \forall \psi \in U(D) \subset L^2(S \times \mathbb{N}, d\mu) \text{ and}$$

$$Uf(H)U^{-1}\psi = f(h)\psi, \ \forall f \in C_\infty(\mathbb{R}), \ \psi \in L^2(S \times \mathbb{N}, d\mu).$$

This Theorem will be a Corollary of the

Theorem 4.20. *Let (H, D) be self-adjoint on $\mathcal{H}$ and $S = \sigma(H) \subset \mathbb{R}$. Further assume that H admits a cyclic vector v. Then, there exists a finite positive measure μ on S and a unitary operator $U : \mathcal{H} \to L^2(S, d\mu) \equiv L^2$ such that if*

$$\begin{aligned} h : S &\to \mathbb{R} \\ x &\mapsto x, \end{aligned}$$

then $\xi \in \mathcal{H}$ belongs to D if and only if $hU\xi \in L^2$ and

$$UHU^{-1}\psi = h\psi \ \forall \psi \in U(D) \subset L^2(S, d\mu).$$

Proof of Theorem 4.20. Let the linear functional $\Phi : C_\infty(\mathbb{R}) \to \mathbb{C}$ be defined by

$$\Phi(f) = \langle v | f(H) v \rangle.$$

The functional calculus shows that $\Phi(\bar{f}) = \overline{\Phi(f)}$. And if $0 \le f \in C_\infty(\mathbb{R})$, then, with $g = \sqrt{f}$, we have

$$\Phi(f) = \|g(H)v\|^2 \ge 0, \ \text{ i.e. } \Phi \text{ is positive.}$$

Thus, by Riesz-Markov Theorem[c], there exists a positive measure on $\mathbb{R}$ such that

---

[c] If X is a locally compact space, any positive linear functional l on $C_\infty(X)$ is of the form $l(f) = \int_X f \, d\mu$, where μ is a (regular) Borel measure with finite total mass.

$$\langle v|f(H)v\rangle = \int_{\mathbb{R}} f(x)d\mu(x), \ \forall f \in C_\infty(\mathbb{R}).$$

Since in case supp $(f) \cap \sigma(H) = \emptyset$, $f(H)$ is zero, we deduce that supp $(\mu) \subseteq S = \sigma(H)$. Also, note that f above belongs to $L^2(S, d\mu)$, since

$$\int |f(x)|^2 d\mu(x) = \langle v|f(H)^* f(H)v\rangle \leq \|f^2\|_\infty \|v\|^2 < \infty.$$

Consider the linear map $T : C_\infty(\mathbb{R}) \to L^2$ such that $Tf = f$. It satisfies for any $f, g \in C_\infty(\mathbb{R})$

$$\begin{aligned}\langle Tg|Tf\rangle_{L^2} &= \int_S \bar{g}(x)f(x)d\mu(x) = \Phi(\bar{g}f)\\ &= \langle v|g^*(H)f(H)v\rangle_{\mathcal{H}} = \langle g(H)v|f(H)v\rangle_{\mathcal{H}}.\end{aligned}$$

Defining

$$M = \{f(H)v \in \mathcal{H} \,|\, f \in C_\infty(\mathbb{R})\},$$

we have existence of an onto isomorphism U such that

$$U : M \to C_\infty(\mathbb{R}) \subseteq L^2 \text{ such that } Uf(H)v = f.$$

Now, M is dense in $\mathcal{H}$ since v is cyclic and $C_\infty(\mathbb{R})$ is dense in L^2, so that U admits a unitary extension from $\mathcal{H}$ to L^2.

Let $f_1, f_2, f \in C_\infty(\mathbb{R})$ and $\psi_i = f_i(H)v \in \mathcal{H}$. Then

$$\begin{aligned}\langle \psi_2|f(H)\psi_1\rangle_{\mathcal{H}} &= \langle f_2(H)v|(ff_1)(H)v\rangle_{\mathcal{H}}\\ &= \int_S \bar{f}_2(x)f(x)f_1(x)d\mu(x) = \langle U\psi_2|fU\psi_1\rangle_{L^2},\end{aligned}$$

where f denotes the obvious multiplication operator. In particular, if $f(x) = r_w(x) = (x-w)^{-1}$, we deduce that for all $\xi \in L^2$ and all $w \notin \mathbb{R}$

$$Ur_w(H)U^{-1}\xi = U(H-w)^{-1}U^{-1}\xi = r_w\xi. \tag{20}$$

Thus, U maps Ran$(H-w)$ to Ran(r_w), i.e. D and $\{\xi \in L^2 \,|\, x\xi(x) \in L^2\}$ are in one to one correspondence.

If $\xi \in L^2$ and $\psi = r_w\xi$, then $\psi \in D(h)$, where $D(h)$ is the domain of the multiplication operator by $h : x \mapsto x$. Then, with (20)

$$UHU^{-1}\psi = UHU^{-1}r_w\xi = UHr_w(H)U^{-1}\xi = wr_w\xi + \xi = xr_w\xi = h\psi.$$

□

Proof of Theorem 4.19. We know $\mathcal{H} = \overline{\oplus_n L_n}$ with cyclic subspaces L_n with vectors v_n. We will assume $\|v_n\| = 1/2^n$, $\forall n \in \mathbb{N}$. By Theorem 4.19, there exist μ_n of mass $\int_S d\mu_n = \|v_n\|^2 = 2^{-2n}$ and unitary operators $U_n : L_n \to L^2(S, d\mu_n)$ such that $H_n = H|_{L_n}$ is unitarily equivalent to the multiplication by x on $L^2(S, d\mu_n)$. Defining μ on $S \times \mathbb{N}$ by imposing $\mu|_{S\times\{n\}} = \mu_n$ and U by $\oplus_n U_n$, we get the result. □

In case $\mathcal{H} = \mathbb{C}^n$ and $H = H^*$ has simple eigenvalues λ_j with associated eigenvectors ψ_j, the measure can be chosen as $\mu = \sum_j \delta(x - \lambda_j)$ and $L^2 = L^2(\mathbb{R}, d\mu) = \mathbb{C}^n$. Note also that $\tilde{\mu} = \sum_j a_j \delta(x - \lambda_j)$, where $a_j > 0$ is as good a measure as μ to represent $\mathbb{C}^n$ as $L^2(\mathbb{R}, d\tilde{\mu})$.

Let us now extend our Spectral Theorem to $\mathcal{B}(\mathbb{R})$, the set of bounded Borel functions on $\mathbb{R}$.

Definition 4.21. *We say that $f_n \in \mathcal{B}(\mathbb{R})$ is* monotonically increasing *to $f \in \mathcal{B}(\mathbb{R})$, if $f_n(x)$ increases monotonically to $f(x)$ for any $x \in \mathbb{R}$.*

Thus, $\|f_n\| = \sup_{x\in\mathbb{R}} |f_n(x)|$ is uniformly bounded in n.

Theorem 4.22. *There exists a unique functional calculus $f \to f(H)$ from $\mathcal{B}(\mathbb{R})$ to $\mathcal{L}(\mathcal{H})$ if one imposes s-$\lim_{n\to\infty} f_n(H) = f(H)$ if $f_n \in \mathcal{B}(\mathbb{R})$ converges monotonically to $f \in \mathcal{B}(\mathbb{R})$.*

Recall that s-lim means limit in the strong sense, i.e. s-lim $A_n = A$ in $\mathcal{L}(\mathcal{H})$ is equivalent to $\lim_n A_n\varphi = A\varphi$, in $\mathcal{H}$, $\forall \varphi \in \mathcal{H}$.

Proof. Consider existence first. By unitary equivalence, we identify $\mathcal{H}$ and $L^2(S \times \mathbb{N}, d\mu)$ and H by multiplication by $h : (x, n) \mapsto x$. We define $f(H)$ for $f \in \mathcal{B}(\mathbb{R})$ by

$$f(H)\psi(x,n) = f(h(x,n))\psi(x,n) \quad \text{on} \quad L^2(S \times \mathbb{N}, d\mu),$$

which is easily shown to satisfy the properties of a functional calculus. Then, by the dominated convergence Theorem, if f_n converges monotonically to f:

$$\lim_{n\to\infty} f_n(H)\psi(x,m) = f(H)\psi(x,m).$$

Uniqueness is shown as follows. Consider two functional calculus with the mentioned properties. Let $\mathcal{C}$ be the subset of $\mathcal{B}(\mathbb{R})$ on which they coincide. We know $C_\infty(\mathbb{R}) \subset \mathcal{C}$ and $\mathcal{C}$ is closed by taking monotone limits. But the smallest set of functions containing $C_\infty(\mathbb{R})$ which is closed under monotone limits is $\mathcal{B}(\mathbb{R})$, see [4]. □

Remark 4.23. It all works the same if one considers functions f_n that converge pointwise to f and such that $\sup_n \|f_n\|_\infty < \infty$

We have the following Corollary concerning the resolvent.

Corollary 4.24. *With the hypotheses and notations above, $\sigma(H)$ is the essential range of h in $L^2(S, d\mu)$ and*

$$\|(H - z)^{-1}\| = \frac{1}{dist\,(z, \sigma(H))}$$

Proof. This follows directly from Theorem 4.20 and our study of multiplication operators. □

Another instance where our previous study of multiplication operators is useful is the case of constant coefficient differential operators on $\mathcal{S}(\mathbb{R}^N)$, the set of Schwartz functions. Such an operator L is defined by a finite sum of the form

$$L = \sum_\alpha a_\alpha D^\alpha, \tag{21}$$

where

$$\alpha = (\alpha_1, \cdots, \alpha_N) \in \mathbb{N}^N, \; a_\alpha \in \mathbb{C}, \; D^{\alpha_j} = \frac{\partial^{\alpha_j}}{\partial x_j^{\alpha_j}}, \; D^\alpha = D^{\alpha_1} \cdots D^{\alpha_N},$$

and L acts on functions in $\mathcal{S}(\mathbb{R}^N) \subset L^2(\mathbb{R}^N)$. This set of functions being invariant under Fourier transformation $\mathcal{F}$, this operator is unitarily equivalent to

$$\mathcal{F}L\mathcal{F}^{-1} = \sum_\alpha a_\alpha (ik)^\alpha, \text{ on } \mathcal{S}(\mathbb{R}^N) \subset L^2(\mathbb{R}^N).$$

The function $\sum_\alpha a_\alpha (ik)^\alpha$ is called the *symbol* of the differential operator. It is not difficult to get the following

Proposition 4.25. *Let L be the differential operator on $\mathbb{R}^n$ with constant coefficients defined in (21). Then L is symmetric iff its symbol is real valued in $\mathbb{R}^N$. In that case, $\bar{L}$ is self-adjoint and*

$$\sigma(\bar{L}) = \overline{\{\sum_\alpha a_\alpha (ik)^\alpha \,|\, k \in \mathbb{R}^N\}}.$$

Let us finally introduce spectral projectors in the general case of unbounded operators.

Theorem 4.26. *Let (H, D) be self-adjoint on $\mathcal{H}$ and (a, b) an open interval. Let f_n be an increasing sequence of non-negative continuous functions on $\mathbb{R}$ with support in (a, b) that converges to $\chi_{(a,b)}$, the characteristic function of (a, b). Then*

$$s\text{-}\lim_n f_n(H) = P_{(a,b)},$$

a canonical orthogonal projector, independent of $\{f_n\}$, that satisfies

$$P_{(a,b)} H \subset H P_{(a,b)}, \text{ and } P_{(a,b)} = 0 \iff (a, b) \cap \sigma(H) = \emptyset$$

Proof. The existence of the limit is ensured by Theorem 4.22 and and its properties are immediate. □

Remark 4.27. The fact that $P_{(a,b)}$ is a projector follows from the identity $\chi_{(a,b)} = \chi^2_{(a,b)}$, which makes $\chi_{(a,b)}$ a projector, when viewed as a multiplication operator. These projectors are called *spectral projectors* and their range $L_{(a,b)} = P_{(a,b)}\mathcal{H}$ are called *spectral subspaces*. These spectral subspaces satisfy

$$L_{(a,b)} \simeq L^2(E_{(a,b)}, d\mu), \text{ where } E_{(a,b)} = \{(x,n) \,|\, a < h(x,n) < b\},$$

and $\simeq$ denotes the unitary equivalence constructed in Theorem 4.19
It is also customary to represent a self-adjoint operator H by the Stieltjes integral

$$H = \int_{-\infty}^{\infty} \lambda dE(\lambda),$$

where $E(\lambda) = P(-\infty, \lambda)$ is projection operator valued. Let us justify this in case H has a cyclic vector, the general case following immediately. By polarization, it is enough to check that for $\xi \in D$

$$\langle \xi | H\xi \rangle_{\mathcal{H}} = \int \lambda d\langle \xi | E(\lambda)\xi \rangle = \int \lambda d\|E(\lambda)\xi\|^2.$$

By unitary equivalence to $L^2(\mathbb{R}, d\mu)$, if $\psi = U\xi$

$$\begin{aligned} d\|E(\lambda)\xi\|^2 &= d\int_{-\infty}^{\infty} |\chi_{(-\infty,\lambda)}(x)\psi(x)|^2 d\mu(x) \\ &= d\int_{-\infty}^{\lambda} |\psi(x)|^2 d\mu(x) = |\psi(\lambda)|^2 d\mu(\lambda). \end{aligned}$$

Hence

$$\langle \xi | H\xi \rangle_{\mathcal{H}} = \int \lambda |\psi(\lambda)|^2 d\mu(\lambda) = \langle \psi | h\psi \rangle_{L^2}.$$

We close this Section about the Spectral Theorem by some results in perturbation theory of unbounded operators.

Definition 4.28. *Let (H, D) and (H_n, D_n) be a sequence of self-adjoint operators on $\mathcal{H}$. We say that $H_n \to H$ in the* norm resolvent sense *if*

$$\lim_{n\to\infty} \|(H_n + i)^{-1} - (H + i)^{-1}\| = 0.$$

The point $i \in \mathbb{C}$ plays no particular role as the following Lemma shows.

Lemma 4.29. *If $z = x + iy \in \mathbb{C} \setminus \mathbb{R}$, and*

$$g(x, y) = \sup_{h\in\mathbb{R}} \left| \frac{h + i}{h - x - iy} \right|,$$

then, there exists a constant c such that

$$\|(H_n - z)^{-1} - (H - z)^{-1}\| \leq c g(x, y) \|(H_n + i)^{-1} - (H + i)^{-1}\|.$$

Proof. Once the identity

$$(H_n - z)^{-1} - (H - z)^{-1} = \\ (H_n + i)(H_n - z)^{-1}((H_n + i)^{-1} - (H + i)^{-1})(H + i)(H - z)^{-1} \tag{22}$$

is established, the Lemma is a consequence of the bound following from the Spectral Theorem

$$\|(H + i)(H - z)^{-1}\| \leq g(x, y).$$

If one doesn't take care of domain issues, (22) is straightforward. We refer to [1] for a careful proof of (22). □

Remark 4.30. If $z \in \rho(H) \cap \rho(H_n)$, the result is similar.

Theorem 4.31. *If $H_n \to H$ in the norm resolvent sense, then*

$$\lim_{n\to\infty} \|f(H_n) - f(H)\| = 0, \ \ \forall f \in C_\infty.$$

Remark 4.32. The result is not generally true if $f \in \mathcal{B}(\mathbb{R})$. Consider spectral projectors, for instance. See also Corollary 4.33 below.

Proof. If $f \in \mathcal{A}$, the definition yields,

$$\begin{aligned} \|f(H) - f(H_n)\| &\leq \frac{1}{\pi}\int_{\mathbb{C}} \left|\frac{\partial \tilde{f}(z)}{\partial \bar{z}}\right| \|(H_n - z)^{-1} - (H - z)^{-1}\| dxdy \\ &\leq \frac{4c}{\pi}\int_{\mathbb{C}} \left|\frac{\partial \tilde{f}(z)}{\partial \bar{z}}\right| g(x, y) dxdy \, \|(H_n + i)^{-1} - (H + i)^{-1}\|. \end{aligned}$$

It is not difficult to see that the last integral is finite, due to the properties of $\frac{\partial \tilde{f}(z)}{\partial \bar{z}}$ and of $g(x, y)$. The convergence in norm is established for $f \in \mathcal{A}$ and the extension of the result to $f \in C_\infty(\mathbb{R})$ comes from the extension of the functional calculus to those f's and by density. □

We have the following spectral consequences.

Corollary 4.33. *If $H_n \to H$ in the norm resolvent sense, we have convergence of the spectrum of H_n to that of H in the following sense:*

$$\begin{aligned} &\lambda \in \mathbb{R} \setminus \sigma(H) \ \Rightarrow \ \lambda \notin \sigma(H_n), \ \ n \textit{ large enough} \\ &\lambda \in \sigma(H) \ \Rightarrow \ \exists \lambda_n \in \sigma(H_n), \ \textit{such that} \ \lim_{n\to\infty} \lambda_n = \lambda. \end{aligned}$$

Proof. If $\lambda \in \mathbb{R} \setminus \sigma(H)$, there exists $f \in C_0^\infty(\mathbb{R})$ whose support is disjoint from $\sigma(H)$ and which is equal to 1 in a neighborhood of λ. Then, Theorem 4.31 implies $\|f(H_n)\| \to 0$ and, in turn, the Spectral Theorem implies $\lambda \notin \sigma(H_n)$ if $\|f(H_n)\| < 1$. Conversely, if $\lambda \in \sigma(H)$, pick an $\epsilon > 0$ and a $f \in C_0^\infty(\mathbb{R})$ such that $f(\lambda) = 1$ and supp $(f) \subset (\lambda - \epsilon, \lambda + \epsilon)$. From $\lim_n \|f(H_n)\| = 1$ follows that $\sigma(H_n) \cap (\lambda - \epsilon, \lambda + \epsilon) \neq \emptyset$, if n is as large as we wish. □

5 Stone's Theorem, Mean Ergodic Theorem and Trotter Formula

The Spectral Theorem allows to prove easily Stone's Theorem, which characterizes one parameter evolution groups which we define below. Such groups are those giving the time evolution of a wave function ψ in Quantum Mechanics governed by the Schrödinger equation

$$i\hbar\frac{\partial}{\partial t}\psi(t) = H\psi(t), \quad \text{with } \psi(0) = \psi_0,$$

where H is the Hamiltonian.

Definition 5.1. *A* one-parameter evolution group on a Hilbert space *is a family* $\{U(t)\}_{t\in\mathbb{R}}$ *of unitary operators satisfying* $U(t+s) = U(t)U(s)$ *for all* $t, s \in \mathbb{R}$ *and* $U(t)$ *is strongly continuous in* t *on* $\mathbb{R}$.

Remark 5.2. It is easy to check that strong continuity at 0 is equivalent to strong continuity everywhere and that weak continuity is equivalent to strong continuity in that setting.

Actually, we have equivalence between the following statements: the map $t \mapsto \langle\varphi|U(t)\psi\rangle$ is measurable for all φ, ψ and $U(t)$ is strongly continuous, see [4] for a proof.

Theorem 5.3. *Let* (A, D) *be self-adjoint on* $\mathcal{H}$ *and* $U(t) = e^{itA}$ *given by functional calculus. Then*

a. $\{U(t)\}_{t\in\mathbb{R}}$ *forms a one parameter evolution group and* $U(t) : D \to D$ *for* $t \in \mathbb{R}$.
b. For any $\psi \in D$,

$$\frac{U(t)\psi - \psi}{t} \to iA\psi \;\; as \;\; t \to 0.$$

c. Conversely,

$$\lim_{t\to 0}\frac{U(t)\psi - \psi}{t} \;\; exists \;\; \Rightarrow \psi \in D.$$

Proof. a) follows from the Functional Calculus and the properties of $x \mapsto f_t(x) = e^{itx}$. Similarly b) is a consequence of the functional calculus (see Theorem 4.22) applied to $x \mapsto g_t(x) = (e^{itx} - 1)/t$ and of the estimate $|e^{ix} - 1| \le |x|$.
c) Define

$$D(B) = \left\{\psi \mid \lim_{t\to 0}\frac{U(t)\psi - \psi}{t} \text{ exists}\right\}, \text{ and } B\psi = \lim_{t\to 0}\frac{U(t)\psi - \psi}{it} \text{ on } D(B). \tag{23}$$

One checks that B is symmetric and b) implies $B \supseteq A$. But $A \subseteq B \subseteq \bar{B}$ and $A = A^*$ is closed, thus $A = B$. □

Remark 5.4. The formula (23) defines the so-called *infinitesimal generator* of the evolution group $U(t)$.

The converse of that result is Stone's Theorem.

Theorem 5.5. *If $\{U(t)\}_{t\in\mathbb{R}}$ forms a one parameter evolution group on $\mathcal{H}$, then there exists (A, D) self-adjoint on $\mathcal{H}$ such that $U(t) = e^{iAt}$.*

Proof. The idea of the proof is to define A as the infinitesimal generator on a set of good vectors and show that A is essentially self-adjoint. Then one shows that $U(t) = e^{i\bar{A}t}$.

Let $f \in C_0^\infty(\mathbb{R})$ and define

$$\varphi_f = \int_{\mathbb{R}} f(t)U(t)\varphi\, dt, \quad \forall \varphi \in \mathcal{H}.$$

Let D be the set of finite linear combinations of such φ_f, with different φ and f.
1) D is dense: Let $j_\epsilon(x) = j(x/\epsilon)/\epsilon$, where $0 \le j \in C_0^\infty$ with support in $[-1,1]$ and $\int j(x)\, dx = 1$. Then, for any φ,

$$\begin{aligned}\|\varphi_{j_\epsilon} - \varphi\| &= \left\| \int j_\epsilon(t)(U(t)\varphi - \varphi)dt \right\| \\ &\le \left(\int_{-\infty}^{\infty} j_\epsilon(t)dt \right) \sup_{|t|\le\epsilon} \|U(t)\varphi - \varphi\| \to 0 \text{ as } \epsilon \to 0.\end{aligned}$$

2) Infinitesimal generator on D: Let $\varphi_f \in D$.

$$\begin{aligned}\frac{(U(s) - \mathbb{1})}{s}\varphi_f &= \int_{\mathbb{R}} f(t)\frac{U(t+s) - U(t)}{s}\varphi dt \\ &= \int_{\mathbb{R}} \frac{f(\tau - s) - f(\tau)}{s} U(\tau)\varphi d\tau \\ &\overset{s\to 0}{\longrightarrow} - \int_{\mathbb{R}} f'(\tau)U(\tau)\varphi d\tau = \varphi_{-f'}.\end{aligned}$$

Hence, we set for $\varphi_f \in D$,

$$A\varphi_f = \frac{1}{i}\varphi_{-f'} = \lim_{t\to 0} \frac{U(t) - \mathbb{1}}{it}\varphi_f$$

and it is easily checked that

$$U(t) : D \to D, \;\; A : D \to D, \;\; \text{and} \;\; U(t)A\varphi_f = AU(t)\varphi_f = \frac{1}{i}\varphi_{-f'(\cdot - t)}.$$

Moreover, for $\varphi_f, \varphi_g \in D$,

$$\langle \varphi_g | A\varphi_f \rangle = \lim_{s\to 0} \left\langle \varphi_f | \frac{U(s) - \mathbb{1}}{is}\varphi_g \right\rangle = \lim_{s\to 0} \left\langle \frac{U(-s) - \mathbb{1}}{-is}\varphi_f | \varphi_g \right\rangle = \langle A\varphi_f | \varphi_g \rangle,$$

so that A is symmetric.
3) A is essentially self-adjoint: Assume there exists $\psi \in D^* = D(A^*)$ such that $A^*\psi = i\psi$. Then, $\forall \varphi \in D$,

$$\frac{d}{dt}\langle\psi|U(t)\varphi\rangle = \langle\psi|iAU(t)\varphi\rangle = i\langle A^*\psi|U(t)\varphi\rangle = \langle\psi|U(t)\varphi\rangle.$$

Hence, solving the differential equation,

$$\langle\psi|U(t)\varphi\rangle = \langle\psi|\varphi\rangle e^t.$$

As $\|U(t)\| = 1$, this implies $\langle\psi|\varphi\rangle = 0$, so that $\psi = 0$ as D is dense. A similar reasoning holds for any $\chi \in \mathrm{Ker}(A^* + i)$, so that A is essentially self-adjoint and $\bar{A}$ is self-adjoint.

4) $U(t) = e^{i\bar{A}t}$: Let $\varphi \in D \subseteq \bar{D} = D(\bar{A})$. On the one hand,

$$e^{i\bar{A}t}\varphi \in \bar{D} \quad \text{and} \quad \frac{d}{dt}e^{i\bar{A}t}\varphi = i\bar{A}e^{i\bar{A}t}\varphi,$$

by b) Theorem 5.3. On the other hand, $U(t)\varphi \in D \subset \bar{D}$ for all t. Thus, introducing $\psi(t) = U(t)\varphi - e^{i\bar{A}t}\varphi$; we compute

$$\psi'(t) = iAU(t)\varphi - i\bar{A}e^{i\bar{A}t}\varphi = i\bar{A}\psi(t), \quad \text{with} \quad \psi(0) = 0,$$

so that $\frac{d}{dt}\|\psi(t)\|^2 \equiv 0$, hence $\psi(t) \equiv 0$. As D is dense, $e^{i\bar{A}t} \equiv U(t)$. □

Examining the above proof, one deduces the following Corollary which can be useful in applications.

Corollary 5.6. *Let (A, D) be self-adjoint on $\mathcal{H}$ et $E \subset D$ be dense. If, for all $t \in \mathbb{R}$, $e^{itA} : E \to E$, then $(A|_E, E)$ is essentially self-adjoint.*

Remark 5.7. *i.* In the situation of the Corollary, one says that E is a *core* for A.
ii. The solution to the following equation, in the strong sense,

$$\frac{d}{dt}\varphi(t) = iA\varphi(t), \ \forall t \in \mathbb{R} \ \varphi(0) = \varphi_0 \in D$$

is unique and is given by $\varphi(t) = e^{iAt}\varphi_0$.

In case A is bounded, the evolution group generated by A can be obtained from the power series of the exponential. This relation remains true in a certain sense when A is unbounded and self-adjoint. Indeed, if φ belongs to the dense set $\cup_{M\geq 0}P(-M,M)\mathcal{H}$, where $P(-M,M)$ denotes the spectral projectors of A, we get

$$\sum_{k=0}^{N}\frac{(itA)^k}{k!}\varphi \to e^{itA}\varphi, \ \text{as} \ N \to \infty.$$

This formula makes sense due to the fact that $\varphi \in \cap_{n\geq 0}D(A^n)$, where $D(A^n)$ is the domain of A^n. Stone's Theorem provides a link between evolution groups and self-adjoint generators, therefore one can expect a relation between essentially self-adjoint operators and the existence of sufficiently many vectors for which the above formula makes sense.

Definition 5.8. *Let A be an operator on a Hilbert space $\mathcal{H}$. A vector φ is called an* analytic vector *if $\varphi \in \cap_{n\geq 0} D(A^n)$ and*

$$\sum_{k=0}^{\infty} \frac{\|A^k \varphi\|}{k!} t^k < \infty, \quad \textit{for some } t > 0.$$

The relation alluded to above is provided by the following criterion for essential self-adjointness.

Theorem 5.9 (Nelson's Analytic Vector Theorem). *Let (A, D) be symmetric on a Hilbert space $\mathcal{H}$. If D contains a total set of analytic vectors, then (A, D) is essentially self-adjoint.*

We refer the reader to [4] for a proof and we proceed by providing a link between the discrete spectrum of the self-adjoint operator H with the evolution operator e^{itH} it generates. This is the so-called

Theorem 5.10 (Mean Ergodic Theorem). *Let P_λ be the spectral projector on an eigenvalue λ of a self-adjoint operator H of domain $D \in \mathcal{H}$. Then*

$$P_\lambda = s - \lim_{t_2 - t_1 \to \infty} \frac{1}{t_2 - t_1} \int_{t_1}^{t_2} e^{itH} e^{-it\lambda} dt.$$

Proof. One can assume $\lambda = 0$ by considering $H - \lambda$ if necessary.
i) If $\varphi \in P_0\mathcal{H}$, then $H\varphi = 0$ and $e^{itH}\varphi = \varphi$, so that

$$\frac{1}{t_2 - t_1} \int_{t_1}^{t_2} e^{itH} \varphi dt = \varphi = P_0 \varphi, \quad \text{for any } t_1, t_2.$$

ii) If $\varphi \in \text{Ran}(H)$, i.e. $\varphi = H\psi$ for some $\psi \in D$, then

$$e^{itH}\varphi = e^{itH} H \psi = -i \frac{d}{dt} e^{itH} \psi,$$

so that we can write

$$\frac{1}{t_2 - t_1} \int_{t_1}^{t_2} e^{itH} \varphi dt = -\frac{i}{t_2 - t_1} (e^{it_2 H} - e^{it_1 H}) \psi \to 0 = P_0 \psi.$$

The result is thus proven for $\text{Ran}(H)$ and $\text{Ran}(P_0)$. As the integral is uniformly bounded, the result is true on $\overline{\text{Ran}(H)}$ as well and we conclude by

$$\mathcal{H} = \overline{\text{Ran}(H)} \oplus \text{Ker}(H) = \overline{\text{Ran}(H)} \oplus P_0 \mathcal{H},$$

which follows from $H = H^*$. □

With a little more efforts, one can prove in the same vein

Theorem 5.11 (von Neumann's Mean Ergodic Theorem). *If V is such that, uniformly in n, $\|V^n\| \leq C$ and P_1 projects on Ker$(V - \mathbb{1})$, then*

$$\frac{1}{N}\sum_{n=0}^{N-1} V^n\varphi \to P_1\varphi, \text{ as } N \to \infty.$$

Remark 5.12. *i.* The projector P_1 is not necessarily self-adjoint.

ii. The projection on Ker$(V - \lambda)$ where $|\lambda| = 1$ is obtained by replacing V by V/λ in the Theorem.

iii. It follows from the assumption that $\sigma(V) \subseteq \{z \,|\, |z| \leq 1\}$, since the spectral radius spr$(V) = \lim_{n\to\infty} \|V^n\|^{1/n} = 1$.

iv. A proof can be found in [5].

Another link between the spectrum of its generator and the behavior of an evolution group arises when a vector is transported away from its initial value at $t = 0$ by the evolution exponentially fast as $|t| \to \infty$.

Proposition 5.13. *Let (H, D) be self-adjoint on $\mathcal{H}$ and assume there exists a normalized vector $\varphi \in \mathcal{H}$ such that, for any $t \in \mathbb{R}$ and for some positive constants A, B,*

$$|\langle\varphi|e^{itH}\varphi\rangle| \leq Ae^{-B|t|}.$$

Then $\sigma(H) = \mathbb{R}$.

Proof. Taking φ as first vector in the decomposition provided in Lemma 4.18, we have by the Spectral Theorem, that on $L^2(d\mu_1)$, the restriction of $L^2(d\mu)$ unitary equivalent to that first cyclic subspace,

$$\varphi \simeq 1, \quad e^{itH}\varphi \simeq e^{itx}1,$$

so that

$$\langle\varphi|e^{itH}\varphi\rangle = \int e^{itx}\, d\mu_1(x) \equiv f(t).$$

This f admits a Fourier transform $\omega \mapsto \hat{f}(\omega)$ which is analytic in a strip $\{\omega \,|\, |\Im\omega| < B\}$. Therefore, we have $d\mu_1(x) = \hat{f}(x)dx$ and the support of $d\mu_1 = \mathbb{R}$. Hence, $\sigma(H) \supset \mathbb{R}$. □

Let us close this Section by a result concerning evolution groups generated by sums of self-adjoint operators.

Theorem 5.14 (Trotter product formula). *Let (A, D_A), (B, D_B) be self-adjoint and $A + B$ be essentially self-adjoint on $D_A \cap D_B$. Then*

$$e^{i(A+B)t} = s\text{-}\lim_{n\to\infty}\left(e^{itA/n}e^{itB/n}\right)^n, \quad \forall t \in \mathbb{R}.$$

If, moreover, A and B are bounded from below,

$$e^{-(A+B)t} = s\text{-}\lim_{n\to\infty}\left(e^{-tA/n}e^{-tB/n}\right)^n, \quad \forall t \geq 0.$$

Remark 5.15. The operator e^{-Ct}, $t \geq 0$ for C self-adjoint and bounded below can be defined via the Spectral Theorem applied to the function f defined as follows: $f(x) = e^{-x}$, if $x \geq x_0$ and $f(x) = 0$ otherwise, with x_0 small enough.

Proof. (partial). We only consider the first assertion under the significantly simplifying hypothesis that $A + B$ is self-adjoint on $D = D_A \cap D_B$. The second assertion is proven along the same lines.

Let $\psi \in D$ and consider

$$s^{-1}(e^{isA}e^{isB} - \mathbb{1})\psi = s^{-1}(e^{isA} - \mathbb{1})\psi + s^{-1}e^{isA}(e^{isB} - \mathbb{1})\psi \to iA\psi + iB\psi$$

as $s \to 0$ and

$$s^{-1}(e^{is(A+B)} - \mathbb{1})\psi \to i(A+B)\psi \text{ as } s \to 0.$$

Thus, setting $K(s) = s^{-1}(e^{isA}e^{isB} - e^{is(A+B)})$, the vector $K(s)\psi \to 0$ as $s \to 0$ or $s \to \infty$, for any $\psi \in D$. Due to the assumed self-adjointness of $A + B$ on D, $A + B$ is closed so that D equipped with the norm

$$\|\psi\|_{A+B} = \|\psi\| + \|(A+B)\psi\|$$

is a Banach space. As $K(s) : D \to \mathcal{H}$ is bounded for each finite s and tends to zero strongly at 0 and ∞, we can apply the uniform boundedness principle or Banach-Steinhaus Theorem [d] to deduce the existence of a constant C so that

$$\|K(s)\psi\| \leq C\|\psi\|_{A+B}, \quad \forall s \in \mathbb{R}, \ \forall \psi \in D.$$

Therefore, on any compact set of D in the $\|\cdot\|_{A+B}$ norm, $K(s)\psi \to 0$ uniformly as $s \to 0$. We know from Theorem 5.3 that $e^{is(A+B)} : D \to D$ and is strongly continuous, thus $\{e^{is(A+B)}\psi \mid s \in [-1,1]\}$ is compact in D for $\|\cdot\|_{A+B}$, for ψ fixed.

Hence

$$t^{-1}(e^{itA}e^{itB} - e^{it(A+B)})e^{is(A+B)}\psi \to 0$$

uniformly in $s \in [-1,1]$ as $t \to 0$. Therefore, writing

$$\left(\left(e^{itA/n}e^{itB/n}\right)^n - \left(e^{it(A+B)/n}\right)^n\right)\psi =$$

$$\sum_{k=0}^{n-1}\left(e^{itA/n}e^{itB/n}\right)^k\left[e^{itA/n}e^{itB/n} - e^{it(A+B)/n}\right]\left(e^{it(A+B)/n}\right)^{n-1-k}\psi,$$

we get that the RHS is bounded in norm by

$$|t| \max_{|s|<t}\left\|\left(\frac{t}{n}\right)^{-1}\left[e^{itA/n}e^{itB/n} - e^{it(A+B)/n}\right]e^{is(A+B)}\psi\right\|.$$

---

[d] If $\mathcal{X}$ and $\mathcal{Y}$ are two Banach spaces, and $\mathcal{F}$ is a family of bounded linear operators from $\mathcal{X}$ to $\mathcal{Y}$ such that for each $x \in \mathcal{X}$, $\{\|Tx\|_{\mathcal{Y}} \mid T \in \mathcal{F}\}$ is bounded, then $\{\|T\| \mid T \in \mathcal{F}\}$ is bounded.

Thus, if $\psi \in D$, we get

$$\left(e^{itA/n}e^{itB/n}\right)^n \psi \to e^{it(A+B)}\psi, \quad \text{as } n \to \infty.$$

The operators being bounded and D being dense, this finishes the proof. □

6 One-Parameter Semigroups

This Section extends some results of the previous one in the sense that unitary groups discussed above are a particular case of semigroups. The setting used in this Section is that of a Banach space denoted by $\mathcal{B}$. One parameter semigroups will be used in the study the time evolution of *Open* Quantum Systems.

Definition 6.1. *Let* $\{S(t)\}_{t\geq 0}$ *be a family of bounded operators defined on* $\mathcal{B}$. *We say that* $\{S(t)\}_{t\geq 0}$ *is a* strongly continuous semigroup *or* C_0 *semigroup if*

1. $S(0) = \mathbb{1}$.
2. $S(t+s) = S(t)S(s)$ *for any* $s, t \geq 0$.
3. $S(t)\varphi$ *is continuous as a function of* t *on* $[0, \infty)$, *for all* $\varphi \in \mathcal{B}$.

Remark 6.2. 3) is equivalent to requiring continuity at 0^+ only.

Definition 6.3. *The* infinitesimal generator *of the semigroup* $\{S(t)\}_{t\geq 0}$, *is the linear operator* (A, D) *defined by*

$$D = \{\varphi \in \mathcal{B} \mid \lim_{t\to 0^+} t^{-1}(S(t) - \mathbb{1})\varphi \text{ exists in } \mathcal{B}\}$$

$$A\varphi = \lim_{t\to 0^+} \frac{(S(t) - \mathbb{1})\varphi}{t}, \quad \varphi \in D.$$

The main properties of semigroups and their generators are listed below.

Proposition 6.4. *Let* $\{S(t)\}_{t\geq 0}$ *be a semigroup of generator A. Then*

a. There exist $\omega \in \mathbb{R}$ *and* $M \geq 1$ *such that* $\|S(t)\| \leq Me^{\omega t}$, *for all* $t \geq 0$.
b. The generator A is closed with dense domain D.
c. For any $t \geq 0$, $\varphi \in D$, *we have* $\int_0^t S(\tau)\varphi d\tau \in D$ *and*

$$A\left(\int_0^t S(\tau)d\tau\right) = S(t)\varphi - \varphi.$$

d. For any $t \geq 0$, $S(t) : D \to D$ *and if* $\varphi \in D$, $t \mapsto S(t)\varphi$ *is in* $C^1([0,\infty))$ *and*

$$\frac{d}{dt}S(t)\varphi = AS(t)\varphi = S(t)A\varphi, \quad t \geq 0.$$

e. If $\{S_1(t)\}_{t\geq 0}$ *and* $\{S_2(t)\}_{t\geq 0}$ *are two* C_0 *semigroups with the same generator A, then* $S_1(t) \equiv S_2(t)$.

Proof. a) By the right continuity at 0 and the Banach-Steinhaus Theorem, there exists $\epsilon > 0$ and $M \geq 1$ such that $\|S(t)\| \leq M$ if $t \in [0, \epsilon]$. For any given $t \geq 0$, there exists $n \in \mathbb{N}$ and $0 \leq \delta < \epsilon$ such that $t = n\epsilon + \delta$, hence

$$\|S(t)\| = \|S(\delta)S(\epsilon)^n\| \leq M^{n+1} \leq MM^{t/\epsilon} \equiv Me^{\omega t}, \text{ with } \omega = \frac{\ln M}{\epsilon} \geq 0.$$

c) For $\varphi \in D$, $t \geq 0$ and any $\epsilon > 0$,

$$\frac{(S(\epsilon) - \mathbb{1})\varphi}{\epsilon} \int_0^t S(\tau)\varphi = \frac{1}{\epsilon} \int_0^t (S(\tau + \epsilon) - S(\tau))\varphi d\tau$$
$$= \frac{1}{\epsilon} \int_t^{t+\epsilon} S(\tau)\varphi - \frac{1}{\epsilon} \int_0^{\epsilon} S(\tau)\varphi$$

which converges to $S(t)\varphi - \varphi$ as $\epsilon \to 0$.
d) For $\varphi \in D$, $t \geq 0$ and any $\epsilon > 0$, we have

$$\frac{(S(\epsilon) - \mathbb{1})\varphi}{\epsilon} S(t)\varphi = S(t) \frac{(S(\epsilon) - \mathbb{1})\varphi}{\epsilon} \varphi \to S(t)A\varphi, \text{ as } \epsilon \to 0. \quad (24)$$

Thus $S(t)\varphi \in D$ and $AS(t) = S(t)A$ on D. As a consequence of (24) , the function $t \mapsto S(t)\varphi$ has a right derivative given by $S(t)A\varphi$ which is continuous on $[0, \infty)$. Therefore, a classical result of analysis shows that the derivative at $t \geq 0$ exists.
b) Let φ and define φ_ϵ for any $\epsilon > 0$ by $\varphi_\epsilon = \frac{1}{\epsilon} \int_0^\epsilon S(\tau)\varphi d\tau$. The vector $\varphi_\epsilon \in D$, by c) and $\varphi_\epsilon \to \varphi$ and $\epsilon \to 0$, so that D is dense. Closedness of A is shown as follows. Let $\{\varphi_n\}_{n \in \mathbb{N}}$ be a sequence of vectors in D, such that $\varphi_n \to \varphi$ and $A\varphi_n \to \psi$, for some φ and ψ. For any $n \in \mathbb{N}$ and $t > 0$, d) implies by integration

$$S(t)\varphi_n - \varphi_n = \int_0^t S(\tau)A\varphi_n d\tau.$$

Taking limits $n \to \infty$ we get $S(t)\varphi - \varphi = \int_0^t S(\tau)\psi d\tau$, therefore

$$\lim_{t \to 0+} t^{-1}S(t)\varphi - \varphi = \psi.$$

In other words, $\psi \in D$ and $\psi = A\varphi$.
e) Finally, for $\varphi \in D$, $t > 0$, we define, for $\tau \in [0, T]$, $\psi(\tau) = S_1(t - \tau)S_2(\tau)\varphi$. In view of c) we are allowed to differentiate w.r.t. τ and we get

$$\frac{d}{d\tau}\psi(\tau) = -S_1(t - \tau)AS_2(\tau)\varphi + S_1(t - \tau)AS_2(\tau)\varphi = 0,$$

hence $\psi(0) = \psi(t)$, i.e. $S_1(t)\varphi = S_2(t)\varphi$. The density of D concludes the proof. □

Remark 6.5. If $\{S(t)\}_{t \geq 0}$ is a semigroup that is continuous in norm, i.e. such that $\|S(t) - \mathbb{1}\| \to 0$ as $t \to 0^+$, it is not difficult to show that (see e.g. [3]) that there exists $A \in \mathcal{L}(\mathcal{B})$ such that $S(t) = e^{At}$ for any $t \geq 0$.

Definition 6.6. *Semigroups characterized by the bound* $\|S(t)\| \leq 1$ *for all* $t \geq 0$ *are called* contraction semigroups .

Remark 6.7. There is no loss of generality in studying contraction semigroups in the sense that if the semigroup $\{S(t)\}_{t\geq 0}$ satisfies the bound a) in the Proposition above, we can consider the new C_0 semigroup $S_1(t) = e^{-\beta t}S(t)$ satisfying $\|S_1(t)\| \leq M$. At the price of a change of the norm of the Banach space, we can turn it into a contraction semigroup. Let

$$|||\varphi||| = \sup_{t\geq 0} \|S_1(t)\varphi\|,$$

such that $\|\varphi\| \leq |||\varphi||| \leq M\|\varphi\|$ and $|||S_1(\tau)\varphi||| \leq |||\varphi|||$.

The characterization of generators of C_0 semigroups is given by the

Theorem 6.8 (Hille Yosida). *Let* (A, D) *be a closed operator with dense domain. The following statements are equivalent:*

1. *The operator* A *generates a* C_0 *semigroup* $\{S(t)\}_{t\geq 0}$ *satisfying*

$$\|S(t)\| \leq Me^{\omega t}, \ t \geq 0$$

2. *The resolvent and resolvent set of* A *are such that for all* $\lambda > \omega$ *and all* $n \in \mathbb{N}^*$

$$\rho(A) \supset (\omega, \infty) \ \ \textit{and} \ \ \|(A-\lambda)^{-n}\| \leq \frac{M}{(\lambda-\omega)^n}$$

3. *The resolvent and resolvent set of* A *are such that for all* $n \in \mathbb{N}^*$

$$\rho(A) \supset \{\lambda \in \mathbb{C} \,|\, \Re\lambda > \omega\} \ \ \textit{and} \ \ \|(A-\lambda)^{-n}\| \leq \frac{M}{(\Re\lambda-\omega)^n}, \ \ \Re\lambda > \omega.$$

Remark 6.9. A proof of this Theorem can be found in [5] or [3]. We simply note here that A generates of a contraction semigroup, $M = 1, \omega = 0$, if and only if $\|(A-\lambda)^{-1}\| \leq \frac{1}{\Re\lambda}$, $\Re\lambda > 0$. This is trivially true if $A = iH$ in a Hilbert space where $H = H^*$, as expected. Moreover, in that case, A and $-A$ satisfy the bound, so that we can construct a group, as we already know, generated by iH.

Let us specialize a little by considering contraction semigroups on a Hilbert space $\mathcal{H}$. We can characterize their generator by roughly saying that their real part in non-positive.

Definition 6.10. *An operator* (A, D) *on a Hilbert space* $\mathcal{H}$ *is called* dissipative *if for any* $\varphi \in D$

$$\langle\varphi|A\varphi\rangle + \langle A\varphi|\varphi\rangle = 2\Re(\langle\varphi|A\varphi\rangle) \leq 0.$$

Proposition 6.11. *Let* A *be the generator of a* C_0 *semigroup* $\{S(t)\}_{t\geq 0}$ *on* $\mathcal{H}$. *Then* $\{S(t)\}_{t\geq 0}$ *is a contraction semigroup iff* A *is dissipative.*

Proof. Assume $S(t)$ is a contraction semigroup and $\varphi \in D$. Consider $f(t) = \langle S(t)\varphi | S(t)\varphi \rangle$. As a function of t, it is differentiable and

$$0 \geq f'(0) = \langle \varphi | A\varphi \rangle + \langle A\varphi | \varphi \rangle.$$

Conversely, if A is dissipative, as we have $S(t) : D \to D$, we get for any $t \geq 0$,

$$f'(t) = \langle S(t)\varphi | S(t)A\varphi \rangle + \langle AS(t)\varphi | S(t)\varphi \rangle \leq 0,$$

so $f(t)$ is monotonically decreasing and $\|S(t)\varphi\| \leq \|\varphi\|$. As D is dense, $\{S(t)\}_{t\geq 0}$ is a contraction semigroup. □

Actually, the notion of dissipative operator can be generalized to the Banach space setting. Moreover, it is still true that dissipative operators and generators of C_0 contraction semigroups are related. This is the content of the Lumer Phillips Theorem stated below.

Let $\mathcal{B}$ be a Banach space and let $\mathcal{B}'$ be its dual. The value of $l \in \mathcal{B}'$ at $\varphi \in \mathcal{B}$ is denoted by $\langle l, \varphi \rangle \in \mathbb{C}$. Let us define for any $\varphi \in \mathcal{B}$ the duality set $F(\varphi) \subset \mathcal{B}'$ by

$$F(\varphi) = \{l \in \mathcal{B}' \mid \langle l, \varphi \rangle = \|\varphi\|^2 = \|l\|^2\}.$$

By the Hahn-Banach Theorem, $F(\varphi) \neq \emptyset$ for any $\varphi \in \mathcal{B}$.

Definition 6.12. *An operator (A, D) on a Banach space $\mathcal{B}$ is called* dissipative *if for any $\varphi \in D$, there exists $l \in F(\varphi)$ such that $\Re(\langle l, A\varphi \rangle) \leq 0$.*

The following characterization of dissipative operators avoiding direct duality considerations can be found in [3]:

Proposition 6.13. *An operator (A, D) is dissipative if and only if*

$$\|(\lambda \mathbb{1} - A)\varphi\| \geq \lambda \|\varphi\|, \ \textit{for all } \varphi \in D \textit{ and all } \lambda > 0.$$

The link between dissipativity and contraction semigroups is provided by the

Theorem 6.14 (Lumer Phillips). *Let (A, D) be an operator with dense domain in a Banach space $\mathcal{B}$.*

a. If A is dissipative and there exists $\lambda_0 > 0$ such that $Ran(\lambda_0 \mathbb{1} - A) = \mathcal{B}$, then A is the generator of a C_0 contraction semigroup.
b. If A is the generator of a C_0 contraction semigroup on $\mathcal{B}$, then one has $Ran(\lambda \mathbb{1} - A) = \mathcal{B}$ for all $\lambda > 0$ and A is dissipative.

The proof of this result can be found in [3], for example.

We close this Section by considerations on the perturbation of semigroups, or more precisely, of their generators. We stick to our Banach space setting for the end of the Section.

We first show that the property of being a generator is stable under perturbation by bounded operators.

Theorem 6.15. *Let A be the generator of a C_0 semigroup $\{S(t)\}_{t\geq 0}$ on $\mathcal{B}$ which satisfies the bound 1) in Theorem 6.8. Then, if B is a bounded operator, the operator $A+B$ generates a C_0 semigroup $\{V(t)\}_{t\geq 0}$ that satisfies the same bound with $M \mapsto M$ and $\omega \mapsto \omega + M\|B\|$. Moreover, if B is replaced by xB, the semigroup generated by $A+xB$ is an entire function of the variable $x \in \mathbb{C}$.*

Proof. If $A+B$ generates a contraction semigroup $V(t)$, it must solve the following differential equation on $D(A+B) = D(A)$

$$\frac{d}{dt}V(t) = (A+B)V(t).$$

Introducing $S(t)$, the solution of that equation must solve

$$V(t) = S(t) + \int_0^t S(t-s)BV(s)ds. \tag{25}$$

Strictly speaking the above integral equation is true on $D(A)$ only, but as $D(A)$ is dense and all operators are bounded, (25) is true on $\mathcal{B}$, as a strong integral. We solve this equation by iteration

$$V(t) = \sum_{n\geq 0} S_n(t), \tag{26}$$

where

$$S_{n+1}(t) = \int_0^t S(t-s)BS_n(s)ds, \quad n = 0, 1, \ldots, \quad \text{and} \quad S_0(t) = S(t).$$

All integrals are strongly continuous and we have the bounds

$$\|S_n(t)\| \leq M^{n+1}\|B\|^n e^{\omega t}t^n/n!$$

which are proven by an easy induction. The starting estimate is true by hypothesis on $S(t)$. Thus we see that (26) is absolutely convergent and satisfies

$$\|V(t)\| \leq Me^{(\omega+M\|B\|)t}.$$

To show that $V(t)$ defined this way is actually generated by $A+B$, we multiply (25) by $e^{-\lambda t}$ and integrate over $[0,\infty]$, assuming $\Re\lambda > \omega + M\|B\|$, to get

$$\begin{aligned} r(\lambda) &= \int_0^\infty e^{-\lambda t}V(t)dt \\ &= \int_0^\infty e^{-\lambda t}S(t)dt + \left(\int_0^\infty e^{-\lambda t}S(t)dt\right) B \left(\int_0^\infty e^{-\lambda t}V(t)dt\right) \\ &= -(A-\lambda)^{-1} - (A-\lambda)^{-1}Br(\lambda). \end{aligned}$$

This yields $(A+B-\lambda)r(\lambda) = -\mathbb{1}$ and, as $\lambda \in \rho(A+B)$ by our choice of λ, it follows by computations we went through already that $r(\lambda) = -(A+B-\lambda)^{-1}$. We finally use Hille Yosida Theorem to conclude. For any $k = 0, 1, 2, \cdots$,

$$\|(A+B-\lambda)^{-k-1}\| = \frac{1}{k!}\left\|\frac{d^k}{d\lambda^k}r(\lambda)\right\| \leq \frac{1}{k!}\int_0^\infty t^k|e^{-\lambda t}|\|V(t)\|dt$$
$$\leq \frac{M}{k!}\int_0^\infty t^k e^{-(\Re\lambda-\omega-M\|B\|)t}dt = M(\Re\lambda-\omega-M\|B\|)^{-k-1},$$

which shows that the resolvent of $A+B$ satisfies the estimate of the point 3) of Hille Yosida's Theorem.

Finally, we note that $V(t)$ has the form of a converging series in powers of B, which proves the last statement. □

More general perturbations of generators of semigroups are allowed under the supplementary hypothesis that both the unperturbed generator and the perturbation generate contraction semigroups.

Theorem 6.16. *Let A and B be generators of contraction semigroups and assume B is relatively bounded w.r.t. A with relative bound smaller that 1/2. Then, $A+B$ generates a contraction semigroup.*

For a proof, see [2], for example.

References

1. E. B. Davies, *Spectral Theory and Differential operators*, Cambridge University Press, 1995.
2. T. Kato, *Perturbation Theory of Linear Operators*, CIM, Springer 1981.
3. A. Pazy, *Semigroups of Linear Operators and Applications to Partial Differential Equations*, Springer, 1983.
4. M. Reed, B. Simon, *Methods of Modern Mathematical Physics*, Vol 1-4, Academic Press, 1971-1978.
5. K. Yosida, *Functional Analysis*, CIM, Springer 1980.

Introduction to Quantum Statistical Mechanics

Alain Joye

Institut Fourier, Université de Grenoble 1,
BP 74, 38402 Saint-Martin-d'Hères Cedex, France
e-mail: alain.joye@ujf-grenoble.fr

This set of lectures is intended to provide a flavor of the physical ideas underlying some of the concepts of Quantum Statistical Mechanics that will be studied in this school devoted to Open Quantum Systems. Although it is quite possible to start with the mathematical definitions of notions such as "bosons", "states", "Gibbs prescription" or "entropy" for example and prove theorems about them, we believe it can be useful to have in mind some of the heuristics that lead to their precise definitions in order to develop some intuition about their properties.

Given the width and depth of the topic, we shall only be able to give a very partial account of some of the key notions of Quantum Statistical Mechanics. Moreover, we do not intend to provide proofs of the statements we make about them, nor even to be very precise about the conditions under which these statements hold. The mathematics concerning these notions will come later. We only aim at giving plausibility arguments, borrowed from physical considerations or based on the analysis of simple cases, in order to give substance to the dry definitions.

Our only hope is that the mathematically oriented reader will benefit somehow from this informal introduction, and that, at worse, he will not be too confused by the many admittedly hand waving arguments provided.

Some of the many general references regarding an aspect or the other of these lectures are provided at the end of these notes.

1 Quantum Mechanics

We provide in this section an introduction to the quantum description of a physical system, starting from the Hamiltonian description of Classical Mechanics. The quantization procedure is illustrated for the standard kinetic plus potential Hamiltonian by means of the usual recipe. A set of postulates underlying the quantum description of systems is introduced and motivated by means of that special though important case. These aspects, and much more, are treated in particular in [2] and [4], for instance.

1.1 Classical Mechanics

Let us recall the Hamiltonian version of Classical Mechanics in the following typical setting, neglecting the geometrical content of the formalism. Consider N particles in $\mathbb{R}^d$ of coordinates $q_k \in \mathbb{R}^d$, masses m_k and momenta $p_k \in \mathbb{R}^d$, $k = 1, \cdots, N$, interacting by means of a potential

$$\begin{aligned} V : \mathbb{R}^{dN} &\to \mathbb{R} \\ q &\mapsto V(q). \end{aligned} \tag{1}$$

The space $\mathbb{R}^{dN}$ of the coordinates $(q_1, q_2, \cdots, q_N)$, with $q_{k,j} \in \mathbb{R}$, $j = 1 \cdots, d$ which we shall sometimes denote collectively by q (and similarly for p), is called the configuration space and the space $\Gamma = \mathbb{R}^{dN} \times \mathbb{R}^{dN} = \mathbb{R}^{2dN}$ of the variables (q, p) is called the phase space. A point (q, p) in phase space characterizes the *state* of the system and the *observables* of the systems, which are the physical quantities one can measure on the system, are given by functions defined on the phase space. For example, the potential is an observable. The Hamiltonian $H : \Gamma \to \mathbb{R}$ of the above system is defined by the observable

$$H(p, q) = \sum_{k=1}^{N} \frac{p_k^2}{2m_k} + V(q_1, q_2, \cdots, q_N), \tag{2}$$
$$\text{where } V(q_1, q_2, \cdots, q_N) = \sum_{i<j} V_{ij}(|q_i - q_j|),$$

which coincides with the sum of the kinetic and potential energies. The equations of motion read for all $k = 1, \cdots, N$ as

$$\dot{q_k} = \frac{\partial}{\partial p_k} H(q, p), \quad \dot{p_k} = -\frac{\partial}{\partial q_k} H(q, p), \quad \text{with } (q(0), p(0)) = (q^0, p^0), \tag{3}$$

where $\frac{\partial}{\partial q_k}$ denotes the gradient with respect to q_k. The equations (3) are equivalent to Newton's equations, with $p_k = m_k \dot{q}_k$,

$$m_k \ddot{q}_k = -\frac{\partial}{\partial q_k} V(q) \text{ with } (q(0), \dot{q}(0)) = (q^0, \dot{q}^0),$$

for all $k = 1, \cdots, N$. In case the Hamiltonian is time independent, i.e. if the potential V is time independent, the total energy E of the system is conserved

$$E = H(q(0), p(0)) \equiv H(q(t), p(t)), \quad \forall t. \tag{4}$$

where $(q(t), p(t))$ are solutions to (3) with initial conditions $(q(0), p(0))$. More generally, a system is said to be Hamiltonian if its equations of motions read as (3) above. We shall essentially only deal with systems governed by Hamiltonians that are time-independent.

These evolution equations are also called *canonical equations of motion*. Changes of coordinates

$$p_k \mapsto P_k, \; q_k \mapsto Q_k, \text{ such that } H(q,p) \mapsto K(Q,P)$$

which conserve the form of the equations of motions, i.e.

$$\dot{Q}_k = \frac{\partial}{\partial P_k} K(Q,P), \; \dot{P}_k = -\frac{\partial}{\partial Q_k} K(Q,P), \quad \text{with } (Q(0), P(0)) = (Q^0, P^0), \tag{5}$$

are called *canonical transformations*. The energy conservation property (4) is just a particular case of time dependence of a particular observable. Assuming the Hamiltonian is time independent, but not necessarily given by (2), the time evolution of any (smooth) observable $B : \Gamma \to \mathbb{R}$ defined on phase space computed along a classical trajectory $B_t(q,p) \equiv B(q(t), p(t))$ is governed by the Liouville equation

$$\frac{d}{dt} B_t(q,p) = L_H B_t(q,p), \quad \text{with } B_0(q,p) = B(q,p), \tag{6}$$

where the linear operator L_H acting on the vector space of observables is given by

$$L_H = \nabla_p H(q,p) \cdot \nabla_q - \nabla_q H(q,p) \cdot \nabla_p, \tag{7}$$

with the obvious notation. Observables which are constant along the trajectories of the system are called *constant of the motions*. Introducing the Lebesgue measure on $\Gamma = \mathbb{R}^{2dN}$,

$$d\mu = \Pi_{k=1}^N dq_k dp_k, \text{ with } dq_k = \Pi_{j=1}^d q_{k,j}, \text{ and } dp_k = \Pi_{j=1}^d p_{k,j},$$

and the Hilbert space $L^2(\Gamma, d\mu)$, one checks that L_H is formally anti self-adjoint on $L^2(\Gamma, d\mu)$, (i.e. antisymmetric on the set of observables in $C_0^\infty(\Gamma)$). Therefore, the formal solution to (6) given by

$$B_t(q,p) = e^{tL_H} B_0(q,p)$$

is such that e^{tL_H} is unitary on $L^2(\Gamma, d\mu)$. Another expression of this fact is *Liouville's Theorem* stating that

$$\frac{\partial(q(t), p(t))}{\partial(q^0, p^0)} \equiv 1,$$

where the LHS above stands for the Jacobian of the transformation $(q^0, p^0) \mapsto (q(t), p(t))$. It is convenient for the quantization procedure to follow to introduce the *Poisson bracket* of observables B, C on $L^2(\Gamma)$ by the definition

$$\{B, C\} = \nabla_q B \cdot \nabla_p C - \nabla_p B \cdot \nabla_q C. \tag{8}$$

For example,

$$\{q_{k,m}, p_{j,n}\} = \delta_{(j,n),(k,m)}, \ \{p_{j,n}, p_{k,m}\} = \{q_{j,n}, q_{k,m}\} = 0, \tag{9}$$

which are particular cases of

$$\{q_{k,m}, F(q,p)\} = \frac{\partial F(q,p)}{\partial p_{k,m}}, \ \{G(q,p), p_{j,n}\} = \frac{\partial G(q,p)}{\partial q_{j,n}}.$$

These brackets fulfill Jacobi's relations,

$$\{A, \{B, C\}\} + \text{circular permutations} = 0 \tag{10}$$

and, as $\{H, B\} = -L_H B$, we can rewrite (6) by means of Poisson brackets as

$$\frac{d}{dt} B_t = -\{H, B_t\}. \tag{11}$$

Therefore, it follows that the Poisson bracket of two constants of the motion is a constant of the motion, though not necessarily independent from the previous ones.

Before we proceed to the quantization procedure, let us introduce another Hamiltonian system we will be interested in later on. It concerns the evolution of N identical particles of mass m and charge e in $\mathbb{R}^3$, interacting with each other and with an external electromagnetic field (E, B).

Let us recall Maxwell's equations for the electromagnetic field

$$\nabla B = 0, \quad \nabla \wedge E = -\frac{\partial B}{\partial t}, \quad \epsilon_0 \nabla E = \rho_e, \quad \nabla \wedge B = \mu_0 j + \frac{1}{c^2}\frac{\partial E}{\partial t}, \tag{12}$$

where ρ_e and j denote the density of charges and of current, respectively, the constant ϵ_0 and μ_0 are characteristics of the vacuum in which the fields propagate, and c is the speed of light. A particle of mass m and charge e in presence of an electromagnetic field obeys the Newtonian equation of motion determined by the Lorentz force

$$m\ddot{q} = e(E + \dot{q} \wedge B). \tag{13}$$

When N charged particles interact with the electromagnetic field, the rule is that each of them becomes a source for the fields and obeys (13), with the densities given by

$$\rho_e(x,t) = \sum_{j=1}^{N} e\delta(x - q_j(t)), \text{ and } j = \sum_{j=1}^{N} e\dot{q}_j(t)\delta(x - q_j(t)). \tag{14}$$

In order to have a Hamiltonian description of this dynamics later, we introduce the scalar potential V and the *vector potential* A associated to the electromagnetic field (E, B). They are defined so that

$$E = -\frac{\partial A}{\partial t} - \nabla V, \quad B = \nabla \wedge A, \tag{15}$$

hence the first two equations (12) are satisfied, and

$$\begin{aligned} &\frac{\partial}{\partial t}(\nabla A) + \Delta V = -\epsilon_0^{-1}\rho_e \\ &\frac{1}{c^2}\frac{\partial^2}{\partial t^2}A - \Delta A + \nabla(\nabla A + \frac{1}{c^2}\frac{\partial V}{\partial t}) = \mu_0 j \end{aligned} \tag{16}$$

yield the last two equations of (12). There is some freedom in the choice of A and V in the sense that the physical fields E and B are unaffected by a change of the type

$$V \mapsto V + \frac{\partial \chi}{\partial t}, \quad A \mapsto A - \nabla \chi,$$

where χ is any scalar function of (x, t). A transformation of this kind is called a *gauge transformation*. This allows in particular to choose the potential vector A so that it satisfies

$$\nabla A = 0, \tag{17}$$

by picking a χ solution to $\Delta\chi = \nabla A$, if (17) is not satisfied. This is the so called *Coulomb gauge* in which (16) reduces to

$$\begin{aligned} &\Delta V = -\epsilon_0^{-1}\rho_e \iff V(x,t) = \frac{1}{4\pi\epsilon_0}\int dy \frac{\rho_e(y,t)}{|x-y|} \\ &\frac{1}{c^2}\frac{\partial^2}{\partial t^2}A - \Delta A = \mu_0 j - \frac{1}{c^2}\frac{\partial \nabla V}{\partial t} \equiv \mu_0 j_T. \end{aligned} \tag{18}$$

The subscript T stands here for transverse, since $\nabla j_T \equiv 0$. Assume we have N particles of identical masses m and identical charges e interacting with the electromagnetic field satisfying (18) with (14) so that

$$V(x,t) = \frac{e}{4\pi\epsilon_0}\sum_{j=1}^{N}\frac{1}{|x - q_j(t)|}. \tag{19}$$

We want to write down a Hamiltonian function which will yield (13) back when we compute the equation of motions for the particles as (3). It is just a matter of computation to show that the following (time-dependent) Hamiltonian fulfills this requirement:

$$H(q,p,t) = \sum_j \frac{1}{2m}|p_j - eA(q,t)|^2 + \frac{1}{8\pi\epsilon_0}\sum_{i\neq j}\frac{e^2}{|q_i - q_j|}. \tag{20}$$

The only thing to note is that when one computes the part of the electric field E that is produced by the ∇V part of (15) at the point $q_j(t)$, the double sum that appears due to the form (19) of V contains an infinite part when the indices take the same value, which is simply ignored .

Remark: It follows also from the above that the momentum p_j of the j'th particle isn't proportional to its velocity, but is given instead by

$$p_j = mv_j + eA(q_j, t).$$

This is an important feature of systems interacting with electromagnetic fields. As noted above, this Hamiltonian is time dependent. We can get a time independent Hamiltonian provided one takes also into account the energy of the field in the Hamiltonian. This new Hamiltonian H^{tot} reads

$$H^{tot} = H(q,p,t) + \frac{1}{2} \int dx \left(\epsilon_0 |\partial A(x,t)/\partial t|^2 + \mu_0^{-1} |\nabla \wedge A(x,t)|^2 \right),$$

where the first term in the integral is the contribution to the electric field that is not provided by the Coulomb potential (19) (which is taken into account in (20)) and the second term is the magnetic energy. It can be shown also that the total energy H^{tot} is conserved.

1.2 Quantization

The Quantum description of a general classical system is given by a set of postulates we list here as **P1** to **P4**. In order to motivate and/or illustrate their meaning, we consider in parallel the typical Hamiltonian (2) to make the link with its quantization by means of the traditional recipe.

P1: *The phase space Γ is replaced by a Hilbert space $\mathcal{H}$ whose scalar product is denoted by $\langle \cdot | \cdot \rangle$, with anti-linearity on the left. The state of the system is characterized by a ray in this space, that is a unit vectors with an arbitrary phase.*

Actually, rays characterize the *pure states* of the system. When we consider Quantum Statistical Mechanics, we will make a distinction between pure states and mixed states that will be introduced then. However, in that section, we will go on talking about states.

In case of our example, $\mathcal{H} = L^2(\mathbb{R}^{dN})$, $\mathbb{R}^{dN}$ being the configuration space. The state of the system is characterized by a normalized complex valued function $\psi(q)$ in $L^2(\mathbb{R}^{dN})$, also called the *wave function* of the system.

P2: *The observables are given by (possibly unbounded) self-adjoint linear operators on $\mathcal{H}$ obtained from their classical counterparts by a quantization procedure.*

The quantization procedure is not always straightforward. In particular, if the phase space has a non trivial topological structure, sophisticated methods of quantizations have to be applied. The link with the corresponding classical observables should be achieved, at a formal level at least, by taking the limit $\hbar \to 0$.

For our example, the formal substitutions

$$p_k \mapsto -i\hbar\nabla_k, \quad q_k \mapsto \text{mult}\, q_k, \quad k = 1, \cdots, N \tag{21}$$

are used. Here mult q_k denotes the operator multiplication by the variable q_k, which we shall simply denote by q_k below, and $\hbar$ is Planck's constant, whose numerical value is about $1.055 \times 10^{-34} J.s.$ In particular, the classical Hamiltonian (2) gives rise to the (formally) self-adjoint operator

$$H = \sum_{k=1}^{N} -\frac{\hbar^2}{2m_k}\Delta_k + V(q_1, q_2, \cdots, q_N) \text{ on } \mathcal{H}, \tag{22}$$

where Δ_k denotes the Laplacian in the variables q_k. This class of operators goes under the name *Schrödinger operators* and plays, for obvious reasons, an important role in Quantum Mechanics. Note that the quantization of observables by the formal rule (21) may need to be precised by a symmetrization procedure due to the non-commutativity of p and q,

$$[p_{j,n}, q_{k,m}] = \frac{\hbar}{i}[\partial_{q_{j,n}}, q_{k,m}] = \frac{\hbar}{i}\mathbb{1}, \tag{23}$$

where $\mathbb{1}$ denotes the identity operator. The symmetrization can be performed by hand in some concrete cases. For example, the dilation operator (for $N = 1$) is the (self-adjoint) quantization of $p \cdot q$ given by

$$p \cdot q \mapsto \frac{-i\hbar}{2}(\nabla_q q + q\nabla_q).$$

Note that in dimension $d = 3$, the angular momentum $x \wedge p$ vector doesn't require symmetrization and yields the operator

$$x \wedge p \mapsto J := -i\hbar q \wedge \nabla_q, \tag{24}$$

whose components satisfy the relations

$$[J_i, J_j] = i\hbar J_k, \quad \text{for } (i, j, k) \text{ a permutation of } (1, 2, 3). \tag{25}$$

The (components of the) angular momentum are unbounded operators which are known to have discrete spectrum, see below.

For a general classical observable $B(q, p)$ (belonging to some reasonable class of smooth real valued functions on $\Gamma \simeq \mathbb{R}^{2d}$, say) the Weyl quantization procedure $B \mapsto B^W$ defined by

$$(B^W\psi)(q) = (2\pi\hbar)^{-d}\int\int B\left(\frac{q+q'}{2}, p\right) e^{i(q-q')\cdot p/\hbar}\psi(q')dq'dp$$

is a good prescription to obtain the corresponding self-adjoint observables. It maps functions of $q \in \mathbb{R}^d$ to the corresponding multiplication operators and polynomials in p_j to the same polynomials in the differential operator $\frac{\hbar}{i}\partial_{q_j}$. Note, however, that there exists other quantization prescriptions that have their own merits.

Also, in other cases, if the geometry of the classical phase space of the system has more structure, the formal operator (22) needs to be supplemented by boundary conditions determined by physical considerations. For example, if the system is confined to a region Λ in configuration space, one customarily provides $\partial\Lambda$ with Dirichlet boundary conditions. In particular, the Hamiltonian of a particle in $\mathbb{R}^d$ confined to a cube Λ whose sides have length L is given by

$$H_\Lambda = -\frac{\hbar^2}{2m}\Delta \text{ plus Dirichlet boundary condition at } \partial\Lambda. \tag{26}$$

P3: *The result of the measure of an observable B on the quantum system characterized by $\psi \in \mathcal{H}$ is an element $b \in \mathbb{R}$ of the spectrum $\sigma(B)$ of the self-adjoint operator B. Moreover, the probability to obtain an element in $(b_1, b_2]$ as the result of this measure on the state ψ is given by*

$$\mathbb{P}_\psi(B \in (b_1, b_2]) = \|P_B((b_1, b_2])\psi\|^2, \tag{27}$$

where $P_B(I)$ denotes the spectral projector of the operator B on the set $I \subset \mathbb{R}$. Furthermore, once a measure of B is performed, and the result yields a value in a set $(b_1, b_2] \subset \mathbb{R}$, the wave function ψ is reduced, i.e. it undergoes the instantaneous change

$$\psi \mapsto \frac{P_B((b_1, b_2])\psi}{\|P_B((b_1, b_2])\psi\|}. \tag{28}$$

Another observable C is said to be compatible with B if B and C commute, i.e. if

$$[P_B(\alpha), P_C(\beta)] = 0, \text{ for any intervals } \alpha, \beta \subseteq \mathbb{R}.$$

This postulate explains the importance of the efforts made by mathematical physicists in order to determine the spectral properties of operators related to Quantum Mechanics. As, in general, the spectrum of a self-adjoint operator is the union of its discrete and continuous components, the result of the measure of an observable may be quantized, even though its classical counterpart may take values in an interval. This justifies the adjective Quantum for the theory.

Several examples of this fact will be discussed in the lectures on the spectral theory of unbounded operators, in particular for Schrödinger operators of the form $-\Delta + V$. Note that due to the Spectral Theorem, the expectation value of an observable B in a state ψ can be written as

$$\mathbb{E}_\psi(B) = \int_{\sigma(B)} b\|P(db)\psi\|^2 = \int_{\sigma(B)} b\langle\psi|P(db)\psi\rangle = \langle\psi|B\psi\rangle.$$

The reduction processes of the wave function (28) after the measurement of B insures that an immediate subsequent measure of B gives a result in the same set

$(b_1, b_2]$ with probability one. The compatibility condition insures that the observables B and C can be simultaneously measured, in the sense that once B and C have been measured with results in the sets α and β respectively, further successive measurements of B and C, in any order, will give results in the same sets with probability one. Or in other words, B and C can be diagonalized simultaneously. This is not the case if B and C do not commute.

Let us introduce some very classical examples as illustrations of the above. The interpretation of the wave function, in the setting where $\Gamma = \mathbb{R}^{2dN}$, is that $|\psi(q)|^2 dq$ is the probability that the system is at point q of configuration space and if $\hat{\psi}(p)$ denotes the Fourier transform of ψ, $|\hat{\psi}(p)|^2 dp$ is the probability that it has momentum p. This is just a particular case of the above rule. Indeed, the operators $q_{k,m}$, $k = 1, \cdots, N$, $m = 1, \cdots, d$ all commute and they have continuous spectra $\mathbb{R}$, as multiplication operators. Hence the interpretation of $|\psi(q)|^2$ follows. That of $|\psi(p)|^2$ is a consequence of the fact that the Fourier transform is unitary on $L^2(\mathbb{R}^{dN})$ which transforms the derivative into a multiplication by the independent variable. Note that (23) shows that $q_{j,n}$ and $p_{j,n}$ cannot be simultaneously determined, an expression of *Heisenberg's uncertainty principle*. Actually, Heisenberg's principle can be put on more quantitative grounds as follows. Let A and B be two self-adjoint operators such that their commutator can be written as

$$[A, B] = iC, \quad \text{where } C = C^*.$$

Then, denoting the variance of A in the state ψ by

$$\Delta_\psi(A)^2 = \langle \psi | (A - \mathbb{E}_\psi(A))^2 \psi \rangle,$$

we get the inequality

$$\Delta_\psi(A)\Delta_\psi(B) \geq \frac{\mathbb{E}_\psi(C)}{2}. \tag{29}$$

Applied to the operators p and q, we get the familiar relation

$$\Delta_\psi(p)\Delta_\psi(q) \geq \frac{\hbar}{2}.$$

Similarly, the components of the angular momentum J_k, $k = 1, 2, 3$ are not compatible, however it follows from (24) that J_k and $J^2 := J_1^2 + J_2^2 + J_3^2$ are compatible observables, for any k. Hence one can measure the third component of the angular momentum and its length simultaneously. The result of such measures belongs to the spectra of these operators which is discrete as we recall here. If we denote by $\mathcal{K}_j$ the eigenspace of J^2 associated with the quantum number j and consider the restriction of J_3 to that subspace, a classical algebraic computation shows that

$$\sigma(J^2) = \{j(j+1) | j \in \mathbb{N}/2\} \text{ and } \sigma(J_3|_{\mathcal{K}_j}) = \{-j, -j+1, \cdots, j-1, j\}. \tag{30}$$

The effect of boundary conditions on the spectrum of operators can be quite dramatic, as the following comparison shows. The Hamiltonian (22) (with $N = 1$ for simplicity) with $V \equiv 0$, the so-called free Hamiltonian $H_0 = -\frac{\hbar^2}{2m}\Delta$ is unitarily equivalent by Fourier transform to a multiplication operator

$$H_0 \simeq \text{mult } k^2 \text{ on } L^2(\mathbb{R}^d).$$

Its spectrum is then $\sigma(H_0) = \mathbb{R}^+$. The spectrum of the operator (26) is easily computed to be

$$\sigma(H_\Lambda) = \left\{ \frac{2\pi^2\hbar^2}{mL^2}(n_1^2 + n_2^2 + \cdots + n_d^2) \mid n_j \in \mathbb{Z} \right\} \tag{31}$$

Another celebrated Hamiltonian is the harmonic oscillator, which will play a prominent role in the quantization of classical fields. In one dimension, this Schrödinger operator reads

$$\frac{p^2}{2m} + \frac{\gamma}{2}q^2, \quad \text{where } \gamma \text{ is a positive constant.}$$

Performing the (canonical) change of operators P and Q by

$$P = (m\gamma)^{-1/4}p, \;\; Q = (m\gamma)^{1/4}q, \;\; \text{so that} \;\; [Q, P] = i\hbar,$$

the operator becomes

$$H_o = \frac{\omega}{2}(P^2 + Q^2), \quad \text{with} \quad \omega = \left(\frac{\gamma}{m}\right)^{1/2}. \tag{32}$$

The spectral analysis of (32) is essentially algebraic once one introduces the *creation* and *annihilation* operators by

$$a^* = \frac{1}{\sqrt{2}}(Q - iP), \;\; a = \frac{1}{\sqrt{2}}(Q + iP), \;\; \text{such that} \;\; [a, a^*] = \mathbb{1}. \tag{33}$$

The operator (32) takes the form

$$H_o = \hbar\omega(a^*a + \frac{1}{2}).$$

Then, defining the vacuum state state $|0\rangle$ as the (normalized) solution to the differential equation

$$a|0\rangle = 0 \iff |0\rangle = \left(\frac{m\omega}{\hbar\pi}\right)^{1/4} e^{-\frac{m\omega}{2\hbar}q^2},$$

one sets by induction

$$|n\rangle = \frac{(a^*)^n}{\sqrt{n!}}|0\rangle, \;\; n = 1, 2, \cdots$$

These vectors are normalized and take the form of a product of polynomials of degree n, known as Hermite polynomials, by the Gaussian $|0\rangle$. One sees easily, using the so-called *canonical commutation relations* (33), that

$$H_o|n\rangle = \hbar\omega\left(n + \frac{1}{2}\right)|n\rangle \;\; \text{since} \;\; a^*|n\rangle = \sqrt{n+1}|n+1\rangle \;\; \text{and} \;\; a|n\rangle = \sqrt{n}|n-1\rangle.$$

These eigenvectors are non-degenerate, such that $\Delta_{|n\rangle}(p)\Delta_{|n\rangle}(q) = \hbar(n+1/2)$, and span $L^2(\mathbb{R})$.

P4: *The time evolution of the system is determined by its Hamiltonian H, the energy observable of the system. There are two equivalent standard descriptions:*
The Schrödinger picture, in which the state ψ evolves in time according to the time-dependent Schrödinger equation in $\mathcal{H}$

$$i\hbar\frac{d}{dt}\psi(t) = H\psi(t), \quad \textit{with} \;\; \psi(0) = \psi. \tag{34}$$

The Heisenberg picture, in which the state is fixed, whereas the observables B evolve in time according to the Heisenberg equation in the space of self-adjoint operators on $\mathcal{H}$

$$i\hbar\frac{d}{dt}B(t) = -[H, B(t)], \quad \textit{with} \;\; B(0) = B. \tag{35}$$

Introducing the unitary evolution group $U(t) = e^{-itH/\hbar}$ (Spectral Theorem again), we get the relation between the Schrödinger and Heisenberg pictures through the identity

$$\mathbb{E}_{\psi(t)}(B) = \mathbb{E}_{\psi}(B(t)), \; \forall t \in \mathbb{R},$$

which follows from

$$\psi(t) = U(t)\psi, \;\; \text{and} \;\; B(t) = U(t)^* B U(t).$$

As a consequence, $\psi(t)$ remains normalized for all times and, since the Hamiltonian H commutes with the evolution group $U(t)$ it generates, the observable energy is constant in time. This is also true for observables which are compatible with H, as (35) shows.

The motivations behind the first order linear evolution equation (34) stem from physical observations leading to the so-called superposition principle implying linearity and from the fact that ψ at time 0 should determine completely the state at any later time t. This equation is the quantum equivalent of (3), whereas (35) is the quantum equivalent of (11).

Applied to our example (2), the relation between (35) and (11) together with (9) and (23) are in keeping with the so-called correspondence principle stating that Poisson brackets are to be replaced by commutators in order to achieve formal quantization in that setting:

$$\{\cdot, \cdot\} \mapsto \frac{[\cdot, \cdot]}{i\hbar}.$$

This yields another motivation for (35).

As an example, consider the case where the Hamiltonian has discrete non-degenerate spectrum $\{E_j\}_{j\in\mathbb{N}}$ with associated eigenvectors $\{\phi_j\}_{j\in\mathbb{N}}$, as is the case if the potential is confining. The time evolution of any initial state ψ reads

$$\psi(t) = \sum_{j\in\mathbb{N}} c_j e^{-itE_j/\hbar}\phi_j, \ \forall t \in \mathbb{R}, \ \text{where } c_j = \langle\phi_j|\psi\rangle. \tag{36}$$

Therefore, the probability of measuring the energy $E_{j_0} \in \sigma(H)$ in the state $\psi(t)$ is given by

$$\begin{aligned}\mathbb{P}_{\psi(t)}(H \in \{E_{j_0}\}) &= \||\phi_{j_0}\rangle\langle\phi_{j_0}|\psi(t)\|^2 \\ &= |\langle\phi_{j_0}|\psi(t)\rangle|^2 = |c_{j_0}e^{-itE_{j_0}/\hbar}|^2 = |c_{j_0}|^2\end{aligned}$$

and is constant. We used the convenient notation $P_H(\{E_j\}) = |\phi_j\rangle\langle\phi_j|$. Similarly, the probability to obtain an energy in a subset $\mathcal{E} = \{E_{j_0}, E_{j_n}, \cdots, E_{j_n}\}$ of the spectrum of H is given by

$$\mathbb{P}_{\psi(t)}(H \in \mathcal{E}) = \left\|\sum_k |\phi_{j_k}\rangle\langle\phi_{j_k}|\psi(t)\rangle\right\|^2 = \sum_k^n |c_{j_k}|^2. \tag{37}$$

Note however, that the sole knowledge of the spectrum of H does not allow in general to get precise information about the evolution of states that are not eigenstates, due to the complicated interferences present in (36). A nice and sometimes useful exception to this rule is the case of *coherent states* for the harmonic oscillator. In the one-dimensional setting used in (32), these normalized states depend on a complex number α and are defined as

$$|\alpha\rangle = e^{-\frac{1}{2}|\alpha|^2}\sum_{n=0}^{\infty}\frac{\alpha^n}{\sqrt{n!}}|n\rangle.$$

They have the properties (which can be checked by means of (33) only)

$$a|\alpha\rangle = \alpha|\alpha\rangle, \ \Delta_{|\alpha\rangle}(p)\Delta_{|\alpha\rangle}(q) = \hbar/2, \ |\alpha\rangle|_{\alpha=0} = |0\rangle.$$

Their explicit expression as functions of $L^2(\mathbb{R})$ reads as

$$|\alpha\rangle = \left(\frac{m\omega}{\pi\hbar}\right)^{1/4}\exp\left(-\frac{1}{2}\alpha(\alpha+\alpha^*) - \frac{m\omega}{2\hbar}q^2 + \left(\frac{2m\omega}{\hbar}\right)^{1/2}\alpha q\right).$$

Then, using $e^{-itH/\hbar}|n\rangle = e^{-i\omega(n+1/2)t}|n\rangle$, we get

$$e^{-itH/\hbar}|\alpha\rangle = e^{-i\omega t/2}|e^{i\omega t}\alpha\rangle.$$

Finally, we note also here that in case the Hamiltonian is time-dependent, (34) gives rise under some regularity conditions to a two-parameter unitary evolution operator $U(t,s)$ on $\mathcal{H}$ satisfying

$$\frac{\partial}{\partial t}U(t,s) = H(t)U(t,s), \ \text{with} \ U(s,s) = \mathbb{1}$$

from which follows the relation

$$U(t,r)U(r,s) = U(t,s), \ \ \forall r,s,t \in \mathbb{R}.$$

In such a case, the evolution operator is no longer an exponential and the future of initial wave functions is usually harder to describe. But again, in case the Hamiltonian is essentially a quadratic form in p and q, with time-dependent coefficients, explicit solutions to the above equation can be obtained, provided the initial conditions are of a coherent state type.

1.3 Fermions and Bosons

So far we have considered systems which have no internal structure or degrees of freedom. Such internal degrees of freedom are introduced by taking a tensor product of the original Hilbert space $\mathcal{H}$ with $\mathcal{K}$, another Hilbert space in which these degrees of freedom live, so that the system is now described by means of the new Hilbert space $\mathcal{H} \otimes \mathcal{K}$. An important internal degree of freedom is the spin of a particle. It is a vector valued operator S in a finite dimensional space $\mathcal{K} \simeq \mathbb{C}^k$ whose components also satisfy the commutation relations (25), and therefore displays the same spectral properties (30). A spin is half-integer or integer, depending whether it is true for the maximal quantum number s of S^2.

Consider now the state of a collection of N identical particles, that is sharing the same physical characteristics like masses, spins, charge, etc. In the framework of our example and slightly abusing notations, it is described by means of a wave function $\psi(q_1, s_1; q_2, s_2; \cdots; q_N, s_N) \in L^2(\mathbb{R}^{dN}) \otimes \mathcal{K}^N$. The fact that the particles are identical is equivalent to saying that all observables $B(q_1, p_1, s_1; \cdots; q_N, p_N, s_N)$ applied to such states are invariant under permutations of their variables (q_j, p_j, s_j). An example of such observable is the kinetic energy part of (22), and it is also true of the potential part of this Hamiltonian due to (2). Therefore, if P_{jk} is the operator whose action is to permute the variables labeled j and k, one has

$$P_{jk}^2 = 1, \ \ P_{jk}^* = P_{jk}, \ \ \text{and} \ \ [P_{jk}, B] = 0, \ \ \forall \ B. \tag{38}$$

Hence, $\sigma(P_{jk}) = \{+1, -1\}$ and the observables and P_{jk} can be diagonalized simultaneously. Thus we can first diagonalize P_{jk} and then describe the observables restricted to the corresponding subspaces of P_{jk}. It is a law of nature that the eigenvalues of P_{jk} are either +1 for all pairs j,k or -1 for all pairs j,k. Therefore, identical particles divide themselves into two distinct sets of particles: those that are invariant under exchange of the variables of their wave function, and those that undergo a sign change under this operations. Particles belonging to the former set are called *bosons* whereas they are called *fermions* if they belong to the latter. As the properties (38) are true for all pairs jk of labels, they are also true for arbitrary permutations $\pi \in \mathcal{P}_N$ in the group of permutations of N elements. In particular, if P_π denotes the permutation of indices corresponding to π, then $P_\pi \psi = \psi$ for bosons and $P_\pi \psi = (-1)^\pi \psi$ for fermions, where $(-1)^\pi$ is the signature of the permutation π. In other words, the above discussion shows that the physical Hilbert spaces for bosons and fermions are not the N-fold tensor product $\mathcal{H}^N$ of the Hilbert space $\mathcal{H}$ of their one particle descriptions, but rather $\mathcal{H}_+^n$ and $\mathcal{H}_-^n$, defined by

$$\mathcal{H}^n_{\pm} = \mathcal{S}_{\pm}\mathcal{H}^n \text{ where } S_{\pm} = \frac{1}{N!}\sum_{\pi\in\mathcal{P}_N}(\pm 1)^{\pi}P_{\pi}.$$

The operators $S_{\pm}$ are easily checked to be orthogonal projectors onto the orthogonal subspace $\mathcal{H}^n_{\pm}$ of $\mathcal{H}^n$ and vectors in these spaces are characterized by the properties described above. These characteristics will have important consequences on the physical properties of collections of such particles. In particular, antisymmetry forbids two independent fermions to be in the same quantum state. Indeed, the sign change induced by exchange of these two particles in the antisymmetrization procedure makes the vector vanish. This goes under the name *Pauli's Principle*.

It turns out that the fermionic or bosonic nature of particles is linked to the properties of their spin. Indeed, it can be shown within the realm of relativistic quantum field theory that fundamental requirements of Physics as micro-causality and Lorentz's invariance imply the so-called *Spin-Statistic Theorem* asserting that particles with half-integer spin are fermions, whereas those with integer spin are bosons. It is an experimental fact that no other statistics is present in nature. Electrons are thus fermions and photons are bosons (although the latter have an internal degree of freedom called helicity, instead of a spin).

2 Quantum Statistical Mechanics

2.1 Density Matrices

We give here, in a particular setting, some heuristics behind the formal definition of *state* (or mixed state) in Quantum Statistical Mechanics which will be used later on. The first approach is based on a time dependent point of view supplemented by a postulate on the behavior of some phases, advocated in [3], for example.

Let us start from (37) which says that in case the Hamiltonian H on $\mathcal{H}$, of a system is time independent, discrete and non-degenerate, *i.e.* when the system is isolated from the rest of the world, its normalized wave function can be written as

$$\psi(t) = \sum_{j\in\mathbb{N}} c_j(t)\phi_j, \tag{39}$$

with explicit time dependent complex valued coefficients $c_j(t)$. The basic postulates of Statistical Mechanics are formulated for isolated systems. However, in practice, one is often interested in a subsystem of the whole system only, so that, certain degrees of freedom are not observed and are incorporated in what we call the rest of the world. Thus Statistical Mechanics effectively deals with systems that interact with the external world, so that the truly isolated system is our initial system plus the rest of the world. The relevant Hilbert space to describe this new system is the tensor product $\mathcal{R}\otimes\mathcal{H}$, where $\mathcal{R}$ is the Hilbert space of the rest of the world, and the corresponding scalar product is the product of the respective scalar products on $\mathcal{R}$ and $\mathcal{H}$. The wave function of this larger isolated system can still be written as

(39), with the proviso that the c_j's are now time dependent elements of $\mathcal{R}$ such that $\sum_j \langle c_j(t)|c_j(t)\rangle_{\mathcal{R}} = 1$, where the subscript specifies what scalar product is used.

Now, if B is an observable acting on the original system only, technically of the form $\mathbb{1} \otimes B$ as an operator on $\mathcal{R} \otimes \mathcal{H}$, the instantaneous expectation value of a set of measurements of this observable on $\psi(t)$ is given by

$$\langle \psi(t)|\, \mathbb{1} \otimes B\psi(t)\rangle_{\mathcal{R}\otimes\mathcal{H}} = \sum_{k,j} \langle c_j(t)|c_k(t)\rangle_{\mathcal{R}} \langle \phi_j|B\phi_k\rangle_{\mathcal{H}}.$$

In an actual experiment, it is rather the time average of the above quantity that is measured, over a time that is large with respect to the "molecular time scale", but short with respect to the resolution of the measurement apparatus. Therefore, the measured quantity is actually

$$\langle B\rangle = \sum_{k,j} \overline{\langle c_j(t)|c_k(t)\rangle_{\mathcal{R}}} \langle \phi_j|B\phi_k\rangle_{\mathcal{H}},$$

where the bar indicates time average.

One postulate concerns the scalar products of the $c_j(t)$'s about which we have minimal knowledge, since it deals with properties of the external world. It is postulated that these scalar products producing interferences are averaged to zero over the time scale on which we observe the system, this is the *Random phases* postulate:

$$\overline{\langle c_j(t)|c_k(t)\rangle_{\mathcal{R}}} = 0, \ \ \forall\ j \neq k.$$

The consequence of this postulate is an effective description of the original system by means of the non-zero scalars

$$\lambda_j = \overline{\langle c_j(t)|c_j(t)\rangle_{\mathcal{R}}}, \quad \text{such that} \quad \sum_j \lambda_j = 1, \tag{40}$$

in the sense that the outcome of a measurement in that framework is given by

$$\langle B\rangle = \sum_j \lambda_j \langle \phi_j|B\phi_j\rangle_{\mathcal{H}}.$$

Therefore one says that the random phases postulate allows to regard the state of the system as an incoherent superposition of eigenstates of H or a *mixed state*.

As a consequence, it is possible to represent the mixed state by a *density matrix*. Let ρ be the linear operator on $\mathcal{H}$ defined by

$$\rho = \sum_j \lambda_j |\phi_j\rangle\langle\phi_j|. \tag{41}$$

It is a positive, trace class operator, of trace one, such that

$$\langle B\rangle = \text{Tr}\ (\rho B).$$

This operator contains all the information we have about the mixed state and allows to compute in a convenient way all expectation values by means of the trace operation.
Note that a pure state χ as defined in the previous section corresponds to the density matrix $\rho_\chi = |\chi\rangle\langle\chi|$. Actually, it is easy to see that a density matrix ρ corresponds to a pure state if and only if it is a rank one projector.

Another approach of mixed states consists in noting that incomplete information about a system always leads to density matrices, without resorting to delicate properties of the time evolution.

A first point of view consists in considering the system as an *ensemble* of true eigenstates, to be considered one at a time, where the relative proportion of the eigenstate ϕ_j is λ_j. The value $\lambda_j \in [0,1]$ is interpreted as the classical probability to get the pure eigenstate ϕ_j in the mixed state. This statistical interpretation $\rho = \sum_j \lambda_j |\phi_j\rangle\langle\phi_j|$ allows to avoid any consideration of effective coupling between the system and the "external world" and makes no use of *a priori* knowledge about the time evolution.

A second interpretation of incomplete knowledge about the system consists in splitting the total Hilbert space it lives in into $\mathcal{R}\otimes\mathcal{H}$, where $\mathcal{R}$ concerns the degrees of freedom that are not known. Therefore, if $\{\varphi_j \otimes \psi_k\}_{j,k}$ denotes an orthonormal basis of $\mathcal{R}\otimes\mathcal{H}$ made out of individual orthonormal bases of $\mathcal{R}$ and $\mathcal{H}$ and ρ is any density matrix on $\mathcal{R}\otimes\mathcal{H}$, one introduces the corresponding reduced density matrix $\rho_\mathcal{H}$ on $\mathcal{H}$ by its matrix elements

$$(\rho_\mathcal{H})_{ij} = \sum_k \langle \varphi_k \otimes \psi_i \,|\rho\, \varphi_k \otimes \psi_j\rangle, \;\; \forall i, j.$$

The matrix $\rho_\mathcal{H}$ is designed so that for any operator of the form $\mathbb{1}\otimes B$, one has

$$\mathrm{Tr}_\mathcal{H}(\rho_\mathcal{H} B) = \mathrm{Tr}\,(\rho\, \mathbb{1}\otimes B),$$

where $\mathrm{Tr}_\mathcal{H}$ denotes the trace in $\mathcal{H}$. This formula is in keeping with our ignorance of the degrees of freedom in $\mathcal{R}$ which are traced out. The point is that one can easily check that if $\rho = |\Psi\rangle\langle\Psi|$ for some $\Psi \in \mathcal{R}\otimes\mathcal{H}$, then, in general, the corresponding reduced density matrix $\rho_\mathcal{H}$ characterizes a mixed state.

Therefore we will adopt from now on the following definition:

A mixed state (or simply state) in Quantum Statistical Mechanics is a positive trace class operator on $\mathcal{H}$ of trace 1.

The time dependence of density matrices is governed by the following equation in the subset of density matrices in $\mathcal{T}(\mathcal{H})$, the linear space of trace class operators on $\mathcal{H}$,

$$i\hbar \frac{d}{dt}\rho = [H, \rho], \;\; \rho(0) = \rho_0. \tag{42}$$

This equation stems from the cyclicity of the trace and the relation which must hold for all t and all observables B

$$\mathrm{Tr}\,(\rho B(t)) = \mathrm{Tr}(\rho(t)B).$$

Note that (42) is in keeping with the evolution of the density matrix of a pure state.

3 Boltzmann Gibbs

So far, we haven't talked about equilibrium properties. The basic postulate in Quantum Statistical Mechanics describes the density matrix of an isolated system that has reached equilibrium.

Assume the energy of the system is known to lie within the range $[E, E+\Delta]$, where $\Delta << E$. The *Equal a priori Probability* postulate, formalizing again our minimal knowledge of the total system, states within the framework where the ϕ_j's represent eigenstates of the Hamiltonian H, that at equilibrium, the λ_j's defining the density matrix ρ_{eq} are given by

$$\lambda_j = \begin{cases} \lambda & \text{if } E < E_j < E + \Delta, \\ 0 & \text{otherwise.} \end{cases} \tag{43}$$

The constant λ is normalized so that the trace of ρ_{eq} is one. Other ways of writing ρ_{eq} are

$$\rho_{eq} = P_H(E, E+\Delta)/\,\mathrm{Tr}\,(P_H(E, E+\Delta)) = \frac{\sum_{E<E_j<E<\Delta} |\phi_j\rangle\langle\phi_j|}{\#\{j \,|\, E_j \in]E, E+\Delta[\}}. \tag{44}$$

Note that as ρ_{eq} is a function of the Hamiltonian H, it is constant in time due to (42), which is what we expect for the equilibrium density matrix.

This above prescription for ρ_{eq} corresponds to the *micro-canonical* ensemble, where it is understood that the system under consideration has fixed energy and fixed number of particles.

In this case, Boltzmann's formula is used to define the *entropy*

The entropy of a system at equilibrium in the microcanonical ensemble reads

$$S(E) = k \ln \Gamma(E), \quad \textit{where} \quad \Gamma(E) = \#\{j \,|\, E_j \in]E, E+\Delta[\} \tag{45}$$

and $k \simeq 1,38 \times 10^{-23} J/K$ *is Boltzmann's constant.*

The quantity $\Gamma(E)$, which is the denominator in (44), gives the number of quantum states that are accessible to the system. The entropy actually depends on other variables such as the volume V of the system, the number N of particles of the system, etc... that we omitted in the notation. The definition (45) makes the bridge between equilibrium Statistical Mechanics and thermodynamics, once the thermodynamic limit is taken. That is, once it is demonstrated that Boltzmann's formula fulfills the following conditions:

a. Extensivity, so that the thermodynamical limit as $V \to \infty$, $N \to \infty$ and $E \to \infty$ exists i.e.

$$\frac{1}{N} S(E,V,N) \to s(e,v), \quad \text{where} \quad E/N \to e, \; V/N \to v, \; S/N \to s$$

where e, s and v are the densities of energy and entropy and v is the specific volume.

b. The fact that if exterior parameters of the system initially at equilibrium are varied in such a way that the system can reach another equilibrium configuration, then the difference of entropies between these configurations is non negative. This is an expression of the second law of thermodynamics which implies that the equilibrium state maximizes the entropy. We'll come back to this point shortly.

Without discussing thermodynamics, we mention that assuming the existence of entropy and that properties a) and b) hold, all thermodynamical quantities can be computed from $S(E,V,N)$ through the definitions:

$$\frac{1}{T} = \left.\frac{\partial S}{\partial E}\right)_{V,N} \quad \text{defines the } \textit{temperature}\text{, so that ,}$$

$U(S,V,N) \equiv E(S,V,N)$ the *internal energy* of the system exists and

$$P = -\left.\frac{\partial U}{\partial V}\right)_{S,N} \quad \text{defines the } \textit{pressure} \text{ whereas}$$

$$\mu = \left.\frac{\partial U}{\partial N}\right)_{S,V} \quad \text{defines the } \textit{chemical potential}.$$

The above definitions yield the familiar differential

$$dU(S,V,N) = TdS - PdV + \mu dN, \tag{46}$$

motivating the physical interpretations of these derivatives. The extensivity property of U, i.e. homogeneity of degree one, associated with (46) implies

$$U = TS - PV + \mu N. \tag{47}$$

In order to complete the picture, let us briefly recall that the first law of thermodynamics asserts that the differential

$$dU = \delta Q - \delta W \quad \text{is exact,}$$

where δQ is amount of heat absorbed by the system and δW is the work done by the system in any transformation. A corollary of the second law of thermodynamics says that the differential

$$dS = \frac{\delta Q}{T} \quad \text{is exact,}$$

relating the experimental notion of heat to entropy. These two statements imply the existence of the functions S and U at equilibrium.

It can be argued that the definition (45) satisfies requirement a), but we shall not provide the argument here. Let us consider b). This last property calls for a variational approach of the entropy. Hence we introduce a more general definition of the entropy of a state by

The entropy of state ρ of a physical system is given by

$$S(\rho) = -k\,Tr\,(\rho \ln(\rho)), \tag{48}$$

where the function $x \mapsto x\ln(x)$ is defined to be zero at $x = 0$.

If $\{\lambda_j\}_{j\in N}$ denotes the set of eigenvalues of the density matrix

$$\rho = \sum_j \lambda_j |\phi_j\rangle\langle\phi_j|,$$

the entropy is given by

$$S(\rho) = -k \sum_j \lambda_j \ln(\lambda_j). \tag{49}$$

Therefore one sees that $S(\rho) \geq 0$ with $S(\rho) = 0$ if and only if $\lambda_j = \delta_{j,k}$ for some k, i.e. ρ corresponds to a pure state. Also, the entropy (48) of the density matrix ρ describing two independent systems defined on $\mathcal{H}_1$ and $\mathcal{H}_2$ is the sum of the individual entropies, as expected. Indeed, in such a case, $\rho = \rho_1 \otimes \rho_2$ on $\mathcal{H} = \mathcal{H}_1 \otimes \mathcal{H}_2$, where ρ_j, $j = 1, 2$ are the density matrices of the individual systems. Using

$$\ln(\rho_1 \otimes \rho_2) = \ln(\rho_1) \otimes \mathbb{1} + \mathbb{1} \otimes \ln(\rho_2),$$

and taking partial traces, one gets $S(\rho) = S(\rho_1) + S(\rho_2)$. More generally, it can be shown that for $\alpha \in [0, 1]$ and arbitrary density matrices ρ_1, ρ_2

$$S(\alpha\rho_1 + (1-\alpha)\rho_2) \geq \alpha S(\rho_1) + (1-\alpha)S(\rho_2),$$

which shows that S is concave as a function on the set of density matrices, i.e. mixing density matrices increases the entropy. And, on the other hand, the entropy is almost convex in the sense

$$S(\alpha\rho_1 + (1-\alpha)\rho_2) \leq \alpha S(\rho_1) + (1-\alpha)S(\rho_2) - \alpha\ln\alpha - (1-\alpha)\ln(1-\alpha).$$

More mathematical properties of the entropy are provided in [1] or [5], for example.

Now, maximizing (49) over the probabilities λ_j's shows that the entropy is maximal when the eigenvalues are all constant. Thus, when the corresponding eigenvectors ϕ_j are those of the Hamiltonian, we get back both the equal a priori probability postulate and Boltzmann's formula.

The micro canonical ensemble is convenient to motivate definitions, but one often prefers to use the *canonical* ensemble for applications. In that setting, the system under consideration interacts with a thermal reservoir whose property is to remain at fixed temperature. Exchanges of energy are allowed between the reservoir and the system, under the constraint that the averaged energy of the system is kept fixed. The maximization of entropy in the canonical ensemble leads to Gibbs prescription for the density matrix, as we now (formally) argue.

Consider the functional over the set of density matrices.

$$\mathcal{F}(\rho) = S(\rho) - k\beta\langle H\rangle_\rho, \tag{50}$$

where $\langle H\rangle_\rho$ denotes the expectation value of the energy computed by means of the density matrix ρ, and β is a Lagrange multiplier associated with the energy constraint.

We need to compute the first variation of $\mathcal{F}$

$$\delta\mathcal{F}(\rho) = \frac{d}{dt}\mathcal{F}(\rho + t\eta)|_{t=0},$$

where the admissible variation η is any trace class operator of zero trace and $\mathrm{Tr}\,(\rho) = 1$. In order to do so, we first justify the following intuitive relation:

If $A(t)$ is a t-dependent self-adjoint operator, such that $A(t) = A(0) + t\eta$, then

$$\frac{d}{dt}\mathrm{Tr}\,(f(A(t))) = \mathrm{Tr}\,(f'(A(t))\eta). \tag{51}$$

Indeed, Hellman-Feynman formula applied to $A(t)$ whose eigenvalues and normalized eigenvectors are denoted by $(a_j(t), \varphi_j(t))$ reads

$$a_j'(t) = \langle\varphi_j(t)|A'(t)\varphi_j(t)\rangle. \tag{52}$$

Thus for any (reasonable) real valued function f,

$$\frac{d}{dt}\mathrm{Tr}\,(f(A(t))) = \sum_j f'(a_j(t))a_j'(t) = \sum_j \langle\varphi_j(t)|f'(A(t))\varphi_j(t)\rangle a_j'(t). \tag{53}$$

In the case under consideration, $A'(t) = \eta$ so that, by (52),

$$a_j'(t) = \langle\varphi_j(t)|\eta\varphi_j(t)\rangle$$

and, using orthonormality of the φ_j's, the RHS of (53) equals

$$\begin{aligned}
&\sum_j \langle\varphi_j(t)|f'(A(t))\varphi_j(t)\rangle\langle\varphi_j(t)|\eta\varphi_j(t)\rangle \\
&= \sum_{j,k} \langle\varphi_j(t)|f'(A(t))\varphi_k(t)\rangle\langle\varphi_k(t)|\eta\varphi_j(t)\rangle \\
&= \sum_j \langle\varphi_j(t)|f'(A(t))\eta\varphi_j(t)\rangle, \quad \text{which yields (51).}
\end{aligned}$$

In our case, we get

$$\delta\mathcal{F}(\rho) = -k\,\mathrm{Tr}\,(\eta(\ln(\rho) + \mathbb{1} + \beta H)),$$

which has to be zero for any admissible η if ρ extremalizes $\mathcal{F}$. In particular, we can choose $\eta = \sum_j \eta_j |\varphi_j\rangle\langle\varphi_j|$ where $\{\varphi_j\}$ are the set of eigenvectors of the self adjoint operator $\ln(\rho) + \mathbb{1} + \beta H$ and $\{\eta_j\}$ are a set of real numbers satisfying $\sum_j \eta_j = 0$. For *that* η, we have

$$\delta\mathcal{F}(\rho) = -k \sum_j \eta_j v_j,$$

where the v_j's denote the eigenvalues of $\ln(\rho) + \mathbb{1} + \beta H$. Choosing above $\eta_0 = -\eta_1 \neq 0$ and $\eta_j = 0$ if $j \geq 2$, we get that $\delta\mathcal{F}(\rho) = 0$ for any η_0 implies $v_0 = v_1$. Iterating, we get $v_0 = v_j$, for any $j \in \mathbb{N}$. Hence

$$\delta\mathcal{F}(\rho) = 0 \ \forall\eta \iff \ln(\rho) + \mathbb{1} + \beta H = v_0\,\mathbb{1}.$$

The constant v_0 will be determined by the normalization of the state. Therefore, exponentiating, we find as extremalizer the *Gibbs distribution*

$$\rho_G = \frac{e^{-\beta H}}{\mathrm{Tr}\,(e^{-\beta H})}. \tag{54}$$

Explicit computations that we do not present here show that the second variation $\delta^2\mathcal{F}$ is negative. Hence we get that the Gibbs prescription yields a maximum of the functional (50). Here the parameter β used to insure constancy of the average energy of the system in that canonical ensemble setting will be identified with the inverse temperature given by the usual formula

$$\beta = \frac{1}{kT}. \tag{55}$$

The normalization of the Gibbs distribution

$$Z := \mathrm{Tr}\,(e^{-\beta H})$$

defines the (canonical) *partition function*. It is related to the internal energy of the system $\langle H\rangle_{\rho_G}$ by

$$\langle H\rangle_{\rho_G} = -\frac{\partial \ln(Z)}{\partial\beta}.$$

Again, in the thermodynamic limit, the partition function of Statistical Mechanics is directly linked to a thermodynamical quantity: the free energy F of the system, defined as $F = U - TS$ (see below), where U is the internal energy $\langle H\rangle_{\rho_G}$ of the system. To substantiate this claim, let us formally compute by means of (54) (assuming the thermodynamic limit and extensivity holds),

$$\begin{aligned} S &= k\beta\langle H\rangle_{\rho_G} + k\ln(Z) \quad \text{so that} \\ F &:= -kT\ln Z \end{aligned} \tag{56}$$

defines the *free energy* in Statistical Mechanics.

Moreover, as a consequence of our variational approach, we get that the free energy of a system is minimized by the equilibrium state, which together with (56) are two familiar properties of the free energy.

More precisely, in thermodynamics, F is a function of (T, V, N), which is the result of its very definition:

The thermodynamical free energy is the Legendre transform of the internal energy $U(S, V, N)$ with respect to the variable S

$$F(T, V, N) = (U - TS)(T, V, N) \tag{57}$$

where $S(T, V, N)$ is computed from $\left.\frac{\partial U}{\partial S}\right)_{V,N} = T$.

This operation allows to trade the entropy variable for the more natural temperature variable. Recall that the Legendre transform of a one variable function $f : x \mapsto f(x)$ is the function $a : p \mapsto a(p)$ defined by

$$a(p) = f(x(p)) - px(p),$$

where $x(p)$ is obtained by inversion of

$$\frac{d}{dx} f(x) = p(x).$$

There is no loss of information in the process as long as the inversion of $f'(x)$ is possible and when f is concave, respectively convex, its Legendre transform is convex, respectively concave. In case f depends on other variables y, one has the identities

$$\frac{\partial}{\partial p} a(p, y) = -x(p, y), \quad \frac{\partial}{\partial y} a(p, y) = \frac{\partial}{\partial y} f(x, y)$$

allowing to recover all thermodynamic quantities from F via the relations

$$\left.\frac{\partial F}{\partial T}\right)_{V,N} = -S, \quad \left.\frac{\partial F}{\partial V}\right)_{T,N} = -P, \quad \left.\frac{\partial F}{\partial N}\right)_{T,V} = \mu.$$

We can now provide a justification of the identification (55) as follows, assuming the thermodynamic limit is taken and extensivity holds. Indeed, by means of that identification, we get by explicit computation on (56)

$$\frac{\partial}{\partial T} F = -k \ln Z - Tk \frac{\mathrm{Tr}\,(\frac{\partial}{\partial \beta} e^{-\beta H})}{Z} \frac{\partial \beta}{\partial T} = -k \ln Z - \frac{\langle H \rangle_{\rho_G}}{T} = -S,$$

which is identical to the first relation above.

Let us present here the classical computation of partition functions associated with independent harmonic oscillators.

Let $\mathcal{H}$ be the Hilbert space spanned by the eigenvectors $\{|n\rangle\}_{n=0,\cdots,\infty}$ of the Hamiltonian H_o in (32) corresponding to the energies $\epsilon_n = \frac{1}{2} + n$, $n = 0, \cdots, \infty$ (we assume $\hbar\omega = 1$, without loss). Working in the canonical ensemble, the partition function of one harmonic oscillator reads

$$Z_1(\beta) = \mathrm{Tr}\,(e^{-\beta H_o}) = e^{-\beta/2} \sum_{n=0}^{\infty} e^{-\beta n} = \frac{1}{e^{\beta/2} - e^{-\beta/2}},$$

the internal energy $U_1(\beta) = \langle H_o \rangle_{\rho_G}$ is given by

$$U(\beta) = -\frac{\partial \ln(Z_1(\beta))}{\partial \beta} = \frac{1}{2} \coth(\beta/2),$$

whereas the free energy $F_1(\beta)$ reads

$$F(\beta) = -kT \ln(Z_1(\beta)) = kT \ln(e^{\beta/2} - e^{-\beta/2}).$$

In case we work with a d-dimensional harmonic oscillator, or, equivalently, with d independent oscillators with the same frequency, we denote by

$$|n_1, n_2, \cdots, n_d\rangle, \qquad n_j \in \mathbb{N}, j = 1, \cdots, d,$$

the eigenvector corresponding to the energy $\frac{d}{2} + n_1 + \cdots + n_d$. Thus, the corresponding partition function reads

$$Z(\beta, d) = e^{-\beta d/2} \sum_{n_1 \geq 0, n_2 \geq 0, \cdots, n_d \geq 0} e^{-\beta(n_1 + \cdots + n_d)} = Z(\beta)^d,$$

so that the internal and free energies and are given by

$$U(\beta, d) = dU(\beta), \quad F_d(\beta, d) = dF(\beta). \tag{58}$$

Going from the micro canonical to the canonical ensemble, we have allowed energy exchanges between the system under consideration and a thermal reservoir. In a similar fashion, we can relax the condition that the number of particles in the system is fixed and allow particles exchanges with the reservoir as well, assuming the their average number only is fixed. This corresponds to working in the *grand canonical ensemble*. As we will see later on, allowing particles exchanges in Quantum Open Systems is essential, in the sense that the statistical properties of these particles, i.e. their bosonic or fermionic nature, have definite physical consequences.

This calls for a precision about the Hilbert space suitable to describe such situations, the so-called *second quantization formalism*. The Hilbert space allowing variable numbers of particles is either the symmetrical or anti-symmetrical *Fock space*, depending on the statistics. These Hilbert spaces will be object of much mathematical care later on, so we will briefly and informally describe here the bosonic and fermionic Fock spaces $\mathcal{F}_\pm(\mathcal{H})$. If $\mathcal{H}$ is the one-particle Hilbert space, the n-fold properly symmetrized tensor product $\mathcal{H}_\pm^n$ is the n-boson or n-fermion subspace. An element Ψ of $\mathcal{F}_\pm(\mathcal{H})$ is a collection $\{\psi(n)\}_{n \in \mathbb{N}}$, where $\psi(n) \in \mathcal{H}_\pm^n$, for all $n > 0$, $\psi(0) \in \mathbb{C} \equiv \mathcal{H}_{\pm}^{0}$, with the obvious linear structure and norm $\|\Psi\|_{\pm}^{2} = \sum_{n} \|\psi(n)\|_{\pm}^{2}$. Observables $\mathbf{B}$ on the Fock space can be constructed as $\mathbf{B} = \sum_{n} B(n)$, where the $B(n)$'s acting on the n-particle subspaces are given (with $B(0) = 0$). In particular, the number operator $\mathbf{N}$ defined by $\mathbf{N}\Psi = \{n\psi(n)\}_{n\in\mathbb{N}}$ has the form $\mathbf{N} = \sum_{n} n\, \mathbb{1}_{\mathcal{H}_{\pm}^{n}}$. Another case is that of *one body operators*. That is when $\mathbf{A} = \sum_{n} A(n)$ with $A(n) = \sum_{j=1}^{n} A_j$, where $A_j = \mathbb{1} \otimes \cdots \mathbb{1} \otimes A \otimes \mathbb{1} \cdots \mathbb{1}$, and A acts on the j'th copy of $\mathcal{H}$.

With these preliminaries behind us, let us assume we are given a Hamiltonian $\mathbf{H}$ with the structure above. The equilibrium state in that framework is obtained by maximization of the entropy, under the constraints that both the average energy and average number of particles are fixed. This leads to the computation of the first and second variations of the functional $\mathcal{G}$ over the density matrices defined by

$$\mathcal{G}(\rho) = S(\rho) - k\beta\langle\mathbf{H}\rangle_{\rho} + k\beta\mu\langle\mathbf{N}\rangle_{\rho},$$

where β and μ are Lagrange multipliers associated with the imposed constraints. They will be identified in the thermodynamic limit, with the inverse temperature and chemical potential, respectively. A maximizing procedure quite similar to the one performed above that we will not detail here yields the extremum

$$\rho_{GC} = \frac{e^{-\beta(\mathbf{H}-\mu\mathbf{N})}}{\mathcal{Z}}$$

where, due to the structure of $\mathbf{H}$ and $\mathbf{N}$, the *grand canonical partition function* $\mathcal{Z}$ can be written as

$$\mathcal{Z} = \sum_{n} e^{\beta\mu n}\, \mathrm{Tr}_{\mathcal{H}_{\pm}^{n}}(e^{-\beta H(n)}).$$

The quantity $z = e^{\beta\mu}$ is also called the *fugacity* and with Z_n the canonical partition function, we can rewrite

$$\mathcal{Z} = \sum_{n} z^{n} Z_{n}.$$

One can also verify that the maximal value of $\mathcal{G}$ is

$$S(\rho_{GC}) - k\beta\langle\mathbf{H}\rangle_{\rho_{GC}} + k\beta\mu\langle\mathbf{N}\rangle_{\rho_{GC}} = k\ln\mathcal{Z}. \tag{59}$$

To make the bridge with thermodynamics, consider the thermodynamical *grand potential* Φ defined by the Legendre transform of F with respect to N, i.e.

$$\Phi(T,V,\mu) = (F - \mu N)(T,V,\mu)$$

where $\frac{\partial F}{\partial N} = \mu$. One can then see by formal manipulations similar to those performed above, assuming the thermodynamic limit and extensivity holds, that Φ is minimal at equilibrium. From (59) and (47), we get that this minimum is given by

$$\Phi = -PV = -kT\ln\mathcal{Z}$$

and we further have the thermodynamical relations

$$\left.\frac{\partial\Phi}{\partial T}\right)_{V,\mu} = -S, \quad \left.\frac{\partial\Phi}{\partial\mu}\right)_{T,V} = -N.$$

The ensemble (microcanonical, canonical or grand canonical) chosen to describe a specific system is largely made according to convenience for the computations. Therefore it is comforting to know that the respective descriptions are all equivalent. This is the statement known as the *equivalence of ensembles* which says that in the thermodynamical limit, one can use either the microcanonical or the canonical ensemble to perform calculations of thermodynamical quantities because the results will agree. Instead of providing a justification of this statement here, we shall be content with the explicit verification of this fact for a system of independent harmonic oscillators considered in the microcanonical and canonical ensembles.

In the microcanonical ensemble, we compute the entropy by means of (45). If we have N independent oscillators each of which has energy levels $j + 1/2$, $j \in \mathbb{N}$, we get for N large,

$$\Gamma(E) \simeq \#\left\{ n_j \in \mathbb{N}, j = 1, \cdots, N, \mid \sum_j n_j = E - N/2 \right\} \simeq \binom{E + N/2}{N},$$

using the combinatoric formula

$$\#\left\{ n_j \in \mathbb{N}, j = 1, \cdots, N, \mid \sum_j n_j = M \in \mathbb{N} \right\} = \binom{M + N - 1}{N - 1}.$$

Hence, by means of Stirling formula, we compute in term of the energy density $e = E/N$,

$$S(E, N) \simeq Nk\left((e + \frac{1}{2})\ln(e + \frac{1}{2}) - (e - \frac{1}{2})\ln(e - \frac{1}{2}) \right) \equiv Ns(e).$$

Therefore, the temperature is determined by

$$\frac{1}{T} = \frac{\partial s(e)}{\partial e} = k\ln\left(\frac{e + \frac{1}{2}}{e - \frac{1}{2}}\right),$$

so that we get the following formula for the energy density

$$e = \frac{1}{2}\left(\frac{e^\beta + 1}{e^\beta - 1}\right) = \frac{1}{2}\coth(\beta/2).$$

For the same system considered in the canonical ensemble, we obtained in (58) with $d = N$,

$$U(\beta, N) = NU(\beta) = N\frac{1}{2}\coth(\beta/2),$$

which yields the same energy density $e = U/N$.

Let us consider now the computation of the grand canonical partition function, in the simple bosonic/fermionic context where particles do not interact with one another.

Consider the normalized vector $|n_0, n_1, n_2, \cdots, n_j, \cdots\rangle_{\pm} \in \mathcal{F}_{\pm}(\mathcal{H})$ in the so-called *occupation number representation* relative to the eigenstates $|n\rangle$ in $\mathcal{H}$ of some nondegenerate Hamiltonian H. This vector consists in a normalized, fully (anti)symmetrized tensor product of states $|n\rangle \in \mathcal{H}$ characterized by n_0 factors $|0\rangle$, n_1 factors $|1\rangle$, $\cdots$ n_j factors $|j\rangle$, etc. The number of particles N in such a state is obviously given by $N = \sum_k n_k$. In case of bosons, $n_j \in \mathbb{N}$ without restriction, whereas in case of fermions, Pauli's principle enforces $n_j \in \{0, 1\}$, for any j. We'll denote by $\mathbb{N}'$ the set of allowed values of the n_j's , depending on the statistics. The collection $\{|n_0, n_1, n_2, \cdots, n_j, \cdots\rangle_{\pm}\}_{n_0 \in \mathbb{N}', \cdots, n_j \in \mathbb{N}'}$ forms an orthonormal basis of $\mathcal{F}_{\pm}(\mathcal{H})$. If one considers only the states with a fixed number of particles, one gets that the set $\{|n_0, n_1, n_2, \cdots, n_j, \cdots\rangle_{\pm} \mid \sum_k n_k = N\}$ forms an orthonormal basis of the subspace $\mathcal{H}_{\pm}^N$.

Let ϵ_n denote the eigenvalue of H corresponding to $|n\rangle$. Then the one body observable $\mathbf{H}$ in $\mathcal{F}_{\pm}(\mathcal{H})$ constructed from H satisfies

$$\mathbf{H}\, |n_0, n_1, n_2, \cdots, n_j, \cdots\rangle_{\pm} = \sum_k n_k \epsilon_k\, |n_0, n_1, n_2, \cdots, n_j, \cdots\rangle_{\pm}.$$

The corresponding physical system consists of a collection of independent fermions or bosons individually driven by the Hamiltonian H. Though quite simple, such systems allow to put forward the effect of the statistics. The canonical partition function $Z_N(\beta)$ of N independent fermions/bosons is

$$Z_N(\beta) = \sum_{\{n_j \mid \sum_j n_j = N\}} e^{-\beta \sum_j n_j \epsilon_j},$$

where the restrictions on the n_j's due to the statistics are implicit in the notation. Hence, with $z = e^{\beta\mu}$,

$$\begin{aligned}\mathcal{Z}(\beta, z) &= \sum_{N \geq 0} z^N \sum_{\{n_j \mid \sum_j n_j = N\}} e^{-\beta \sum_j n_j \epsilon_j} = \sum_{N \geq 0} \sum_{\{n_j \mid \sum_j n_j = N\}} \prod_j (z e^{-\beta \epsilon_j})^{n_j} \\ &= \prod_j \left[\sum_n (z e^{-\beta \epsilon_j})^n \right] = \begin{cases} \prod_j (1 - z e^{-\beta \epsilon_j})^{-1} & \text{for bosons,} \\ \prod_j (1 + z e^{-\beta \epsilon_j}) & \text{for fermions.} \end{cases}\end{aligned}$$

In particular, we compute

$$\langle N \rangle_{\rho_{GC}} = z \frac{\partial}{\partial z} \ln(\mathcal{Z}(\beta, z)) = \begin{cases} \sum_j \dfrac{z e^{-\beta \epsilon_j}}{1 - z e^{-\beta \epsilon_j}} & \text{for bosons,} \\ \sum_j \dfrac{z e^{-\beta \epsilon_j}}{1 + z e^{-\beta \epsilon_j}} & \text{for fermions,} \end{cases}$$

which allows to determine z. Similarly, the average occupation numbers can be obtained as

$$\langle n_j \rangle_{\rho GC} = -\frac{1}{\beta}\frac{\partial}{\partial \epsilon_j} \ln(\mathcal{Z}(\beta, z)) = \frac{z e^{-\beta \epsilon_j}}{1 \mp z e^{-\beta \epsilon_j}} \quad \begin{cases} \text{for bosons,} \\ \text{for fermions.} \end{cases}$$

Therefore, we get the expected relation

$$\langle N \rangle_{\rho GC} = \sum_j \langle n_j \rangle_{\rho GC},$$

where we clearly see the effects of the statistics and temperature on the the average occupation numbers.

References

1. O. Bratteli, D. W. Robinson, *Operator Algebras and Quantum Statistical Mechanics II*, Texts and Monographs in Physics, Springer, New York, Heidelberg, Berlin, 1981
2. J. Glimm, A. Jaffe, *Quantum Physics*, Springer, New York, Heidelberg, Berlin, 1981
3. K. Huang, *Statistical Mechanics*, J. Wiley & Sons, New York, London, Sydney, 1963
4. Ph. A. Martin, F. Rothen, *Many -body Problems and Quantum Field Theory*, Texts and Monographs in Physics, Springer, 2nd Edition, 2004
5. B. Simon, *The Statistical Mechanics of Lattice Gases*, Princeton Series in Physics, Princeton New-Jersey, 1993

Elements of Operator Algebras and Modular Theory

Stéphane Attal

Institut Camille Jordan, Université Claude Bernard Lyon 1,
21 av Claude Bernard, 69622 Villeurbanne Cedex, France,
e-mail: attal@math.univ-lyon1.fr
url: http://math.univ-lyon1.fr/~attal

1 Introduction

1.1 Discussion

Techniques and tools coming from operator algebras, that is, C^*-algebras or von Neumann algebras, are central in all the approaches of open quantum system theory. They are intensively used in most of the courses of these three volumes.

They are essential in the Hamiltonian approach for they provide the setup for describing equilibrium states in quantum statistical mechanics, the so-called K.M.S. states (see C.-A. Pillet's course in this volume), for describing the observable algebras of free Bose and Fermi gas (see M. Merkli's course in this volume), for computing the standard Liouvillian of quantum dynamical systems (see C.-A. Pillet's course in this volume) which is the starting point of ergodic theory on these systems.

In the Markovian approach, they are essential tools for proving the Stinespring and the Lindblad theorems, which are the mathematical foundation of the so-called quantum master equations (see R. Rebolledo's course in the second volume), for developing the theory of quantum Markov processes, their dilations and their qualitative behavior (see F. Fagnola's course in the second volume and Fagnola-Rebolledo's course in the third volume).

In the third volume of this series ("Recent developments"), there is no course which does not make a heavy use of all the techniques of von Neumann algebras, states, representations, modular theory.

The aim of this course is to give a basic introduction to this theory. Writing such a course is a challenge, for these theories are difficult, deep and subtle. It suffices to have a look to the most well-known references ([1], [2], [3], [4], [5], [6], [7], [8],) to understand that we here enter into a heavy theory.

Hundreds of volumes have been written on operator algebras, it has been a life work of numerous famous mathematicians to try to understand them, to classify them. It is not our task here to enter into all these details, we just want to browse the most basic elements, the tools that appear everywhere. We try to give as many proof as possible, we only skip the very long and difficult ones (at least, we give the ideas). This course has to be considered as a very first step in the theory, a tool box for the other courses. Readers who are interested in more details or more developments are encouraged to enter into the classical literature that we quoted above.

Among the huge literature on the subject, we have selected few references at the end of this course.

The two volumes by Bratteli and Robinson,[1] and [2], are fundamental references for these three volumes, for they really develop the theory of operator algebras in perspective with applications in quantum statistical mechanics. They are the only references in our list which connect operator algebras with physics. Furthermore, their exposition is concise and pedagogical.

The series of volumes by Kadison and Ringrose,[4], [5] and [6], are sorts of bibles on operator algebras. They provide a very complete exposition on all the old and modern theory of operator algebras. For example they completely treat the classification theory of von Neumann algebra, which we do not treat here.

Sakai's book [8] is a well-known reference on the basic elements of C^* and von Neumann algebras. It is a concise and dense book, it should maybe not considered as a beginner reference. It does not treat modular theory.

Pedersen's book [7] is very complete and more pedagogical. Dixmier's book [3] is an older reference, far from the modern pedagogical and notational standards, but it has influenced a whole generation of mathematicians.

1.2 Notations

All the vector spaces, Hilbert spaces, algebras, are here supposed to be defined on the field of complex numbers $\mathbb{C}$.

On a Hilbert space the scalar product $< \cdot\,, \cdot >$ is linear in the right variable and antilinear in the left one. The notation for the scalar product does not refer to the underlying space, there should not be any possible confusion.

In the same way, the norm of any normed space is denoted by $||\cdot||$, unless it is necessary for the comprehension to specify the underlying space.

On any Hilbert space, the identity operator is denoted by I, without precising the associated space. The algebra of bounded operators on a Hilbert space $\mathcal{H}$ is denoted by $\mathcal{B}(\mathcal{H})$.

In any normed space, $B(x, r)$ denotes the closed ball with center x and radius r.

2 C^*-algebras

The two main operator algebras are C^*-algebras and von Neumann algebras. They can be represented as sub-algebras of bounded operator algebras $\mathcal{B}(\mathcal{H})$, with different topologies. But they also admit abstract definitions, without reference to any particular representation. This is with the abstract theory we start with, representation theorems come later.

2.1 First definitions

A *C^*-algebra* is an algebra $\mathcal{A}$ equipped with an involution $A \mapsto A^*$ and a norm $||\cdot||$ satisfying, for all $A, B \in \mathcal{A}$, all $\lambda, \mu \in \mathbb{C}$:

i) $A^{**} = A$

ii) $(\lambda A + \mu B)^* = \overline{\lambda} A^* + \overline{\mu} B^*$

iii) $(AB)^* = B^* A^*$

i') $||A||$ is always positive and $||A|| = 0$ if and only if $A = 0$

ii') $||\lambda A|| = |\lambda| \, ||A||$

iii') $||A + B|| \leq ||A|| + ||B||$

iv') $||AB|| \leq ||A|| \, ||B||$

i") $\mathcal{A}$ is complete for $||\cdot||$

ii") $||AA^*|| = ||A||^2$.

An algebra with an involution as above satisfying i), ii) and iii) is called a $*$-*algebra*. An algebra satisfying all the conditions above but where ii") is replaced by

ii") $||A^*|| = ||A||$

is called a *Banach algebra*.

The basic examples of C^*-algebras are:

1) $\mathcal{A} = \mathcal{B}(\mathcal{H})$, the algebra of bounded operators on a Hilbert space $\mathcal{H}$. The involution is the usual adjoint mapping and the norm is the usual operator norm:

$$||A|| = \sup_{||f||=1} ||Af||$$

.

2) $\mathcal{A} = \mathcal{K}(\mathcal{H})$, the algebra of compact operators on $\mathcal{H}$. It is a sub-C^*-algebra of $\mathcal{B}(\mathcal{H})$.

3) $\mathcal{A} = C_0(X)$, the space of continuous functions vanishing at infinity on a locally compact space X. Recall that a function f is *vanishing at infinity* if for every $\varepsilon > 0$ there exists a compact $K \subset X$ such that $|f| < \varepsilon$ outside of K. The involution on $\mathcal{A}$ is the complex conjugation $\overline{f}$ and the norm is

$$||f|| = \sup_{x\in X} |f(x)|.$$

We will see later that these examples are more than basic : every commutative C^*-algebra is of the form 3), and every C^*-algebra is a sub-algebra of a type 1) example.

Proposition 2.1. *On a C^*-algebra $\mathcal{A}$ we have $||A^*|| = ||A||$ for all $A \in \mathcal{A}$.*

Proof. We have $||A||^2 = ||A^*A|| \leq ||A^*||\,||A||$ and thus $||A|| \leq ||A^*||$. Inverting the role of A and A^* gives the result. □

An element I of a C^*-algebra $\mathcal{A}$ is a *unit* if

$$IA = AI = A$$

for all $A \in \mathcal{A}$.

If a unit exists it is unique and norm 1 (unless $\mathcal{A} = \{0\}$). But it may not always exists. Indeed, in the example $\mathcal{K}(\mathcal{H})$ there exists a unit if and only if $\mathcal{H}$ is finite dimensional. In the example $C_0(X)$ there exists a unit if and only if X is compact.

But if a C^*-algebra does not contain a unit one can easily add one as follows. Consider the vector space $\mathcal{A}' = \mathcal{A} \oplus \mathbb{C}$ and provide it with the product

$$(A, \lambda)(B, \mu) = (AB + \lambda B + \mu A, \lambda\mu),$$

with the involution

$$(A,\lambda)^* = (A^*, \overline{\lambda})$$

and with the norm

$$||(A,\lambda)|| = \sup_{||B||=1} ||AB + \lambda B|| .$$

Equipped this way $\mathcal{A}'$ is a C^*-algebra. It admits a unit $(0,1)$. The algebra $\mathcal{A}$ identifies to the subset of elements of the form $(A,0)$. The only delicate point is to check that $||(A,\lambda)|| = 0$ if and only if $A = 0$ and $\lambda = 0$. One can assume that $\lambda \neq 0$ for if not we are in $\mathcal{A}$. Hence one can assume that $\lambda = 1$. We have

$$||B - AB|| \le ||B||\,||(-A,1)|| .$$

Thus if $||(-A,1)|| = 0$ then $B = AB$ for all $B \in \mathcal{A}$. Applying the involution gives $B = BA^*$ for all $B \in \mathcal{A}$. In particular $A^* = AA^* = A$ and thus $B = AB = BA$. This means A is a unit and contradicts the assumption.

Note that the above definition of the norm in $\mathcal{A}'$ comes from the fact that in any C^*-algebra we have

$$||A|| = \sup_{||B||=1} ||AB|| .$$

Indeed, there is obviously an inequality $\ge$ between the two terms above. The equality is obtained by considering $B = A^*/\,||A||$.

2.2 Spectral analysis

Let $\mathcal{A}$ be a C^*-algebra with unit I. An element A of $\mathcal{A}$ is *invertible* if there exists an element A^{-1} of $\mathcal{A}$ such that

$$A^{-1}A = AA^{-1} = I.$$

One calls *resolvent set* of A the set

$$\rho(A) = \{\lambda \in \mathbb{C};\, \lambda I - A \text{ is invertible}\}.$$

We put

$$\sigma(A) = \mathbb{C} \setminus \rho(A)$$

and call it the *spectrum* of A.

If $|\lambda| > ||A||$ then the series

$$\frac{1}{\lambda}\sum_n \left(\frac{A}{\lambda}\right)^n$$

is normally convergent and its sum is equal to $(\lambda I - A)^{-1}$. This implies that $\sigma(A)$ is included in $B(0, ||A||)$.

Furthermore, if λ_0 belongs to $\rho(A)$ and if $\lambda \in \mathbb{C}$ is such that $|\lambda - \lambda_0| < ||\lambda_0 I - A||$, then the series

$$(\lambda_0 I - A)^{-1} \sum_n \left(\frac{\lambda_0 - \lambda}{\lambda_0 I - A} \right)^n$$

normally converges to $(\lambda I - A)^{-1}$. In particular we have proved that:

1) the set $\rho(A)$ is open
2) the mapping $\lambda \mapsto (\lambda I - A)^{-1}$ is analytic on $\rho(A)$
3) the set $\sigma(A)$ is compact.

We define

$$r(A) = \sup\{|\lambda| \,;\, \lambda \in \sigma(A)\},$$

the *spectral radius* of A.

Theorem 2.2. *We have for all $A \in \mathcal{A}$*

$$r(A) = \lim_n \|A^n\|^{1/n} = \inf_n \|A^n\|^{1/n} \leq \|A\| .$$

In particular the above limit always exists and $\sigma(A)$ is never empty.

Proof. Let n be fixed and let $|\lambda| > \|A^n\|^{1/n}$. Every integer m can be written $m = pn + q$ with p, q integers and $q < n$. Thus we have

$$\begin{aligned} \sum_m \left\| \left(\frac{A}{\lambda} \right)^m \right\| &= \sum_m \left\| \left(\frac{A}{\lambda} \right)^{pn+q} \right\| \leq \sum_m \left(\frac{\|A^n\|}{|\lambda|^n} \right)^p \left(\frac{\|A\|}{|\lambda|} \right)^q \\ &\leq \left(1 + \frac{\|A\|}{|\lambda|} + \ldots + \left(\frac{\|A\|}{|\lambda|} \right)^{n-1} \right) \sum_p \left(\frac{\|A^n\|}{|\lambda|^n} \right)^p < \infty. \end{aligned}$$

Thus the series

$$\frac{1}{\lambda} \sum_m \left(\frac{A}{\lambda} \right)^m$$

converges and its sum is equal to $(\lambda I - A)^{-1}$. This proves that $r(A) \leq \|A^n\|^{1/n}$ and thus $r(A) \leq \liminf_n \|A^n\|^{1/n}$.

Let us now prove that $r(A) \geq \limsup_n \|A^n\|^{1/n}$. If we have

$$r(A) < \limsup_n \|A^n\|^{1/n}$$

then consider the open set

$$\mathcal{O} = \{\lambda \in \mathbb{C}; r(A) < |\lambda| < \limsup_n \|A^n\|^{1/n}\}.$$

On $\mathcal{O}$ all the operators $\lambda I - A$ are invertible, thus so are the operators $I - \frac{1}{\lambda} A$. The mapping $\lambda \mapsto (I - \frac{1}{\lambda} A)^{-1}$ is analytic on $\mathcal{O}$ and its Taylor series $\sum_n (\frac{A}{\lambda})^n$ converges. But the convergence radius of the series $\sum_n z^n A^n$ is exactly

$$(\limsup_n ||A^n||^{1/n})^{-1}.$$

This would mean

$$\frac{1}{|\lambda|} < (\limsup_n ||A^n||^{1/n})^{-1}$$

which contradicts the fact that $\lambda \in \mathcal{O}$. We have proved the first part of the theorem.

If $r(A) > 0$ then it is clear that $\sigma(A)$ is not empty. It remains to consider the case $r(A) = 0$. But note that if 0 belongs to $\rho(A)$ this means that A is invertible and $1 = ||A^n A^{-n}|| \leq ||A^n|| \, ||A^{-n}||$. In particular, $1 \leq ||A^n||^{1/n} \, ||A^{-n}||^{1/n}$. Passing to the limit, we get $r(A) > 0$. Thus if $r(A) = 0$ we must have $0 \in \sigma(A)$. In any case $\sigma(A)$ is non empty. □

Corollary 2.3. *A C^*-algebra $\mathcal{A}$ with unit and all of which elements, except 0, are invertible is isomorphic to $\mathbb{C}$.*

Proof. If $A \in \mathcal{A}$ its spectrum $\sigma(A)$ is non empty. Thus there exists a $\lambda \in \mathbb{C}$ such that $\lambda I - A$ is not invertible. This means $\lambda I - A = 0$ and $A = \lambda I$. □

All the above results made use of the fact that we considered a C^*-algebra with unit. If $\mathcal{A}$ is a C^*-algebra without unit and if $\widetilde{\mathcal{A}}$ is its natural extension with unit, then the notion of spectrum and resolvent set are extended as follows. The *spectrum* of $A \in \mathcal{A}$ is its spectrum as an element of $\widetilde{\mathcal{A}}$. We extend the notion of resolvent set in the same way.

An element A of a C^*-algebra $\mathcal{A}$ is
normal if $A^*A = AA^*$,
self-adjoint if $A = A^*$.
If $\mathcal{A}$ contains a unit, then an element $A \in \mathcal{A}$ is
isometric if $A^*A = I$,
unitary if $A^*A = AA^* = I$.

Theorem 2.4. *Let $\mathcal{A}$ be a C^*-algebra with unit.*

a) If A is normal then $r(A) = ||A||$.
b) If A is self-adjoint then $\sigma(A) \subset [-||A||\, , ||A||]$.
c) If A is isometric then $r(A) = 1$.
d) If A is unitary then $\sigma(A) \subset \{\lambda \in \mathbb{C}; |\lambda| = 1\}$.
e) For all $A \in \mathcal{A}$ we have $\sigma(A^*) = \overline{\sigma(A)}$ and $\sigma(A^{-1}) = \sigma(A)^{-1}$.
f) For every polynomial function P we have

$$\sigma(P(A)) = P(\sigma(A)).$$

g) For any two $A, B \in \mathcal{A}$ we have

$$\sigma(AB) \cup \{0\} = \sigma(BA) \cup \{0\}.$$

Proof. a) If A is normal then

$$\left\|A^{2^n}\right\|^2 = \left\|A^{2^n} A^{*2^n}\right\| = \left\|(AA^*)^{2^n}\right\| = \left\|(AA^*)^{2^{n-1}} (AA^*)^{2^{n-1}}\right\|$$
$$= \left\|(AA^*)^{2^{n-1}}\right\|^2 = \ldots = \|AA^*\|^{2^n} = \|A\|^{2^{n+1}}.$$

It is now easy to conclude with Theorem 2.2

b) We only have to prove that the spectrum of any self-adjoint element of $\mathcal{A}$ is a subset of $\mathbb{R}$. Let $\lambda = x + iy$ be an element of $\sigma(A)$, with x, y real. We have $x + i(y + t) \in \sigma(A + itI)$. But

$$\|A + itI\|^2 = \|(A + itI)(A - itI)\| = \left\|A^2 + t^2 I\right\| \le \|A\|^2 + t^2.$$

This implies

$$|x + i(y + t)|^2 = x^2 + (y + t)^2 \le \|A\|^2 + t^2$$

or else

$$2yt \le \|A\|^2 - x^2 - y^2$$

for all t. This means $y = 0$.

c) If A is isometric then

$$\|A^n\|^2 = \|A^{*n} A^n\| = \left\|A^{*n-1} A^{n-1}\right\| = \ldots = \|A^* A\| = \|I\| = 1.$$

d) Assume e) is proved. Then if A is unitary we have

$$\sigma(A) = \overline{\sigma(A^*)} = \overline{\sigma(A^{-1})} = \overline{\sigma(A)}^{-1}.$$

This and c) imply that $\sigma(A)$ is included in the unit circle.

e) The property $\sigma(A^*) = \overline{\sigma(A)}$ is obvious. For the other identity we write $\lambda I - A = \lambda A(A^{-1} - \lambda^{-1} I)$ and $\lambda^{-1} I - A^{-1} = \lambda^{-1} A^{-1}(A - \lambda I)$.

f) Note that if $B = A_1 \ldots A_n$ in $\mathcal{A}$, where all the A_i are two by two commuting, we have that B is invertible if and only if each A_i is invertible. Now choose α and $\alpha_1, \ldots \alpha_n$ in $\mathbb{C}$ such that

$$P(x) - \lambda = \alpha \prod_i (x - \alpha_i).$$

In particular we have

$$P(A) - \lambda I = \alpha \prod_i (A - \alpha_i I).$$

As a consequence $\lambda \in \sigma(P(A))$ if and only if $\alpha_i \in \sigma(A)$ for a i. But as $P(\alpha_i) = \lambda$ this exactly means that λ belongs to $\sigma(P(A))$ if and only if λ belongs to $P(\sigma(A))$.

g) If λ belong to $\rho(BA)$ then

$$(\lambda I - AB)(I + A(\lambda I - BA)^{-1} B) = \lambda I.$$

This proves that $\lambda I - AB$ is invertible on the right, with possible exception of $\lambda = 0$. The invertibility on the left is obtained in a similar way. This proves one inclusion. The converse inclusion is obtained exchanging the role of A and B. □

Theorem 2.5. *The norm which makes a $*$-algebra being a C^*-algebra, when it exists, is unique.*

Proof. By the above results we have

$$||A||^2 = ||AA^*|| = r(AA^*)$$

for AA^* is always normal. But $r(AA^*)$ depends only on the algebraic structure of $\mathcal{A}$. □

Proposition 2.6. *The set of invertible elements of a C^*-algebra $\mathcal{A}$ with unit is open and the mapping $A \mapsto A^{-1}$ is continuous on this set.*

Proof. If A is invertible and if B is such that $||B - A|| < \left|\left|A^{-1}\right|\right|^{-1}$ then $B = A(I - A^{-1}(A - B))$ is invertible for

$$r(A^{-1}(A - B)) \leq \left|\left|A^{-1}(A - B)\right|\right| < 1$$

and thus $I - A^{-1}(A - B)$ is invertible. The open character is proved. Let us now show the continuity. If $||B - A|| < 1/2 \left|\left|A^{-1}\right|\right|^{-1}$ then

$$\begin{aligned} ||B^{-1} - A^{-1}|| &= \left|\left| \sum_{n=0}^{\infty} \left(A^{-1}(A - B)\right)^n A^{-1} - A^{-1} \right|\right| \\ &\leq \sum_{n=1}^{\infty} ||A^{-1}(A-B)||^n \, ||A^{-1}|| \\ &\leq \frac{\left|\left|A^{-1}\right|\right|^2 ||A - B||}{1 - ||A^{-1}(A - B)||} \\ &\leq 2\left|\left|A^{-1}\right|\right|^2 ||A - B|| . \end{aligned}$$

This proves the continuity. □

In the following, we denote by $\mathbb{1}$ the constant function equal to 1 on $\mathbb{C}$ and by id_E the function $\lambda \mapsto \lambda$ on $E \subset \mathbb{C}$.

A $*$-*algebra morphism* is a linear mapping $\Pi : \mathcal{A} \to \mathcal{B}$, between two $*$-algebras $\mathcal{A}$ and $\mathcal{B}$, such that $\Pi(A^*B) = \Pi(A)^*\Pi(B)$ for all $A, B \in \mathcal{A}$. A C^*-*algebra morphism* is $*$-algebra morphism Π between two C^*-algebras $\mathcal{A}$ and $\mathcal{B}$, such that $||\Pi(A)||_{\mathcal{B}} = ||A||_{\mathcal{A}}$, for all $A \in \mathcal{A}$.

Theorem 2.7 (Functional calculus). *Let $\mathcal{A}$ be a C^*-algebra with unit. Let A be a self-adjoint element in $\mathcal{A}$. Let $C(\sigma(A))$ be the C^*-algebra of continuous functions on $\sigma(A)$. Then there is a unique morphism of C^*-algebra*

$$\begin{aligned} C(\sigma(A)) &\longrightarrow \mathcal{A} \\ f &\longmapsto f(A) \end{aligned}$$

which sends the function $\mathbb{1}$ on I and the function $\mathrm{id}_{\sigma(A)}$ on A.

Furthermore we have

$$\sigma(f(A)) = f(\sigma(A)) \tag{1}$$

for all $f \in C(\sigma(A))$.

Proof. When f is a polynomial function the application $f \mapsto f(A)$ is well-defined and isometric for

$$||f(A)|| = \sup\{|\lambda|\,;\lambda \in \sigma(f(A))\} = \sup\{|\lambda|\,;\lambda \in f(\sigma(A))\} = ||f||\,.$$

Thus it extends to an isometry on $C(\sigma(A))$ by Weierstrass theorem. The extension is easily seen to be a morphism also. The only delicate point to check is the identity (1). Let $\mu \in f(\sigma(A))$, with $\mu = f(\lambda)$. Let $(f_n)_{n\in\mathbb{N}}$ be a sequence of polynomial functions converging to f. The sequence $(f_n(\lambda)I - f_n(A))_{n\in\mathbb{N}}$ converges to $\mu I - f(A)$. As none of the $f_n(\lambda)I - f_n(A)$ is invertible then $\mu I - f(A)$is not either (Proposition 1.6). Thus $f(\sigma(A)) \subset \sigma(f(A))$. Finally, if $\mu \in \mathbb{C} \setminus f(\sigma(A))$ then let $g(t) = (\mu - f(t))^{-1}$. Then g belongs to $C(\sigma(A))$ and $g(A) = (\mu I - f(A))^{-1}$. Thus μ belongs to $\mathbb{C} \setminus \sigma(f(A))$. □

An element A of a C^*-algebra $\mathcal{A}$ is *positive* if it is self-adjoint and its spectrum is included in $\mathbb{R}^+$.

Theorem 2.8. *Let A be an element of $\mathcal{A}$. The following assertions are equivalent.*

i) A is positive.

ii) (if $\mathcal{A}$ contains a unit) A is self-adjoint and $||tI - A|| \leq t$ for some $t \geq ||A||$.

iii) (if $\mathcal{A}$ contains a unit) A is self-adjoint and $||tI - A|| \leq t$ for all $t \geq ||A||$.

iv) $A = B^*B$ for a $B \in \mathcal{A}$.

v) $A = C^2$ for a self-adjoint $C \in \mathcal{A}$.

Proof. Let us first prove that i) implies iii). If i) is satisfied then $tI - A$ is a normal operator and

$$||tI - A|| = \sup\{|\lambda|\,;\lambda \in \sigma(tI - A)\} = \sup\{|\lambda - t|\,;\lambda \in \sigma(A)\} \leq t.$$

This gives iii).

Obviously iii) implies ii). Let us prove that ii) implies i). If ii) is satisfied and if $\lambda \in \sigma(A)$ then $t - \lambda \in \sigma(tI - A)$ and with the same computation as above $|t - \lambda| \leq ||tI - A|| \leq t$. But as $\lambda \leq t$ we must have $\lambda \geq 0$. This proves i). We have proved that the first 3 assertions are equivalent.

We have that v) implies iv) obviously. In order to show that i) implies v) it suffices to consider $C = \sqrt{A}$ (using the functional calculus of Theorem 2.7 and identity (1)). It remains to prove that iv) implies i). Let $f_+(t) = t \vee 0$ and $f_-(t) = (-t) \vee 0$. Let $A_+ = f_+(A)$ and $A_- = f_-(A)$ (note that when iv) holds true then A is automatically self-adjoint and thus accepts the functional calculus of Theorem 2.7). We have $A = A_+ - A_-$ and the elements A_+ and A_- are positive by (1). Furthermore the identity $f_+f_- = 0$ implies $A_+A_- = 0$. We have

$$(BA_-)^*(BA_-) = A_-(A_+ - A_-)A_- = -A_-^3.$$

In particular $-(BA_-)^*(BA_-)$ is positive.

Writing $BA_- = S + iT$ with S and T self-adjoint gives

$$(BA_-)(BA_-)^* = -(BA_-)^*(BA_-) + 2(S^2 + T^2).$$

In particular, as the equivalence established between i), ii) and iii) proves it easily, the set of positive elements of $\mathcal{A}$ is a cone, thus the element $(BA_-)(BA_-)^*$ is positive. As a consequence

$$\sigma((BA_-)(BA_-)^*) \subset [0, ||B||\,||A_-||].$$

But by Theorem 1.4 g) we must also have $\sigma((BA_-)^*(BA_-)) \subset [0, ||B||\,||A_-||]$. In particular $\sigma(-A_-^3) \subset [0, ||B||^2||A_-||^2]$. This implies $\sigma(A_-^3) = \{0\}$ and $||A_-^3|| = 0 = ||A_-||^3$. That is $A_- = 0$. □

This notion of positivity defines an order on elements of $\mathcal{A}$, by saying that $U \geq V$ in $\mathcal{A}$ if $U - V$ is a positive element of $\mathcal{A}$.

Proposition 2.9. *Let U, V be self-adjoint elements of $\mathcal{A}$ such that $U \geq V \geq 0$. Then*

i) $W^*UW \geq W^*VW \geq 0$ for all $W \in \mathcal{A}$;

ii) $(V + \lambda I)^{-1} \geq (U + \lambda I)^{-1}$ for all $\lambda \geq 0$.

Proof. i) is obvious from Theorem 2.8.

ii) As we have $U + \lambda I \geq V + \lambda I$, then by i) we have

$$(V + \lambda I)^{-1/2}(U + \lambda I)V + \lambda I)^{-1/2} \geq I.$$

Now, note that if W is self-adjoint and $W \geq I$ then $\sigma(W) \subset [1, +\infty[$ and $\sigma(W^{-1}) \subset [0, 1]$. In particular $W^{-1} \leq I$. This argument applied to the above inequality shows that

$$(V + \lambda I)^{1/2}(U + \lambda I)^{-1}V + \lambda I)^{1/2} \leq I.$$

Multiplying both sides by $(V + \lambda I)^{-1/2}$ gives the result. □

2.3 Representations and states

Note that a $*$-algebra morphism is always positive, that is, it maps positive elements to positive elements. Indeed we have $\Pi(A^*A) = \Pi(A)^*\Pi(A)$.

Theorem 2.10. *If Π is a morphism between two C^*-algebras $\mathcal{A}$ and $\mathcal{B}$ then Π is continuous, with norm smaller than 1. Furthermore the range of Π is a sub-C^*-algebra of $\mathcal{B}$.*

Proof. If A is self-adjoint then so is $\Pi(A)$ and thus

$$||\Pi(A)|| = \sup\{|\lambda|\,;\lambda \in \sigma(\Pi(A))\}.$$

But it is easy to see that $\sigma(\Pi(A))$ is included in $\sigma(A)$ and consequently

$$||\Pi(A)|| \leq \sup\{|\lambda|\,;\lambda \in \sigma(A)\} = ||A||\,.$$

For a general A we have

$$||\Pi(A)||^2 = ||\Pi(A^*A)|| \leq ||A^*A|| = ||A||^2.$$

We have proved the first part of the theorem.

For proving the second part we reduce the problem to the case where $\ker \Pi = \{0\}$. If this is not the case, following the Appendix subsection 2.5, we consider the quotient of $\mathcal{A}$ by the two-sided closed ideal $\ker \Pi : \mathcal{A}_\Pi = \mathcal{A}/\ker \Pi$ which is a C^*-algebra. We can thus assume $\ker \Pi = \{0\}$. Let $\mathcal{B}_\Pi$ be the image of Π, it is sufficient to prove that it is closed. Consider the inverse morphism Π^{-1} from $\mathcal{B}_\Pi$ onto $\mathcal{A}$. As previously, for A self-adjoint in $\mathcal{A}$ we have

$$||A|| = \left|\left|\Pi^{-1}(\Pi(A))\right|\right| \leq ||\Pi(A)|| \leq ||A||\,.$$

Thus Π^{-1} and Π are isometric and one concludes easily. □

A *representation* of a C^*-algebra $\mathcal{A}$ is a pair $(\mathcal{H}, \Pi)$ made of a Hilbert space $\mathcal{H}$ and a morphism Π from $\mathcal{A}$ to $\mathcal{B}(\mathcal{H})$. The representation is *faithful* if $\ker \Pi = \{0\}$.

Proposition 2.11. *Let $(\mathcal{H}, \Pi)$ be a representation of a C^*-algebra $\mathcal{A}$. Then the following assertions are equivalent.*

i) Π is faithful.

ii) $||\Pi(A)|| = ||A||$ for all $A \in \mathcal{A}$.

iii) $\Pi(A) > 0$ if $A > 0$.

Proof. We have already seen that i) implies ii), in the proof above. Let us prove that ii) implies iii). If $A > 0$ then $||A|| > 0$ and thus $||\Pi(A)|| > 0$ and $\Pi(A) \neq 0$. As we already know that $\Pi(A) \geq 0$, we conclude that $\Pi(A) > 0$. Finally, assume iii) is satisfied. If B belongs to $\ker \Pi$ and $B \neq 0$ then $\Pi(B^*B) = 0$. But $||B^*B|| = ||B||^2 > 0$ and thus $B^*B > 0$. Which is contradictory and ends the proof. □

Clearly we have not yet discussed the existence of representations for C^*-algebras. The key tool for this existence theorem is the notion of *state*.

A linear form ω on $\mathcal{A}$ is *positive* if $\omega(A^*A) \geq 0$ for all $A \in \mathcal{A}$. Note that for such positive linear form one can easily prove a Cauchy-Schwarz inequality:

$$|\omega(B^*A)|^2 \leq \omega(B^*B)\,\omega(A^*A),$$

with the same proof as for the usual Cauchy-Schwarz inequality.

Proposition 2.12. *Let ω be a linear form on $\mathcal{A}$, a C^*-algebra with unit. Then the following assertions are equivalent.*

i) ω is positive.

ii) ω is continuous with $||\omega|| = \omega(I)$.

Proof. By Theorem 2.8 ii), recall that a self-adjoint element A of $\mathcal{A}$, with $||A|| = 1$, is positive if and only if $||(I - A)|| \leq 1$. In particular, for any $A \in \mathcal{A}$, we have that $||A^*A||\, I - A^*A$ is positive.

If i) is satisfied then $\omega(A^*A) \leq ||A^*A||\, \omega(I)$. By Cauchy-Schwarz we have

$$|\omega(A)| \leq \omega(I)^{1/2} \, |\omega(A^*A)|^{1/2} \leq ||A^*A||^{1/2} \, \omega(I) = ||A|| \, \omega(I). \tag{2}$$

This proves ii).

Conversely, if ii) is satisfied. One can assume $\omega(I) = 1$. Let A be a self-adjoint element of $\mathcal{A}$. Write $\omega(A) = \alpha + i\beta$ for some α, β real. For every $\lambda \in I\!R$ we have

$$||A + i\lambda I||^2 = ||A^2 + \lambda^2 I|| = ||A||^2 + \lambda^2.$$

Thus we have

$$\beta^2 + 2\lambda\beta + \lambda^2 \leq \left|\alpha^2 + i(\beta + \lambda)\right|^2 = |\omega(A + i\lambda I)|^2 \leq ||A||^2 + \lambda^2.$$

This implies that $\beta = 0$ and $\omega(A)$ is real. Consider now A positive, with $||A|| = 1$. We have

$$|1 - \omega(A)| = |\omega(I - A)| \leq ||I - A|| \leq I.$$

Thus $\omega(A)$ is positive. □

When the C^*-algebra $\mathcal{A}$ does not contain a unit, the norm property $||\omega|| = \omega(I)$ above has to be replaced by

$$||\omega|| = \lim_{\alpha} \omega(E_\alpha^2)$$

for an approximate unit (E_α) in $\mathcal{A}$ (cf subsection 2.5), we do not develop the proof in this case.

We call *state* any positive linear form on $\mathcal{A}$ such that $||\omega|| = 1$. We need an existence theorem for states.

Theorem 2.13. *Let A be any element of $\mathcal{A}$. Then there exists a state ω on $\mathcal{A}$ such that $\omega(A^*A) = ||A||^2$.*

Proof. On the space $\mathcal{B} = \{\alpha I + \beta A^*A; \alpha, \beta \in \mathbb{C}\}$ we define the linear form

$$f(\alpha I + \beta A^*A) = \alpha + \beta ||A||^2.$$

One easily checks that $||f|| = 1$. By Hahn-Banach we extend f to the whole of $\mathcal{A}$ into a norm 1 continuous linear form ω. By the previous proposition ω is a state. □

We now turn to the construction of a representation which is going to be fundamental for us, the so called Gelfand-Naimark-Segal construction (G.N.S. construction). Indeed, note that if $(\mathcal{H}, \Pi)$ is a representation of a C^*-algebra $\mathcal{A}$ and if Ω is any norm 1 vector of $\mathcal{H}$, then the mapping

$$\omega(A) = <\Omega\,,\, \Pi(A)\Omega>$$

clearly defines a state on $\mathcal{A}$. The G.N.S. construction proves that any C^*-algebra with a state can be represented this way.

Theorem 2.14 (G.N.S. representation). *Let $\mathcal{A}$ be a C^*-algebra with unit and ω be a state on $\mathcal{A}$. Then there exists a Hilbert space $\mathcal{H}_\omega$, a representation Π_ω of $\mathcal{A}$ in $\mathcal{B}(\mathcal{H}_\omega)$ and a unit vector Ω_ω of $\mathcal{H}_\omega$ such that*

$$\omega(A) = <\Omega_\omega\,,\, \Pi_\omega(A)\Omega_\omega>$$

for all A and such that the space $\{\Pi_\omega(A)\Omega_\omega; A \in \mathcal{A}\}$ is dense in $\mathcal{H}_\omega$. Such a representation is unique up to unitary isomorphism.

Proof. Let $V_\omega = \{A \in \mathcal{A}; \omega(A^*A) = 0\}$. The set V_ω is a left ideal for if $A \in V_\omega$ and $B \in \mathcal{A}$ then

$$0 \leq \omega((AB)^*AB) \leq ||B||^2\omega(A^*A) = 0.$$

We consider the quotient space $\mathcal{A}/V_\omega$. On $\mathcal{A}/V_\omega$ we define

$$<[A]\,,\,[B]> = \omega(B^*A)$$

(We leave to the reader to check that this definition is consistent, in the sense that $\omega(B^*A)$ only depends on the equivalence classes of A and B). It is a positive sesquilinear form which makes $\mathcal{A}/V_\omega$ a pre-Hilbert space. Let $\mathcal{H}_\omega$ be the its closure. We put

$$\begin{aligned} L_A\,:\, \mathcal{A}/V_\omega &\to \mathcal{A}/V_\omega \\ [B] &\mapsto [AB]. \end{aligned}$$

We have

$$<L_A[B]\,,\, L_A[B]> = \omega(B^*A^*AB) \leq ||A||^2\omega(B^*B)$$

for $C \mapsto \omega(B^*CB)$ is a positive linear form equal to $\omega(B^*B)$ on $C = I$. In particular $<L_A[B]\,,\, L_A[B]> \leq ||A||^2 <[B]\,,\,[B]>$. One can extend L_A into a bounded operator $\Pi_\omega(A)$ on $\mathcal{H}_\omega$. If we put $\Omega_\omega = [I]$ then the construction is finished.

Let us check uniqueness. If $(\mathcal{H}', \Pi', \Omega')$ is another such triple, we have

$$\begin{aligned} <\Pi_\omega(B)\Omega_\omega\,,\, \Pi_\omega(A)\Omega_\omega> &= <\Omega_\omega\,,\, \Pi_\omega(B^*A)\Omega_\omega> = \omega(B^*A) \\ &= <\Omega'\,,\, \Pi'(B^*A)\Omega'> = <\Pi'(B)\Omega'\,,\, \Pi'(A)\Omega'>. \end{aligned}$$

The unitary isomorphism is thus defined by $U\,:\, \Pi_\omega(A)\Omega_\omega \mapsto \Pi'(A)\Omega'$. □

This G.N.S. representation theorem gives the fundamental representation theorem for C^*-algebras.

Theorem 2.15. *Let $\mathcal{A}$ be a C^*-algebra . Then $\mathcal{A}$ is isomorphic to a sub-C^*-algebra of $\mathcal{B}(\mathcal{H})$ for some Hilbert space $\mathcal{H}$.*

Proof. For every state ω we have the G.N.S. representation $(\mathcal{H}_\omega, \Pi_\omega, \Omega_\omega)$. Put $\mathcal{H} = \oplus_\omega \mathcal{H}_\omega$ and $\Pi = \oplus_\omega \Pi_\omega$ where the direct sums run over the set of all states on $\mathcal{A}$.

For every $A \in \mathcal{A}$ there exists a state ω_A such that $||\Pi_{\omega_A}(A)|| = ||A||$ (Theorem 2.13). But we have $||\Pi(A)|| \geq ||\Pi_{\omega_A}(A)|| = ||A||$. Thus we get $||\Pi(A)|| = ||A||$ by Theorem 2.10. This means that Π is faithful by Proposition 2.11. In particular $\mathcal{A}$ is isomorphic to $\Pi(\mathcal{A})$ which is, by Theorem 2.10 a sub-C^*-algebra of $\mathcal{B}(\mathcal{H})$. □

2.4 Commutative C^*-algebras

We have shown the very important characterization of C^*-algebras, namely they are exactly the closed $*$-sub-algebras of bounded operators on Hilbert space. We dedicate this last section to prove the (not very useful for us but) interesting characterization of commutative C^*-algebras .

Let $\mathcal{A}$ be a commutative C^*-algebra . A *character* on $\mathcal{A}$ is a linear form χ on $\mathcal{A}$ satisfying

$$\chi(AB) = \chi(A)\chi(B)$$

for all $A, B \in \mathcal{A}$. On then calls *spectrum* of $\mathcal{A}$ the set $\sigma(\mathcal{A})$ of all characters on $\mathcal{A}$.

Proposition 2.16. *Every character is positive.*

Proof. If necessary, we extend the C^*-algebra $\mathcal{A}$ to $\widetilde{\mathcal{A}}$ so that it contains a unit I. A character χ on $\mathcal{A}$ then extends to a character on $\widetilde{\mathcal{A}}$ by $\chi(\lambda I + A) = \lambda + \chi(A)$. Thus we may assume that $\mathcal{A}$ contains a unit I.

Let $A \in \mathcal{A}$ and $\lambda \notin \sigma(A)$. Then there exists $B \in \mathcal{A}$ such that $(\lambda I - A)B = I$. Thus $\chi(\lambda I - A)\chi(B) = (\lambda\chi(I) - \chi(A))\chi(B) = \chi(I) = 1$. This implies in particular that $\lambda \neq \chi(A)$. We have proved that $\chi(A)$ always belong to $\sigma(A)$. In particular $\chi(A^*A)$ is always positive. □

As a corollary every character is a state and thus is continuous. The set $\sigma(\mathcal{A})$ is a subset of $\mathcal{A}^*$, the dual of $\mathcal{A}$.

Theorem 2.17. *Let $\mathcal{A}$ be a commutative C^*-algebra and X be the spectrum of $\mathcal{A}$ endowed with the $*$-weak topology of $\mathcal{A}^*$. Then X is a Hausdorff locally compact set; it is compact if and only if $\mathcal{A}$ admits a unit.*

Furthermore $\mathcal{A}$ is isomorphic to the C^*-algebra $C_0(X)$ of continuous functions on X which vanish at infinity.

Proof. Let $\omega_0 \in X$. Let A positive be such that $\omega_0(A) > 0$. One can assume $\omega_0(A) > 1$. Let $K = \{\omega \in X; \omega(A) > 1\}$. It is an open neighborhood of ω_0. Its closure $\overline{K}$ is included into $\{\omega \in X; \omega(A) \geq 1\}$. The latest set is closed and included in the unit ball of $\mathcal{A}^*$ which is compact. Thus X is locally compact.

If $\mathcal{A}$ contains a unit I, then the same argument applied to $A = 2I$ shows that X is compact.

Now, for all $A \in \mathcal{A}$ we put $\widehat{A}(\omega) = \omega(A)$. Then $\widehat{A}$ is a continuous complex function and $A \mapsto \widehat{A}$ is a morphism. Furthermore

$$\left|\left|\widehat{A}\right|\right|^2 = \sup_{\omega \in X} \left|\widehat{A}(\omega)\right|^2 = \sup_{\omega \in X} \left|\widehat{A^* A}(\omega)\right| = ||A||^2$$

for it exists an ω such that $|\omega(A^*A)| = ||A||$. Thus $A \mapsto \widehat{A}$ is an isomorphism.

The set $K_\varepsilon = \{\omega \in X; \omega(A) > \varepsilon\}$ is $*$-weakly compact and thus $\widehat{A}$ belong to $C_0(X)$. Finally $\widehat{A}$ separates the points of X, thus by Stone-Weierstrass theorem, the mapping $\widehat{A}$ gives the whole of $C_0(X)$. □

2.5 Appendix

This is an appendix of the C^*-algebra section, on Quotient algebras and approximate identities. It is not necessary at first reading.

A subspace $\mathcal{J}$ of a C^*-algebra $\mathcal{A}$ is a *left ideal* if for all $J \in \mathcal{J}$ and all $A \in \mathcal{A}$ then JA belongs to $\mathcal{J}$. In the same way one obviously defines *right ideals* and *two-sided ideals*.

If $\mathcal{J}$ is a two-sided, self-adjoint ideal of $\mathcal{A}$, one can easily define the quotient algebra $\mathcal{A}/\mathcal{J}$ by the usual rules:

$$\begin{aligned} &\text{i)}\lambda[X] + \mu[Y] = [\lambda X + \mu Y], \\ &\text{ii)}[X]\,[Y] = [XY], \\ &\text{iii)}[X]^* = [X^*], \end{aligned}$$

where $[X] = \{X + J; J \in \mathcal{J}\}$ is the equivalence class of $X \in \mathcal{A}$ modulo $\mathcal{J}$. We leave to the reader to check the consistency of the above definitions.

We now define a norm on $\mathcal{A}/\mathcal{J}$ by

$$||[X]|| = \inf\{||X + J||\, ; J \in \mathcal{J}\}.$$

The true difficulty is to check that the above norm is a C^*-algebra norm. For this aim we need the notion of approximate identity.

If $\mathcal{J}$ is a left ideal of $\mathcal{A}$ then an *approximate identity* or *approximate unit* in $\mathcal{J}$ is a generalized sequence $(e_\alpha)_a$ of positive elements of $\mathcal{J}$ satisfying i) $||e_\alpha|| \leq 1$, ii) $\alpha \leq \beta$ implies $e_\alpha \leq e_\beta$, iii) $\lim_\alpha ||Xe_\alpha - X|| = 0$ for all $X \in \mathcal{J}$.

Proposition 2.18. *Every left ideal $\mathcal{J}$ of a C^*-algebra $\mathcal{A}$ possesses an approximate unit.*

Proof. Let $\mathcal{J}_+$ be the set of positive elements of $\mathcal{J}$. For each $J \in \mathcal{J}_+$ put

$$e_J = J(I + J)^{-1} = I - (I + J)^{-1}.$$

It is a generalized sequence, it is increasing by Proposition 1.9 and $||e_J|| \leq 1$. Let us now fix $X \in \mathcal{J}$. For every $n \in \mathbb{N}$ there exists a $J \in \mathcal{J}_+$ such that $J \geq nX^*X$. Thus

$$(X - Xe_J)^*(X - Xe_J) = (I - e_J)X^*X(I - e_J) \leq \frac{1}{n}(I - e_J)J(I - e_J)$$

by Proposition 1.9. It suffices to prove that

$$\sup_{J \in \mathcal{J}_+} ||J(I - e_J)^2|| < \infty.$$

But note that $J(I-e_J)^2 = J(I+J)^{-2}$ and using the functional calculus this reduces to the obvious remark that $\lambda/(1 + \lambda^2)$ is bounded on $\mathbb{R}^+$. □

We can now prove the main result of the appendix.

Theorem 2.19. *If $\mathcal{J}$ is a closed, self-adjoint, two-sided ideal of a C^*-algebra $\mathcal{A}$, then the quotient algebra $\mathcal{A}/\mathcal{J}$, equipped with the quotient norm, is a C^*-algebra.*

Proof. Let us first show that

$$||[X]|| = \lim_\alpha ||e_\alpha X - X||$$

for all $X \in \mathcal{J}$. By definition of the quotient we obviously have

$$||[X]|| \leq \lim_\alpha ||e_\alpha X - X|| \, .$$

As $\sigma(e_\alpha) \subset [0, 1]$ we have $\sigma(I - e_\alpha) \subset [0, 1]$ and $||I - e_\alpha|| \leq 1$. This implies

$$||(X + e_\alpha X) + (Y + e_\alpha Y)|| = ||(I - e_\alpha)(X + Y)|| \leq ||X + Y|| \, .$$

In particular $\limsup_\alpha ||(X + e_\alpha X)|| \leq ||X + Y||$ for every $Y \in \mathcal{J}$. This proves our claim.

Now we have

$$\begin{aligned} ||[X]||^2 &= \lim_\alpha ||X - e_\alpha X||^2 = \lim_\alpha ||(X^* - X^* e_\alpha)(X - e_\alpha X)|| \\ &= \lim_\alpha ||(I - e_\alpha)(X^*X + Y^*)(I - e_\alpha)|| \\ &\leq ||X^*X + Y|| \end{aligned}$$

for every $Y \in \mathcal{J}$. This implies

$$||[X]||^2 \leq ||[X]^*[X]||$$

and thus the result. □

3 von Neumann algebras

3.1 Topologies on $\mathcal{B}(\mathcal{H})$

As every C^*-algebra is a sub-$*$-algebra of some $\mathcal{B}(\mathcal{H})$, closed for the operator norm topology (or *uniform topology*), then it inherits new topologies, which are weaker.

On $\mathcal{B}(\mathcal{H})$ we define the *strong topology* to be the locally convex topology defined by the semi-norms $P_x(A) = ||Ax||$, $x \in \mathcal{H}$, $A \in \mathcal{B}(\mathcal{H})$. This is to say that a base of neighborhood is formed by the sets

$$V(A; x_1, \ldots, x_n; \varepsilon) = \{B \in \mathcal{B}(\mathcal{H}); ||(B - A)x_i|| < \varepsilon, i = 1, \ldots, n\}.$$

On $\mathcal{B}(\mathcal{H})$ we define the *weak topology* to be the locally convex topology defined by the semi-norms $P_{x,y}(A) = |<x\,,\, Ay>|$, $x, y \in \mathcal{H}$, $A \in \mathcal{B}(\mathcal{H})$. This is to say that a base of neighborhood is formed by the sets

$$V(A; x_1, \ldots, x_n; y_1, \ldots, y_n; \varepsilon) = \{B \in \mathcal{B}(\mathcal{H}); |<x_i\,,\, (B - A)y_j>| < \varepsilon, \\ i, j = 1, \ldots, n\}.$$

Proposition 3.1.

i) The weak topology is weaker than the strong topology which is itself weaker than the uniform topology. Once $\mathcal{H}$ is infinite dimensional then these comparisons are strict.

ii) A linear form on $\mathcal{B}(\mathcal{H})$ is strongly continuous if and only if it is weakly continuous.

iii) The strong and the weak closure of any convex subset of $\mathcal{B}(\mathcal{H})$ coincide.

Proof. i) All the comparisons are obvious in the large sense. To make the difference in infinite dimension assume that $\mathcal{H}$ is separable with orthonormal basis $(e_n)_{n\in\mathbb{N}}$. Let P_n be the orthogonal projection onto the space generated by $e_1, \ldots, e_n$. The sequence $(P_n)_{n\in\mathbb{N}}$ converges strongly to I but not uniformly. Furthermore, consider the unilateral shift $S : e_i \mapsto e_{i+1}$. Then S^k converges weakly to 0 when k tends to $+\infty$ but not strongly.

ii) Let $\Psi : \mathcal{B}(\mathcal{H}) \to \mathbb{C}$ be a strongly continuous linear form. Then there exists $x_1, \ldots, x_n \in \mathcal{H}$ such that

$$|\Psi(B)| \leq \sum_{i=1}^{n} ||Bx_i||$$

for all $B \in \mathcal{B}(\mathcal{H})$ (classical result on locally convex topologies, not proved here). On $\mathcal{B}(\mathcal{H})^n$ let P be the semi-norm defined by

$$P(A_1, \ldots, A_n) = \sum_{i=1}^{n} ||A_i x_i||\,.$$

On the diagonal of $\mathcal{B}(\mathcal{H})^n$ we define the linear form $\widetilde{\Psi}$ by $\widetilde{\Psi}(A,\ldots,A) = \Psi(A)$. We then have $\left|\widetilde{\Psi}(A,\ldots,A)\right| \leq P(A,\ldots,A)$. By Hahn-Banach, there exists a linear form Ψ on $\mathcal{B}(\mathcal{H})^n$ which extends $\widetilde{\Psi}$ and such that

$$|\Psi(A_1,\ldots,A_n)| \leq P(A_1,\ldots,A_n).$$

Let Ψ_k be the linear form on $\mathcal{B}(\mathcal{H})$ defined by

$$\Psi_k(A) = \Psi(0,\ldots,0,A,0,\ldots,0). \qquad (A \text{ is at the } k\text{-th place})$$

Then $|\Psi_k(A)| \leq ||Ax_k||$ for every A. Every vector $y \in \mathcal{H}$ can be written as Ax_k for some $A \in \mathcal{B}(\mathcal{H})$. The linear form $Ax_k \mapsto \Psi_k(A)$ is thus well-defined and continuous on $\mathcal{H}$. By Riesz theorem there exists a $y_k \in \mathcal{H}$ such that $\Psi_k(A) = <y_k\,,\,Ax_k>$. We have proved that

$$\Psi(A) = \sum_{i=1}^{n} <y_k\,,\,Ax_k>.$$

Thus Ψ is weakly continuous.

iii) is an easy consequence of ii) and of the geometric form of Hahn-Banach theorem. □

Another topology is of importance for us, the *σ-weak topology*. It is the one determined by the semi-norms

$$p_{(x_n)_{n\in\mathbb{N}},(y_n)_{n\in\mathbb{N}}}(A) = \sum_{n=0}^{\infty} |<x_n\,,\,Ay_n>|$$

where $(x_n)_{n\in\mathbb{N}}$ and $(y_n)_{n\in\mathbb{N}}$ run over all sequences in $\mathcal{H}$ such that

$$\sum_n ||x_n||^2 < \infty \quad \text{and} \quad \sum_n ||y_n||^2 < \infty.$$

Let $\mathcal{T}(\mathcal{H})$ denote the Banach space of trace class operators on $\mathcal{H}$, equipped with the trace norm $||H||_1 = \mathrm{tr}\,|H|$, where $|H| = \sqrt{H^*H}$.

Theorem 3.2. *The Banach space $\mathcal{B}(\mathcal{H})$ is the topological dual of $\mathcal{T}(\mathcal{H})$ thanks to the duality*

$$(A,T) \mapsto tr(AT),$$

$A \in \mathcal{B}(\mathcal{H})$, $T \in \mathcal{T}(\mathcal{H})$. Furthermore the ∗-weak topology on $\mathcal{B}(\mathcal{H})$ associated to this duality is the σ-weak topology.

Proof. The inequality $|\mathrm{tr}(AT)| \leq ||A||\,||T||_1$ proves that $\mathcal{B}(\mathcal{H})$ is included in the topological dual of $\mathcal{T}(\mathcal{H})$. Conversely, let ω be an element of the dual of $\mathcal{T}(\mathcal{H})$. Consider the rank one operators $E_{\xi,\nu} = |\xi><\nu|$. One easily checks that

$||E_{\xi,\nu}||_1 = ||\xi||\,||\nu||$. Thus $|\omega(E_{\xi,\nu})| \leq ||\omega||\,||\xi||\,||\nu||$. By Riesz theorem there exists an operator $A \in \mathcal{B}(\mathcal{H})$ such that $\omega(E_{\xi,\nu}) = <\nu\,,\,A\xi>$. The linear form $\mathrm{tr}(A\,\cdot)$ then coincides with ω on rank one projectors. One concludes that they coincide on $\mathcal{T}(\mathcal{H})$ by density of finite rank operators. This proves the announced duality.

The $*$-weak topology associated to this duality is defined by the seminorms

$$P_T(A) = |\mathrm{tr}(AT)|$$

where T runs over $\mathcal{T}(\mathcal{H})$. But every trace class operator T writes

$$T = \sum_{n=0}^{\infty} \lambda_n\,|\,\xi_n ><\,\nu_n|$$

for some orthonormal systems $(\nu_n)_{n\in\mathbb{N}}$, $(\xi_n)_{n\in\mathbb{N}}$ and some absolutely summable sequence of complex numbers $(\lambda_n)_{n\in\mathbb{N}}$. Thus

$$\mathrm{tr}(AT) = \sum_{n=0}^{\infty} \lambda_n < \nu_n\,,\,A\xi_n >$$

and the seminorms P_T are equivalent to those defining the σ-weak topology. □

Corollary 3.3. *Every σ-weakly continuous linear form on $\mathcal{B}(\mathcal{H})$ is of the form*

$$A \mapsto tr(AT)$$

for some $T \in \mathcal{T}(\mathcal{H})$.

We can now give the first definition of a von Neumann algebra.

A *von Neumann algebra* is a C^*-algebra acting on $\mathcal{H}$ which contains the unit I of $\mathcal{B}(\mathcal{H})$ and which is weakly (strongly) closed.

Of course the whole of $\mathcal{B}(\mathcal{H})$ is the first example of a von Neumann algebra.

Another example, which is actually the archetype of commutative von Neumann algebra, is obtained when considering a locally compact measured space (X,μ), with a σ-finite measure μ. The $*$-algebra $L^\infty(X,\mu)$ acts on $\mathcal{H} = L^2(X,\mu)$ by multiplication. The C^*-algebra $C_0(X)$ also acts on $\mathcal{H}$. But every function $f \in L^\infty(X,\mu)$ is almost sure limit of a sequence $(f_n)_{n\in\mathbb{N}}$ in $C_0(X)$. By dominated convergence, the space $L^\infty(X,\mu)$ is included in the weak closure of $C_0(X)$. But as $L^\infty(X,\mu)$ is also equal to its weak closure, we have that $L^\infty(X,\mu)$ is the weak closure of $C_0(X)$. We have proved that $L^\infty(X,\mu)$ is a von Neumann algebra and we have obtained it as the weak closure of some C^*-algebra.

3.2 Commutant

Let $\mathcal{M}$ be a subset of $\mathcal{B}(\mathcal{H})$. We put

$$\mathcal{M}' = \{B \in \mathcal{B}(\mathcal{H}); BM = MB \text{ for all } M \in \mathcal{M}\}.$$

The space $\mathcal{M}'$ is called the *commutant* of $\mathcal{M}$. We also define

$$\mathcal{M}'' = (\mathcal{M}')', \ldots, \mathcal{M}^{(n)} = (\mathcal{M}^{(n-1)})', \ldots$$

Proposition 3.4. *For every subset $\mathcal{M}$ of $\mathcal{B}(\mathcal{H})$ we have*

i) $\mathcal{M}'$ is weakly closed;
ii) $\mathcal{M}' = \mathcal{M}''' = \mathcal{M}^{(5)} = \ldots$
and $\mathcal{M} \subset \mathcal{M}'' = \mathcal{M}^{(4)} = \ldots$

Proof. i) If $(A_n)_{n\in\mathbb{N}}$ is a sequence in $\mathcal{M}'$ which converges weakly to A in $\mathcal{B}(\mathcal{H})$ then for all $B \in \mathcal{M}$ and all $x, y \in \mathcal{H}$ we have

$$\begin{aligned} &|< x\,, (AB - BA)y > | \leq \\ &\quad |< x\,, (A - A_n)By >| + |< x\,, B(A - A_n)y >| \xrightarrow{n \to \infty} 0. \end{aligned}$$

Thus A belongs to $\mathcal{M}'$.

ii) If B belongs to $\mathcal{M}'$ and A belongs to $\mathcal{M}$ then $AB = BA$, thus A belongs to $(\mathcal{M}')' = \mathcal{M}''$. This proves the inclusion $\mathcal{M} \subset \mathcal{M}''$. But note that if $\mathcal{M}_1 \subset \mathcal{M}_2$ then clearly $\mathcal{M}_2' \subset \mathcal{M}_1'$. Applying this to the previous inclusion gives $\mathcal{M}''' \subset \mathcal{M}'$. But as $\mathcal{M}'''$ is also equal to $(\mathcal{M}')''$ we should also have the converse inclusion to hold true. This means $\mathcal{M}' = \mathcal{M}'''$. We now conclude easily. □

Proposition 3.5. *Let $\mathcal{M}$ be a self-adjoint subset of $\mathcal{B}(\mathcal{H})$. Let $\mathcal{E}$ be a closed subspace of $\mathcal{H}$ and P be the orthogonal projector onto $\mathcal{E}$. Then $\mathcal{E}$ is invariant under $\mathcal{M}$ (in the sense $M\mathcal{E} \subset \mathcal{E}$ for all $M \in \mathcal{M}$) if and only if $P \in \mathcal{M}'$.*

Proof. The space $\mathcal{E}$ is invariant under $M \in \mathcal{M}$ if and only $MP = PMP$. Thus if $\mathcal{E}$ is invariant under $\mathcal{M}$ we have $MP = PMP$ for all $M \in \mathcal{M}$. Applying the involution on this equality and using the fact that $\mathcal{M}$ is self-adjoint, gives $PM = PMP$ for all $M \in \mathcal{M}$. Finally $PM = MP$ for all $M \in \mathcal{M}$ and P belongs to $\mathcal{M}'$. The converse is obvious. □

Theorem 3.6 (von Neumann density theorem). *Let $\mathcal{M}$ be a sub-$*$-algebra of $\mathcal{B}(\mathcal{H})$ which contains the identity I. Then $\mathcal{M}$ is weakly (strongly) dense in $\mathcal{M}''$.*

Proof. f Let $B \in \mathcal{M}''$. Let $x_1, \ldots, x_n \in \mathcal{H}$. Let

$$V = \{A \in \mathcal{B}(\mathcal{H}); ||(A - B)x_i|| < \varepsilon,\ i = 1, \ldots, n\}$$

be a strong neighborhood of B. It is sufficient to show that V intersects $\mathcal{M}$. One can assume B to be self-adjoint as it can always be decomposed as a linear combination of two self-adjoint operators which also belong to $\mathcal{M}''$.

Let $\widetilde{\mathcal{H}} = \oplus_{i=1}^n \mathcal{H}$ and $\pi : \mathcal{B}(\mathcal{H}) \to \mathcal{B}(\widetilde{\mathcal{H}})$ be given by $\pi(A) = \oplus_{i=1}^n A$. Let $x = (x_1, \dots, x_n) \in \widetilde{\mathcal{H}}$. Let P be the orthogonal projection from $\widetilde{\mathcal{H}}$ onto the closure of $\pi(\mathcal{M})x = \{\pi(A)x; A \in \mathcal{M}\} \subset \widetilde{\mathcal{H}}$. By Proposition 2.5 we have that P belongs to $\pi(\mathcal{M})'$.

If one identifies $\mathcal{B}(\widetilde{\mathcal{H}})$ to $M_n(\mathcal{B}(\mathcal{H}))$ it is easy to see that $\pi(\mathcal{M})' = M_n(\mathcal{M}')$ and $\pi(\mathcal{M}'') \subset M_n(\mathcal{M}')'$ (be aware that the prime symbols above are relative to different operator spaces!).

This means that $\pi(B)$ belong to $\pi(\mathcal{M}'') \subset M_n(\mathcal{M}')' = \pi(\mathcal{M})''$. In particular B commutes with $P \in \pi(\mathcal{M})'$. This means that the space $\overline{\pi(\mathcal{M})x}$ is invariant under $\pi(B)$. In particular

$$\pi(B)\,(\pi(I)x) = \begin{pmatrix} Bx_1 \\ \vdots \\ Bx_n \end{pmatrix}$$

belongs to $\overline{\pi(\mathcal{M})x}$. This means that there exists a $A \in \mathcal{M}$ such that $||(B - A)x_i||$ is small for all $i = 1, \dots n$. Thus A belongs to $\mathcal{M} \cap V$. □

As immediate corollary we have a characterization of von Neumann algebras.

Corollary 3.7 (Bicommutant theorem). *Let $\mathcal{M}$ be a sub-$*$-algebra of $\mathcal{B}(\mathcal{H})$ which contains I. Then the following assertions are equivalent.*

i) $\mathcal{M}$ is weakly (strongly) closed.

ii) $\mathcal{M} = \mathcal{M}''$.

As I always belong to $\mathcal{M}''$, we have that a C^*-algebra $\mathcal{M} \subset \mathcal{B}(\mathcal{H})$ is a von Neumann algebra if and only if $\mathcal{M} = \mathcal{M}''$.

3.3 Predual, normal states

Let $\mathcal{M}$ be a von Neumann algebra. Put $\mathcal{M}_1 = \{M \in \mathcal{M}; ||M|| \leq 1\}$. Note that the weak topology and the σ-weak topology coincide on $\mathcal{M}_1$. Hence $\mathcal{M}_1$ is a weakly closed subset of the unit ball of $\mathcal{B}(\mathcal{H})$ which is weakly compact. Thus $\mathcal{M}_1$ is weakly compact.

We denote by $\mathcal{M}_*$ the space of weakly (σ-weakly) linear forms on $\mathcal{M}$ which are continuous on $\mathcal{M}_1$. The space $\mathcal{M}_*$ is called the *predual* of $\mathcal{M}$, for a reason that will appear clear in next proposition. If Ψ belongs to $\mathcal{M}_*$ then $\Psi(\mathcal{M}_1)$ is compact in $\mathbb{C}$, thus Ψ is norm continuous. Thus $\mathcal{M}_*$ is a subspace of $\mathcal{M}^*$ the topological dual of $\mathcal{M}$.

Proposition 3.8.

i) $\mathcal{M}_*$ is closed in $\mathcal{M}^*$, it is thus a Banach space.

ii) $\mathcal{M}$ is the dual of $\mathcal{M}_*$.

Proof. i) Let $(f_n)_{n\in\mathbb{N}}$ be a sequence in $\mathcal{M}_*$ which converges to a f in $\mathcal{M}^*$, that is

$$\sup_{||A||=1} |f_n(A) - f(A)| \longrightarrow_{n\to\infty} 0.$$

We want to show that f belongs to $\mathcal{M}_*$, that is f is weakly continuous on $\mathcal{M}_1$. Let $(A_n)_{n\in\mathbb{N}}$ be a sequence in $\mathcal{M}_1$ which converges weakly to $A \in \mathcal{M}_1$. Then

$$\begin{aligned}|f(A_n) - f(A)| &\leq |f(A_n) - f_m(A_n)| + |f_m(A) - f(A)| + |f_m(A_n) - f_m(A)|\\ &\leq 2 \sup_{||B||=1} |f_m(B) - f(B)| + |f_m(A_n) - f_m(A)|\\ &\to_{n\to\infty} 2 \sup_{||B||=1} |f_m(B) - f(B)|\\ &\to_{m\to\infty} 0.\end{aligned}$$

This proves i).

ii) For a $A \in \mathcal{M}$ we put

$$||A||_{du} = \sup_{\omega\in\mathcal{M}_*;||\omega||=1} |\omega(A)|$$

the norm of A for the duality announced in the statement of ii). Clearly we have $||A||_{du} \leq ||A||$.

For $x, y \in \mathcal{H}$ we denote by $\omega_{x,y}$ the linear form $A \mapsto <y\,,\, Ax>$ on $\mathcal{B}(\mathcal{H})$ and $\omega_{x,y|\mathcal{M}}$ the restriction of $\omega_{x,y}$ to $\mathcal{M}$. We have

$$||A|| = \sup_{||x||=||y||=1} |<y\,,\, Ax>| \leq \sup_{\omega=\omega_{x,y};||\omega||=1} |\omega(A)| \leq ||A||_{du}\,.$$

Thus $\mathcal{M}$ is indeed identified linearly and isometrically to a subspace of $(\mathcal{M}_*)^*$. We just have to prove that this identification is onto.

Let ϕ be a continuous linear form on $\mathcal{M}_*$. Let $\phi'(x,y) = \phi(\omega_{x,y|\mathcal{M}})$. Then ϕ' is a continuous sesquilinear form on $\mathcal{H}$, it is thus of the form $\phi'(x,y) = <y\,,\, Ax>$ for some $A \in \mathcal{B}(\mathcal{H})$.

If T' is a self-adjoint element of $\mathcal{M}'$ then $\omega_{T'x,y|\mathcal{M}} = \omega_{x,T'y|\mathcal{M}}$ and

$$<AT'x\,,\, y> = <T'Ax\,,\, y>$$

for all $x, y \in \mathcal{H}$. Thus A belong to $\mathcal{M}'' = \mathcal{M}$.

As $\omega_{x,y}(A) = <y\,,\, Ax> = \phi'(x,y) = \phi(\omega_{x,y|\mathcal{M}})$ then the image of A in $(\mathcal{M}_*)^*$ coincides with ϕ at least on the $\omega_{x,y}$. Now, it remains to show that this is sufficient for A and ϕ to coincide everywhere. That is, we have to prove that an element a of $(\mathcal{M}_*)^*$ which vanishes on all the $\omega_{x,y}$ is null. But all the elements of $\mathcal{M}_*$ are linear forms ω of the form $\omega(A) = \mathrm{tr}(\rho A)$ for some trace class operator ρ. As every trace class operator ρ writes as

$$\rho = \sum_n \lambda_n\, |x_n\rangle\langle y_n|$$

for some orthonormal basis $(x_n)_{n\in I\!N}$ and $(y_n)_{n\in I\!N}$ and some summable sequence $(\lambda_n)_{n\in I\!N}$, we have that

$$\omega = \sum_n \lambda_n \, \omega_{x_n, y_n}$$

where the series above is convergent in $\mathcal{M}_*$. One concludes easily. □

The two main examples of von Neumann algebra have well-known preduals. Indeed, if $\mathcal{M} = \mathcal{B}(\mathcal{H})$ then $\mathcal{M}_* = \mathcal{T}(\mathcal{H})$ the space of trace class operators.

If $\mathcal{M} = L^\infty(X, \mu)$ then $\mathcal{M}_* = L^1(X, \mu)$.

Theorem 3.9 (Sakai theorem). *A C^*-algebra is a von Neumann algebra if and only if it is the dual of some Banach space.*

Admitted. □

A state on a von Neumann algebra $\mathcal{M}$ is called *normal* if it is σ-weakly continuous. The following characterization is now straightforward.

Theorem 3.10. *On a von Neumann algebra $\mathcal{M}$, for a state ω on $\mathcal{M}$, the following assertions are equivalent.*

i) The state ω is normal

ii) There exists a positive, trace class operator ρ on $\mathcal{H}$ such that $\mathrm{tr}\rho = 1$ *and*

$$\omega(A) = \mathrm{tr}(\rho A)$$

for all $A \in \mathcal{M}$.

4 Modular theory

4.1 The modular operators

The starting point here is a pair $(\mathcal{M}, \omega)$, where $\mathcal{M}$ is a von Neumann algebra acting on some Hilbert space, ω is a normal faithful state on $\mathcal{M}$. Recall that ω is then of the form

$$\omega(A) = \mathrm{tr}(\rho A)$$

for a positive nonsingular ρ, with $\mathrm{tr}\rho = 1$.

Let us consider the G.N.S. representation of $(\mathcal{M}, \omega)$. That is, a triple $(\mathcal{H}, \Pi, \Omega)$ such that

i) Π is a morphism from $\mathcal{M}$ to $\mathcal{B}(\mathcal{H})$.

ii) $\omega(A) = <\Omega\,,\, \Pi(A)\Omega>$

iii) $\Pi(\mathcal{M})\Omega$ is dense in $\mathcal{H}$.

From now on, we omit to mention the representation Π and identify $\mathcal{M}$ and $\mathcal{M}'$ with $\Pi(\mathcal{M})$ and $\Pi(\mathcal{M}')$. We thus write $\omega(A) = <\Omega\,,\, A\Omega>$.

Proposition 4.1. *The vector Ω is cyclic and separating for $\mathcal{M}$ and $\mathcal{M}'$.*

Proof. Ω is cyclic for $\mathcal{M}$ by iii) above. Let us see that it is separating for $\mathcal{M}$. If $A \in \mathcal{M}$ is such that $A\Omega = 0$ then $\omega(A^*A) = 0$, but as ω is faithful this implies $A = 0$.

Let us now see that these properties of Ω on $\mathcal{M}$ imply the same ones on $\mathcal{M}'$. If A' belongs to $\mathcal{M}'$ and $A'\Omega = 0$ then $A'B\Omega = BA'\Omega = 0$ for all $B \in \mathcal{M}$. Thus A' vanishes on a dense subspace of $\mathcal{H}$, it is thus the null operator. This proves that Ω is separating for $\mathcal{M}'$.

Finally, let P' be the orthogonal projector onto the space $\mathcal{M}'\Omega$. As it is the projection onto a $\mathcal{M}'$-invariant space, it belongs to $(\mathcal{M}')' = \mathcal{M}$. But $P\Omega = \Omega$ and thus $(I - P)\Omega = 0$. As Ω is separating for $\mathcal{M}$ this implies $I - P = 0$ and Ω is cyclic for $\mathcal{M}'$. □

As a consequence the (**anti-linear**) operators

$$\begin{aligned} S_0 \,:\, \mathcal{M}\Omega &\longrightarrow \mathcal{M}\Omega \\ A\Omega &\longmapsto A^*\Omega \end{aligned}$$

$$\begin{aligned} F_0 \,:\, \mathcal{M}'\Omega &\longrightarrow \mathcal{M}'\Omega \\ B\Omega &\longmapsto B^*\Omega \end{aligned}$$

are well-defined (by the separability of Ω) on dense domains.

Proposition 4.2. *The operators S_0 and F_0 are closable and $\overline{F}_0 = S_0^*$, $\overline{S}_0 = F_0^*$.*

Proof. For all $A \in \mathcal{M}$, $B \in \mathcal{M}'$ we have

$$<B\Omega\,,\,S_0A\Omega> = <B\Omega\,,\,A^*\Omega> = <A\Omega\,,\,B^*\Omega> = <A\Omega\,,\,F_0B\Omega>.$$

This proves that $F_0 \subset S_0^*$ and $S_0 \subset F_0^*$. The operators S_0 and F_0 are thus closable.

Let us show that $\overline{F}_0 = S_0^*$. Actually it is sufficient to show that $S_0^* \subset \overline{F}_0$. Let $x \in \mathrm{Dom}\, S_0^*$ and $y = S_0^*x$. For any $A \in \mathcal{M}$ we have

$$<A\Omega\,,\,y> = <A\Omega\,,\,S_0^*x> = <x\,,\,S_0A\Omega> = <x\,,\,A^*\Omega>.$$

If we define the operators Q_0 and Q_0^+ by

$$\begin{aligned} Q_0 &\,:\, A\Omega \longmapsto Ax \\ Q_0^+ &\,:\, A\Omega \longmapsto Ay \end{aligned}$$

we then have

$$\begin{aligned} <B\Omega\,,\,Q_0A\Omega> &= <B\Omega\,,\,Ax> = <A^*B\Omega\,,\,x> \\ &= <y\,,\,B^*A\Omega> = <By\,,\,A\Omega> \\ &= <Q_0^+B\Omega\,,\,A\Omega>. \end{aligned}$$

This proves that $Q_0^+ \subset Q_0^*$ and Q_0 is closable. Let $Q = \overline{Q}_0$. Note that we have

$$Q_0 AB\Omega = ABx = AQ_0 B\Omega.$$

This proves that $Q_0 A = AQ_0$ on $\operatorname{Dom} Q_0$ and thus $AQ \subset QA$ for all $A \in \mathcal{M}$. This means that Q is *affiliated* to $\mathcal{M}'$, that is, it fails from belonging to $\mathcal{M}'$ only by the fact it is an unbounded operator; but every bounded function of Q is thus in $\mathcal{M}'$. In particular, if $Q = U\,|Q|$ is the polar decomposition of Q then U belongs to $\mathcal{M}'$ and the spectral projections of $|Q|$ also belong to $\mathcal{M}'$.

Let $E_n = \mathbb{1}_{[0,n]}(|Q|)$. The operator $Q_n = UE_n\,|Q|$ thus belongs to $\mathcal{M}'$ and

$$\begin{aligned} Q_n\Omega &= UE_n\,|Q|\,\Omega = UE_nU^*U\,|Q|\,\Omega \\ &= UE_nU^*Q_0\Omega = UE_nU^*x. \end{aligned}$$

Furthermore we have

$$Q_n^*\Omega = E_n\,|Q|\,U^*\Omega = E_nQ_0^+\Omega = E_ny.$$

This way UE_nU^*x belongs to $\operatorname{Dom} F_0$ and $F_0(UE_nU^*x) = E_ny$. But E_n tends to I and UU^* is the orthogonal projector onto $\operatorname{Ran} Q$, which contains x.

Finally, we have proved that $x \in \operatorname{Dom} \overline{F}_0$ and $\overline{F}_0 x = y = S_0^* x$. That is, $S_0^* \subset \overline{F}_0$.

The other case is treated similarly. □

We now put $S = \overline{S}_0$ and $F = \overline{F}_0$.

Lemma 4.3. *We have*

$$S = S^{-1}.$$

Proof. Let $z \in \operatorname{Dom} S^*$. We have

$$< S_0 A\Omega\,,\, S^*z > = < A^*\Omega\,,\, S_0^*z > = < z\,,\, S_0A^*\Omega > = < z\,,\, A\Omega >.$$

Thus S^*z belongs to $\operatorname{Dom} S_0^* = S^*$ and $(S^*)^2 z = z$.

Let $y \in \operatorname{Dom} S$ and $z \in \operatorname{Dom} S^*$, we have $S^*z \in \operatorname{Dom} S^*$ and

$$< S^*z\,,\, Sy > = < y\,,\, (S^*)^2 z > = < y\,,\, z >.$$

This means that Sy belongs to $\operatorname{Dom} S^{**} = \operatorname{Dom} S$ and $S^2 y = S^{**}Sy = y$.

We have proved that $\operatorname{Dom} S^2 = \operatorname{Dom} S$ and $S^2 = I$ on $\operatorname{Dom} S$. □

We had proved in Proposition 3.2 that $F = S^*$. Thus the operators FS and SF are (self-adjoint) positive. The operators F and S have their range equal to their domain, they are invertible and equal to their inverse.

Let $\Delta = FS = S^*S$. Then Δ is invertible, with inverse $\Delta^{-1} = SF = SS^*$.

As S, Δ and thus $\Delta^{1/2}$ have a dense range then the partial anti-isometry J such that

$$S = J(S^*S)^{1/2}$$

(polar decomposition of S) is an anti-isometry from $\mathcal{H}$ to $\mathcal{H}$.

Furthermore

$$S = J\Delta^{1/2} = (SS^*)^{1/2}J = \Delta^{-1/2}J.$$

Let x belong to Dom S. Then

$$x = S^2x = J\Delta^{1/2}\Delta^{-1/2}Jx = J^2x$$

and thus $J^2 = I$. Note the following relations

$$S = J\Delta^{1/2}$$
$$F = S^* = \Delta^{1/2}J$$
$$\Delta^{-1} = J\Delta J.$$

The operator Δ has a spectral measure (E_λ). Thus the operator $\Delta^{-1} = J\Delta J$ has the spectral measure $(JE_\lambda J)$. Let f be a bounded Borel function, we have

$$\begin{aligned}
< f(\Delta^{-1})x\,,\, x> &= \int \overline{f}(\lambda)\, d< JE_\lambda Jx\,,\, x> \\
&= \int \overline{f}(\lambda)\, d< Jx\,,\, E_\lambda Jx> \\
&= \int \overline{f}(\lambda)\, d< E_\lambda Jx\,,\, Jx> \\
&= < f(\Delta)Jx\,,\, Jx> \\
&= < Jx\,,\, \overline{f}(\Delta)Jx> \\
&= < J\overline{f}(\Delta)Jx\,,\, x>.
\end{aligned}$$

This proves

$$f(\Delta^{-1}) = J\overline{f}(\Delta)J.$$

In particular

$$\Delta^{it} = J\Delta^{it}J$$
$$\Delta^{it}J = J\Delta^{it}.$$

Finally note that $S\Omega = F\Omega = \Omega$ and thus $\Delta\Omega = FS\Omega = \Omega$ which finally gives

$$\Delta^{1/2}\Omega = \Omega.$$

Let us now summarize the situation we have already described.

Theorem 4.4. *There exists an anti-unitary operator J from $\mathcal{H}$ to $\mathcal{H}$ and an (unbounded) invertible, positive operator Δ such that*

$$\begin{aligned}\Delta = FS,\ \Delta^{-1} = SF,\ J^2 = I \\ S = J\Delta^{1/2} = \Delta^{-1/2}J \\ F = J\Delta^{-1/2} = \Delta^{1/2}J \\ J\Delta^{it} = \Delta^{it}J \\ J\Omega = \Delta\Omega = \Omega.\end{aligned}$$

The operator Δ is called the *modular operator* and J is the *modular conjugation*.

It is interesting to note the following. If the state ω were *tracial*, that is, $\omega(AB) = \omega(BA)$ for all A, B, we would have

$$||S_0 A\Omega||^2 = ||A^*\Omega||^2 = < A^*\Omega\,,\, A^*\Omega > = \omega(AA^*) = \omega(A^*A) = ||A\Omega||^2.$$

Thus S_0 would be an isometry and

$$\begin{aligned}S = J = F \\ \Delta = I.\end{aligned}$$

4.2 The modular group

Let $A, B, C \in \mathcal{M}$. We have

$$SASBC\Omega = SAC^*B^*\Omega = BCA^*\Omega = BSAC^*\Omega = BSASC\Omega.$$

This proves that B and SAS commute. Thus SAS is affiliated to $\mathcal{M}'$.

Let us assume for a moment that Δ **is bounded**. In that case the operators $\Delta^{-1} = J\Delta J$, S and F are also bounded.

We have seen that

$$\begin{aligned}S\mathcal{M}S \subset \mathcal{M}' \\ F\mathcal{M}'F \subset \mathcal{M}.\end{aligned}$$

This way we have

$$\begin{aligned}\Delta\mathcal{M}\Delta^{-1} &= \Delta^{1/2}JJ\Delta^{1/2}\mathcal{M}\Delta^{-1/2}JJ\Delta^{-1/2} \\ &= FS\mathcal{M}SF \subset F\mathcal{M}'F \subset \mathcal{M}.\end{aligned}$$

We also have

$$\Delta^n\mathcal{M}\Delta^{-n} \subset \mathcal{M}$$

for all $n \in \mathbb{N}$.

For any $A \in \mathcal{M}$, $A' \in \mathcal{M}'$, the function

$$f(z) = ||\Delta||^{-2z} < \phi\,,\, [\Delta^z A \Delta^{-z}, A']\psi >$$

is analytic on $\mathbb{C}$. It vanishes for $z = 0, 1, 2, ...$

As $||\Delta^{-1}|| = ||J\Delta J|| = ||\Delta||$ we have

$$|f(z)| = O\left(||\Delta||^{-2\Re z} (||\Delta||^{|\Re z|})^2\right) = O(1)$$

when $\Re z > 0$.

By Carlson's theorem we have $f(z) = 0$ for all $z \in \mathbb{C}$. Thus

$$\Delta^z \mathcal{M} \Delta^{-z} \subset \mathcal{M}'' = \mathcal{M}$$

for all $z \in \mathbb{C}$. But

$$\mathcal{M} = \Delta^z(\Delta^{-z}\mathcal{M}\Delta^z)\Delta^{-z} \subset \Delta^z \mathcal{M} \Delta^{-z}$$

and finally

$$\Delta^z \mathcal{M} \Delta^{-z} = \mathcal{M}.$$

Furthermore

$$J\mathcal{M}J = J\Delta^{1/2}\mathcal{M}\Delta^{-1/2}J = S\mathcal{M}S \subset \mathcal{M}'$$
$$J\mathcal{M}'J = J\Delta^{-1/2}\mathcal{M}\Delta^{1/2}J = F\mathcal{M}F \subset \mathcal{M}.$$

We have proved

$$J\mathcal{M}J = \mathcal{M}'.$$

The results we have obtained here are fundamental and extend to the case when Δ is unbounded. This is what the following theorem says. We do not prove it as it implies pages of difficult analytic considerations. We hope that the above computations make it credible.

Theorem 4.5 (Tomita-Takesaki's theorem). *In any case we have*

$$J\mathcal{M}J = \mathcal{M}'$$
$$\Delta^{it}\mathcal{M}\Delta^{-it} = \mathcal{M}.$$

Put

$$\sigma_t(A) = \Delta^{it} A \Delta^{-it}.$$

This defines a one parameter group of automorphisms of $\mathcal{M}$.

Proposition 4.6. *We have, for all $A, B \in \mathcal{M}$*

$$\omega(A\sigma_t(B)) = \omega(\sigma_{t+i}(B)A). \tag{1}$$

Proof.

$$
\begin{aligned}
<\Omega\,,\, A\Delta^{it}B\Delta^{-it}\Omega> &= <\Delta^{-it}A^*\Omega\,,\, B\Omega> \\
&= <\Delta^{-it-1/2}A^*\Omega\,,\, \Delta^{1/2}B\Omega> \\
&= <\Delta^{-it-1}\Delta^{1/2}A^*\Omega\,,\, \Delta^{1/2}B\Omega> \\
&= <J\Delta^{-it+1}J\Delta^{1/2}A^*\Omega\,,\, \Delta^{1/2}B\Omega> \\
&= <J\Delta^{1/2}B\Omega\,,\, \Delta^{-it+1}J\Delta^{1/2}A^*\Omega> \\
&= <B^*\Omega\,,\, \Delta^{-it+1}A\Omega> \\
&= <\Omega\,,\, B\Delta^{-i(t+i)}A\Omega> \\
&= <\Omega\,,\, \Delta^{i(t+i)}B\Delta^{-i(t+i)}A\Omega> \\
&= \omega(\sigma_{t+i}(B)A).
\end{aligned}
$$

□

It is interesting to relate the above equality with the following result.

Proposition 4.7. *Let ω be a state of the form*

$$\omega(A) = tr(\rho A)$$

on $\mathcal{B}(\mathcal{K})$ for some trace-class positive ρ with $tr\rho = 1$. Let (σ_t) be the following group of automorphisms of $\mathcal{B}(\mathcal{K})$:

$$\sigma_t(A) = e^{itH}Ae^{-itH}$$

for some self-adjoint operator H on $\mathcal{K}$. Then the following assertions are equivalent.

i) For all $A, B \in \mathcal{B}(\mathcal{K})$, all $t \in \mathbb{R}$ and a fixed $\beta \in \mathbb{R}$ we have

$$\omega(A\sigma_t(B)) = \omega(\sigma_{t-\beta i}(B)A).$$

ii) ρ is given by

$$\rho = \frac{1}{Z}e^{-\beta H},$$

where $Z = tr(\exp(-\beta H))$.

Proof. ii) implies i): We compute directly

$$
\begin{aligned}
\omega(A\sigma_t(B)) &= \frac{1}{Z}\mathrm{tr}(e^{-\beta H}Ae^{itH}Be^{-itH}) \\
&= \frac{1}{Z}\mathrm{tr}(Ae^{itH}Be^{(-it-\beta)H}) \\
&= \frac{1}{Z}\mathrm{tr}(Ae^{-\beta H}e^{(it+\beta)H}Be^{(-it-\beta)H}) \\
&= \frac{1}{Z}\mathrm{tr}(e^{-\beta H}e^{(it+\beta)H}Be^{(-it-\beta)H}A) \\
&= \omega(\sigma_{t-\beta i}(B)A).
\end{aligned}
$$

i) implies ii): We have

$$\begin{aligned}\mathrm{tr}(AB\rho) &= \mathrm{tr}(\rho AB) = \omega(AB)\\ &= \omega(\sigma_{-\beta i}(B)A) = \mathrm{tr}(\rho e^{\beta H} B e^{-\beta H} A) = \mathrm{tr}(A\rho e^{\beta H} B e^{-\beta H}).\end{aligned}$$

As this is valid for any A we conclude that

$$B\rho = \rho e^{\beta H} B e^{-\beta H}$$

for all B. This means

$$B\left(\rho e^{\beta H}\right) = \left(\rho e^{\beta H}\right) B.$$

As this is valid for all B we conclude that $\rho \exp(\beta H)$ is a multiple of the identity. This gives ii). □

Another very interesting result to add to Proposition 3.6 is that the modular group is the only one to perform the relation (1).

Theorem 4.8. *$\sigma_.$ is the only automorphism group to satisfy (1) on $\mathcal{M}$ for the given state ω.*

Proof. Let $\tau_.$ be another automorphism group on $\mathcal{M}$ which satisfies (1). Define the operators U_t by

$$U_t A\Omega = \tau_t(A)\Omega.$$

Then U_t is unitary for

$$\begin{aligned}\|U_t A\Omega\|^2 &= <\tau_t(A)\Omega\,,\,\tau_t(A)\Omega> = <\Omega\,,\,\tau_t(A^*A)\Omega>\\ &= \omega(\tau_t(A^*A)) = \omega(\tau_{t+i}(I)A^*A)\\ &= \omega(A^*A) = \|A\Omega\|^2.\end{aligned}$$

The family $U_.$ is clearly a group, it is thus of the form $U_t = \exp itM$ for a self adjoint operator M.

Note that $U_t\Omega = \Omega$ and thus $M\Omega = 0$.

Let A, B be entire elements for $\tau_.$, then the relation $\omega(\tau_i(B)A) = \omega(AB)$ implies

$$\begin{aligned}<B^*\Omega\,,\,\Delta A\Omega> &= <\Delta^{1/2}B^*\Omega\,,\,J J\Delta^{1/2}A\Omega>\\ &= <A^*\Omega\,,\,B\Omega>\\ &= \omega(AB)\\ &= \omega(\tau_i(B)A)\\ &= <\Omega\,,\,e^{-M}Be^{M}A\Omega>\\ &= <B^*\Omega\,,\,e^{M}A\Omega>.\end{aligned}$$

This means

$$\Delta = e^M$$

and $\tau = \sigma$. □

4.3 Self-dual cone and standard form

We put

$$\mathcal{P} = \overline{\{AJAJ\Omega; A \in \mathcal{M}\}}.$$

Proposition 4.9.

i) $\mathcal{P} = \overline{\Delta^{1/4}\mathcal{M}_+\Omega} = \overline{\Delta^{-1/4}\mathcal{M}'_+\Omega}$ *and thus* $\mathcal{P}$ *is a convex cone.*

ii) $\Delta^{it}\mathcal{P} = \mathcal{P}$ *for all t.*

iii) If f *is of positive type then* $f(\log \Delta)\mathcal{P} \subset \mathcal{P}$.

iv) If $\xi \in \mathcal{P}$ *then* $J\xi = \xi$.

v) If $A \in \mathcal{M}$ *then* $AJAJ\mathcal{P} \subset \mathcal{P}$.

Proof. i) Let $\mathcal{M}_0$ be the $*$-algebra of elements of $\mathcal{M}$ which are *entire* for the modular group σ. (that is, $t \mapsto \sigma_t(A)$ admits an analytic extension). We shall admit here that $\mathcal{M}_0$ is σ-weakly dense in $\mathcal{M}$.

For every $A \in \mathcal{M}_0$ we have

$$\begin{aligned}\Delta^{1/4}AA^*\Omega &= \sigma_{-i/4}(A)\sigma_{i/4}(A)^*\Omega \\ &= \sigma_{-i/4}(A)J\Delta^{1/2}\sigma_{i/4}(A)\Omega \\ &= \sigma_{-i/4}(A)J\sigma_{-i/4}(A)J\Omega \\ &= BJBJ\Omega\end{aligned}$$

where $B = \sigma_{-i/4}(A)$. By $\sigma_{-i/4}(\mathcal{M}_0) = \mathcal{M}_0$ and by the density of $\mathcal{M}_0$ in $\mathcal{M}$ we have

$$BJBJ\Omega \in \overline{\Delta^{1/4}\mathcal{M}_+\Omega} \subset \overline{\Delta^{1/4}\overline{\mathcal{M}_+\Omega}}$$

for all $B \in \mathcal{M}$. Thus

$$\mathcal{P} \subset \overline{\Delta^{1/4}\mathcal{M}_+\Omega} \subset \overline{\Delta^{1/4}\overline{\mathcal{M}_+\Omega}}.$$

Conversely, $\mathcal{M}_0^+\Omega$ is dense in $\overline{\mathcal{M}_+\Omega}$. Let $\psi \in \overline{\mathcal{M}_+\Omega}$. There exists a sequence $(A_n) \subset \mathcal{M}_0^+$ such that $A_n\Omega \to \psi$. We know by the above that $\Delta^{1/4}A_n\Omega$ belongs to $\mathcal{P}$. But

$$J\Delta^{1/2}A_n\Omega = A_n\Omega \to \psi = J\Delta^{1/2}\psi$$

and thus

$$\left|\left|\Delta^{1/4}(\psi - A_n\Omega)\right|\right|^2 = <\psi - A_n\Omega\,,\, \Delta^{1/2}(\psi - A_n\Omega)> \to 0.$$

Thus $\Delta^{1/4}\psi$ belongs to $\mathcal{P}$ and $\overline{\Delta^{1/4}\overline{\mathcal{M}_+\Omega}} \subset \mathcal{P}$.

This proves the first equality of i). The second one is treated exactly in the same way.

ii) We have

$$\Delta^{it}\Delta^{1/4}\mathcal{M}_+\Omega = \Delta^{1/4}\Delta^{it}\mathcal{M}_+\Omega = \Delta^{1/4}\sigma_t(\mathcal{M}_+)\Omega = \Delta^{1/4}\mathcal{M}_+\Omega.$$

iii) If f is of positive type then f is the Fourier transform of some positive, finite, Borel measure μ on $\mathbb{R}$. In particular

$$f(\log \Delta) = \int \Delta^{it} \, d\mu(t).$$

One conclues with ii) now.

iv) $JAJAJ\Omega = JAJA\Omega = AJAJ\Omega$.

v) $AJAJBJBJ\Omega = ABJAJJBJ\Omega = ABJABJ\Omega$. □

Theorem 4.10.

i) P is self-dual, *that is $\mathcal{P} = \mathcal{P}^\vee$ where*

$$\mathcal{P}^\vee = \{x \in \mathcal{H}; < y \, , \, x > \geq 0, \forall y \in \mathcal{P}\}.$$

ii) $\mathcal{P}$ is pointed, *that is,*

$$\mathcal{P} \cap (-\mathcal{P}) = \{0\}.$$

iii) If $J\xi = \xi$ then ξ admits a unique decomposition as $\xi = \xi_1 - \xi_2$ with $\xi_1, \xi_2 \in \mathcal{P}$ and ξ_1 orthogonal to ξ_2.

iv) The linear span of $\mathcal{P}$ is the whole of $\mathcal{H}$.

Proof. i) If $A \in \mathcal{M}_+$ and $A' \in \mathcal{M}'_+$ then

$$< \Delta^{1/4} A\Omega \, , \, \Delta^{-1/4} A'\Omega > = < A\Omega \, , \, A'\Omega > = < \Omega \, , \, A^{1/2} A' A^{1/2} \Omega > \geq 0.$$

Thus $\mathcal{P}$ is included in $\mathcal{P}^\vee$.

Conversely, if $\xi \in \mathcal{P}^\vee$, that is $< \xi \, , \, \nu > \geq 0$ for all $\nu \in \mathcal{P}$, we put

$$\xi_n = f_n(\log \Delta)\xi$$

where $f_n(x) = \exp(-x^2/2n^2)$. Then ξ_n belongs to $\cap_{\alpha \in \mathbb{C}} \mathrm{Dom}\, \Delta^\alpha$ and ξ_n converges to ξ. We know that $f_n(\log \Delta)\nu$ belongs to $\mathcal{P}$ and thus

$$< \xi_n \, , \, \nu > = < \xi \, , \, f_n(\log \Delta)\nu > \geq 0.$$

Let $A \in \mathcal{M}_+$ then $\Delta^{1/4} A\Omega$ belongs to $\mathcal{P}$ and

$$< \Delta^{1/4} \xi_n \, , \, A\Omega > = < \xi_n \, , \, \Delta^{1/4} A\Omega > \geq 0.$$

Thus $\Delta^{1/4}\xi_n$ belongs to $\overline{\mathcal{M}_+\Omega}^\vee$ which coincides with $\overline{\mathcal{M}'_+\Omega}$ (admitted). This finally gives that ξ_n belongs to $\Delta^{-1/4}\overline{\mathcal{M}'_+\Omega} \subset \mathcal{P}$. This proves i).

ii) If $\xi \in \mathcal{P} \cap (-\mathcal{P}) = \mathcal{P} \cap (-\mathcal{P}^\vee)$ then $< \xi \, , \, -\xi > \geq 0$ and $\xi = 0$.

iii) If $J\xi = \xi$ then, as $\mathcal{P}$ is convex and closed, there exists a unique $\xi_1 \in \mathcal{P}$ such that

$$||\xi - \xi_1|| = \inf\{||\xi - \nu|| \, ; \nu \in \mathcal{P}\}.$$

We put $\xi_2 = \xi_1 - \xi$. Let $\nu \in \mathcal{P}$ and $\lambda > 0$. Then $\xi_1 + \lambda\nu$ belongs to $\mathcal{P}$ and

$$||\xi - \xi_1||^2 \leq ||\xi_1 + \lambda\nu - \xi||^2.$$

That is $||\xi_2||^2 \leq ||\xi_2 + \lambda\nu||^2$, or else $\lambda^2||\nu||^2 + 2\lambda\Re< \xi_2 \, , \, \nu > \geq 0$. This implies that $\Re< \xi_2 \, , \, \nu >$ is positive. But as $J\xi_2 = \xi_2$ and $J\nu = \nu$ then

$$< \xi_2 \, , \, \nu > = < J\xi_2 \, , \, J\nu > = \overline{< \xi_2 \, , \, \nu >}.$$

That is $< \xi_2 \, , \, \nu > \geq 0$ and $\xi_2 \in \mathcal{P}^\vee = \mathcal{P}$.

iv) If ξ is orthogonal to the linear span of $\mathcal{P}$ then ξ belongs to $\mathcal{P}^\vee = \mathcal{P}$. thus $< \xi \, , \, \xi > = 0$ and $\xi = 0$. □

Theorem 4.11 (Universality).

1) If $\xi \in \mathcal{P}$ then ξ is cyclic for $\mathcal{M}$ if and only if it is separating for $\mathcal{M}$.

2) If $\xi \in \mathcal{P}$ is cyclic for $\mathcal{M}$ then $J_\xi, \mathcal{P}_\xi$ associated to $(\mathcal{M}, \xi)$ satisfy

$$J_\xi = J \quad \text{and} \quad \mathcal{P}_\xi = \mathcal{P}.$$

Proof. 1) If ξ is cyclic for $\mathcal{M}$ then $J\xi$ is cyclic for $\mathcal{M}' = J\mathcal{M}J$ and thus $\xi = J\xi$ is separating for $\mathcal{M}$. And conversely.

2) Define as before (the closed version of)

$$\begin{aligned} S_\xi \, &: A\xi \longmapsto A^*\xi \\ F_\xi \, &: A'\xi \longmapsto A'^*\xi. \end{aligned}$$

We have

$$\begin{aligned} JF_\xi JA\xi &= JF_\xi JAJ\xi \\ &= J(JAJ)^*\xi \\ &= A^*\xi \\ &= S_\xi A\xi. \end{aligned}$$

This proves that $S_\xi \subset JF_\xi J$. By a symmetric argument $F_\xi \subset JS_\xi J$ and thus $JS_\xi = F_\xi J$.

Note that

$$(JS_\xi)^* = S_\xi^* J = F_\xi J = JS_\xi.$$

This means that JS_ξ is self-adjoint. Let us prove that it is positive. We have

$$< A\xi \, , \, JS_\xi A\xi > = < A\xi \, , \, JA^*\xi > = < \xi \, , \, A^*JA^*\xi >$$

which is a positive quantity for ξ and A^*JA^*J belong to $\mathcal{P}$. This proves the positivity of JS_ξ.

We have

$$S_\xi = J_\xi \Delta_\xi^{1/2} = J(JS_\xi).$$

By uniqueness of the polar decomposition we must have $J = J_\xi$.

Finally, we have that $\mathcal{P}_\xi$ is generated by the $AJ_\xi AJ_\xi \xi = AJAJ\xi$. But as ξ belongs to $\mathcal{P}$ we have that $AJAJ\xi$ belongs to $\mathcal{P}$ and thus $\mathcal{P}_\xi \subset \mathcal{P}$. Finally, $\mathcal{P} = \mathcal{P}^\vee \subset \mathcal{P}_\xi^\vee = \mathcal{P}_\xi$ and $\mathcal{P} = \mathcal{P}_\xi$. □

The following theorem is very useful and powerful, but its proof is very long, tedious and cannot be summarized, thus we prefer not enter into it and give the result as it is (cf [B-R], p. 108-117).

For every $\xi \in \mathcal{P}$ one can define a particular normal positive form

$$\omega_\xi(A) = <\xi\,,\,A\xi>$$

on $\mathcal{M}$. That is, $\omega_\xi \in \mathcal{M}_{*+}$.

Theorem 4.12. *1) For every $\omega \in \mathcal{M}_{*+}$ there exists a unique $\xi \in \mathcal{P}$ such that*

$$\omega = \omega_\xi .$$

2) The mapping $\xi \longmapsto \omega_\xi$ is an homeomorphism and

$$||\xi - \nu||^2 \le ||\omega_\xi - \omega_\nu||^2 \le ||\xi - \nu||\; ||\xi + \nu|| .$$

We denote by $\omega \longmapsto \xi(\omega)$ the inverse mapping of $\xi \longmapsto \omega_\xi$.

Corollary 4.13. *There exists a unique unitary representation*

$$\alpha \in Aut(\mathcal{M}) \longmapsto U_\alpha$$

of the group of $*$-automorphisms of $\mathcal{M}$ on $\mathcal{H}$, such that
i) $U_\alpha A U_\alpha^* = \alpha(A)$, for all $A \in \mathcal{M}$,
ii) $U_\alpha \mathcal{P} \subset \mathcal{P}$ and, moreover,

$$U_\alpha \xi(\omega) = \xi(\alpha^{-1*}(\omega))$$

for all $\omega \in \mathcal{M}_{*+}$ and where $(\alpha^*\omega)(A) = \omega(\alpha(A))$.
iii) $[U_\alpha, J] = 0$.

Proof. Let $\alpha \in \mathrm{Aut}(\mathcal{M})$. Let $\xi \in \mathcal{P}$ be the representant of the state

$$A \longmapsto <\Omega\,,\,\alpha^{-1}(A)\Omega>.$$

That is,

$$<\xi\,,\,A\xi> = <\Omega\,,\,\alpha^{-1}(A)\Omega>.$$

In particular ξ is separating for $\mathcal{M}$ and hence cyclic. Define the operator

$$UA\Omega = \alpha(A)\xi.$$

We have

$$||UA\Omega||^2 = <\xi\,,\,\alpha(A^*A)\xi> = <\Omega\,,\,A^*A\Omega> = ||A\Omega||^2.$$

Thus U is unitary. In particular

$$U^* A\xi = \alpha^{-1}(A)\Omega.$$

Now, for $A, B \in \mathcal{M}$ we have

$$UAU^*B\xi = UA\alpha^{-1}(B)\Omega = \alpha(A\alpha^{-1}(B))\xi = \alpha(A)B\xi$$

and

$$\alpha(A) = UAU^*.$$

We have proved the existence of the unitary representation. Note that

$$\begin{aligned} SU^*A\xi &= S\alpha^{-1}(A)\Omega \\ &= \alpha^{-1}(A)^*\Omega \\ &= \alpha^{-1}(A^*)\Omega \\ &= U^*A^*\xi \\ &= U^*S_\xi A\xi. \end{aligned}$$

Hence by closure

$$J\Delta^{1/2}U^* = U^*J_\xi\Delta_\xi^{1/2} = U^*J\Delta_\xi^{1/2}.$$

That is

$$UJU^*U\Delta^{1/2}U^* = J\Delta_\xi^{1/2}.$$

By uniqueness of the polar decomposition we must have $UJU^* = J$. This gives iii). For $A \in \mathcal{M}$ we have

$$UAJAJ\Omega = \alpha(A)J\alpha(A)J\xi.$$

Since ξ belongs to $\mathcal{P}$ we deduce

$$U\mathcal{P} = \mathcal{P}.$$

If $\phi \in \mathcal{M}_{*+}$ we have

$$\begin{aligned} <U\xi(\phi)\,,\, AU\xi(\phi)> &= <\xi(\phi)\,,\, U^*AU\xi(\phi)> \\ &= <\xi(\phi)\,,\, \alpha^{-1}(A)\xi(\phi)> \\ &= \phi(\alpha^{-1}(A)) \\ &= (\alpha^{-1^*}(\phi))(A) \\ &= <\xi(\alpha^{-1^*}(\phi))\,,\, A\xi(\alpha^{-1^*}(\phi))>. \end{aligned}$$

By uniqueness of the representing vector in $\mathcal{P}$

$$U(\alpha)\xi(\phi) = \xi(\alpha^{-1^*}(\phi)).$$

This gives ii) and also the uniqueness of the unitary representation. □

References

1. O. Bratteli, D.W. Robinson, *Operator Algebras and Quantum Statistical Mechanics 1. C*- and W*-Algebras. Symmetry Groups. Decomposition of States*, Texts and Monographs in Physics, 2nd ed. 1987, Springer Verlag.
2. O. Bratteli, D.W. Robinson, *Operator Algebras and Quantum Statistical Mechanics 2. Equilibrium States. Models in Quantum Statistical Mechanics*, Texts and Monographs in Physics, 2nd ed. 1987, Springer Verlag.
3. J. Dixmier, *Les C^*-algèbres et leurs représentations*, Gauthier-Villars, Paris, 1964.
4. R.V. Kadison and J.R. Ringrose, *Fundamentals of the theory of operator algebras (I)*, Acad. Press, 1983.
5. R.V. Kadison and J.R. Ringrose, *Fundamentals of the theory of operator algebras (II)*, Acad. Press, 1986.
6. R.V. Kadison and J.R. Ringrose, *Fundamentals of the theory of operator algebras (III)*, Acad. Press, 1991.
7. G.K. Pedersen, *C^*-algebras and their automorphism groups*, London Mathematical Society Monographs, Academic Press, 1989.
8. S. Sakai, *C^*-algebras and W^*-algebras*, Ergebnisse der Mathematik und ihrer Grenzgebiete 60, Springer Verlag.

Quantum Dynamical Systems

Claude-Alain Pillet

CPT-CNRS (UMR 6207), Université du Sud, Toulon-Var,
BP 20132, 83957 La Garde Cedex, France
e-mail: pillet@univ-tln.fr

1 Introduction

Many problems of classical and quantum physics can be formulated in the mathematical framework of dynamical systems. Within this framework ergodic theory provides a probabilistic interpretation of dynamics which is suitable to study the statistical properties of the evolution of a mechanical system over large time scales. The conceptual foundation of ergodic theory is intimately related to the birth of statistical mechanics and goes back to Boltzmann and Gibbs. It started to develop as a mathematical theory with the pioneering works of von Neumann [48] and Birkhoff [10]. It is now a beautiful cross-disciplinary part of Mathematics with numerous connections to analysis and probability, geometry, algebra, number theory, combinatorics...

The Koopman–von Neumann approach to ergodic theory [34] provides an effective way to translate ergodic properties of dynamical systems into spectral properties of some associated linear operator (which I shall call Liouvillean). The resulting spectral approach to dynamics is particularly well adapted to the study of open systems. During the last decade, this spectral approach has been successfully applied to the problem of return to equilibrium [27, 8, 16, 22, 21] and to the construction of non-equilibrium steady states for quantum open systems [29]. The reader should consult [30] for an introduction to these problems.

The aim of this lecture is to provide a short introduction to quantum dynamical systems and their ergodic properties with particular emphasis on the quantum Koopman–von Neumann spectral theory. However, I shall not discuss the spectral analysis of the resulting Liouvillean operators. The interested reader should consult the above mentioned references. For other approaches based on scattering ideas see [40, 6, 26].

Ergodic theory also played an important role in the development of the algebraic approach to quantum field theory and quantum statistical mechanics, mainly in connection with the analysis of symmetries. Most of the results obtained in this framework rely on some kind of asymptotic abelianness hypothesis which is often inappropriate in the context of dynamical systems. The reader should consult Chapter 4 and in particular Sections 4.3 and the corresponding notes in [11] for an introduction to the results.

I have assumed that the reader is familiar with the material covered in the first Lectures of this Volume [33, 32, 7]. Besides that, the notes are reasonably self-contained and most of the proofs are given. Numerous examples should provide the reader with a minimal toolbox to construct basic models of quantum open systems.

These notes are organized as follows.

Section 2 is an extension of Subsection 3.3 in Lecture [7]. It consists of two parts. In Subsections 2.1, I review some topological properties of von Neumann algebras. I also introduce the notions of support and central support of a normal state. Subsection 2.2 explores some elementary consequences of the GNS construction.

In Section 3, I briefly review some basic facts of the ergodic theory of classical dynamical systems. The discussion is centered around two simple properties: ergodicity and mixing. As a motivation for the transposition of these concepts to quantum mechanics I discuss the classical Koopman–von Neumann approach in Subsection 3.2. General references for Section 3 are [15], [35], [49] and [5].

The main part of these notes is contained in Section 4 which deals with the ergodic theory of quantum systems. The basic concepts of the algebraic theory of quantum dynamics – C^*- and W^*-dynamical systems and their invariant states – are introduced in Subsections 4.1–4.3.

In Subsection 4.4, I define a more general notion of quantum dynamical system. The GNS construction provides an efficient way to bring such systems into normal form. This normal form plays an essential role in quantum ergodic theory. In particular it allows to define a Liouvillean which will be the central object of the quantum Koopman–von Neumann theory. In Subsection 4.5, I introduce the related notion of standard form of a quantum dynamical system.

The ergodic properties – ergodicity and mixing – of quantum dynamical systems are defined and studied in Subsection 4.6. The quantum Koopman–von Neumann spectral theory is developed in Subsection 4.7.

In many physical applications, and in particular in simple models of open systems, the dynamics is constructed by coupling elementary subsystems. Perturbation theory provides a powerful tool to analyze such models. In Subsection 4.8, I discuss a simple adaptation of the Dyson-Schwinger-Tomonaga time dependent perturbation theory – which played an important role in the early development of quantum electrodynamics – to C^*- and W^*-dynamical systems.

General references for the material in Section 4 are Chapters 2.5, 3.1 and 3.2 in [11] as well as [47], [46] and [44]. More examples of dynamical systems can be found in [9].

Among the invariant states of a C^*- or W^*-dynamical system, the KMS states introduced in Section 5 form a distinguished class, from the physical as well as from the mathematical point of view. On the physical side, KMS states play the role of thermodynamical equilibrium states. As such, they are basic building blocks for an important class of models of open quantum systems where the reservoirs are at thermal equilibrium. On the mathematical side, KMS states appear naturally in the modular structure associated with faithful normal states (or more generally semi-finite weight) on von Neumann algebras. They are thus intimately connected with their mathematical structure. This tight relation between dynamics and the structure of the observable algebra is one of the magical feature of quantum mechanics which has no classical counterpart.

In Subsection 5.2, I discuss the perturbation theory of KMS states. For bounded perturbations, the theory is due to Araki [2] and [4] (see also Section 5.4.1 in [12]). Extensions to unbounded perturbations have been developed in [43],[19] and [17]. This subject being very technical, I only give some plausibility argument and state the main results without proofs.

Acknowledgments. I wish to thank Jan Dereziński and Vojkan Jakšić for fruitful discussions related to the material covered by these notes. I am particularly grateful to Stephan De Bièvre for his constructive comments on an early version of the manuscript.

2 The State Space of a C^*-algebras

This section is a complement to Lecture [7] and contains a few additions that are needed to develop the ergodic theory of quantum dynamical systems. It consists of two parts.

In the first part I present the basic properties of normal states on a von Neumann algebra $\mathfrak{M}$. In particular I discuss the connection between the σ-weak topology on $\mathfrak{M}$ and its algebraic structure. I also introduce the very useful concepts of support and central support of a normal state.

The second part of the section deals with the GNS construction and its consequences on the structure of the state space of a C^*-algebra: enveloping von Neumann algebra and folium of a state, relative normality and orthogonality of states. I also discuss the special features of the GNS representation associated with a normal state.

The material covered by this section is standard and the reader already familiar with the above concepts may skip it.

Warning: All C^*-algebras in this lecture have a unit I.

2.1 States

A linear functional ω on a C^*-algebra $\mathfrak{A}$ is positive if $\omega(A^*A) \geq 0$ for all $A \in \mathfrak{A}$. Such a functional is automatically continuous and $\|\omega\| = \omega(I)$ (see Proposition 5 in Lecture [7]). If μ, ν are two positive linear functionals such that $\nu - \mu$ is a positive linear functional then we write $\mu \leq \nu$.

A state is a normalized ($\|\omega\| = 1$) positive linear functional. A state ω is *faithful* if $\omega(A^*A) = 0$ implies $A = 0$.

Denote by $\mathfrak{A}_1^\star$ the unit ball of the Banach space dual $\mathfrak{A}^\star$. By the Banach-Alaoglu theorem, $\mathfrak{A}_1^\star$ is compact in the weak-$\star$ topology. The set of all states on $\mathfrak{A}$ is given by

$$E(\mathfrak{A}) = \{\omega \in \mathfrak{A}_1^\star \,|\, \omega(A^*A) \geqslant 0 \text{ for all } A \in \mathfrak{A}\},$$

and it immediately follows that it is a weak-$\star$ compact, convex subset of $\mathfrak{A}^\star$.

Normal States

Recall (Subsection 3.1 in Lecture [7]) that the σ-weak topology on a von Neumann algebra $\mathfrak{M} \subset \mathfrak{B}(\mathcal{H})$ is the locally convex topology generated by the semi-norms

$$A \mapsto \sum_{n\in\mathbb{N}} |(\psi_n, A\phi_n)|,$$

for sequences $\psi_n, \phi_n \in \mathcal{H}$ such that $\sum_n \|\psi_n\|^2 < \infty$ and $\sum_n \|\phi_n\|^2 < \infty$. The σ-strong and σ-strong* topologies are defined similarly by the semi-norms

$$A \mapsto \left(\sum_{n \in \mathbb{N}} \|A\psi_n\|^2 \right)^{1/2}, \quad A \mapsto \left(\sum_{n \in \mathbb{N}} \|A\psi_n\|^2 + \sum_{n \in \mathbb{N}} \|A^*\psi_n\|^2 \right)^{1/2},$$

with $\sum_n \|\psi_n\|^2 < \infty$. Note that, except when $\mathcal{H}$ is finite dimensional, these topologies are not first countable. Therefore, the use of nets (directed sets) is mandatory.

As a Banach space, the von Neumann algebra $\mathfrak{M}$ is the dual of the space $\mathfrak{M}_\star$ of all σ-weakly continuous linear functionals on $\mathfrak{M}$. In particular, the predual $\mathfrak{M}_\star$ is a norm-closed subspace of the dual $\mathfrak{M}^* = (\mathfrak{M}_\star)^{**}$.

Exercise 2.1. Show that the σ-strong* topology is stronger than the σ-strong topology which is stronger than the σ-weak topology. Show also that the σ-strong (resp. σ-weak) topology is stronger than the strong (resp. weak) topology and that these two topologies coincide on norm bounded subsets.

Exercise 2.2. Adapt the proof of Proposition 8.ii in Lecture [7] to show that a linear functional ω on $\mathfrak{M}$ is σ-weakly continuous if and only if it is σ-strongly continuous. Using Corollary 2 in Lecture [7] and the Hahn-Banach theorem show that ω is σ-weakly continuous if and only if there exists a trace class operator T on $\mathcal{H}$ such that $\omega(A) = \mathrm{Tr}(TA)$ for all $A \in \mathfrak{M}$.

The von Neumann density theorem (Theorem 13 in Lecture [7]) asserts that a $*$-subalgebra $\mathcal{D} \subset \mathfrak{B}(\mathcal{H})$ containing I is dense in $\mathcal{D}''$ in the weak and strong topologies. In fact one can prove more (see for example Corollary 2.4.15 in [11]).

Theorem 2.3. *(Von Neumann density theorem) A $*$-subalgebra $\mathcal{D} \subset \mathfrak{B}(\mathcal{H})$ containing I is σ-strongly* dense in $\mathcal{D}''$.*

Thus, any element A of the von Neumann algebra generated by $\mathfrak{D}$ is the σ-strong* limit of a net A_α in $\mathfrak{D}$. By Exercise 2.1 the net A_α also approximates A in the σ-strong, σ-weak, strong and weak topologies. In particular, $\mathfrak{D}''$ coincides with the closure of $\mathfrak{D}$ in all these topologies. It is often useful to approximate A by a *bounded* net in $\mathfrak{D}$. That this is also possible is a simple consequence of the following theorem (see Theorem 2.4.16 in [11]).

Theorem 2.4. *(Kaplansky density theorem) Let $\mathcal{D} \subset \mathfrak{B}(\mathcal{H})$ be a $*$-subalgebra and denote by $\overline{\mathfrak{D}}$ its weak closure. Then $\mathfrak{D}_r \equiv \{A \in \mathfrak{D} \mid \|A\| \le r\}$ is σ-strongly* dense in $\overline{\mathfrak{D}}_r \equiv \{A \in \overline{\mathfrak{D}} \mid \|A\| \le r\}$ for any $r > 0$.*

Recall also that a self-adjoint element A of a C^*-algebra $\mathfrak{A}$ is positive if its spectrum is a subset of $[0, \infty[$ (see Theorem 5 in Lecture [7]). This definition induces a partial order on the set of self-adjoint elements of $\mathfrak{A}$: $A \le B$ if and only if $B - A$ is positive. Moreover, one writes $A < B$ if $A \le B$ and $A \ne B$. The relation $\le$ is clearly a purely algebraic concept, *i.e.*, it is independent on the action of $\mathfrak{A}$ on some Hilbert space. However, if $\mathfrak{A}$ acts on a Hilbert space $\mathcal{H}$ then its positive elements are characterized by the fact that $(\psi, A\psi) \ge 0$ for all ψ in a dense subspace of $\mathcal{H}$. In particular, $A \le B$ if and only if $(\psi, A\psi) \le (\psi, B\psi)$ for all ψ in such a subspace.

Let A_α be a bounded increasing net of self-adjoint elements of $\mathfrak{B}(\mathcal{H})$, *i.e.*, such that $A_\alpha \geq A_\beta$ for $\alpha \succ \beta$ and $\sup_\alpha \|A_\alpha\| < \infty$. Then, for any $\psi \in \mathcal{H}$, one has $\sup_\alpha(\psi, A_\alpha\psi) < \infty$ and since a self-adjoint element A of $\mathfrak{B}(\mathcal{H})$ is completely determined by its quadratic form $\psi \mapsto (\psi, A\psi)$, there exists a unique $A \in \mathfrak{B}(\mathcal{H})$ such that

$$(\psi, A\psi) \equiv \lim_\alpha(\psi, A_\alpha\psi) = \sup_\alpha(\psi, A_\alpha\psi).$$

It follows immediately from this definition that $A = \sup_\alpha A_\alpha$. Since one has $0 \leq A - A_\alpha \leq A$, the estimate

$$\begin{aligned}\|(A - A_\alpha)\psi\|^2 &= (\psi, (A - A_\alpha)^2\psi)\\ &\leq \|A - A_\alpha\| \, \|(A - A_\alpha)^{1/2}\psi\|^2\\ &\leq \|A\| \, (\psi, (A - A_\alpha)\psi),\end{aligned}$$

further shows that A_α converges strongly to A. Moreover, since the net A_α is bounded, one also has $\lim_\alpha A_\alpha = A$ in the σ-strong and σ-weak topologies (Exercise 2.1). In particular, if $\mathfrak{M} \subset \mathfrak{B}(\mathcal{H})$ is a von Neumann algebra and $A_\alpha \in \mathfrak{M}$, then $A \in \mathfrak{M}$. Finally, we note that if $B \in \mathfrak{M}$ then

$$\sup_\alpha(B^* A_\alpha B) = B^*(\sup_\alpha A_\alpha)B. \tag{1}$$

Definition 2.5. *A positive linear functional ω on a von Neumann algebra $\mathfrak{M}$ is called normal if, for all bounded increasing net A_α of self-adjoint elements of $\mathfrak{M}$, one has*

$$\omega(\sup_\alpha A_\alpha) = \sup_\alpha \omega(A_\alpha).$$

In particular, a normal state is a normalized, normal, positive linear functional.

Remark 2.6. This definition differs from the one given in Section 3.3 of Lecture [7]. However, Theorem 2.7 below shows that these two definitions are equivalent.

Note that the concept of normality only depends on the partial order relation $\leq$ and hence on the algebraic structure of $\mathfrak{M}$. Since by Exercise 2.2 any σ-weakly continuous linear functional on $\mathfrak{M}$ is of the form $A \mapsto \mathrm{Tr}(TA)$ for some trace class operator T, it is a finite linear combination of positive, σ-weakly continuous linear functionals (because T is a linear combination of 4 positive trace class operators). Thus, the following theorem characterizes the σ-weak topology on $\mathfrak{M}$ in a purely algebraic way.

Theorem 2.7. *A positive linear functional on a von Neumann algebra is normal if and only if it is σ-weakly continuous.*

Proof. If ω is a σ-weakly continuous positive linear functional on the von Neumann algebra $\mathfrak{M}$ and A_α a bounded increasing net of self-adjoint elements of $\mathfrak{M}$ one has, in the σ-weak topology,

$$\omega(\sup_\alpha A_\alpha) = \omega(\lim_\alpha A_\alpha) = \lim_\alpha \omega(A_\alpha) = \sup_\alpha \omega(A_\alpha).$$

Hence, ω is normal.

To prove the reverse statement let ω be a normal positive linear functional and consider the set

$$\mathcal{A} \equiv \{A \in \mathfrak{M} \,|\, 0 \leq A \leq I, \omega_A \in \mathfrak{M}_\star\},$$

where $\omega_A(X) \equiv \omega(XA)$. If $0 \leq B \leq A \leq I$ then the Cauchy-Schwarz inequality,

$$\begin{aligned}|\omega_A(X) - \omega_B(X)|^2 &= |\omega(X(A-B))|^2 \\ &\leq \omega(X(A-B)X^*)\omega(A-B) \\ &\leq \|X\|^2\omega(A-B),\end{aligned}$$

yields that

$$\|\omega_A - \omega_B\|^2 \leq \omega(A) - \omega(B). \tag{2}$$

Let A_α be an increasing net in $\mathcal{A}$ and set $A \equiv \sup_\alpha A_\alpha$. One clearly has $0 \leq A_\alpha \leq A \leq I$ and since ω is normal Equ. (2) shows that ω_{A_α} converges in norm to ω_A. $\mathfrak{M}_\star$ being a norm-closed subspace of $\mathfrak{M}^\star$ one has $\omega_A \in \mathfrak{M}_\star$ and we conclude that $A \in \mathcal{A}$. Thus, $\mathcal{A}$ is inductively ordered and by Zorn's lemma there exists a maximal element $N \in \mathcal{A}$. We set $M \equiv I - N$ and note that if $\omega(M) = 0$ then Equ. (2) shows that $\omega = \omega_N \in \mathfrak{M}_\star$.

To conclude the proof we assume that $\omega(M) > 0$ and show that this leads to a contradiction. Since $M > 0$ we can pick $\psi \in \mathcal{H}$ such that $\omega(M) < (\psi, M\psi)$. Consider an increasing net B_α in the set

$$\mathcal{B} \equiv \{B \in \mathfrak{M} \,|\, 0 \leq B \leq M, \omega(B) \geq (\psi, B\psi)\},$$

and let $B \equiv \sup_\alpha B_\alpha$. Then $0 \leq B_\alpha \leq B \leq M$ and since ω is normal

$$\omega(B) = \sup_\alpha \omega(B_\alpha) \geq \sup_\alpha (\psi, B_\alpha\psi) = (\psi, B\psi),$$

shows that $B \in \mathcal{B}$. Hence $\mathcal{B}$ is inductively ordered. Let S be a maximal element of $\mathcal{B}$. Remark that $M \notin \mathcal{B}$ since $\omega(M) < (\psi, M\psi)$. This means that

$$T \equiv M - S > 0. \tag{3}$$

Next we note that if $0 \leq B \leq T$ and $\omega(B) \geq (\psi, B\psi)$ then $B + S \in \mathcal{B}$ and the maximality of S yields that $B = 0$. It follows that $0 \leq B \leq T$ implies $\omega(B) \leq (\psi, B\psi)$. Since for any $B \in \mathfrak{M}$ such that $\|B\| \leq 1$ one has

$$TB^*BT \leq T^2 \leq T,$$

we can conclude that

$$\omega(TB^*BT) \leq (\psi, TB^*BT\psi) = \|BT\psi\|^2.$$

By Cauchy-Schwarz inequality we further get

$$|\omega_T(B)|^2 = |\omega(IBT)|^2 \leq \omega(I)\omega(TB^*BT) \leq \|BT\psi\|^2.$$

This inequality extends by homogeneity to all $B \in \mathfrak{M}$ and shows that ω_T is σ-strongly continuous and hence, by Exercise 2.2, σ-weakly continuous. Finally, we note that $\omega_{N+T} = \omega_N + \omega_T \in \mathfrak{M}_\star$ and by Equ. (3),

$$N < N + T = N + (M - S) = I - S \leq I,$$

a contradiction to the maximality of N. $\square$

Thus, the set of normal states on a von Neumann algebra $\mathfrak{M}$ coincides with the set of σ-weakly continuous states and with the set of σ-strongly continuous states. It is given by

$$N(\mathfrak{M}) = \mathfrak{M}_\star \cap E(\mathfrak{M}) \subset E(\mathfrak{M}),$$

and is clearly a norm closed subset of $E(\mathfrak{M})$. If $\mathfrak{M}$ acts on the Hilbert space $\mathcal{H}$ then, according to Exercise 2.2, a normal state ω on $\mathfrak{M}$ is described by a *density matrix, i.e.,* a non-negative trace class operator ρ on $\mathcal{H}$ such that $\mathrm{Tr}\,\rho = 1$ and $\omega(A) = \mathrm{Tr}(\rho A)$.

Lemma 2.8. *Let $\mathfrak{M} \subset \mathfrak{B}(\mathcal{H})$ be a von Neumann algebra. The set of vector states $V(\mathfrak{M}) \equiv \{(\Psi, (\,\cdot\,)\Psi) \mid \Psi \in \mathcal{H}, \|\Psi\| = 1\}$ is total in $N(\mathfrak{M})$ i.e., finite convex linear combinations of elements of $V(\mathfrak{M})$ are norm dense in $N(\mathfrak{M})$.*

Proof. We first note that if $\mu, \nu \in N(\mathfrak{M})$ are given by density matrices ρ, σ then

$$|\mu(A) - \nu(A)| = |\,\mathrm{Tr}((\rho - \sigma)A)| \leq \|\rho - \sigma\|_1 \, \|A\|,$$

where $\|T\|_1 \equiv \mathrm{Tr}(T^*T)^{1/2}$ denotes the trace norm. Hence $\|\mu - \nu\| \leq \|\rho - \sigma\|_1$. Let $\mu \in N(\mathfrak{M})$ and ρ a corresponding density matrix. Denote by

$$\rho = \sum_n p_n \psi_n(\psi_n, \cdot),$$

its spectral decomposition, *i.e.*, $(\psi_n, \psi_k) = \delta_{n,k}$, $0 < p_n \leq 1$, $\sum_n p_n = 1$. From the trace norm estimate

$$\left\| \rho - \sum_{n=1}^{N-1} p_n \psi_n(\psi_n, \cdot) \right\|_1 = \left\| \sum_{n \geq N} p_n \psi_n(\psi_n, \cdot) \right\|_1 = \sum_{n \geq N} p_n \equiv q_N,$$

it follows that

$$\left\| \mu - \left(\sum_{n=1}^{N-1} p_n \mu_n + q_N \mu_N \right) \right\| \leq 2q_N,$$

where $\mu_n = (\psi_n, (\,\cdot\,)\psi_n) \in V(\mathfrak{M})$. Since $\lim_N q_N = 0$ we conclude that finite convex linear combinations of vector states are norm dense in $N(\mathfrak{M})$. $\square$

Exercise 2.9. (Complement to Lemma 2.8) Let $\mathcal{D} \subset \mathcal{H}$ be a dense subspace. Show that the set of vector states $V_{\mathcal{D}}(\mathfrak{M}) \equiv \{(\Psi, (\cdot)\Psi) \mid \Psi \in \mathcal{D}, \|\Psi\| = 1\}$ is total in $N(\mathfrak{M})$.

Exercise 2.10. Show that a net A_α in a von Neumann algebra $\mathfrak{M}$ converges σ-weakly (resp. σ-strongly) to 0 if and only if, for all $\omega \in N(\mathfrak{M})$ one has $\lim_\alpha \omega(A_\alpha) = 0$ (resp. $\lim_\alpha \omega(A_\alpha^* A_\alpha) = 0$).

Lemma 2.11. *Let $\mathfrak{M}$, $\mathfrak{N}$ be von Neumann algebras. A $*$-morphism $\phi : \mathfrak{M} \to \mathfrak{N}$ is σ-weakly continuous if and only if it is σ-strongly continuous.*

Proof. Suppose that ϕ is σ-weakly continuous and that the net A_α converges σ-strongly to 0. By Exercise 2.10, $A_\alpha^* A_\alpha$ converges σ-weakly to 0. It follows that $\phi(A_\alpha)^*\phi(A_\alpha) = \phi(A_\alpha^* A_\alpha)$ converges σ-weakly to zero and hence, by Exercise 2.10 again, that $\phi(A_\alpha)$ converges σ-strongly to 0.

Suppose now that ϕ is σ-strongly continuous. Since any $\omega \in N(\mathfrak{N})$ is σ-strongly continuous, so is the state $\omega \circ \phi$. This means that $\omega \circ \phi$ is σ-weakly continuous for all $\omega \in N(\mathfrak{N})$ and hence that ϕ itself is σ-weakly continuous. □

Corollary 2.12. *A $*$-isomorphism between two von Neumann algebras is σ-weakly and σ-strongly continuous.*

Proof. Let $\mathfrak{M}$, $\mathfrak{N}$ be von Neumann algebras and $\phi : \mathfrak{M} \to \mathfrak{N}$ a $*$-isomorphism. If A_α is a bounded increasing net of self-adjoint elements in $\mathfrak{M}$ then so is $\phi(A_\alpha)$ in $\mathfrak{N}$. Set $A \equiv \sup_\alpha A_\alpha$. Since ϕ preserves positivity one has $\phi(A_\alpha) \leq \phi(A)$ and hence $\sup_\alpha \phi(A_\alpha) \leq \phi(A)$. Moreover, since ϕ is surjective there exists $B \in \mathfrak{M}$ such that $\sup_\alpha \phi(A_\alpha) = \phi(B)$ and

$$\phi(A_\alpha) \leq \phi(B) \leq \phi(A),$$

holds for all α. These inequalities and the injectivity of ϕ further yield

$$A_\alpha \leq B \leq A,$$

for all α. Thus, we conclude that $B = A$, that is,

$$\sup_\alpha \phi(A_\alpha) = \phi(\sup_\alpha A_\alpha). \tag{4}$$

By Theorem 2.7, any $\omega \in N(\mathfrak{N})$ is normal and Equ. (4) yields

$$\sup_\alpha \omega(\phi(A_\alpha)) = \omega(\sup_\alpha \phi(A_\alpha)) = \omega(\phi(\sup_\alpha A_\alpha)),$$

which shows that $\omega \circ \phi$ is normal and hence σ-weakly continuous. It follows that ϕ itself is σ-weakly continuous and, by Lemma 2.11, σ-strongly continuous. □

Functional Calculus

Let $\mathfrak{A}$ be a C^*-algebra and $A \in \mathfrak{A}$ a self-adjoint element. By Theorem 4 in Lecture [7] there is a unique $*$-morphism $\pi_A : C(\sigma(A)) \to \mathfrak{A}$ such that $\pi_A(f) = A$ if $f(x) = x$. Accordingly, if f is continuous we write $f(A) \equiv \pi_A(f)$.

When dealing with a von Neumann algebra $\mathfrak{M} \subset \mathfrak{B}(\mathcal{H})$, it is necessary to extend this morphism to a larger class of functions. This can be done with the help of Theorem 7 and Remark 10 in Lecture [32]. Let $A \in \mathfrak{B}(\mathcal{H})$ be self-adjoint and denote by $\mathcal{B}(\mathbb{R})$ the $*$-algebra of bounded Borel functions on $\mathbb{R}$. Then there exists a unique $*$-morphism $\Pi_A : \mathcal{B}(\mathbb{R}) \to \mathfrak{B}(\mathcal{H})$ such that

(i) $\Pi_A(f) = f(A)$ if $f \in C(\sigma(A))$.
(ii) If $f, f_n \in \mathcal{B}(\mathbb{R})$ are such that $\lim_n f_n(x) = f(x)$ for all $x \in \mathbb{R}$ and $\sup_{n,x\in\mathbb{R}} |f_n(x)| < \infty$ then $\Pi_A(f_n) \to \Pi_A(f)$ strongly.

Again we write $f(A) \equiv \Pi_A(f)$ for $f \in \mathcal{B}(\mathbb{R})$. Thus, if $A \in \mathfrak{M}$ is self-adjoint then $f(A) \in \mathfrak{M}$ for any continuous function f. More generally, assume that $f \in \mathcal{B}(\mathcal{R})$ can be approximated by a sequence f_n of continuous functions such that (ii) holds. Since $\mathfrak{M}$ is strongly closed it follows that $f(A) \in \mathfrak{M}$.

In particular, if χ_I denotes the characteristic function of an interval $I \subset \mathbb{R}$ then $\chi_I(A)$ is the spectral projection of A for the interval I and one has $\chi_I(A) \in \mathfrak{M}$.

The Support of a Normal State

Exercise 2.13. Let P_α be an increasing net of orthogonal projections of the Hilbert space $\mathcal{H}$. Denote by P the orthogonal projection on the smallest closed subspace of $\mathcal{H}$ containing all the subspaces Ran P_α. Show that

$$\mathrm{s} - \lim_\alpha P_\alpha = \sup_\alpha P_\alpha = P.$$

Exercise 2.14. Let P and Q be two orthogonal projections on the Hilbert space $\mathcal{H}$. Denote by $P \wedge Q$ the orthogonal projection on $\mathrm{Ran}\, P \cap \mathrm{Ran}\, Q$ and by $P \vee Q$ the orthogonal projection on $\mathrm{Ran}\, P + \mathrm{Ran}\, Q$.

i. Show that $I - P \vee Q = (I - P) \wedge (I - Q)$.
ii. Show that $P \vee Q \le P + Q \le I + P \wedge Q$.
iii. Set $T \equiv PQP$ and show that $\mathrm{Ran}\, P \cap \mathrm{Ran}\, Q = \mathrm{Ker}(I - T)$. Mimic the proof of Theorem 3.13 to show that

$$P \wedge Q = \mathrm{s} - \lim_{n\to\infty} T^n.$$

iv. Show that if $\mathfrak{M}$ is a von Neumann algebra on $\mathcal{H}$ and if $P, Q \in \mathfrak{M}$ then $P \wedge Q \in \mathfrak{M}$ and $P \vee Q \in \mathfrak{M}$.
v. Show that if $\omega \in E(\mathfrak{M})$ then $\omega(P \vee Q) = 0$ if and only if $\omega(P) = \omega(Q) = 0$ and $\omega(P \wedge Q) = 1$ if and only if $\omega(P) = \omega(Q) = 1$.

Let ω be a normal state on the von Neumann algebra $\mathfrak{M}$. We denote by $\mathfrak{M}^P$ the set of orthogonal projections in $\mathfrak{M}$. Exercise 2.13 shows that the non-empty set $\mathfrak{P}_\omega \equiv \{P \in \mathfrak{M}^P \mid \omega(P) = 0\}$ is inductively ordered: any increasing net P_α in $\mathfrak{P}_\omega$ has a least upper bound

$$\sup_\alpha P_\alpha = \mathrm{s} - \lim_\alpha P_\alpha \in \mathfrak{P}_\omega.$$

By Zorn's lemma, $\mathfrak{P}_\omega$ has a maximal element $\bar{P}_\omega$. For any $P \in \mathfrak{P}_\omega$ one has $P \vee \bar{P}_\omega \geq \bar{P}_\omega$ and, by exercise 2.14, $P \vee \bar{P}_\omega \in \mathfrak{P}_\omega$. The maximality of $\bar{P}_\omega$ yields $P \vee \bar{P}_\omega = \bar{P}_\omega$ from which we can conclude that $P \leq \bar{P}_\omega$. Thus, one has $\bar{P}_\omega = \sup \mathfrak{P}_\omega$. The complementary projection

$$s_\omega \equiv I - \bar{P}_\omega = \inf\{P \in \mathfrak{M}^P \mid \omega(P) = 1\},$$

is called the *support* of ω. For any normal state ω and any $A \in \mathfrak{M}$ one has, by the Cauchy-Schwarz inequality

$$|\omega(A(I - s_\omega))| \leq \omega(AA^*)^{1/2}\omega(I - s_\omega)^{1/2} = 0,$$

from which we conclude that $\omega(A) = \omega(As_\omega) = \omega(s_\omega A)$.

Exercise 2.15. Show that $\omega(A^*A) = 0$ if and only if $As_\omega = 0$. Conclude that the state ω is *faithful* if and only if $s_\omega = I$.

Hint: if $As_\omega \neq 0$ there exists $\epsilon > 0$ and a non-zero $P_\epsilon \in \mathfrak{M}^P$ such that $sA^*As \geq \epsilon P_\epsilon$ and $P_\epsilon = P_\epsilon s_\omega$.

Exercise 2.16. Let $\mathfrak{M} \subset \mathfrak{B}(\mathcal{H})$ be a von Neumann algebra. If $\mathcal{K} \subset \mathcal{H}$ is a vector subspace, denote by $[\mathcal{K}]$ the orthogonal projection on its closure $\bar{\mathcal{K}}$. Use Proposition 10 in Lecture [7] to show that, for any subset $\mathcal{M} \subset \mathcal{H}$, $[\mathfrak{M}'\mathcal{M}] \in \mathfrak{M}$.

Lemma 2.17. *Let $\mathfrak{M} \subset \mathfrak{B}(\mathcal{H})$ be a von Neumann algebra. The support of the state $\omega \in N(\mathfrak{M})$ is given by*

$$s_\omega = [\mathfrak{M}' \operatorname{Ran} \rho],$$

where ρ is any density matrix on $\mathcal{H}$ such that $\omega(A) = \mathrm{Tr}(\rho A)$ for all $A \in \mathfrak{M}$. In particular, the support of the vector state $\omega_\Phi(A) = (\Phi, A\Phi)$ is given by

$$s_{\omega_\Phi} = [\mathfrak{M}'\Phi].$$

ω_Φ is faithful if and only if Φ is cyclic for $\mathfrak{M}'$.

Proof. Set $P = [\mathfrak{M}' \operatorname{Ran} \rho]$ and note that $P \in \mathfrak{M}$ by Exercise 2.16. On the one hand $I \in \mathfrak{M}'$ implies $\operatorname{Ran} \rho \subset \mathfrak{M}' \operatorname{Ran} \rho$ and hence $P \operatorname{Ran} \rho = \operatorname{Ran} \rho$. Thus, $P\rho = \rho$ and $\omega(P) = \mathrm{Tr}(P\rho) = \mathrm{Tr}(\rho) = 1$, from which we conclude that $P \geq s_\omega$. On the other hand,

$$0 = \omega(I - s_\omega) = \mathrm{Tr}(\rho(I - s_\omega)) = \mathrm{Tr}(\rho^{1/2}(I - s_\omega)\rho^{1/2}) = \|(I - s_\omega)\rho^{1/2}\|_2^2,$$

yields $(I - s_\omega)\rho = 0$. It follows that $s_\omega \operatorname{Ran} \rho = \operatorname{Ran} \rho$ and

$$s_\omega \mathfrak{M}' \operatorname{Ran} \rho = \mathfrak{M}' s_\omega \operatorname{Ran} \rho = \mathfrak{M}' \operatorname{Ran} \rho,$$

implies $s_\omega P = P$, that is $P \leq s_\omega$. □

The Central Support of a Normal State

The *center* of a von Neumann algebra $\mathfrak{M}$ is the Abelian von Neumann subalgebra $\mathfrak{Z}(\mathfrak{M}) \equiv \mathfrak{M} \cap \mathfrak{M}'$. One easily sees that $\mathfrak{Z}(\mathfrak{M}) = (\mathfrak{M} \cup \mathfrak{M}')'$ so that $\mathfrak{Z}(\mathfrak{M})' = \mathfrak{M} \vee \mathfrak{M}'$ is the smallest von Neumann algebra containing $\mathfrak{M}$ and $\mathfrak{M}'$. The elementary proof of the following lemma is left to the reader.

Lemma 2.18. *Assume that $\mathfrak{M}$ and $\mathfrak{N}$ are two von Neumann algebras and let ϕ : $\mathfrak{M} \to \mathfrak{N}$ be a $*$-morphism.*

(i) If ϕ is surjective then $\phi(\mathfrak{Z}(\mathfrak{M})) \subset \mathfrak{Z}(\mathfrak{N})$.
(ii) If ϕ is injective then $\phi^{-1}(\mathfrak{Z}(\mathfrak{N})) \subset \mathfrak{Z}(\mathfrak{M})$.
(iii) If ϕ is bijective then $\phi(\mathfrak{Z}(\mathfrak{M})) = \mathfrak{Z}(\mathfrak{N})$.

$\mathfrak{M}$ is a *factor* if $\mathfrak{Z}(\mathfrak{M}) = \mathbb{C}I$ or equivalently $\mathfrak{M} \vee \mathfrak{M}' = \mathfrak{B}(\mathcal{H})$.

The *central support* of a normal state ω on $\mathfrak{M}$ is the support of its restriction to the center of $\mathfrak{M}$,

$$z_\omega \equiv \inf\{P \in \mathfrak{Z}(\mathfrak{M}) \cap \mathfrak{M}^P \,|\, \omega(P) = 1\}. \tag{5}$$

For any normal state ω one clearly has $0 < s_\omega \leq z_\omega \leq I$ and hence $\omega(A) = \omega(Az_\omega) = \omega(z_\omega A)$ for all $A \in \mathfrak{M}$. The state ω is *centrally faithful* if $z_\omega = I$. Lemma 2.17 shows that the central support of the vector state ω_Φ is

$$z_{\omega_\Phi} = [\mathfrak{M} \vee \mathfrak{M}'\Phi]. \tag{6}$$

In particular, if Φ is cyclic for $\mathfrak{M}$ or $\mathfrak{M}'$ or if $\mathfrak{M}$ is a factor, then ω_Φ is centrally faithful. More generally one has

Lemma 2.19. *Let ω be a normal state on the von Neumann algebra $\mathfrak{M} \subset \mathfrak{B}(\mathcal{H})$. If s_ω is the support of ω, its central support is given by*

$$z_\omega = [\mathfrak{M} s_\omega \mathcal{H}].$$

Proof. Set $\mathcal{K} \equiv \overline{\mathfrak{M} s_\omega \mathcal{H}}$ and denote by P the orthogonal projection on $\mathcal{K}$. We first claim that $P \in \mathfrak{Z}(\mathfrak{M})$. Using Proposition 10 in [7] and the relation $\mathfrak{Z}(\mathfrak{M})' = (\mathfrak{M} \cup \mathfrak{M}')''$, this follows from the fact that $\mathfrak{M} s_\omega \mathcal{H}$ and hence $\mathcal{K}$ are invariant under $\mathfrak{M}$ and $\mathfrak{M}'$. Next we note that $\operatorname{Ran} s_\omega \subset \mathcal{K}$ implies $P \geq s_\omega$ and hence $\omega(P) = 1$. Finally, if $Q \in \mathfrak{M}^P \cap \mathfrak{Z}(\mathfrak{M})$ is such that $\omega(Q) = 1$, Cauchy-Schwarz inequality yields that $\omega(A(I - Q)) = 0$ for all $A \in \mathfrak{M}$. Exercise 2.15 further leads to $(I - Q)As_\omega = A(I - Q)s_\omega = 0$ for all $A \in \mathfrak{M}$. This shows that $(I - Q)P = 0$, *i.e.*, $Q \geq P$ and Equ. (5) yields that $P = z_\omega$. □

By Corollary 2.12, a $*$-automorphism τ of a von Neumann algebra is automatically continuous in the σ-weak topology. In particular $\omega \circ \tau$ is a normal state for any normal state ω. It immediately follows from the definitions that $s_{\omega\circ\tau} = \tau^{-1}(s_\omega)$ and $z_{\omega\circ\tau} = \tau^{-1}(z_\omega)$.

2.2 The GNS Representation

Let $\mathfrak{A}$ be a C^*-algebra and $\omega \in E(\mathfrak{A})$. Throughout these notes I shall use the standard notation $(\mathcal{H}_\omega, \pi_\omega, \Omega_\omega)$ for the GNS representation of $\mathfrak{A}$ associated to the state ω (Theorem 8 in Lecture [7]).

Enveloping von Neumann Algebra and Folium of a State

Since Ω_ω is cyclic for $\pi_\omega(\mathfrak{A})$,

$$\hat{\omega}(A) \equiv (\Omega_\omega, A\Omega_\omega),$$

defines a centrally faithful normal state on the von Neumann algebra $\pi_\omega(\mathfrak{A})''$.

By the von Neumann density theorem, $\pi_\omega(\mathfrak{A})$ is σ-weakly dense in $\pi_\omega(\mathfrak{A})''$ and hence we have a canonical injection

$$\begin{aligned} N(\pi_\omega(\mathfrak{A})'') &\to E(\mathfrak{A}) \\ \tilde{\nu} &\hookrightarrow \pi_\omega^\star(\tilde{\nu}) = \tilde{\nu} \circ \pi_\omega. \end{aligned}$$

Thus, we can identify $N(\pi_\omega(\mathfrak{A})'')$ with a subset $N(\mathfrak{A}, \omega)$ of $E(\mathfrak{A})$. Explicitly, $\nu \in N(\mathfrak{A}, \omega)$ if and only if there exists a density matrix ρ on $\mathcal{H}_\omega$ and a corresponding normal state $\tilde{\nu}$ on $\pi_\omega(\mathfrak{A})''$ such that

$$\nu(A) = \tilde{\nu} \circ \pi_\omega(A) = \mathrm{Tr}(\rho \pi_\omega(A)).$$

Definition 2.20. *Let $\mathfrak{A}$ be a C^*-algebra and $\omega \in E(\mathfrak{A})$.*

(i) $\mathfrak{A}_\omega \equiv \pi_\omega(\mathfrak{A})'' \subset \mathfrak{B}(\mathcal{H}_\omega)$ is the enveloping von Neumann algebra of $\mathfrak{A}$ associated to ω.
(ii) $N(\mathfrak{A}, \omega) \subset E(\mathfrak{A})$ is the folium of the state ω. It is the image under $\pi_\omega^\star$ of the set of states on $\pi_\omega(\mathfrak{A})$ which have a unique normal extension to the enveloping algebra $\mathfrak{A}_\omega$. Its elements are said to be normal relative to ω, or simply ω-normal.

Note that $\hat{\omega}$ is the unique normal extension of the state $\pi_\omega(A) \mapsto \omega(A)$ from $\pi_\omega(\mathfrak{A})$ to its weak closure $\mathfrak{A}_\omega$. By a slight abuse of language I shall say that $\hat{\omega}$ is the normal extension of ω to $\mathfrak{A}_\omega$. Similarly, if $\nu = \tilde{\nu} \circ \pi_\omega \in N(\mathfrak{A}, \omega)$ I shall say that $\tilde{\nu}$ is the normal extension of ν to $\mathfrak{A}_\omega$. I further denote by $s_{\nu|\omega}$ the support of $\tilde{\nu}$ and by $z_{\nu|\omega}$ its central support. Abusing notation, I also set $s_\omega \equiv s_{\omega|\omega} = s_{\hat{\omega}}$.

Definition 2.21. *Let ω, ν be two states on the C^*-algebra $\mathfrak{A}$.*

(i) ν, ω are quasi-equivalent, written $\nu \approx \omega$, if $N(\mathfrak{A}, \nu) = N(\mathfrak{A}, \omega)$.
(ii) They are orthogonal, written $\nu \perp \omega$, if $\lambda\mu \leq \nu$ and $\lambda\mu \leq \omega$ for some $\mu \in E(\mathfrak{A})$ and $\lambda \geq 0$ implies $\lambda = 0$.
(iii) They are disjoint if $N(\mathfrak{A}, \nu) \cap N(\mathfrak{A}, \omega) = \varnothing$.

The GNS Representation of a Normal State

In this subsection we study the special features of the GNS representation associated to a normal state ω on the von Neumann algebra $\mathfrak{M} \subset \mathfrak{B}(\mathcal{H})$.

The first result relates the central support z_ω of the state ω to the kernel of the $*$-morphism π_ω. Before stating this relation let me make the following remark.

Remark 2.22. If $P \in \mathfrak{Z}(\mathfrak{M})$ is an orthogonal projection then $Q \equiv I - P \in \mathfrak{Z}(\mathfrak{M})$ and since $\mathfrak{M}P\mathcal{H} = P\mathfrak{M}\mathcal{H} \subset P\mathcal{H}$ and $\mathfrak{M}Q\mathcal{H} = Q\mathfrak{M}\mathcal{H} \subset Q\mathcal{H}$, any element of $\mathfrak{M}$ can be written, according to the orthogonal decomposition $\mathcal{H} = P\mathcal{H} \oplus Q\mathcal{H}$, as a 2×2-matrix

$$A = \begin{pmatrix} B & 0 \\ 0 & C \end{pmatrix},$$

where $B \in \mathfrak{B}(P\mathcal{H})$ and $C \in \mathfrak{B}(Q\mathcal{H})$. Using the injection

$$P\mathfrak{M} \ni \begin{pmatrix} B & 0 \\ 0 & 0 \end{pmatrix} \hookrightarrow B \in \mathfrak{B}(P\mathcal{H}),$$

we can identify $P\mathfrak{M}$ with a von Neumann algebra on $P\mathcal{H}$ and similarly for $Q\mathfrak{M}$. We then write $\mathfrak{M} = P\mathfrak{M} \oplus Q\mathfrak{M}$, and say that $\mathfrak{M}$ is the direct sum of the von Neumann algebras $P\mathfrak{M}$ and $Q\mathfrak{M}$. Of course the same argument applies to the commutant and we also have $\mathfrak{M}' = P\mathfrak{M}' \oplus Q\mathfrak{M}'$. It follows immediately that $P\mathfrak{M}' = (P\mathfrak{M})'$ and $Q\mathfrak{M}' = (Q\mathfrak{M})'$ as von Neumann algebras on $P\mathcal{H}$ and $Q\mathcal{H}$. With the same interpretation we can write $\mathfrak{Z}(\mathfrak{M}) = P\mathfrak{Z}(\mathfrak{M}) \oplus Q\mathfrak{Z}(\mathfrak{M})$ and $\mathfrak{Z}(P\mathfrak{M}) = P\mathfrak{Z}(\mathfrak{M})$, $\mathfrak{Z}(Q\mathfrak{M}) = Q\mathfrak{Z}(\mathfrak{M})$.

Lemma 2.23. *If* $\omega \in N(\mathfrak{M})$ *then* $\operatorname{Ker}(\pi_\omega) = (I - z_\omega)\mathfrak{M}$. *In particular,* π_ω *is faithful (i.e., is a $*$-isomorphism from* $\mathfrak{M}$ *onto* $\pi_\omega(\mathfrak{M})$*) if and only if* ω *is centrally faithful. More generally, the map*

$$\hat{\pi}_\omega : Az_\omega \mapsto \pi_\omega(A),$$

defines a $*$-isomorphism from the von Neumann algebra $z_\omega\mathfrak{M}$ *onto* $\pi_\omega(\mathfrak{M})$ *such that, for all* $A \in \mathfrak{M}$ *and all* $B \in \pi_\omega(\mathfrak{M})$ *one has* $\pi_\omega \circ \hat{\pi}_\omega^{-1}(B) = B$ *and* $\hat{\pi}_\omega^{-1} \circ \pi_\omega(A) = z_\omega A$.

Proof. For $A, B \in \mathfrak{M}$ one has

$$\|\pi_\omega(A(I - z_\omega))\pi_\omega(B)\Omega_\omega\|^2 = \omega(B^* A^* AB(I - z_\omega)) = 0.$$

Since $\pi_\omega(\mathfrak{M})\Omega_\omega$ is dense in $\mathcal{H}_\omega$ one concludes that $\pi_\omega(A(I - z_\omega)) = 0$, *i.e.*, $\mathfrak{M}(I - z_\omega) \subset \operatorname{Ker}(\pi_\omega)$.

To prove the reverse inclusion note that $\pi_\omega(A) = 0$ implies that

$$\omega(B^* A^* AB) = \|\pi_\omega(A)\pi_\omega(B)\Omega_\omega\|^2 = 0,$$

for all $B \in \mathfrak{M}$. Exercise 2.15 further gives $ABs_\omega = 0$ for all $B \in \mathfrak{M}$ and Lemma 2.19 yields $Az_\omega = 0$, *i.e.*, $A = A(I - z_\omega)$. The proof of the last statement of the lemma is easy and left to the reader □

Corollary 2.24. *Let $\mathfrak{M}$ be a von Neumann algebra and $\omega, \nu \in N(\mathfrak{M})$. Then ν is ω-normal if and only if $s_\nu \leq z_\omega$.*

Proof. Suppose that $\nu = \hat{\nu} \circ \pi_\omega$ for some $\hat{\nu} \in N(\mathfrak{M}_\omega)$. By Lemma 2.23 we have

$$\nu(I - z_\omega) = \hat{\nu}(\pi_\omega(I - z_\omega)) = \hat{\nu}(0) = 0.$$

Thus, $\nu(z_\omega) = 1$ from which we conclude that $s_\nu \leq z_\omega$.

Suppose now that $s_\nu \leq z_\omega$ and set $\hat{\nu} \equiv \nu \circ \hat{\pi}_\omega^{-1}$. Since $\hat{\pi}_\omega$ is a $*$-isomorphism $\hat{\nu}$ is normal. Moreover, from $\nu(A) = \nu(z_\omega A)$ we conclude that

$$\hat{\nu} \circ \pi_\omega(A) = \nu(\hat{\pi}_\omega^{-1}(\pi_\omega(A))) = \nu(z_\omega A) = \nu(A).$$

□

The continuity properties of π_ω follow from the simple lemma:

Lemma 2.25. *The map π_ω is normal i.e., for any bounded increasing net A_α of self-adjoint elements of $\mathfrak{M}$ one has*

$$\sup_\alpha \pi_\omega(A_\alpha) = \pi_\omega(\sup_\alpha A_\alpha).$$

Proof. For any $B \in \mathfrak{M}$ one has $\omega(B^* A_\alpha B) = (\pi_\omega(B)\Omega_\omega, \pi_\omega(A_\alpha)\pi_\omega(B)\Omega_\omega)$ and Equ. (1) allows us to write

$$\begin{aligned}(\pi_\omega(B)\Omega_\omega, \sup_\alpha \pi_\omega(A_\alpha)\pi_\omega(B)\Omega_\omega) &= \sup_\alpha(\pi_\omega(B)\Omega_\omega, \pi_\omega(A_\alpha)\pi_\omega(B)\Omega_\omega)\\ &= \sup_\alpha \omega(B^* A_\alpha B)\\ &= \omega(B^*(\sup_\alpha A_\alpha)B)\\ &= (\pi_\omega(B)\Omega_\omega, \pi_\omega(\sup_\alpha A_\alpha)\pi_\omega(B)\Omega_\omega).\end{aligned}$$

Since $\pi_\omega(\mathfrak{M})\Omega_\omega$ is dense in $\mathcal{H}_\omega$ the claim follows. □

Exercise 2.26. Prove the following lemma using Lemma 2.25 and following the proof of Corollary 2.12.

Lemma 2.27. *If $\omega \in N(\mathfrak{M})$ then π_ω is σ-weakly and σ-strongly continuous.*

Lemma 2.28. *If $\omega \in N(\mathfrak{M})$ then $\pi_\omega(\mathfrak{M})$ is a von Neumann algebra in $\mathfrak{B}(\mathcal{H}_\omega)$, i.e., $\mathfrak{M}_\omega = \pi_\omega(\mathfrak{M})$.*

Proof. Since $\pi_\omega(\mathfrak{M}) = \hat{\pi}_\omega(\mathfrak{M}z_\omega)$ we can assume, without loss of generality, that π_ω is faithful and hence isometric (Proposition 4 in Lecture [7]). Let $B \in \pi_\omega(\mathfrak{M})''$. By the Kaplansky density theorem there exists a net A_α in $\mathfrak{M}$ such that $\|\pi_\omega(A_\alpha)\| \leq \|B\|$ and $\pi_\omega(A_\alpha)$ converges σ-weakly to B. Since

$$\|A_\alpha\| = \|\pi_\omega(A_\alpha)\| \leq \|B\|,$$

it follows from the Banach-Alaoglu theorem that there exists a subnet A_β of the net A_α which converges σ-weakly to some $A \in \mathfrak{M}$. Since π_ω is σ-weakly continuous one has

$$\pi_\omega(A) = \lim_\beta \pi_\omega(A_\beta) = \lim_\alpha \pi_\omega(A_\alpha) = B,$$

and hence $B \in \pi_\omega(\mathfrak{M})$. □

Lemma 2.29. *If $\omega \in N(\mathfrak{M})$ then*

$$N(\mathfrak{M}, \omega) = \{\nu \in N(\mathfrak{M}) \mid s_\nu \leq z_\omega\} \subset N(\mathfrak{M}).$$

In particular, if ω is centrally faithful then $N(\mathfrak{M}, \omega) = N(\mathfrak{M})$.

Proof. By Lemma 2.27, if $\omega \in N(\mathfrak{M})$ then π_ω is σ-weakly continuous. Hence $N(\mathfrak{M}, \omega) \subset N(\mathfrak{M})$ and Corollary 2.24 apply. □

As an application of the above results let us prove the following characterization of the relative normality of two states on a C^*-algebra.

Theorem 2.30. *Let $\mathfrak{A}$ be a C^*-algebra and $\omega, \mu \in E(\mathfrak{A})$. Denote the induced GNS representations by $(\mathcal{H}_\omega, \pi_\omega, \Omega_\omega)$, $(\mathcal{H}_\mu, \pi_\mu, \Omega_\mu)$ and the corresponding enveloping von Neumann algebras by $\mathfrak{A}_\omega$, $\mathfrak{A}_\mu$. Then $\mu \in N(\mathfrak{A}, \omega)$ if and only if there exists a σ-weakly continuous $*$-morphism $\pi_{\mu|\omega} : \mathfrak{A}_\omega \to \mathfrak{A}_\mu$ such that $\pi_\mu = \pi_{\mu|\omega} \circ \pi_\omega$. If this is the case then the following also hold.*

(i) $N(\mathfrak{A}, \mu) \subset N(\mathfrak{A}, \omega)$.
(ii) If $\nu \in N(\mathfrak{A}, \mu)$ has the normal extension $\tilde{\nu}$ to $\mathfrak{A}_\mu$ then $\tilde{\nu} \circ \pi_{\mu|\omega}$ is its normal extension to $\mathfrak{A}_\omega$.
(iii) $\pi_{\mu|\omega}$ is σ-strongly continuous.
(iv) $\hat{\pi}_{\mu|\omega} : z_{\mu|\omega} A \mapsto \pi_{\mu|\omega}(A)$ is a $*$-isomorphism from $z_{\mu|\omega}\mathfrak{A}_\omega$ onto $\mathfrak{A}_\mu$.
(v) $\mathrm{Ker}\, \pi_{\mu|\omega} = (I - z_{\mu|\omega})\mathfrak{A}_\omega$
(vi) $\pi_{\mu|\omega}(s_{\mu|\omega}) = s_\mu$.
(vii) If $\tilde{\mu}$ denotes the normal extension of μ to $\mathfrak{A}_\omega$ then $(\mathcal{H}_\mu, \pi_{\mu|\omega}, \Omega_\mu)$ is the induced GNS representation of $\mathfrak{A}_\omega$. In particular $\mathfrak{A}_\mu = (\mathfrak{A}_\omega)_{\tilde{\mu}}$.

Proof. Let $\hat{\mu}(A) \equiv (\Omega_\mu, A\Omega_\mu)$ be the normal extension of μ to $\mathfrak{A}_\mu$. If the morphism $\pi_{\mu|\omega}$ exists then one has $\mu = \hat{\mu} \circ \pi_\mu = (\hat{\mu} \circ \pi_{\mu|\omega}) \circ \pi_\omega$. Since $\pi_{\mu|\omega}$ is σ-weakly continuous $\hat{\mu} \circ \pi_{\mu|\omega} \in N(\mathfrak{A}_\omega)$ and we conclude that $\mu \in N(\mathfrak{A}, \omega)$.

Assume now that $\mu \in N(\mathfrak{A}, \omega)$, *i.e.*, that $\mu = \tilde{\mu} \circ \pi_\omega$ with $\tilde{\mu} \in N(\mathfrak{A}_\omega)$. Consider the GNS representation $(\mathcal{K}, \phi, \Psi)$ of $\mathfrak{A}_\omega$ induced by $\tilde{\mu}$. By Lemma 2.27 ϕ is σ-weakly and σ-strongly continuous. Since by the von Neumann density theorem $\pi_\omega(\mathfrak{A})$ is σ-strongly dense in $\mathfrak{A}_\omega$, we get

$$\mathcal{K} = \overline{\phi(\mathfrak{A}_\omega)\Psi} = \overline{\phi(\pi_\omega(\mathfrak{A}))\Psi}.$$

Finally, for any $A \in \mathfrak{A}$ one has

$$\mu(A) = \tilde{\mu}(\pi_\omega(A)) = (\Psi, \phi(\pi_\omega(A))\Psi),$$

and we can conclude that $(\mathcal{K}, \phi \circ \pi_\omega, \Psi)$ is a GNS representation of $\mathfrak{A}$ induced by μ. By the unicity, up to unitary equivalence, of such representations there exists a unitary $U : \mathcal{K} \to \mathcal{H}_\mu$ such that $\pi_\mu(A) = U\phi(\pi_\omega(A))U^*$ and $\Omega_\mu = U\Psi$. Set $\pi_{\mu|\omega}(A) \equiv U\phi(A)U^*$ then $(\mathcal{H}_\mu, \pi_{\mu|\omega}, \Omega_\mu)$ is another GNS representation of $\mathfrak{A}_\omega$ induced by $\tilde{\mu}$ and $\pi_\mu = \pi_{\mu|\omega} \circ \pi_\omega$. Since $\pi_{\mu|\omega}$ is also σ-weakly and σ-strongly continuous Lemma 2.28 and the von Neumann density theorem yield

$$(\mathfrak{A}_\omega)_{\tilde{\mu}} = \pi_{\mu|\omega}(\mathfrak{A}_\omega) = \pi_{\mu|\omega}(\pi_\omega(\mathfrak{A}))'' = \pi_\mu(\mathfrak{A})'' = \mathfrak{A}_\mu.$$

This proves the existence of $\pi_{\mu|\omega}$ with Properties (iii) and (vii).

To prove Properties (i) and (ii) let $\nu \in N(\mathfrak{A}, \mu)$ and denote by $\tilde{\nu}$ its normal extension to $\mathfrak{A}_\mu$. One has $\nu = \tilde{\nu} \circ \pi_\mu = \tilde{\nu} \circ \pi_{\mu|\omega} \circ \pi_\omega$ and it follows that $\nu \in N(\mathfrak{A}, \omega)$ and that its normal extension to $\mathfrak{A}_\omega$ is $\tilde{\nu} \circ \pi_{\mu|\omega}$. Properties (iv) and (v) follow directly from Lemma 2.23. By Lemma 2.25, $\pi_{\mu|\omega}$ is normal and hence

$$\begin{aligned}
\pi_{\mu|\omega}(s_{\mu|\omega}) &= \pi_{\mu|\omega}\left(\inf\{P \in \mathfrak{A}_\omega^P \mid \tilde{\mu}(P) = 1\}\right) \\
&= \inf\{\pi_{\mu|\omega}(P) \mid P \in \mathfrak{A}_\omega^P, \hat{\mu}(\pi_{\mu|\omega}(P)) = 1\} \\
&= \inf\{Q \in \mathfrak{A}_\mu^P \mid \hat{\mu}(Q) = 1\} \\
&= s_\mu,
\end{aligned}$$

proves Property (vi). □

3 Classical Systems

3.1 Basics of Ergodic Theory

Let X be a measurable space, *i.e.*, a set equipped with a σ-field $\mathcal{F}$. A *dynamics* on X is a family of maps $\varphi_t : X \to X$, indexed by a *time* t running in $\mathbb{R}$, such that $(x, t) \mapsto \varphi_t(x)$ is measurable and the group properties

$$\varphi_0(x) = x, \quad \varphi_t \circ \varphi_s = \varphi_{t+s},$$

hold. In particular, the map φ_t is one to one with inverse φ_{-t}. Given $x \in X$, we set $x_t = \varphi_t(x)$ and call $(x_t)_{t\in\mathbb{R}} \subset X$ the *orbit* or *trajectory* starting at x. *Observables* are bounded measurable functions $f : X \to \mathbb{C}$.

Instead of considering individual orbits, think of the initial configuration x as a random variable distributed according to a probability measure[1] μ on X. Then $(x_t)_{t\in\mathbb{R}}$ becomes a stochastic process and we denote by $\mathbb{E}_\mu$ the corresponding mathematical expectation. If f is an observable, $\mathbb{E}_\mu[f(x_0)] = \mu(f) = \int f \mathrm{d}\mu$ is the expectation of f at time zero. Its expectation at time t is $\mu_t(f) \equiv \mathbb{E}_\mu[f(x_t)] = \mu(f \circ \varphi_t)$ which defines the evolution μ_t of the measure μ. A measure μ is called *invariant* if

[1] All measures in this section are probabilities, so I use the words measure and probability interchangeably.

the corresponding process is stationary, *i.e.*, if $\mu_t = \mu$ for all t. Such an invariant measure describes a stationary regime of the system, or an *equilibrium state.*

Invariant probabilities may fail to exist (see Exercise I.8.6 in [35]) but most physical systems of interest have a lot of them. For example, if $x \in X$ is periodic of period T, then

$$\mu_x \equiv \frac{1}{T}\int_0^T \delta_{x_t}\,\mathrm{d}t,$$

is invariant, supported by the orbit of x. Under additional topological assumptions, one can prove that there is at least one invariant measure.

Exercise 3.1. Let M be a compact metric space and φ_t a continuous dynamics on M, *i.e.*, assume that the map $(t,x) \mapsto \varphi_t(x)$ is continuous. Recall the following facts: The set of continuous observables $C(M)$ is a separable Banach space. Its dual $C(M)^\star$ is the set of Baire measures on M. The set of Baire probabilities is a compact, metrizable subset of $C(M)^\star$ for the weak-$\star$ topology. Every Baire measure has a unique regular Borel extension.

Show that for any Baire probability μ on M, there exists a sequence $T_n \to +\infty$, such that the weak-$\star$ limit

$$\mu^+ = \mathrm{w}^\star\!-\!\lim_n \frac{1}{T_n}\int_0^{T_n} \mu_t\,\mathrm{d}t,$$

exists and defines an invariant probability for φ_t on M.

Definition 3.2. *A classical dynamical system is a triple (X, φ_t, μ) where X, the phase space of the system, is a measurable space, φ_t a dynamics on X and μ an invariant measure for φ_t.*

If (X, φ_t, μ) is a classical dynamical system, then $U^t f \equiv f \circ \varphi_t$ defines a group of isometries on each Banach space $L^p(X, \mathrm{d}\mu)$. Indeed, for any $f \in L^p(X, \mathrm{d}\mu)$ we have $U^0 f = f \circ \varphi_0 = f$,

$$U^t U^s f = U^t(f \circ \varphi_s) = (f \circ \varphi_s) \circ \varphi_t = f \circ (\varphi_s \circ \varphi_t) = f \circ \varphi_{t+s} = U^{t+s} f,$$

and

$$\|U^t f\|_p^p = \mu(|f \circ \varphi_t|^p) = \mu(|f|^p \circ \varphi_t) = \mu_t(|f|^p) = \mu(|f|^p) = \|f\|^p.$$

A function $f \in L^p(X, \mathrm{d}\mu)$ is invariant if $U^t f = f$ for all t. A measurable set $A \subset X$ is invariant modulo μ if its characteristic function χ_A is invariant, that is

$$0 = |\chi_A - U^t \chi_A| = |\chi_A - \chi_{\varphi_t^{-1}(A)}| = \chi_{A\Delta\varphi_t^{-1}(A)},$$

which is equivalent to $\mu(A\Delta\varphi_t^{-1}(A)) = 0$.

Ergodic theory deals with the study of invariant measures and their connections with the large time behavior of dynamical system. The cornerstone of ergodic theory is the following, so called *Birkhoff or individual ergodic theorem.*

Theorem 3.3. *Let (X, φ_t, μ) be a classical dynamical system. Then for any $f \in L^1(X, \mathrm{d}\mu)$, the two limits*

$$(\mathbb{P}_\mu f)(x) = \lim_{T\to\infty} \frac{1}{T} \int_0^T f \circ \varphi_{\pm t}(x)\,\mathrm{d}t,$$

exist and coincide for μ-almost all $x \in X$. They define a linear contraction $\mathbb{P}_\mu$ on $L^1(X, \mathrm{d}\mu)$ with the following properties:

(i) $\mathbb{P}_\mu f \geq 0$ if $f \geq 0$,
(ii) $\mathbb{P}_\mu^2 = \mathbb{P}_\mu$,
(iii) $U^t\mathbb{P}_\mu = \mathbb{P}_\mu U^t = \mathbb{P}_\mu$ for all t,
(iv) $\mu(f) = \mu(\mathbb{P}_\mu f)$,
(v) if $g \in L^1(X, \mathrm{d}\mu)$ is invariant, then $\mathbb{P}_\mu gf = g\mathbb{P}_\mu f$. In particular, $\mathbb{P}_\mu g = g$.

Remark 3.4. $\mathbb{P}_\mu$ is a conditional expectation with respect to the σ-field of invariant sets modulo μ.

Even in the case of a smooth (continuous or analytic) dynamical system, the conditional expectation $\mathbb{P}_\mu f(x)$ can display a weird dependence on the starting point x, reflecting the complexity of the orbits. Of special interest are the measures μ for which $\mathbb{P}_\mu f(x)$ is μ-almost surely independent of x.

Definition 3.5. *The dynamical system (X, φ_t, μ) is called ergodic if, for all $f \in L^1(X, \mathrm{d}\mu)$, one has $\mathbb{P}_\mu f(x) = \mu(f)$ for μ-almost all $x \in X$. In this case, we also say that μ is an ergodic measure for φ_t, or that the dynamics φ_t is ergodic for μ.*

Proposition 3.6. *The following propositions are equivalent.*

(i) (X, φ_t, μ) is ergodic.
(ii) For any measurable set A invariant modulo μ, one has $\mu(A) \in \{0, 1\}$.
(iii) For any invariant $f \in L^1(X, \mathrm{d}\mu)$ one has $f = \mu(f)$.
(iv) For any μ-absolutely continuous probability ρ one has

$$\lim_{T\to\infty} \frac{1}{T} \int_0^T \rho_t(f)\,\mathrm{d}t = \mu(f),$$

for all $f \in L^\infty(X, \mathrm{d}\mu)$.

Proof. Let us first show that (i), (ii) and (iii) are equivalent.

(i)⇒(ii). If A is invariant modulo μ then its indicator function χ_A is invariant. Property (v) of Theorem 3.3 shows that $\chi_A = \mathbb{P}_\mu \chi_A$. If μ is ergodic we further have $\mathbb{P}_\mu \chi_A = \mu(A)$. Therefore, $\chi_A = \mu(A)$ and we conclude that $\mu(A) \in \{0, 1\}$.

(ii)⇒(iii). If $f \in L^1(X, \mathrm{d}\mu)$ is invariant so are its real and imaginary parts. Without loss of generality we can assume that f is real valued. Then the set $\{x \mid f(x) > a\}$ is invariant modulo μ. Therefore, the distribution function $F_f(a) = \mu(\{x \mid f(x) > a\})$ takes values in the set $\{0, 1\}$ from which we conclude that f is constant μ-almost everywhere.

(iii)⇒(i). If $f \in L^1(X, \mathrm{d}\mu)$ then $\mathbb{P}_\mu f$ is invariant by property (iii) of Theorem 3.3. Therefore, $\mathbb{P}_\mu f = \mu(\mathbb{P}_\mu f)$ and by property (iv) of Theorem 3.3 this is equal to $\mu(f)$.

Now we consider property (iv). Denote by $g \in L^1(X, \mathrm{d}\mu)$ the Radon-Nikodym derivative of ρ with respect to μ. Then, for any $f \in L^\infty(X, \mathrm{d}\mu)$, Fubini Theorem and the invariance of μ give

$$\begin{aligned}\frac{1}{T}\int_0^T \rho_t(f)\,\mathrm{d}t &= \frac{1}{T}\int_0^T \left(\int g(x) f\circ\varphi_t(x)\,\mathrm{d}\mu(x)\right)\mathrm{d}t \\ &= \int \left(\frac{1}{T}\int_0^T g\circ\varphi_{-t}(x)\,\mathrm{d}t\right) f(x)\,\mathrm{d}\mu(x).\end{aligned}$$

Birkhoff Theorem together with Lebesgue dominated convergence Theorem further lead to

$$\rho^+(f) \equiv \lim_{T\to\infty} \frac{1}{T}\int_0^T \rho_t(f)\,\mathrm{d}t = \mu((\mathbb{P}_\mu g) f).$$

Therefore, if μ is ergodic, we obtain $\rho^+(f) = \mu(\mu(g)f) = \mu(f)$. Reciprocally, if $\rho^+ = \mu$ for all μ-absolutely continuous probabilities ρ, we can conclude that for all $g \in L^1(X, \mathrm{d}\mu)$ such that $\mu(g) = 1$ one has $\mathbb{P}_\mu g = 1 = \mu(g)$. This clearly implies that μ is ergodic. □

The following *ergodic decomposition theorem* shows that ergodic measures are elementary building blocks of invariant measures.

Theorem 3.7. *Let M be a compact metric space and φ_t a continuous dynamics on M. Then the set $E_\varphi(M)$ of invariant probabilities for φ_t is a non-empty, convex, weak$-\star$ compact subset of the dual $C(M)^\star$. Denote by $X_\varphi(M)$ the set of extremal points[2] of $E_\varphi(M)$:*

(i) $\mu \in X_\varphi(M)$ if and only if it is ergodic for φ_t.
(ii) If $\mu, \nu \in X_\varphi(M)$ and $\mu \neq \nu$, then μ and ν are mutually singular.
(iii) For any $\mu \in E_\varphi(M)$ there exists a probability measure ρ on $X_\varphi(M)$ such that

$$\mu = \int \nu\,\mathrm{d}\rho(\nu).$$

The relevance of an ergodic measure μ in the study of the large time behavior of the system comes from the fact that the *time average* of an observable along a generic orbit is given by the *ensemble average* described by μ:

$$\lim_{T\to\infty} \frac{1}{T}\int_0^T f(x_t)\,\mathrm{d}t = \mu(f),$$

[2] μ is extremal if it can not be expressed as a non-trivial convex linear combination of two distinct measures: $\mu = \alpha\mu_1 + (1-\alpha)\mu_2$ with $\alpha \in]0, 1[$ and $\mu_i \in E_\varphi(M)$ implies $\mu_1 = \mu_2 = \mu$.

for μ-almost all x. Note however that, depending on the nature of μ, generic orbits may fill a very small portion of the phase space (as shown by the above example of a periodic orbit). Here small refers to some additional feature that the physical problem may induce on the space X. For example if X is a smooth manifold, small could mean of measure zero with respect to some (any) Riemannian volume on X (see [39] for a more detailed discussion of this important point).

A more precise information on the large time asymptotics of the system is given by the following mixing condition.

Definition 3.8. *A dynamical system (X, φ_t, μ) is mixing if, for any μ-absolutely continuous measure ρ and all observables $f \in L^\infty(X, \mathrm{d}\mu)$, one has*

$$\lim_{t\to\infty} \rho_t(f) = \mu(f).$$

One also says that μ is mixing, or φ_t is mixing for μ.

If we think of an invariant measure μ as describing an equilibrium state of the system then μ is mixing if all initial measures ρ which are not too far from equilibrium, *i.e.*, which are absolutely continuous with respect to μ, converge to μ as $t \to \infty$. For this reason, the mixing condition is often referred to as *return to equilibrium.*

If μ is mixing, it is obviously ergodic from Property (iv) in Theorem 3.6. Here is another proof. If A is an invariant set modulo μ such that $\mu(A) > 0$, then $\rho(f) = \mu(f\chi_A)/\mu(A)$ defines a μ-absolutely continuous invariant measure with $\rho(A) = 1$. Therefore, if μ is mixing we have

$$1 = \rho(A) = \rho_t(A) = \lim_{t\to\infty} \rho_t(A) = \mu(A).$$

From Theorem 3.6 we conclude that μ is ergodic. Thus mixing implies ergodicity, but the reverse is not true in general.

Exercise 3.9. Let $X = \mathbb{R}/\mathbb{Z}$ be the one dimensional torus, μ the measure induced on X by the Lebesgue measure on $\mathbb{R}$ and $\varphi_t(x) = x + t \pmod 1$. Show that (X, φ_t, μ) is ergodic but not mixing.

Exercise 3.10. Show that (X, φ_t, μ) is mixing if and only if, for all measurable subsets $A, B \subset X$ one has

$$\lim_{t\to\infty} \mu(\varphi_t^{-1}(A) \cap B) = \mu(A)\,\mu(B).$$

Ergodicity and mixing are only two elements of the so called ergodic hierarchy which contains many other properties like exponential mixing, K-system... The interested reader should consult the general references given in the Introduction.

3.2 Classical Koopmanism

Ergodicity and mixing can be quite difficult to prove in concrete situations. One of the more powerful tools to do it is the Koopman–von Neumann spectral theory that I shall now introduce. The Koopman space of the dynamical system (X, φ_t, μ) is the Hilbert space $\mathcal{H} = L^2(X, \mathrm{d}\mu)$ on which the Koopman operators U^t are defined by

$$U^t f \equiv f \circ \varphi_t.$$

In the following, I shall always assume that the Koopman space is separable. This is the case for example if X is a (locally) compact metric space equipped with its natural Borel structure.

Lemma 3.11. *(Koopman Lemma) If $\mathcal{H}$ is separable, U^t is a strongly continuous group of unitary operators on $\mathcal{H}$.*

Proof. We have already shown that U^t is a group of isometries on $\mathcal{H}$. Since $U^t U^{-t} = I$ we have $\mathrm{Ran}\, U^t = \mathcal{H}$ and therefore U^t is unitary. Finally, since $t \mapsto (f, U^t g)$ is measurable and $\mathcal{H}$ is separable, it follows from a well known result of von Neumann (Theorem VIII.9 in [41]) that the map $t \mapsto U^t$ is strongly continuous. □

Remark 3.12. The separability condition is satisfied if X is a finite dimensional manifold. In infinite dimensional cases it can often be replaced by a weak continuity assumption. Indeed, if the map $t \mapsto \int f(x) f(\varphi_t(x))\, \mathrm{d}\mu(x)$ is continuous for all $f \in L^2(X, \mathrm{d}\mu)$, then U^t is strongly continuous since

$$\|U^t f - f\|^2 = 2(\|f\|^2 - Re(f, U^t f)). \tag{7}$$

By the Stone Theorem, there exists a self-adjoint operator L on $\mathcal{H}$ such that

$$U^t = \mathrm{e}^{-itL}.$$

We call L the *Liouvillean* of the system. Note that $1 \in \mathcal{H}$ and $U^t 1 = 1$ for all t, from which we conclude that $1 \in \mathrm{D}(L)$ and $L1 = 0$. In other words, 0 is always an eigenvalue of the Liouvillean, with the associated eigenfunction 1. The connection between $\mathrm{Ker}\, L$ and ergodic theory is the content of the following *von Neumann or mean ergodic theorem.*

Theorem 3.13. *Let $U^t = \mathrm{e}^{-itA}$ be a strongly continuous group of unitaries on a Hilbert space $\mathcal{H}$ and denote by P the orthogonal projection on* Ker *A. Then, for any $f \in \mathcal{H}$,*

$$\lim_{T \to \infty} \frac{1}{T} \int_0^T U^t f\, \mathrm{d}t = Pf,$$

holds in $\mathcal{H}$.

Proof. Since U^t is strongly continuous,

$$\langle f \rangle_T = \frac{1}{T} \int_0^T U^t f\, \mathrm{d}t$$

is well defined as a Riemann integral. We first remark that for $f \in \mathrm{Ran}\, A$, we have

$$U^t f = U^t A g = i\partial_t U^t g$$

for some $g \in \mathrm{D}(A)$. Hence an explicit integration gives

$$\lim_{T\to\infty} \langle f\rangle_T = \lim_{T\to\infty} \frac{i}{T}(U^T - I)g = 0.$$

Using the simple estimate $\|\langle u\rangle_T - \langle v\rangle_T\| \le \|u-v\|$, this result immediately extends to all $f \in \overline{\mathrm{Ran}\, A} = \mathrm{Ker}\, A^\perp$. Since for $f \in \mathrm{Ker}\, A$ we have $\langle f\rangle_T = f$, we get for arbitrary $f \in \mathcal{H}$

$$\lim_{T\to\infty} \langle f\rangle_T = \lim_{T\to\infty} \langle Pf\rangle_T + \lim_{T\to\infty} \langle (I-P)f\rangle_T = Pf.$$

□

Theorem 3.14. *(Koopman Ergodicity Criterion) A dynamical system is ergodic if and only if 0 is a simple eigenvalue of its Liouvillean L.*

Proof. Let $f \in \mathrm{Ker}\, L$, then f is an invariant function in $L^1(X, \mathrm{d}\mu)$ and, by Theorem 3.6, if μ is ergodic we must have $f = \mu(f)1$. Thus, $\mathrm{Ker}\, L$ is one dimensional.

Assume now that $\mathrm{Ker}\, L$ is one dimensional. Let A be an invariant set modulo μ. It follows that $\chi_A \in \mathcal{H}$ is invariant and hence belongs to $\mathrm{Ker}\, L$. Thus, we have $\chi_A = \mu(A)1$, from which we may conclude that $\mu(A) \in \{0, 1\}$. Ergodicity of μ follows from Theorem 3.6. □

Theorem 3.15. *(Koopman Mixing Criterion) A dynamical system is mixing if and only if*

$$\mathrm{w}-\lim_{t\to\infty} U^t = (1, \cdot)1. \tag{8}$$

In particular, if the spectrum of the Liouvillean L is purely absolutely continuous on $\{1\}^\perp$, *then the system is mixing.*

Proof. Assume first that the system is mixing, and set

$$\mathcal{H}_{1+} = \{g \in \mathcal{H} \,|\, g \ge 0, \mu(g) = 1\}.$$

Since any $g \in \mathcal{H}_{1+}$ is the Radon-Nikodym derivative of some μ-absolutely continuous probability ρ, we get for any $f \in L^\infty(X, \mathrm{d}\mu)$

$$(g, U^t f) = \rho_t(f) \to \mu(f) = (g, 1)(1, f), \tag{9}$$

as $t \to \infty$. Since any $g \in \mathcal{H}$ is a finite linear combination of elements of $\mathcal{H}_{1+}$, Equ. (9) actually holds for all $g \in \mathcal{H}$ and $f \in L^\infty(X, \mathrm{d}\mu)$. Finally, since both side of Equ. (9) are $\mathcal{H}$-continuous in f uniformly in t, Equ. (8) follows from the fact that $L^\infty(M, \mathrm{d}\mu)$ is dense in $\mathcal{H}$.

The reverse statement is proved in an analogous way. Suppose ρ is a μ-absolutely continuous probability and denote by g its Radon-Nikodym derivative. Assuming for a while that $g \in \mathcal{H}$, we get from Equ. (8) that $\lim_{t\to\infty} \rho_t(f) = \mu(f)$ for all

$f \in L^\infty(M, \mathrm{d}\mu)$. Since $\rho_t(f)$ is L^1-continuous in g, uniformly in t, the desired result follows from the fact that $\mathcal{H}$ is dense in $L^1(M, \mathrm{d}\mu)$.

To prove the last statement, we first remark that to obtain Equ. (8) it suffices to show that

$$\lim_{t\to\infty} (f, U^t f) = 0,$$

for all $f \in \{1\}^\perp$ (use the orthogonal decomposition $\mathcal{H} = \mathbb{C} \cdot 1 \oplus \{1\}^\perp$ and the polarization identity). If L has purely absolutely continuous spectrum on $\{1\}^\perp$ then its spectral measure associated to $f \in \{1\}^\perp$ can be written as $d\nu_f(\lambda) = g(\lambda)\,\mathrm{d}\lambda$ for some function $g \in L^1(\mathbb{R})$. Therefore,

$$(f, U^t f) = \int \mathrm{e}^{it\lambda} g(\lambda)\,\mathrm{d}\lambda \to 0,$$

as $t \to \infty$ by the Riemann-Lebesgue Lemma. □

Note that $(f, U^t f)$ is the Fourier transform of the spectral measure of L associated to f. It is clear from the above proof that a dynamical system is mixing if and only if the Fourier transform of the spectral measure of its Liouvillean associated to any vector in $\{1\}^\perp$ vanishes at infinity. See Subsection 2.5 in Lecture [25].

Exercise 3.16. On the n-dimensional torus $\mathbb{T}^n = \mathbb{R}^n/\mathbb{Z}^n$, consider the dynamics $\varphi_t(x) = x + t\omega \pmod 1$ where $\omega = (\omega_1, \cdots, \omega_n) \in \mathbb{R}^n$. Show that the (Haar) measure $d\mu(x) = dx_1 \cdots dx_n$ is invariant and compute the Liouvillean L of $(\mathbb{T}^n, \varphi_t, \mu)$. Determine the spectrum of L and discuss the ergodic properties of the system.

4 Quantum Systems

In the traditional description of quantum mechanics, a quantum system is completely determined by its Hilbert space $\mathcal{H}$ and its Hamiltonian H, a self-adjoint operator on $\mathcal{H}$. The Hilbert space $\mathcal{H}$ determines both, the set of observables and the set of states of the system. The Hamiltonian H specifies its dynamics.

An observable is a bounded linear operator on $\mathcal{H}$. A states is specified by a "wave function", a unit vector $\Psi \in \mathcal{H}$, or more generally by a "density matrix", a non-negative, trace class operator ρ on $\mathcal{H}$ with $\mathrm{Tr}\,\rho = 1$. The state associated with the density matrix ρ is the linear functional

$$A \mapsto \rho(A) \equiv \mathrm{Tr}(\rho A), \tag{10}$$

on the set of observables. As a state, a unit vector Ψ is equivalent to the density matrix $\rho = (\Psi, \cdot)\Psi$. Such density matrices are characterized by $\rho^2 = \rho$ and the corresponding states

$$A \mapsto (\Psi, A\Psi),$$

are called vector states.

The role played by quantum mechanical states is somewhat similar to the role played by probability distributions in classical dynamical systems. However, since there is no quantum mechanical phase space, there is nothing like a trajectory and no state corresponding to the Dirac measures δ_x of classical dynamical systems.

To any self-adjoint observable A, a state ρ associate a probability measure on the spectrum of A

$$\mathrm{d}\rho_A(a) = \rho(E_A(\mathrm{d}a)),$$

where $E_A(\cdot)$ is the projection valued spectral measure of A given by the spectral theorem. The measure ρ_A specifies the probability distribution for the outcome of a measure of the observable A. In particular, the expectation value of A in the state ρ is

$$\int a\,\mathrm{d}\rho_A(a) = \rho(A).$$

If two or more self-adjoint observables $A, B, \ldots$, commute, they have a joint spectral measure and a similar formula gives the joint probability distribution for the simultaneous measurement of $A, B, \ldots$ However, if A and B do not commute, it is not possible to measure them simultaneously and there is no joint probability distribution for them (see also Lecture [33] in this Volume and [13] or any textbook on quantum mechanics for more details).

By the spectral decomposition theorem for compact operators, a density matrix ρ can be written as

$$\rho = \sum_j p_j(\Psi_j, \cdot)\Psi_j, \tag{11}$$

where the Ψ_j are eigenvectors of ρ and form an orthonormal basis of $\mathcal{H}$ and the coefficients p_j are the corresponding eigenvalues which satisfy

$$0 \leq p_j \leq 1, \qquad \sum_j p_j = 1.$$

Thus, a general state is a convex linear combination of vector states. p_j is the probability for the system to be in the vector state associated to Ψ_j. In the physics literature, such a state is sometimes called statistical mixture (or incoherent superposition) of the states Ψ_j with the statistical weights p_j.

A wave function Ψ evolves according to the Schrödinger equation of motion

$$i\partial_t\Psi_t = H\Psi_t.$$

Since H is self-adjoint, the solution of this equation, with initial value $\Psi_0 = \Psi$, is given by

$$\Psi_t = \mathrm{e}^{-itH}\Psi.$$

According to the decomposition (11), a density matrix ρ evolves as

$$\rho_t = \mathrm{e}^{-itH}\rho\,\mathrm{e}^{itH},$$

and thus, satisfies the quantum Liouville equation

$$\partial_t\rho_t = i[\rho_t, H].$$

The expectation value of an observable A at time t is then

$$\rho_t(A) = \mathrm{Tr}(\rho_t A).$$

This is the so-called Schrödinger picture of quantum dynamics. Since the cyclicity of the trace implies that

$$\mathrm{Tr}(\rho_t A) = \mathrm{Tr}(\mathrm{e}^{-itH}\rho\,\mathrm{e}^{itH}A) = \mathrm{Tr}(\rho\,\mathrm{e}^{itH}A\,\mathrm{e}^{-itH}),$$

we can alternatively keep the state fixed and let observables evolve according to

$$A_t = \mathrm{e}^{itH}A\,\mathrm{e}^{-itH}.$$

Such time evolved observables satisfy the Heisenberg equation of motion

$$\partial_t A_t = i[H, A_t].$$

We obtain in this way the Heisenberg picture of quantum dynamics. Since

$$\rho_t(A) = \rho(A_t),$$

the Schrödinger and the Heisenberg pictures are obviously equivalent.

For systems with a finite number of degrees of freedom, this Hilbert space description of a quantum system is good enough because it is essentially unique (a precise form of this statement is the Stone–von Neumann uniqueness theorem, see Lecture [36]). However, for systems with an infinite number of degrees of freedom, (*i.e.*, quantum fields) this is no more the case. An intrinsic description, centered around the C^*-algebra of observables becomes more convenient.

4.1 C^*-Dynamical Systems

Definition 4.1. *A C^*-dynamical system is a pair $(\mathfrak{A}, \tau^t)$ where $\mathfrak{A}$ is a C^*-algebra with a unit and τ^t a strongly continuous group of $*$-automorphisms of $\mathfrak{A}$.*

Since $\tau^t((z - A)^{-1}) = (z - \tau^t(A))^{-1}$, a $*$-automorphism τ^t clearly preserves the spectrum. Hence, given the relation between the norm and the spectral radius (see the proof of Theorem 3 in Lecture [7]), it is isometric and in particular norm continuous. Strong continuity of τ^t means that, for any $A \in \mathfrak{A}$, the map $t \mapsto \tau^t(A)$ is continuous in the norm topology of $\mathfrak{A}$.

From the general theory of strongly continuous semi-groups, there exists a densely defined, norm closed linear operator δ on $\mathfrak{A}$ such that $\tau^t = \mathrm{e}^{t\delta}$. Since $\tau^t(1) = 1$, if follows that $1 \in \mathrm{D}(\delta)$ and $\delta(1) = 0$. Differentiation of the identities $\tau^t(AB) = \tau^t(A)\tau^t(B)$ and $\tau^t(A^*) = \tau^t(A)^*$ for $A, B \in \mathrm{D}(\delta)$ further show that the generator δ is a $*$-derivation as defined by the following

Definition 4.2. *Let $\mathfrak{A}$ be a $*$-algebra and $\mathfrak{D} \subset \mathfrak{A}$ a subspace. A linear operator $\delta : \mathfrak{D} \to \mathfrak{A}$ is called $*$-derivation if*

(i) $\mathfrak{D}$ *is a* $*$*-subalgebra of* $\mathfrak{A}$.
(ii) $\delta(AB) = \delta(A)B + A\delta(B)$ *for all* $A, B \in \mathfrak{D}$.
(iii) $\delta(A^*) = \delta(A)^*$ *for all* $A \in \mathfrak{D}$.

Generators of strongly continuous groups of $*$-automorphisms are characterized by the following simple adaptation of the Hille-Yosida Theorem (see [11], Theorem 3.2.50).

Proposition 4.3. *Let* $\mathfrak{A}$ *be a* C^**-algebra with a unit. A densely defined, closed operator* δ *on* $\mathfrak{A}$ *generates a strongly continuous group of* $*$*-automorphisms of* $\mathfrak{A}$ *if and only if:*

(i) δ *is a* $*$*-derivation.*
(ii) $\mathrm{Ran}(\mathrm{Id} + \lambda\delta) = \mathfrak{A}$ *for all* $\lambda \in \mathbb{R} \setminus \{0\}$.
(iii) $\|A + \lambda\delta(A)\| \geqslant \|A\|$ *for all* $\lambda \in \mathbb{R}$ *and* $A \in \mathrm{D}(\delta)$.

If the C^*-algebra $\mathfrak{A}$ acts on a Hilbert space $\mathcal{H}$ then a dynamical group τ^t can be constructed from a group of unitary operators U^t on $\mathcal{H}$

$$\tau^t(A) = U^t A U^{t*}.$$

Such $*$-automorphisms are called *spatial*. They are particularly pleasant since it is possible to lift most of their analysis to the Hilbert space $\mathcal{H}$ itself.

Example 4.4. (Finite quantum systems) Consider the quantum system with a finite number of degrees of freedom determined by the data $\mathcal{H}$ and H as described at the beginning of this section. On the C^*-algebra $\mathfrak{A} = \mathfrak{B}(\mathcal{H})$ the dynamics is given by

$$\tau^t(A) = \mathrm{e}^{itH} A \,\mathrm{e}^{-itH}.$$

Exercise 4.5. Show that the group τ^t is strongly continuous if and only if H is bounded.

Thus, if τ^t is strongly continuous it is automatically uniformly continuous and its generator is the bounded $*$-derivation

$$\delta(A) = i[H, A],$$

on $\mathfrak{B}(\mathcal{H})$. More specific examples are:

1. N-levels systems: $\mathcal{H} = \mathbb{C}^N$ and, without loss of generality, the Hamiltonian is a $N \times N$ diagonal matrix $H_{ij} = \epsilon_i \delta_{ij}$.
2. Lattice quantum systems: $\mathcal{H} = l^2(\mathbb{Z}^d)$ and the Hamiltonian is of tight-binding type

$$(H\psi)(x) = \frac{1}{2d} \sum_{|x-y|=1} \psi(y) + V(x)\psi(x),$$

where $|x|$ denotes the Euclidean norm of $x = (x_1, \cdots, x_d) \in \mathbb{Z}^d$ and $V \in l^\infty(\mathbb{Z}^d)$. See Example 4 in Subsection 5.2 of Lecture [6] for a continuation of this example.

Example 4.6. (The ideal Fermi gas) Let $\mathfrak{h}$ be a Hilbert space and $\Gamma_-(\mathfrak{h})$ the Fermionic Fock space over $\mathfrak{h}$. Recall that

$$\Gamma_-(\mathfrak{h}) = \bigoplus_{n\in\mathbb{N}} \Gamma_-^{(n)}(\mathfrak{h}),$$

where $\Gamma_-^{(0)}(\mathfrak{h}) = \mathbb{C}$ and $\Gamma_-^{(n)}(\mathfrak{h}) = \mathfrak{h}\wedge\mathfrak{h}\wedge\cdots\wedge\mathfrak{h}$ is the n-fold totally anti-symmetric tensor product of $\mathfrak{h}$. For $f \in \mathfrak{h}$, the action of the Fermionic *creation operator* $a^*(f)$ on a Slater determinant $f_1 \wedge f_2 \wedge \cdots \wedge f_n \in \Gamma_-^{(n)}(\mathfrak{h})$ is defined by

$$a^*(f) f_1 \wedge f_2 \wedge \cdots \wedge f_n = \sqrt{n+1}\, f_1 \wedge f_2 \wedge \cdots \wedge f_n \wedge f, \tag{12}$$

and is extended by linearity to the dense subspace

$$\Gamma_{\text{fin}\,-}(\mathfrak{h}) = \bigcup_{k\in\mathbb{N}} \bigoplus_{n=0}^{k} \Gamma_-^{(n)}(\mathfrak{h}),$$

of $\Gamma_-(\mathfrak{h})$. The *annihilation operators* are defined in a similar way starting from

$$a(f) f_1 \wedge f_2 \wedge \cdots \wedge f_n = \sqrt{n} \sum_{k=1}^{n} (f, f_k) f_1 \wedge \cdots f_{k-1} \wedge f_{k+1} \cdots \wedge f_n.$$

A simple calculation shows that these operators satisfy $a^*(f)\psi = a(f)^*\psi$ as well as the *Canonical Anti-commutation Relations* (CAR)

$$\{a(f), a^*(g)\}\psi = (f, g)\psi,$$
$$\{a^*(f), a^*(g)\}\psi = 0,$$

for all $f, g \in \mathfrak{h}$ and $\psi \in \Gamma_{\text{fin}\,-}(\mathfrak{h})$. It immediately follows that

$$\|a(f)\psi\|^2 + \|a^*(f)\psi\|^2 = \|f\|^2\|\psi\|^2,$$

from which we conclude that the closures of $a(f)$ and $a^*(f)$ are bounded operators on $\Gamma_-(\mathfrak{h})$. If we denote these extensions by the same symbols then $a^*(f) = a(f)^*$ and the canonical anti-commutation relations hold for all $\psi \in \Gamma_-(\mathfrak{h})$.

Exercise 4.7. Prove that $\|a(f)\| = \|a^*(f)\| = \|f\|$ (Hint: Use the CAR to compute $(a^*(f)a(f))^2$ and the C^*-property of the norm). Compute the spectrum of $a(f)$, $a^*(f)$ and $a^*(f)a(f)$.

Let $\mathfrak{u}$ be a subspace of $\mathfrak{h}$, not necessarily closed, and denote by $\mathrm{CAR}(\mathfrak{u})$ the C^*-algebra generated by the family $\{a(f) \,|\, f \in \mathfrak{u}\}$, *i.e.*, the norm closure in $\mathfrak{B}(\Gamma_-(\mathfrak{h}))$ of the linear span of monomials $a^*(f_1)\cdots a^*(f_n)a(f_{n+1})\cdots a(f_m)$ with $f_j \in \mathfrak{u}$. It follows from the CAR that this algebra has a unit. To any self-adjoint operator h on $\mathfrak{h}$ we can associate the second quantization $U^t = \Gamma(\mathrm{e}^{iht})$ on the Fock space. By definition

$$U^t f_1 \wedge \cdots \wedge f_n \equiv \mathrm{e}^{ith} f_1 \wedge \cdots \wedge \mathrm{e}^{ith} f_n,$$

and therefore U^t is a strongly continuous unitary group on $\Gamma_-(\mathfrak{h})$. Denote by τ^t the corresponding group of spatial $*$- automorphisms of $\mathfrak{B}(\Gamma_-(\mathfrak{h}))$.

Exercise 4.8. Show that the generator $H = \mathrm{d}\Gamma(h)$ of U^t is bounded if and only if h is trace class.

Hence, by Exercise 4.5, if h is not trace class then τ^t is not strongly continuous on $\mathfrak{B}(\Gamma_-(\mathfrak{h}))$. Let me show that its restriction to $\mathrm{CAR}(\mathfrak{u})$ is strongly continuous.

It follows from Equ. (12) that

$$\tau^t(a^*(f)) = U^t a^*(f) U^{t*} = a^*(\mathrm{e}^{ith} f).$$

Therefore, for any monomial $A = a^\#(f_1)\cdots a^\#(f_m)$, where $a^\#$ stands for either a or a^*, we get a telescopic expansion

$$\begin{aligned}\tau^t(A) - A \\ = \sum_{k=1}^{m} a^\#(\mathrm{e}^{ith} f_1)\cdots a^\#(\mathrm{e}^{ith} f_{k-1}) a^\#(\mathrm{e}^{ith} f_k - f_k) a^\#(f_{k+1})\cdots a^\#(f_m),\end{aligned}$$

which, together with Exercise 4.7, leads to the estimate

$$\|\tau^t(A) - A\| \leqslant m \left(\max_{1\leqslant k\leqslant m} \|f_k\|\right)^{m-1} \max_{1\leqslant k\leqslant m} \|\mathrm{e}^{ith} f_k - f_k\|.$$

We conclude that $\lim_{t\to 0}\|\tau^t(A) - A\| = 0$ for all such monomials and hence for arbitrary polynomials. Since these polynomials are norm dense in $\mathrm{CAR}(\mathfrak{u})$ and τ^t is isometric, the result follows from an $\varepsilon/3$-argument.

The simplest non-trivial example of strongly continuous group of $*$-automorphisms of $\mathrm{CAR}(\mathfrak{u})$ is the gauge group ϑ^t obtained by setting $h = I$, *i.e.*,

$$\vartheta^t(a^\#(f)) = a^\#(\mathrm{e}^{it} f). \tag{13}$$

The corresponding operator $N \equiv \mathrm{d}\Gamma(I)$ is the *number operator.*

The one particle Hilbert space for an infinite d-dimensional Fermi gas is $\mathfrak{h} = L^2(\mathbb{R}^d, \mathrm{d}x)$. The non-relativistic Hamiltonian is $h = -\Delta$. To any compact region $\Lambda \Subset \mathbb{R}^d$ one can associate the Hilbert space $\mathfrak{h}_\Lambda = L^2(\Lambda, \mathrm{d}x)$ which is canonically embedded in $\mathfrak{h}$. Accordingly, the corresponding C^*-algebra $\mathrm{CAR}(\mathfrak{h}_\Lambda)$ can be identified with a subalgebra of $\mathrm{CAR}(\mathfrak{h})$. Elements of

$$\mathfrak{A}_{\mathrm{loc}} = \bigcup_{\Lambda \Subset \mathbb{R}^d} \mathrm{CAR}(\mathfrak{h}_\Lambda),$$

are called local and we say that $A \in \mathrm{CAR}(\mathfrak{h}_\Lambda)$ is supported by Λ.

Exercise 4.9. Show that $\mathfrak{A}_{\mathrm{loc}}$ is a dense $*$-subalgebra of $\mathrm{CAR}(\mathfrak{h})$.

From a physical point of view, local observables supported by disjoint subsets should be simultaneously measurable, *i.e.*, they should commute. However, it follows from the fact that

$$\{AB, C'\} = A\{B, C'\} - \{A, C'\}B,$$

that $\mathrm{CAR}(\mathfrak{h}_\Lambda)$ and $\mathrm{CAR}(\mathfrak{h}_{\Lambda'})$ anti-commute if $\Lambda \cap \Lambda' = \varnothing$ and will generally not commute. Thus the full CAR algebra is in a sense too big. To obtain an algebra fulfilling the above locality requirement, one introduces the so called even subalgebra $\mathrm{CAR}^+(\mathfrak{h})$ generated by monomials of even degrees in a and a^*. It follows from

$$[AB, C'D'] = \{A, C'\}D'B + A\{B, C'\}D' - C'\{D', A\}B - AC'\{B, D'\},$$

that two elements of this subalgebra supported by disjoint subset commute.

Another way to characterize the even subalgebra is to consider the $*$-morphism of $\mathrm{CAR}(\mathfrak{h})$ defined by $\theta(a(f)) = -a(f)$. Then one clearly has

$$\mathrm{CAR}^+(\mathfrak{h}) = \{A \in \mathrm{CAR}(\mathfrak{h}) | \theta(A) = A\}.$$

Since τ^t commutes with θ, it leaves $\mathrm{CAR}^+(\mathfrak{h})$ invariant and $(\mathrm{CAR}^+(\mathfrak{h}), \tau^t)$ is a C^*-dynamical system which describes an ideal Fermi gas.

When the C^*-algebra $\mathfrak{A}$ does not act naturally on a Hilbert space, the situation is more involved and Banach space techniques must be used. I will only mention a simple example based on the powerful technique of analytic vectors (see Chapter 3.1 in [11], Section 5 in [32] and Chapter 6.2 in [12] for a more systematic exposition of these techniques).

Definition 4.10. *Let T be an operator on a Banach space $\mathcal{B}$. A vector $x \in \mathcal{B}$ is called analytic for T if $x \in \cap_{n\in\mathbb{N}} \mathrm{D}(T^n)$ and if there exists $\rho_x > 0$ such that the power series*

$$\sum_{n=0}^{\infty} \frac{z^n}{n!} \|T^n x\|$$

defines an analytic function in $\{z \in \mathbb{C} \,|\, |z| < \rho_x\}$.

Note that the set of analytic vectors of an operator T is a subspace. If x is analytic for T one can define

$$\mathrm{e}^{tT} x \equiv \sum_{n=0}^{\infty} \frac{t^n}{n!} T^n x,$$

for $|t| < \rho_x$. Moreover, if T is closed and $t < \rho_x$ then it is easy to show that $\mathrm{e}^{tT} x$ is analytic for T and a simple manipulation of norm convergent series yields that $\mathrm{e}^{sT} \mathrm{e}^{tT} x = \mathrm{e}^{(s+t)T} x$ as long as $|s| + |t| < \rho_x$. If the subspace of analytic vectors is dense in $\mathcal{B}$ and if one can prove that the linear operator e^{tT} defined in this way is bounded (this is usually done with the help of some dissipativity estimate), then it extends to all of $\mathcal{B}$ and using the group property its definition can be extended to all $t \in \mathbb{R}$.

Example 4.11. (Quantum spin systems) Let Γ be an infinite lattice (for example $\Gamma = \mathbb{Z}^d$, with $d \geqslant 1$) and to each $x \in \Gamma$ associate a copy $\mathfrak{h}_x$ of a finite dimensional Hilbert space $\mathfrak{h}$. For finite subsets $\Lambda \subset \Gamma$ set $\mathfrak{h}_\Lambda \equiv \otimes_{x\in\Lambda} \mathfrak{h}_x$ and define the local C^*-algebra

$$\mathfrak{A}_\Lambda \equiv \mathfrak{B}(\mathfrak{h}_\Lambda).$$

If $\Lambda \subset \Lambda'$, the natural injection $A \mapsto A \otimes I_{\mathfrak{h}_{\Lambda' \setminus \Lambda}}$ allows to identify $\mathfrak{A}_\Lambda$ with a subalgebra of $\mathfrak{A}_{\Lambda'}$. Therefore, a C^*-norm can be defined on $\mathfrak{A}_{\text{loc}} \equiv \cup_{\Lambda \subset \Gamma} \mathfrak{A}_\Lambda$, the union being over finite subsets of Γ. Denote by $\mathfrak{A}$ the C^*-algebra obtained as norm closure of $\mathfrak{A}_{\text{loc}}$. Each local algebra $\mathfrak{A}_\Lambda$ is identified with the corresponding subalgebra of $\mathfrak{A}$. C^*-algebras of this type are called uniformly hyperfinite (UHF). Interpreting $\mathfrak{h}$ as the Hilbert space of a single spin, $\mathfrak{A}$ describes the observables of a quantum spin system on the lattice Γ.

An interaction is a map $X \mapsto \phi(X)$ which, to any finite subset X of the lattice Γ, associates a self-adjoint element $\phi(X)$ of $\mathfrak{A}_X$ describing the interaction energy of the degrees of freedom inside X. For example, in the spin interpretation, if $X = \{x\}$ then $\phi(X)$ is the energy of the spin at x due to the coupling of its magnetic moment with an external magnetic field. If $X = \{x, y\}$ then $\phi(X)$ is the coupling energy due to the pair interaction between the corresponding magnetic moments.

Given an interaction ϕ, the local Hamiltonian for a finite region $\Lambda \subset \Gamma$ is the self-adjoint element of $\mathfrak{A}_\Lambda$ given by

$$H_\Lambda \equiv \sum_{X \subset \Lambda} \phi(X),$$

and a C^*-dynamical system is defined on $\mathfrak{A}$ by

$$\tau_\Lambda^t(A) \equiv \mathrm{e}^{itH_\Lambda} A \,\mathrm{e}^{-itH_\Lambda}.$$

Assume that the interaction ϕ has sufficiently short range, more precisely that

$$\|\phi\|_\sigma \equiv \sup_{x \in \Gamma} \sum_{X \ni x} \|\phi(X)\| \mathrm{e}^{2\sigma|X|} < \infty,$$

for some $\sigma > 0$, where $|X|$ denotes the cardinality of the subset X. We shall show that the limit

$$\tau^t(A) \equiv \lim_{\Lambda \uparrow \Gamma} \tau_\Lambda^t(A), \tag{14}$$

exists for all $A \in \mathfrak{A}$, $t \in \mathbb{R}$ and defines a strongly continuous group of $*$-morphisms.

The generator of τ_Λ^t is the bounded derivation

$$\delta_\Lambda(A) = i[H_\Lambda, A],$$

hence we have a norm convergent expansion

$$\tau_\Lambda^t(A) = \sum_{n=0}^\infty \frac{t^n}{n!} \delta_\Lambda^n(A). \tag{15}$$

For $A \in \mathfrak{A}_{\Lambda_0}$, we can further write

$$\delta_\Lambda^n(A) = \sum_{X_1, \cdots, X_n \subset \Lambda} i[\phi(X_n), i[\phi(X_{n-1}), \cdots, i[\phi(X_1), A] \cdots]].$$

Since local algebras corresponding to disjoint subsets of Γ commute, this sum can be restricted by the condition $X_j \cap \Lambda_{j-1} \neq \varnothing$ with $\Lambda_{j-1} = \Lambda_0 \cup X_1 \cdots \cup X_{j-1}$. We proceed to estimate the norm of $\delta_\Lambda^n(A)$

$$\begin{aligned}
\|\delta_\Lambda^n(A)\| &\leqslant \sum_{x_1 \in \Lambda_0} \sum_{X_1 \ni x_1} \sum_{x_2 \in \Lambda_1} \sum_{X_2 \ni x_2} \cdots \sum_{x_n \in \Lambda_{n-1}} \sum_{X_n \ni x_n} 2^n \|A\| \prod_{i=1}^n \|\phi(X_i)\| \\
&\leqslant 2^n \|A\| \, |\Lambda_0| \sup_{x_1} \sum_{X_1 \ni x_1} |\Lambda_1| \sup_{x_2} \sum_{X_2 \ni x_2} \cdots |\Lambda_{n-1}| \sup_{x_n} \sum_{X_n \ni x_n} \prod_{i=1}^n \|\phi(X_i)\| \\
&\leqslant 2^n \|A\| \sup_{x_1, \cdots, x_n} \sum_{X_1 \ni x_1} \cdots \sum_{X_n \ni x_n} (|\Lambda_0| + |X_1| + \cdots + |X_n|)^n \prod_{i=1}^n \|\phi(X_i)\|.
\end{aligned}$$

Using the inequality $\mathrm{e}^{2\sigma x} \geqslant (2\sigma x)^n / n!$ with $x = |\Lambda_0| + |X_1| + \cdots + |X_n|$ we finally get the following uniform estimate in Λ

$$\|\delta_\Lambda^n(A)\| \leq \frac{n!}{\sigma^n} \mathrm{e}^{2\sigma|\Lambda_0|} \|A\| \|\phi\|_\sigma^n. \tag{16}$$

From this we conclude that

$$\delta^{(n)}(A) = \lim_\Lambda \delta_\Lambda^n(A) = \sum_{X_1, \cdots, X_n} i[\phi(X_n), i[\phi(X_{n-1}), \cdots, i[\phi(X_1), A] \cdots]],$$

exists and satisfies the same estimate (16). It is also clear from this argument that

$$\lim_\Lambda \delta^{(n)}(\delta_\Lambda^k(A)) = \delta^{(n+k)}(A). \tag{17}$$

Introducing (16) into the expansion (15), we conclude that the limit

$$\tau^t(A) = \lim_\Lambda \tau_\Lambda^t(A) = \sum_{n=0}^\infty \frac{t^n}{n!} \delta^{(n)}(A), \tag{18}$$

exists for $|t| < \sigma / \|\phi\|_\sigma$.

Since τ_Λ^t is isometric from any local algebra $\mathfrak{A}_{\Lambda_0}$ into $\mathfrak{A}$, it follows that the limit τ^t is norm continuous and extends by continuity to all of $\mathfrak{A}$. Furthermore, as a norm limit of $*$-morphisms, τ^t is a $*$-morphism.

For $|s| + |t| < \sigma / \|\phi\|_\sigma$ and $A \in \mathfrak{A}_{\mathrm{loc}}$ the continuity of τ^t yields

$$\lim_\Lambda \tau^t(\tau_\Lambda^s(A)) = \tau^t(\tau^s(A)),$$

while Equ. (17) and the expansions (15)(18) lead, after a simple manipulation to

$$\lim_\Lambda \tau^t(\tau_\Lambda^s(A)) = \tau^{t+s}(A).$$

Thus τ^t satisfies the local group property, and in particular is a $*$-isomorphism. Finally τ^t can be extended to a group by setting $\tau^{nt_0+t} = (\tau^{t_0})^n \circ \tau^t$ and relation (14) then extends to all $t \in \mathbb{R}$ and $A \in \mathfrak{A}$.

Example 4.12. (Continuous classical dynamical system) Let M be a compact metric space and φ_t a continuous dynamics on M. Then the space $C(M)$ of continuous functions on M is a commutative C^*-algebra with a unit on which the map $\tau^t(f) = f \circ \varphi_t$ defines a strongly continuous group of $*$-automorphisms.

4.2 W^*-Dynamical Systems

We have seen in Example 4.4 that even the simplest quantum mechanical system leads to a dynamics which is not strongly continuous when its Hamiltonian is unbounded. Thus, the notion of C^*-dynamical system is too restrictive for our purposes and we need to consider weaker topologies on the algebra of observables.

Let $\mathfrak{M} \subset \mathfrak{B}(\mathcal{H})$ be a von Neumann algebra. A group of $*$-automorphisms of $\mathfrak{M}$ is σ-weakly continuous if for all $A \in \mathfrak{M}$ the map $t \mapsto \tau^t(A)$ is continuous in the σ-weak topology. This means that for any $A \in \mathfrak{M}$ and any trace class operator T on $\mathcal{H}$ the map $t \mapsto \mathrm{Tr}(T\tau^t(A))$ is continuous.

Definition 4.13. *A W^*-dynamical system is a pair $(\mathfrak{M}, \tau^t)$ where $\mathfrak{M}$ is a von Neumann algebra acting on a Hilbert space $\mathcal{H}$ and τ^t is a σ-weakly continuous group of $*$-automorphisms of $\mathfrak{M}$.*

Remark 4.14. More generally, $\mathfrak{M}$ could be an abstract W^*-algebra, *i.e.*, a C^*-algebra which is the dual Banach space of a Banach space $\mathfrak{M}_\star$, and τ^t a weak-$\star$ continuous group of $*$-automorphisms. Since by Sakai theorem (see [42], Theorem 1.16.7) a W^*-algebra is $*$-isomorphic to a von Neumann subalgebra of $\mathfrak{B}(\mathcal{H})$ for some Hilbert space $\mathcal{H}$, I will only consider this particular situation. The predual $\mathfrak{M}_\star$ is then canonically identified with the quotient $\mathcal{L}^1(\mathcal{H})/\mathfrak{M}^\perp$ where $\mathcal{L}^1(\mathcal{H})$ is the Banach space of trace class operators on $\mathcal{H}$ and

$$\mathfrak{M}^\perp = \{T \in \mathcal{L}^1(\mathcal{H}) \mid \mathrm{Tr}(TA) = 0 \text{ for all } A \in \mathfrak{M}\},$$

the annihilator of $\mathfrak{M}$. The weak-$\star$ topology on $\mathfrak{M}$ induced by $\mathfrak{M}_\star$ coincides with the σ-weak topology and elements of the predual corresponds to σ-weakly continuous linear functionals on $\mathfrak{M}$.

We have seen in Subsection 2.1 (Corollary 2.12) that a $*$-automorphism of a von Neumann algebra is σ-weakly continuous. A σ-weakly continuous group of σ-weakly continuous linear operators is characterized by its generator, as in the strongly continuous case (see Chapter 3 in [11] for details). Thus, one can write $\tau^t = \mathrm{e}^{t\delta}$ and there is a characterization, parallel to Proposition 4.3, of W^*-dynamical systems (see Theorem 3.2.51 in [11]).

Proposition 4.15. *Let $\mathfrak{M}$ be a von Neumann algebra. A σ-weakly densely defined and closed linear operator δ on $\mathfrak{M}$ generates a σ-weakly continuous group of $*$-automorphisms of $\mathfrak{M}$ if and only if:*

(i) δ is a $*$-derivation and $1 \in \mathrm{D}(\delta)$.
(ii) $\mathrm{Ran}(\mathrm{Id} + \lambda\delta) = \mathfrak{M}$ for all $\lambda \in \mathbb{R} \setminus \{0\}$.

(iii) $\|A + \lambda\delta(A)\| \geqslant \|A\|$ *for all* $\lambda \in \mathbb{R}$ *and* $A \in \mathrm{D}(\delta)$.

Example 4.16. (Finite quantum systems, continuation of Example 4.4) Consider now the case of an unbounded Hamiltonian H on the Hilbert space $\mathcal{H}$. Clearly $\tau^t(A) = \mathrm{e}^{itH} A\, \mathrm{e}^{-itH}$ defines a group of $*$-automorphisms of the von Neumann algebra $\mathfrak{M} = \mathfrak{B}(\mathcal{H})$. For any unit vectors $\Phi, \Psi \in \mathcal{H}$ the function

$$t \mapsto (\Phi, \tau^t(A)\Psi) = (\mathrm{e}^{-itH}\Phi, A\mathrm{e}^{-itH}\Psi)$$

is continuous and uniformly bounded in t by $\|A\|$. If T is a trace class operator on $\mathcal{H}$ then one has $T = \sum_n t_n(\Phi_n, \cdot\,)\Psi_n$ with $\|\Psi_n\| = \|\Phi_n\| = 1$ and $\sum_n |t_n| < \infty$. It follows that

$$t \mapsto \mathrm{Tr}(T\tau^t(A)) = \sum_n t_n(\Phi_n, \tau^t(A)\Psi_n),$$

is continuous (as a uniformly convergent series of continuous functions). Thus τ^t is σ-weakly continuous and $(\mathfrak{M}, \tau^t)$ is a W^*-dynamical system.

Example 4.17. (Ideal Bose gas) Let $\mathfrak{h}$ be a Hilbert space and $\Gamma_+(\mathfrak{h})$ the Bosonic Fock space over $\mathfrak{h}$ (see Section 2 in Lecture [36]). For $f \in \mathfrak{h}$, denote by

$$\Phi(f) = \frac{1}{\sqrt{2}}(a^*(f) + a(f)),$$

the self-adjoint Segal field operator. Since $\Phi(f)$ is unbounded for $f \neq 0$, it is more convenient to use the unitary Weyl operators $W(f) = \mathrm{e}^{i\Phi(f)}$. They define a projective representation of the additive group $\mathfrak{h}$ on $\Gamma_+(\mathfrak{h})$ satisfying the *Weyl relations*

$$W(f)W(g) = \mathrm{e}^{-i\,\mathrm{Im}(f,g)/2}W(f+g). \tag{19}$$

To any self-adjoint operator h on $\mathfrak{h}$ we can, as in the Fermionic case, associate the second quantized strongly continuous unitary group $U^t = \Gamma(\mathrm{e}^{ith})$ on $\Gamma_+(\mathfrak{h})$. The action of this group on Weyl operators is given by

$$U^t W(f) U^{t*} = W(\mathrm{e}^{ith} f). \tag{20}$$

It follows from Equ. (19) that, for $f \neq 0$ and $g = i\theta f/\|f\|^2$,

$$W(g)^* W(f) W(g) = \mathrm{e}^{-i\theta} W(f).$$

This shows that the spectrum of $W(f)$ is the full unit circle and therefore

$$\|W(f) - W(g)\| = \|W(f-g) - \mathrm{e}^{-i\,\mathrm{Im}(f,g)/2}\| = 2,$$

for $f \neq g$. This makes clear that in the Bosonic framework there is no chance to obtain a strongly continuous group from Equ. (20). On the other hand it follows from Exercise 4.16 that the von Neumann algebra

$$\mathfrak{W}(\mathfrak{h}) = \{W(f) \,|\, f \in \mathfrak{h}\}'',$$

together with the group

$$\tau^t(A) = U^t A U^{t*}, \tag{21}$$

form a W^*-dynamical system.

Exercise 4.18. Show that the system $\{W(f) \mid f \in \mathfrak{h}\}$ is irreducible, *i.e.*, that

$$\{W(f) \mid f \in \mathfrak{h}\}' = \mathbb{C}I.$$

Conclude that $\mathfrak{W}(\mathfrak{h}) = \mathfrak{B}(\Gamma_+(\mathfrak{h}))$. (Hint: see Subsection 2.4 in Lecture [36])

4.3 Invariant States

Definition 4.19. *If τ^t is a group of $*$-automorphisms on a C^*-algebra $\mathfrak{A}$, a state μ on $\mathfrak{A}$ is called τ^t-invariant if $\mu \circ \tau^t = \mu$ for all $t \in \mathbb{R}$. We denote by $E(\mathfrak{A}, \tau^t) \subset E(\mathfrak{A})$ the set of τ^t-invariant states.*

As in the case of classical dynamical systems, invariant states play a important role in the study of quantum dynamics.

Theorem 4.20. *Let τ^t be a group of $*$-automorphisms of the C^*-algebra $\mathfrak{A}$. If there exists a state ω on $\mathfrak{A}$ such that the function $t \mapsto \omega(\tau^t(A))$ is continuous for all $A \in \mathfrak{A}$ then $E(\mathfrak{A}, \tau^t)$ is a non-empty, convex and weak-$\star$ compact subset of $\mathfrak{A}^*$. In particular, these conclusions hold if $(\mathfrak{A}, \tau^t)$ is a C^*- or W^*-dynamical system.*

Proof. To show that $E(\mathfrak{A}, \tau^t)$ is not empty, we follow the strategy of Exercise 3.1. For all $A \in \mathfrak{A}$ consider the expression

$$\omega_T(A) \equiv \frac{1}{T} \int_0^T \omega \circ \tau^s(A)\, \mathrm{d}s.$$

Since the function $s \mapsto \omega \circ \tau^s(A)$ is continuous, the integral is well defined and we clearly have $\omega_T \in E(\mathfrak{A})$ for all $T > 0$. Since $E(\mathfrak{A})$ is weak-$\star$ compact, the net $(\omega_T)_{T>0}$ has a weak-$\star$ convergent subnet. The formula

$$\omega_T(\tau^t(A)) = \omega_T(A) - \frac{1}{T} \int_0^t \omega \circ \tau^t(A)\, \mathrm{d}s + \frac{1}{T} \int_T^{T+t} \omega \circ \tau^s(A)\, \mathrm{d}s,$$

leads to the estimate

$$|\omega_T(\tau^t(A)) - \omega_T(A)| \leqslant 2\|A\| \frac{|t|}{T},$$

from which it follows that the limit of any convergent subnet of $(\omega_T)_{T>0}$ is τ^t-invariant.

It is clear that the set of τ^t-invariant states is convex and weak-$\star$ closed. □

Definition 4.21. *If τ^t is a group of $*$-automorphisms of the von Neumann algebra $\mathfrak{M}$ we denote by $N(\mathfrak{M}, \tau^t) \equiv E(\mathfrak{M}, \tau^t) \cap N(\mathfrak{M})$ the set of normal τ^t-invariant states.*

It immediately follows from the last paragraph in Subsection 2.1 that if $\omega \in N(\mathfrak{M}, \tau^t)$ then $\tau^t(s_\omega) = s_\omega$ and $\tau^t(z_\omega) = z_\omega$.

We note that for a W^*-dynamical system $(\mathfrak{M}, \tau^t)$ the compactness argument used in the proof of Theorem 4.20 breaks down if we replace $E(\mathfrak{M})$ by $N(\mathfrak{M})$. There is no general existence result for *normal invariant states* of W^*-dynamical systems. In fact, there exists W^*-dynamical systems without normal invariant states (see Exercise 4.34). However, we shall see below that any invariant state of a C^*- or W^*-dynamical system can be described as a normal invariant state of some associated W^*-dynamical system. For this reason, normal invariant states play an important role in quantum dynamics.

4.4 Quantum Dynamical Systems

Definition 4.22. *If $\mathfrak{C}$ is a C^*-algebra and τ^t a group of $*$-automorphisms of $\mathfrak{C}$ we define*

$$\mathcal{E}(\mathfrak{C},\tau^t) \equiv \{\mu \in E(\mathfrak{C},\tau^t) \,|\, t \mapsto \mu(A^*\tau^t(A)) \text{ is continuous for all } A \in \mathfrak{C}\}.$$

If $\mu \in \mathcal{E}(\mathfrak{C},\tau)$ we say that $(\mathfrak{C},\tau^t,\mu)$ is a quantum dynamical system.

Example 4.23. If $(\mathfrak{A},\tau^t)$ is a C^*-dynamical system then $\mathcal{E}(\mathfrak{A},\tau) = E(\mathfrak{A},\tau)$ and $(\mathfrak{A},\tau^t,\mu)$ is a quantum dynamical system for any τ^t-invariant state μ.

Example 4.24. If $(\mathfrak{M},\tau^t)$ is a W^*-dynamical system then $N(\mathfrak{A},\tau) \subset \mathcal{E}(\mathfrak{A},\tau)$ and $(\mathfrak{M},\tau^t,\mu)$ is a quantum dynamical system for any τ^t-invariant normal state μ.

Exercise 4.25. Show that $\mathcal{E}(\mathfrak{C},\tau^t)$ is a convex, norm closed subset of $E(\mathfrak{C},\tau^t)$.

In this subsection we shall study the GNS representation of a quantum dynamical system. We start with the following extension of the GNS construction.

Lemma 4.26. *Let $(\mathfrak{C},\tau^t,\mu)$ be a quantum dynamical system and denote the GNS representation of $\mathfrak{C}$ associated to μ by $(\mathcal{H}_\mu,\pi_\mu,\Omega_\mu)$. Then there exists a unique self-adjoint operator L_μ on $\mathcal{H}_\mu$ such that*

(i) $\pi_\mu(\tau^t(A)) = \mathrm{e}^{itL_\mu}\pi_\mu(A)\mathrm{e}^{-itL_\mu}$ for all $A \in \mathfrak{C}$ and $t \in \mathbb{R}$.
(ii) $L_\mu\Omega_\mu = 0$.

Proof. For fixed $t \in \mathbb{R}$ one easily checks that $(\mathcal{H}_\mu,\pi_\mu\circ\tau^t,\Omega_\mu)$ is a GNS representation of $\mathfrak{C}$ associated to μ. By unicity of the GNS construction there exists a unitary operator U_μ^t on $\mathcal{H}_\mu$ such that, for any $A \in \mathfrak{C}$, one has

$$U_\mu^t\pi_\mu(A)\Omega_\mu = \pi_\mu(\tau^t(A))\Omega_\mu, \tag{22}$$

and in particular

$$U_\mu^t\Omega_\mu = \Omega_\mu. \tag{23}$$

For $t,s \in \mathbb{R}$ we have

$$U_\mu^tU_\mu^s\pi_\mu(A)\Omega_\mu = U_\mu^t\pi_\mu(\tau^s(A))\Omega_\mu = \pi_\mu(\tau^{t+s}(A))\Omega_\mu = U_\mu^{t+s}\pi_\mu(A)\Omega_\mu,$$

and the cyclic property of Ω_μ yields that U_μ^t is a unitary group on $\mathcal{H}_\mu$. Using Equ. (7) it follows from the continuity of the map

$$t \mapsto (\pi_\mu(A)\Omega_\mu, U^t_\mu \pi_\mu(A)\Omega_\mu) = \mu(A^*\tau^t(A)),$$

that U^t_μ is a strongly continuous. By Stone theorem $U^t_\mu = \mathrm{e}^{itL_\mu}$ for some self-adjoint operator L_μ and property (ii) follows from Equ. (23).

Finally, for $A, B \in \mathfrak{C}$ we get

$$U^t_\mu \pi_\mu(A)\pi_\mu(B)\Omega_\mu = \pi_\mu(\tau^t(A))\pi_\mu(\tau^t(B))\Omega_\mu = \pi_\mu(\tau^t(A))U^t_\mu \pi_\mu(B)\Omega_\mu,$$

and property (i) follows from the cyclic property of Ω_μ.

To prove the uniqueness of L_μ note that Equ. (22) uniquely determines U^t_μ and that conditions (i) and (ii) imply that e^{itL_μ} satisfies (22). □

Recall from subsection 2.2 that $\hat{\mu}(A) = (\Omega_\mu, A\Omega_\mu)$ defines a centrally faithful normal state on the enveloping von Neumann algebra $\mathfrak{C}_\mu = \pi_\mu(\mathfrak{C})''$. Moreover, by property (i) of Lemma 4.26, the σ-weakly continuous group of $*$-automorphisms of $\mathfrak{B}(\mathcal{H}_\mu)$ defined by

$$\hat{\tau}^t_\mu(A) \equiv \mathrm{e}^{itL_\mu} A \mathrm{e}^{-itL_\mu}, \tag{24}$$

leaves $\pi_\mu(\mathfrak{C})$ and hence its σ-weak closure $\mathfrak{C}_\mu$ invariant. Thus, $(\mathfrak{C}_\mu, \hat{\tau}^t_\mu)$ is a W^*-dynamical system. By property (ii) of Lemma 4.26, $\hat{\mu}$ is $\hat{\tau}^t_\mu$-invariant.

Definition 4.27. *A quantum dynamical system $(\mathfrak{C}, \tau^t, \mu)$ is in normal form if*

(i) $\mathfrak{C}$ is a von Neumann algebra on a Hilbert space $\mathcal{H}$.
(ii) $\tau^t(A) = \mathrm{e}^{itL} A \mathrm{e}^{-itL}$ for some self-adjoint operator L on $\mathcal{H}$.
(iii) $\mu(A) = (\Omega, A\Omega)$ for some unit vector $\Omega \in \mathcal{H}$.
(iv) $\overline{\mathfrak{C}\Omega} = \mathcal{H}$.
(v) $L\Omega = 0$.

We denote by $(\mathfrak{C}, \mathcal{H}, L, \Omega)$ such a system.

The above considerations show that to any quantum dynamical system $(\mathfrak{C}, \tau^t, \mu)$ we can associate a quantum dynamical system in normal form $(\mathfrak{C}_\mu, \mathcal{H}_\mu, L_\mu, \Omega_\mu)$.

Definition 4.28. *$(\pi_\mu, \mathfrak{C}_\mu, \mathcal{H}_\mu, L_\mu, \Omega_\mu)$ is the normal form of the quantum dynamical system $(\mathfrak{C}, \tau^t, \mu)$. The operator L_μ is its μ-Liouvillean.*

The normal form of a quantum dynamical system is uniquely determined, up to unitary equivalence.

Definition 4.29. *Two quantum dynamical systems $(\mathfrak{C}, \tau^t, \mu)$, $(\mathfrak{D}, \sigma^t, \nu)$ are isomorphic if there exists a $*$-isomorphism $\phi : \mathfrak{C} \to \mathfrak{D}$ such that $\phi \circ \tau^t = \sigma^t \circ \phi$ for all $t \in \mathbb{R}$ and $\mu = \nu \circ \phi$.*

Exercise 4.30. Show that two isomorphic quantum dynamical systems share the same normal forms.

Remark 4.31. If $\omega \in \mathcal{E}(\mathfrak{C}, \tau^t)$ and $\mu \in N(\mathfrak{C}, \omega) \cap E(\mathfrak{C}, \tau^t)$ then $\mu = \tilde{\mu} \circ \pi_\omega$ for some $\tilde{\mu} \in N(\mathfrak{C}_\omega)$ and therefore

$$\mu(A^* \tau^t(A)) = \tilde{\mu}(\pi_\omega(A)^* \pi_\omega(\tau^t(A))) = \tilde{\mu}(\pi_\omega(A)^* \mathrm{e}^{itL_\omega} \pi_\omega(A) \mathrm{e}^{-itL_\omega}),$$

is a continuous function of $t \in \mathbb{R}$. It follows that $\mu \in \mathcal{E}(\mathfrak{C}, \tau^t)$. Denote by $\hat{\tau}_\omega^t$ the W^*-dynamics on $\mathfrak{C}_\omega$ generated by the ω-Liouvillean L_ω. Since μ is τ^t-invariant one has $\tilde{\mu}(\hat{\tau}_\omega^t(A)) = \tilde{\mu}(A)$ for all $A \in \pi_\omega(\mathfrak{C})$ and by continuity $\tilde{\mu}$ is $\hat{\tau}_\omega^t$-invariant. Let $\pi_{\mu|\omega} : \mathfrak{C}_\omega \to \mathfrak{C}_\mu$ be the $*$-morphism of Theorem 2.30. From the identity

$$\begin{aligned} \hat{\tau}_\mu^t \circ \pi_{\mu|\omega} \circ \pi_\omega &= \hat{\tau}_\mu^t \circ \pi_\mu \\ &= \pi_\mu \circ \tau^t = \pi_{\mu|\omega} \circ \pi_\omega \circ \tau^t \\ &= \pi_{\mu|\omega} \circ \hat{\tau}_\omega^t \circ \pi_\omega, \end{aligned}$$

it follows by σ-weak continuity that

$$\hat{\tau}_\mu^t \circ \pi_{\mu|\omega} = \pi_{\mu|\omega} \circ \hat{\tau}_\omega^t, \tag{25}$$

and since $z_{\mu|\omega}$ is invariant under $\hat{\tau}_\omega^t$

$$\hat{\tau}_\mu^t \circ \hat{\pi}_{\mu|\omega} = \hat{\pi}_{\mu|\omega} \circ \hat{\tau}_\omega^t. \tag{26}$$

We conclude that $\hat{\pi}_{\mu|\omega}$ is an isomorphism between the quantum dynamical systems $(z_{\mu|\omega}\mathfrak{C}_\omega, \hat{\tau}_\omega^t, \tilde{\mu})$ and $(\mathfrak{C}_\mu, \hat{\tau}_\mu^t, \hat{\mu})$.

The normal form turns out to be a very useful tool in the study of quantum dynamics since it provides a unifying framework in which both C^*- and W^*-systems can be handled on an equal footing.

Example 4.32. (Finite quantum systems) Let $(\mathfrak{M}, \tau^t)$ be the C^*- or W^*-dynamical system constructed from the Hilbert space $\mathcal{H}$ and the Hamiltonian H, *i.e.*, $\mathfrak{M} \equiv \mathfrak{B}(\mathcal{H})$ and $\tau^t(A) \equiv \mathrm{e}^{itH} A \mathrm{e}^{-itH}$ (Examples 4.4 and 4.16). A density matrix ρ on $\mathcal{H}$ such that $\mathrm{e}^{-itH} \rho \mathrm{e}^{itH} = \rho$ defines a normal, τ^t-invariant state $\mu(A) = \mathrm{Tr}(\rho A)$. If H has non-empty point spectrum, such states are easily obtained as mixtures of eigenstates of H. $(\mathfrak{M}, \tau^t, \mu)$ is a quantum dynamical system. Its normal form can be described in the following way. Set $\mathcal{G} \equiv \overline{\mathrm{Ran}\,\rho}$, denote by $\iota : \mathcal{G} \hookrightarrow \mathcal{H}$ the canonical injection and by $\mathcal{L}^2(\mathcal{G}, \mathcal{H})$ the set of Hilbert-Schmidt operators from $\mathcal{G}$ to $\mathcal{H}$. Then

$$\mathcal{H}_\mu \equiv \mathcal{L}^2(\mathcal{G}, \mathcal{H}),$$

is a Hilbert space with inner product $(X, Y) \equiv \mathrm{Tr}(X^* Y)$. Since $P \equiv \iota\iota^*$ is the orthogonal projection of $\mathcal{H}$ on $\mathcal{G}$, a simple calculation shows that $\Omega_\mu \equiv \rho^{1/2}\iota$ is a unit vector in $\mathcal{H}_\mu$. Setting

$$\pi_\mu(A)X = AX,$$

for all $X \in \mathcal{H}_\mu$, the map $A \mapsto \pi_\mu(A)$ defines a faithful representation of $\mathfrak{M}$ on $\mathcal{H}_\mu$. Since $(X, \pi_\mu(A)\Omega_\mu) = \mathrm{Tr}((\rho^{1/2}\iota X^*)A)$, one immediately checks that $\pi_\mu(\mathfrak{M})\Omega_\mu$

is dense in $\mathcal{H}_\mu$. Finally a simple calculation shows that for all $A \in \mathfrak{M}$ one has $\mu(A) = (\Omega_\mu, \pi_\mu(A)\Omega_\mu)$.

Since μ is normal and centrally faithful[3] the enveloping von Neumann algebra is $\pi_\mu(\mathfrak{M})$ and $N(\mathfrak{M}, \mu) = N(\mathfrak{M})$. Let $\bar{\cdot}$ denote an arbitrary complex conjugation on the Hilbert space $\mathcal{G}$. Then $\varphi \otimes \bar{\psi} \mapsto \varphi(\psi, \cdot\,)$ extends to a unitary map $U : \mathcal{H} \otimes \mathcal{G} \to \mathcal{H}_\mu$ such that $\pi_\mu(A) = U(A \otimes I)U^*$. Thus the enveloping von Neumann algebra $\pi_\mu(\mathfrak{M})$ is unitarily equivalent to $\mathfrak{M} \otimes I$.

Exercise 4.33. Show that the μ-Liouvillean of the preceding example is given by

$$\mathrm{e}^{itL_\mu} X = \mathrm{e}^{itH} X \mathrm{e}^{-itH'},$$

where H' is the restriction of H to $\mathcal{G}$. What is the spectrum of L_μ ?

Exercise 4.34. Show that if H has purely continuous spectrum, there is no trace class operator commuting with H, hence no normal invariant state.

Example 4.35. (Ideal Bose gas, continuation of Example 4.17) Let $\mathcal{D} \subset \mathfrak{h}$ be a subspace and denote by $\mathrm{CCR}(\mathcal{D})$ the C^*-algebra generated by the Weyl system $W(\mathcal{D}) \equiv \{W(f) \mid f \in \mathcal{D}\}$, *i.e.*, the norm closure of the linear span of $W(\mathcal{D})$. Since we can replace $\mathfrak{h}$ by $\overline{\mathcal{D}}$ we may assume, without loss of generality, that $\mathcal{D}$ is dense in $\mathfrak{h}$.

Let $\mathcal{H}$ be a Hilbert space and $\pi : W(\mathcal{D}) \to \mathfrak{B}(\mathcal{H})$ be such that $\pi(W(f))$ is unitary and

$$\pi(W(f))\pi(W(g)) = \mathrm{e}^{-i\,\mathrm{Im}(f,g)/2}\pi(W(f+g)), \tag{27}$$

for all $f, g \in \mathcal{D}$. Then one can show (see Theorem 6 in Lecture [36]) that π has a unique extension to an injective $*$-morphism from $\mathrm{CCR}(\mathcal{D})$ into $\mathfrak{B}(\mathcal{H})$. Thus, a representation of $\mathrm{CCR}(\mathcal{D})$ is completely determined by its restriction to $W(\mathcal{D})$.

A representation $(\mathcal{H}, \pi)$ of $\mathrm{CCR}(\mathcal{D})$ is *regular* if the map $\lambda \mapsto \pi(W(\lambda f))$ is strongly continuous for all $f \in \mathcal{D}$. Regular representations are physically appealing since by Stone theorem there exists a self-adjoint operator $\Phi_\pi(f)$ on $\mathcal{H}$ such that $\pi(W(f)) = \mathrm{e}^{i\Phi_\pi(f)}$ for all $f \in \mathcal{D}$. The operator $\Phi_\pi(f)$ is the Segal field operator in the representation π. The corresponding creation and annihilation operators are obtained by linear combination, for example $a_\pi(f) = (\Phi_\pi(f) + i\Phi_\pi(if))/\sqrt{2}$. A state ω on $\mathrm{CCR}(\mathcal{D})$ is called regular if its GNS representation is regular.

Since the finite linear combinations of elements of $W(\mathcal{D})$ are norm dense in $\mathrm{CCR}(\mathcal{D})$, a state ω on this C^*-algebra is completely determined by its characteristic function

$$\mathcal{D} \ni f \mapsto S_\omega(f) \equiv \omega(W(f)).$$

Clearly, if the function $\lambda \mapsto S_\omega(\lambda f)$ is continuous for all $f \in \mathcal{D}$, the state ω is regular and we denote by $\Phi_\omega(f)$, $a_\omega(f)$ and $a^*_\omega(f)$ the corresponding operators. The state ω is said to be C^n, C^∞ respectively analytic if the function $\lambda \mapsto S_\omega(\lambda f)$ has this smoothness near $\lambda = 0$. If ω is an analytic state, it is easy to see that $S_\omega(\lambda f)$

[3] Because $\mathfrak{M}$ is a factor.

is actually analytic in an open strip around the real axis and that Ω_ω is an analytic vector for all field operators $\Phi_\omega(f)$. The characteristic function $S_\omega(f)$ and therefore the state ω itself are then completely determined by the derivatives $\partial_\lambda^n S_\omega(\lambda f)|_{\lambda=0}$, or equivalently by the family of correlation functions

$$W_{m,n}(g_1,\cdots,g_m;f_1,\cdots,f_n) = (\Omega_\omega, a^*_\omega(g_m)\cdots a^*_\omega(g_1)a_\omega(f_1)\cdots a_\omega(f_n)\Omega_\omega).$$

Characteristic functions of regular states on $\mathrm{CCR}(\mathcal{D})$ are characterized by the the following result (see Lecture [36] for a proof).

Theorem 4.36. *(Araki-Segal) A map $S:\mathcal{D}\to\mathbb{C}$ is the characteristic function of a regular state ω on* $\mathrm{CCR}(\mathcal{D})$ *if and only if*

(i) $S(0)=1$.
(ii) The function $\lambda\mapsto S(\lambda f)$ is continuous for all $f\in\mathcal{D}$.
(iii) For all integer $n\geqslant 2$, all $f_1,\cdots,f_n\in\mathcal{D}$ and all $z_1,\cdots,z_n\in\mathbb{C}$ one has

$$\sum_{j,k=1}^n S(f_j-f_k)\,\mathrm{e}^{-i\,\mathrm{Im}(f_j,f_k)/2}\,\overline{z}_j z_k \geqslant 0.$$

If the subspace $\mathcal{D}$ is invariant under the one-particle dynamics, *i.e.*, if $\mathrm{e}^{-ith}\mathcal{D}\subset\mathcal{D}$ for all $t\in\mathbb{R}$, then it follows from Equ. (20) that the group τ^t defined by Equ. (21) leaves $W(\mathcal{D})$ and hence its closed linear span $\mathrm{CCR}(\mathcal{D})$ invariant. Thus, τ^t is a group of $*$-automorphisms of $\mathrm{CCR}(\mathcal{D})$. Note that $(\mathrm{CCR}(\mathcal{D}),\tau^t)$ is neither a C^*-dynamical (τ^t is not strongly continuous by Example 4.17), nor a W^*-dynamical system ($\mathrm{CCR}(\mathcal{D})$ is not a von Neumann algebra).

Suppose that the map $S:\mathcal{D}\to\mathbb{C}$ satisfies the conditions of Theorem 4.36 and

(iv) $S(\mathrm{e}^{ith}f)=S(f)$ for all $f\in\mathcal{D}$ and $t\in\mathbb{R}$.
(v) $\lim_{t\to0} S(\mathrm{e}^{ith}f-f)=1$ for all $f\in\mathcal{D}$.

Then the corresponding regular state ω is τ^t-invariant and

$$\omega(W(f)^*\tau^t(W(f)))=S(\mathrm{e}^{ith}f-f),$$

is continuous at $t=0$. By the Cauchy-Schwarz inequality the same is true of $\omega(W(g)^*\tau^t(W(f)))$ for any $f,g\in\mathcal{D}$. Since the linear span of $W(\mathcal{D})$ is norm dense in $\mathrm{CCR}(\mathcal{D})$ the group property of τ^t allows to conclude that for all $A\in\mathrm{CCR}(\mathcal{D})$ the function $t\mapsto\omega(A^*\tau^t(A))$ is continuous. Therefore, $\omega\in\mathcal{E}(\mathrm{CCR}(\mathcal{D}),\tau^t)$ and $(\mathrm{CCR}(\mathcal{D}),\tau^t,\omega)$ is a quantum dynamical system.

Definition 4.37. *A state on* $\mathrm{CCR}(\mathcal{D})$ *is called quasi-free if its characteristic function takes the form*

$$S(f)=\mathrm{e}^{-\frac14\|f\|^2-\frac12\rho[f]}, \tag{28}$$

where ρ is a closable non-negative quadratic form on $\mathcal{D}$.

We shall denote by ω_ρ the quasi-free state characterized by Equ. (28). We shall also use the symbol ρ to denote the non-negative self-adjoint operator associated with the quadratic form ρ. Thus, one has $\mathcal{D} \subset \mathrm{D}(\rho^{1/2})$ and $\rho[f] = \|\rho^{1/2} f\|^2$ for $f \in \mathcal{D}$.

The quasi-free state ω_ρ is clearly regular and analytic. If $\mathrm{e}^{-ith}\rho\,\mathrm{e}^{ith} = \rho$ for all $t \in \mathbb{R}$ then condition (iv) is satisfied and ω_ρ is τ^t-invariant. Moreover, since

$$\rho[\mathrm{e}^{ith} f - f] = 4\|\sin(th/2)\rho^{1/2} f\|^2$$

condition (v) is also satisfied and $\omega_\rho \in \mathcal{E}(\mathrm{CCR}(\mathcal{D}), \tau^t)$.

To describe the normal form of $(\mathrm{CCR}(\mathcal{D}), \tau^t, \omega_\rho)$ let $\mathfrak{g} \equiv \overline{\mathrm{Ran}\,\rho} \subset \mathfrak{h}$, denote by $\iota : \mathfrak{g} \hookrightarrow \mathfrak{h}$ the canonical injection and set

$$\mathcal{H}_{\omega_\rho} \equiv \mathcal{L}^2(\Gamma_+(\mathfrak{g}), \Gamma_+(\mathfrak{h})),$$

the set of Hilbert-Schmidt operators from $\Gamma_+(\mathfrak{g})$ to $\Gamma_+(\mathfrak{h})$ with scalar product $(X, Y) \equiv \mathrm{Tr}(X^*Y)$. For $g \in \mathfrak{g}$ denote by $W'(g)$ the Weyl operator in $\Gamma_+(\mathfrak{g})$.

For $f \in \mathcal{D}$ define

$$\pi_{\omega_\rho}(W(f)) : X \mapsto W((I+\rho)^{1/2} f) X W'(\iota^* \rho^{1/2} f)^*.$$

Using the CCR (19), one easily checks that $\pi_{\omega_\rho}(W(f))$ is unitary and satisfies Equ. (27). Thus, π_{ω_ρ} has a unique extension to a representation of $\mathrm{CCR}(\mathcal{D})$ in $\mathcal{H}_{\omega_\rho}$. Denote by Ω, Ω' the Fock vacua in $\Gamma_+(\mathfrak{h})$, $\Gamma_+(\mathfrak{g})$ and set $\Omega_{\omega_\rho} = \Omega(\Omega', \cdot)$.

Exercise 4.38. Show that Ω_{ω_ρ} is cyclic for $\pi_{\omega_\rho}(\mathfrak{C})$. (Hint: For $X = \Psi(\Phi, \cdot)$ with $\Psi = a^*(f_n) \cdots a^*(f_1)\Omega$ and $\Phi = a'^*(g_m) \cdots a'^*(g_1)\Omega'$ and $f \in \mathfrak{h}$ compute $\partial_\lambda \pi_{\omega_\rho}(W(\lambda f))X|_{\lambda=0}$ and $\partial_\lambda \pi_{\omega_\rho}(W(i\lambda f))X|_{\lambda=0}$.)

A simple calculation shows that

$$\omega_\rho(W(f)) = (\Omega_{\omega_\rho}, \pi_{\omega_\rho}(W(f))\Omega_{\omega_\rho}).$$

Finally, the ω_ρ-Liouvillean is given by

$$\mathrm{e}^{itL_{\omega_\rho}} X \equiv \Gamma(\mathrm{e}^{ith}) X \Gamma(\mathrm{e}^{-ith'}),$$

where h' is the restriction of h to $\mathfrak{g}$. The representation of $\mathrm{CCR}(\mathcal{D})$ obtained in this way is called *Araki-Woods representation* [1]. We refer the reader to [36] for a more detailed discussion and to [18] for a thorough exposition of the representation theory of canonical commutation relations.

4.5 Standard Forms

Recall that a subset $\mathcal{C}$ of a Hilbert space $\mathcal{H}$ is a cone if $t\psi \in \mathcal{C}$ for all $t \geq 0$ and $\psi \in \mathcal{C}$. The dual of a cone $\mathcal{C}$ is the closed cone

$$\widehat{\mathcal{C}} \equiv \{\psi \in \mathcal{H} \,|\, (\phi, \psi) \geq 0 \text{ for any } \phi \in \mathcal{C}\}.$$

A cone $\mathcal{C}$ is self-dual if $\widehat{\mathcal{C}} = \mathcal{C}$. A self-dual cone is automatically closed.

Definition 4.39. *A von Neumann algebra $\mathfrak{M} \subset \mathfrak{B}(\mathcal{H})$ is said to be in standard form if there exist a anti-unitary involution J on $\mathcal{H}$ and a self-dual cone $\mathcal{C} \subset \mathcal{H}$ such that:*

(i) $J\mathfrak{M}J = \mathfrak{M}'$.
(ii) $J\Psi = \Psi$ for all $\Psi \in \mathcal{C}$.
(iii) $AJA\mathcal{C} \subset \mathcal{C}$ for all $A \in \mathfrak{M}$.
(iv) $JAJ = A^*$ for all $A \in \mathfrak{M} \cap \mathfrak{M}'$.

We shall denote by $(\mathfrak{M}, \mathcal{H}, J, \mathcal{C})$ a von Neumann algebra in standard form.

Theorem 4.40. *Any von Neumann algebra $\mathfrak{M}$ has a faithful representation $(\mathcal{H}, \pi)$ such that $\pi(\mathfrak{M})$ is in standard form. Moreover, this representation is unique up to unitary equivalence.*

Sketch of the proof. If $\mathfrak{M}$ is separable, *i.e.*, if any family of mutually orthogonal projections in $\mathfrak{M}$ is finite or countable then $\mathfrak{M}$ has a normal faithful state ω (Proposition 2.5.6 in [11]). The associated GNS construction provides a faithful representation $(\mathcal{H}_\omega, \pi_\omega, \Omega_\omega)$ of $\mathfrak{M}$ and it follows from Tomita-Takesaki theory (Chapter 4 in Lecture [7]) that $\pi_\omega(\mathfrak{M})$ is in standard form. The anti-unitary involution J is the modular conjugation and the self-dual cone $\mathcal{C}$ is the natural cone $\overline{\{AJA\Omega_\omega \,|\, A \in \mathfrak{M}_\omega\}}$ (see Section 4.3 in Lecture [7]). This construction applies in particular to any von Neumann algebra over a separable Hilbert space which is the case most often encountered in physical applications.

In the general case, the construction is similar to the above one, substituting faithful normal states with faithful normal semi-finite weights. The general theory of standard forms was developed by Haagerup [23] following the works of Araki [3] and Connes [14] (see also [45], where standard forms are called hyper-standard). □

Von Neumann algebras in standard form have two important properties which are of crucial importance in the study of quantum dynamical systems. The first one concerns normal states.

Theorem 4.41. *Let $(\mathfrak{M}, \mathcal{H}, J, \mathcal{C})$ be a von Neumann algebra in standard form and for any unit vector $\Psi \in \mathcal{H}$ denote by $\omega_\Psi \in N(\mathfrak{M})$ the corresponding vector state $\omega_\Psi(A) = (\Psi, A\Psi)$. Then the map*

$$\begin{aligned} \{\Phi \in \mathcal{C} \,|\, \|\Phi\| = 1\} &\to N(\mathfrak{M}) \\ \Psi &\mapsto \omega_\Psi \end{aligned}$$

is an homeomorphism (for the norm topologies). In particular, for any normal state ν on $\mathfrak{M}$, there is a unique unit vector $\Psi_\nu \in \mathcal{C}$ such that $\nu = \omega_{\Psi_\nu}$. We call Ψ_ν the standard vector representative of ν. Moreover, for any unit vectors $\Psi, \Phi \in \mathcal{C}$ one has:

(i) $\|\Psi - \Phi\|^2 \leqslant \|\omega_\Psi - \omega_\Phi\| \leqslant \|\Psi - \Phi\| \|\Psi + \Phi\|$.
(ii) Ψ is cyclic for $\mathfrak{M} \iff \Psi$ is cyclic for $\mathfrak{M}' \iff \omega_\Psi$ is faithful.
(iii) More generally $\overline{\mathfrak{M}\Psi} = J\overline{\mathfrak{M}'\Psi}$.

Proof. I shall not prove the fact that $\Psi \mapsto \omega_\Psi$ is an homeomorphism and the first inequality in (i) since this requires rather long and involved arguments. A proof can be found for example in [11]. The Second inequality in (i) follows from the polarization identity. The first equivalence in (ii) is a special case of (iii) which is a direct consequence of (i) and (ii) in Definition 4.39. The second equivalence in (ii) is a special case of the last statement of Lemma 2.17. □

The second important property of von Neumann algebras in standard form has to do with the unitary implementation of $*$-automorphisms. To formulate this property let me introduce the following definition.

Definition 4.42. *Let $(\mathfrak{M}, \mathcal{H}, J, \mathcal{C})$ be a von Neumann algebra in standard form. A unitary operator U on $\mathcal{H}$ is called standard if the following holds:*

(i) $U\mathcal{C} \subset \mathcal{C}$.
(ii) $U^*\mathfrak{M}U = \mathfrak{M}$.

Obviously, the set of standard unitaries form a subgroup of the unitary group of $\mathcal{H}$. It is not hard to see that this subgroup is closed in the strong topology of $\mathfrak{B}(\mathcal{H})$. The following is essentially a rewriting of Corollary 4 in Lecture [7].

Theorem 4.43. *Let $(\mathfrak{M}, \mathcal{H}, J, \mathcal{C})$ be a von Neumann algebra in standard form. Denote by $\mathcal{U}_s$ the group of standard unitaries of $\mathcal{H}$ equipped with the strong topology and by $\mathrm{Aut}(\mathfrak{M})$ the group of $*$-automorphisms of $\mathfrak{M}$ with the topology of pointwise σ-weak convergence.*

For any $U \in \mathcal{U}_s$ denote by τ_U the corresponding spatial $*$-automorphism $\tau_U(A) = UAU^*$. Then the map

$$\begin{aligned} \mathcal{U}_s &\to \mathrm{Aut}(\mathfrak{M}) \\ U &\mapsto \tau_U \end{aligned}$$

is an homeomorphism. In particular, for any $*$-automorphism σ of $\mathfrak{M}$ there is a unique standard unitary U such that $\sigma = \tau_U$. We call U the standard implementation of σ. Moreover, for any $U \in \mathcal{U}_s$ one has:

(i) $[U, J] = 0$.
(ii) $U\mathfrak{M}'U^* = \mathfrak{M}'$.
(iii) $U^*\Psi_\omega = \Psi_{\omega\circ\tau_U}$ for all $\omega \in N(\mathfrak{M})$.

Definition 4.44. *A quantum dynamical system $(\mathfrak{C}, \tau^t, \mu)$ is in standard form if*

(i) $\mathfrak{C}$ is a von Neumann algebra in standard form $(\mathfrak{C}, \mathcal{H}, J, \mathcal{C})$.
(ii) $(\mathfrak{C}, \tau^t)$ is a W^*-dynamical system.
(iii) $\mu \in N(\mathfrak{C}, \tau^t)$.

Suppose that $(\mathfrak{C}, \tau^t, \mu)$ is in standard form. By Theorem 4.43, τ^t has a standard implementation U^t for each $t \in \mathbb{R}$. Since $t \mapsto \tau^t(A)$ is σ-weakly continuous, U^t is strongly continuous. Therefore, there exists a self-adjoint operator L on $\mathcal{H}$ such that $U^t = e^{itL}$. By Theorem 4.41, there exists a unique vector $\Phi \in \mathcal{C}$ such that $\mu(A) = (\Phi, A\Phi)$. Moreover, it follows from properties (i)-(iii) of Theorem 4.43 that

$$JL + LJ = 0 \tag{29}$$

$$e^{itL}\mathfrak{C}'e^{-itL} = \mathfrak{C}', \tag{30}$$

$$L\Psi_\omega = 0, \tag{31}$$

for all $t \in \mathbb{R}$ and all $\omega \in N(\mathfrak{C}, \tau^t)$.

Definition 4.45. *We denote by $(\mathfrak{C}, \mathcal{H}, J, \mathcal{C}, L, \Phi)$ a quantum dynamical system in standard form and we call L its standard Liouvillean.*

As an immediate consequence of this definition we have the

Proposition 4.46. *If the quantum dynamical system $(\mathfrak{C}, \tau^t, \mu)$ is in standard form $(\mathfrak{C}, \mathcal{H}, J, \mathcal{C}, L, \Phi)$, its standard Liouvillean is the unique self-adjoint operator L on $\mathcal{H}$ such that, for all $t \in \mathbb{R}$ and all $A \in \mathfrak{C}$ one has*
(i) $e^{-itL}\mathcal{C} \subset \mathcal{C}$.
(ii) $e^{itL}Ae^{-itL} = \tau^t(A)$.

Let $(\mathfrak{C}, \tau^t, \mu)$ be a quantum dynamical system, $(\pi_\mu, \mathfrak{C}_\mu, \mathcal{H}_\mu, L_\mu, \Omega_\mu)$ its normal form and $(\mathcal{H}, \pi)$ a standard representation of $\mathfrak{C}_\mu$. Then $\mathfrak{M} \equiv \pi(\mathfrak{C}_\mu)$ is a von Neumann algebra in standard form $(\mathfrak{M}, \mathcal{H}, J, \mathcal{C})$. Note that since π is faithful it is σ-weakly continuous by Corollary 2.12. The same remark apply to its inverse $\pi^{-1} : \mathfrak{M} \to \mathfrak{C}_\mu$.

Let $\eta \equiv \pi \circ \pi_\mu$ be the induced representation of $\mathfrak{C}$ in $\mathcal{H}$. Since $\pi_\mu(\mathfrak{C})$ is σ-weakly dense in $\mathfrak{C}_\mu$ we have $\mathfrak{M} = \eta(\mathfrak{C})''$. For $A \in \mathfrak{M}$ we define

$$\sigma^t(A) \equiv \pi(e^{itL_\mu}\pi^{-1}(A)e^{-itL_\mu}).$$

It follows that for $A \in \mathfrak{C}$ one has

$$\begin{aligned}\sigma^t(\eta(A)) &= \pi(e^{itL_\mu}\pi^{-1}(\eta(A))e^{-itL_\mu}) \\ &= \pi(e^{itL_\mu}\pi_\mu(A)e^{-itL_\mu}) \\ &= \pi(\pi_\mu(\tau^t(A))) \\ &= \eta(\tau^t(A)).\end{aligned}$$

The group σ^t defines a W^*-dynamical system on $\mathfrak{M}$. It has a standard implementation with standard Liouvillean L. Finally we remark that $\omega \equiv (\Omega_\mu, \pi^{-1}(\cdot)\Omega_\mu) \in N(\mathfrak{M}, \sigma^t)$ satisfies $\omega \circ \eta = \mu$. Denote by $\Phi \in \mathcal{C}$ its standard vector representative. It follows that $(\mathfrak{M}, \sigma^t, \omega)$ is in standard form.

Definition 4.47. *We say that $(\eta, \mathfrak{M}, \mathcal{H}, J, \mathcal{C}, L, \Phi)$ is the standard form of the quantum dynamical system $(\mathfrak{C}, \tau^t, \mu)$ and that L is its standard Liouvillean.*

If $(\pi_1, \mathfrak{D}_1, \mathcal{K}_1, M_1, \Omega_1)$ and $(\pi_2, \mathfrak{D}_2, \mathcal{K}_2, M_2, \Omega_2)$ are two normal forms of the quantum dynamical system $(\mathfrak{C}, \tau^t, \mu)$ then, by the unicity of the GNS representation, there exists a unitary $U : \mathcal{K}_1 \to \mathcal{K}_2$ such that $\pi_2 = U\pi_1 U^*$, $\Omega_2 = U\Omega_1$ and $M_2 = UM_1U^*$. Let $(\eta_1, \mathcal{H}_1)$ and $(\eta_2, \mathcal{H}_2)$ be standard representations of $\mathfrak{D}_1$ and $\mathfrak{D}_2$ and denote by $(\eta_1 \circ \pi_1, \mathfrak{M}_1, \mathcal{H}_1, J_1, \mathcal{C}_1, L_1, \Phi_1)$ and $(\eta_2 \circ \pi_2, \mathfrak{M}_2, \mathcal{H}_2, J_2, \mathcal{C}_2, L_2, \Phi_2)$ the corresponding standard forms of $(\mathfrak{C}, \tau^t, \mu)$. Since $(\eta_2(U \cdot U^*), \mathcal{H}_2)$ is another standard representation of $\mathfrak{D}_1$, there exists a unitary $V : \mathcal{H}_1 \to \mathcal{H}_2$ such that $\eta_2(UAU^*) = V\eta_1(A)V^*$. It follows that $\eta_2 \circ \pi_2 = V\eta_1(U^*\pi_2 U)V^* = V\eta_1 \circ \pi_1 V^*$. Thus, the standard form and in particular the standard Liouvillean of a quantum dynamical system are uniquely determined, up to unitary equivalence.

The next proposition elucidate the relation between the μ-Liouvillean and the standard Liouvillean. It also shows that in many applications (see Lemma 5.11 below) the two coincide.

Proposition 4.48. *Let $(\mathfrak{C}, \tau^t, \mu)$ be a quantum dynamical system with normal form $(\pi_\mu, \mathfrak{C}_\mu, \mathcal{H}_\mu, L_\mu, \Omega_\mu)$. If the state $\hat{\mu} \equiv (\Omega_\mu, (\cdot)\Omega_\mu)$ is faithful on $\mathfrak{C}_\mu$ then this von Neumann algebra is in standard form $(\mathfrak{C}_\mu, \mathcal{H}_\mu, J, \mathcal{C})$ and the standard form of $(\mathfrak{C}, \tau^t, \mu)$ is given by $(\pi_\mu, \mathfrak{C}_\mu, \mathcal{H}_\mu, J, \mathcal{C}, L_\mu, \Omega_\mu)$. In particular L_μ is its standard Liouvillean.*

Proof. If $\hat{\mu}$ is faithful then, as mentioned in the sketch of the proof of Theorem 4.40, $\mathfrak{C}_\mu$ is in standard form. J is the modular conjugation associated to Ω_μ and $\mathcal{C}$ is the associated natural cone. In particular, Ω_μ is the standard vector representative of $\hat{\mu}$. If L is the standard Liouvillean then $L\Omega_\mu = 0$ by Equ. (31) hence L is the μ-Liouvillean. □

More generally, the relation between μ-Liouvillean and standard Liouvillean is given by the following result.

Proposition 4.49. *Let $(\mathfrak{C}, \tau^t, \mu)$ be a quantum dynamical system with standard form $(\eta, \mathfrak{M}, \mathcal{H}, J, \mathcal{C}, L, \Phi)$. Then the subspace $\mathcal{K} \equiv \overline{\mathfrak{M}\Phi} \subset \mathcal{H}$ reduces the standard Liouvillean L and the μ-Liouvillean L_μ is (unitarily equivalent to) the restriction of L to $\mathcal{K}$. In particular, one has $\sigma(L_\mu) \subset \sigma(L)$.*

Proof. We reconstruct a normal form of $(\mathfrak{C}, \tau^t, \mu)$ out of its standard form. Recall that $\eta : \mathfrak{C} \to \mathfrak{M}$ is a representation such that $\mathfrak{M} = \eta(\mathfrak{C})''$, $\mu(A) = (\Phi, \eta(A)\Phi)$ and $\eta(\tau^t(A)) = \mathrm{e}^{itL}\eta(A)\mathrm{e}^{-itL}$. Denote by $k : \mathcal{K} \hookrightarrow \mathcal{H}$ the canonical injection. Since $\mathcal{K}$ is invariant under $\mathfrak{M}$, $\phi(A) \equiv k^*\eta(A)k$ defines a representation of $\mathfrak{C}$ on $\mathcal{K}$. By the von Neumann density theorem $\eta(\mathfrak{C})$ is σ-strongly dense in $\mathfrak{M}$ and one has $\overline{\phi(\mathfrak{C})\Phi} = \overline{k^*\eta(\mathfrak{C})\Phi} = \overline{k^*\mathfrak{M}\Phi} = \mathcal{K}$. Finally $\mu(A) = (\Phi, \eta(A)\Phi) = (\Phi, \phi(A)\Phi)$. Thus $(\mathcal{K}, \phi, \Phi)$ is the required GNS representation. Since $\mathrm{e}^{itL}\eta(A)\Phi = \eta(\tau^t(A))\Phi$ the subspace $\mathcal{K}$ reduces L and the restriction of L to $\mathcal{K}$ is the μ-Liouvillean. □

If $(\mathfrak{C}, \tau^t, \omega)$ is a quantum dynamical system and μ is a τ^t-invariant ω-normal state on $\mathfrak{C}$ then $(\mathfrak{C}, \tau^t, \mu)$ is also a quantum dynamical system. The standard Liouvilleans of these two systems are not independent. Their relation is explicited in the next Theorem.

Theorem 4.50. *Let $(\mathfrak{C}, \tau^t, \omega)$ be a quantum dynamical system with standard form $(\eta, \mathfrak{M}, \mathcal{H}, J, \mathcal{C}, L, \Phi)$ and μ a τ^t-invariant ω-normal state. Denote by $P \in \mathfrak{Z}(\mathfrak{M})$ the central support of the normal extension of μ to $\mathfrak{M}$. The standard Liouvillean of the quantum dynamical system $(\mathfrak{C}, \tau^t, \mu)$ is (unitarily equivalent to) the restriction of L to the subspace $P\mathcal{H}$. In particular, its spectrum is contained in $\sigma(L)$.*

Proof. We construct a standard form $(\eta_\mu, \mathfrak{M}_\mu, \mathcal{H}_\mu, J_\mu, \mathcal{C}_\mu, L_\mu, \Phi_\mu)$ of $(\mathfrak{C}, \tau^t, \mu)$. Since μ is ω-normal there exists $\tilde{\mu} \in N(\mathfrak{C}_\omega)$ with central support $z_{\mu|\omega}$ such that $\mu = \tilde{\mu} \circ \pi_\omega$. There is also a $*$-isomorphism $\pi : \mathfrak{C}_\omega \to \mathfrak{M}$ such that $\pi \circ \pi_\omega = \eta$. The normal extension of μ to $\mathfrak{M}$ is $\bar{\mu} \equiv \tilde{\mu} \circ \pi^{-1}$ so that $\mu = \bar{\mu} \circ \eta$. We denote its standard vector representative by Φ_μ. It follows from Lemma 6 that $P = [\mathfrak{M} \vee \mathfrak{M}' \Phi_\mu] = \pi(z_{\mu|\omega})$ and in particular that $\Phi_\mu \in \mathcal{H}_\mu \equiv P\mathcal{H}$. By Remark 2.22 one has $\mathfrak{M} = P\mathfrak{M} \oplus (I - P)\mathfrak{M}$ and we can identify $\mathfrak{M}_\mu \equiv P\mathfrak{M}$ with a von Neumann algebra on $\mathcal{H}_\mu$. Clearly $\eta_\mu(A) \equiv P\eta(A)$ defines a representation of $\mathfrak{C}$ in $\mathcal{H}_\mu$. By Property (iv) of Definition 4.39 one has $JPJ = P$ so that the subspace $\mathcal{H}_\mu$ reduces J. It follows that the restriction J_μ of J to this subspace is an anti-unitary involution. Finally, $\mathcal{C}_\mu \equiv P\mathcal{C}$ is a cone in $\mathcal{H}_\mu$ and

$$\begin{aligned}\hat{\mathcal{C}}_\mu &= \{\psi \in \mathcal{H}_\mu \,|\, (\psi, P\phi) \geq 0 \text{ for all } \phi \in \mathcal{C}\}\\ &= \{\psi \in \mathcal{H}_\mu \,|\, (P\psi, \phi) \geq 0 \text{ for all } \phi \in \mathcal{C}\}\\ &= \{\psi \in \mathcal{H}_\mu \,|\, (\psi, \phi) \geq 0 \text{ for all } \phi \in \mathcal{C}\}\\ &= \mathcal{H}_\mu \cap \mathcal{C}.\end{aligned}$$

Let me show that $\mathcal{H}_\mu \cap \mathcal{C} = \mathcal{C}_\mu$. On the one hand $P = JPJ$ yields $P = P^2 = PJPJ$ and hence $P\mathcal{C} = PJPJ\mathcal{C} = PJP\mathcal{C} \subset \mathcal{C}$ by Properties (ii) and (iii) of Definition 4.39. But since $P\mathcal{C} \subset \mathcal{H}_\mu$ one concludes that $P\mathcal{C} \subset \mathcal{H}_\mu \cap \mathcal{C}$. On the other hand if $\psi \in \mathcal{H}_\mu \cap \mathcal{C}$ then $\psi = P\psi \in \mathcal{C}$ and hence $\psi \in P\mathcal{C}$ proving the claim and hence the fact that $\mathcal{C}_\mu$ is self-dual.

We have to show that Properties (i)-(iv) of Definition 4.39 are satisfied. By Remark 2.22, $\mathfrak{M}_\mu' = P\mathfrak{M}'$ and Property (i) follows from $J_\mu \mathfrak{M}_\mu J_\mu = JP\mathfrak{M}PJ = PJ\mathfrak{M}JP = P\mathfrak{M}'P = \mathfrak{M}_\mu'$. To prove Property (ii) note that for $\psi \in \mathcal{C}_\mu \subset \mathcal{C}$ we have $J_\mu \psi = J\psi = \psi$. For $A \in \mathfrak{M}_\mu$ we further have $A = AP = PA = PAP$ and hence $AJ_\mu A\mathcal{C}_\mu = P(APJAP\mathcal{C}) \subset P\mathcal{C} = \mathcal{C}_\mu$ which proves Property (iii). By Remark 2.22, $\mathfrak{Z}(\mathfrak{M}_\mu) = P\mathfrak{Z}(\mathfrak{M})$ so that if $A \in \mathfrak{Z}(\mathfrak{M}_\mu)$ we have $J_\mu A J_\mu = JPAPJ = (PAP)^* = PA^*P = A^*$ and Property (iv) is verified.

By Theorem 2.30, $\hat{\pi}_{\mu|\omega} : z_{\mu|\omega}\mathfrak{C}_\omega \to \mathfrak{C}_\mu$ is a $*$-isomorphism. Hence $\phi \equiv \pi \circ \hat{\pi}_{\mu|\omega}^{-1}$ is a faithful representation of $\mathfrak{C}_\mu$ in $\mathcal{H}_\mu$ such that $\phi(\mathfrak{C}_\mu) = \pi(z_{\mu|\omega}\mathfrak{C}_\omega) = P\mathfrak{M} = \mathfrak{M}_\mu$ and $(\mathfrak{M}_\mu, \mathcal{H}_\mu, J_\mu, \mathcal{C}_\mu)$ is indeed a standard form of $\mathfrak{C}_\mu$. To determine the standard Liouvillean note that since $\bar{\mu}$ is an invariant state its central support P is also invariant

$$\mathrm{e}^{itL} P \mathrm{e}^{-itL} = P.$$

It follows that P reduces the standard Liouvillean L. By Equ. (26), we have $\hat{\pi}_{\mu|\omega}^{-1} \circ \hat{\tau}_\mu^t = \hat{\tau}_\omega^t \circ \hat{\pi}_{\mu|\omega}^{-1}$ from which we derive

$$\phi \circ \hat{\tau}_\mu^t(A) = \pi \circ \hat{\tau}_\omega^t \circ \hat{\pi}_{\mu|\omega}^{-1}(A) = \mathrm{e}^{itL}\phi(A)\mathrm{e}^{-itL}.$$

Finally, since $e^{itL}\mathcal{C}_\mu = e^{itL}P\mathcal{C} = Pe^{itL}\mathcal{C} = P\mathcal{C} = \mathcal{C}_\mu$ we conclude that the standard Liouvillean of $\hat{\tau}_\mu^t$ is the restriction of L to $\mathcal{H}_\mu$. □

Example 4.51. (Standard form of a finite quantum system) Consider the quantum dynamical system $(\mathfrak{M}, \tau^t, \mu)$ of Example 4.32 and suppose that the Hilbert space $\mathcal{H}$ is separable. To construct a standard form let ρ_0 be a density matrix on $\mathcal{H}$ such that $\operatorname{Ker}\rho_0 = \{0\}$. It follows that $\overline{\operatorname{Ran}\rho_0} = \mathcal{H}$ and Lemma 2.17 yields that the normal state $\omega_0(A) \equiv \operatorname{Tr}(\rho_0 A)$ is faithful. The GNS representation of $\mathfrak{M}$ corresponding to ω_0 is given, according to Example 4.32, by

$$\mathcal{H}_{\omega_0} = \mathcal{L}^2(\mathcal{H}),$$

$$\pi_{\omega_0}(A)X = AX,$$

$$\Omega_{\omega_0} = \rho_0^{1/2}.$$

By Lemma 2.23 this representation is faithful and by Lemma 2.28 the corresponding enveloping von Neumann algebra is $\mathfrak{M}_{\omega_0} = \pi_{\omega_0}(\mathfrak{M})$. It follows that $\hat{\omega}_0$ is faithful on $\mathfrak{M}_{\omega_0}$ and hence, by Proposition 4.48, that $\mathfrak{M}_{\omega_0}$ is in standard form. Indeed, one easily checks that $J : X \mapsto X^*$ and $\mathcal{C} \equiv \{X \in \mathcal{L}^2(\mathcal{H}) \,|\, X \geq 0\}$ satisfy all the conditions of Definition 4.39.

By Proposition 4.46, the standard Liouvillean L is given by

$$e^{itL}X = e^{itH}Xe^{-itH},$$

since this is a unitary implementation of the dynamics

$$e^{itL}\pi_{\omega_0}(A)e^{-itL}X = e^{itH}(Ae^{-itH}Xe^{itH})e^{-itH} = \pi_{\omega_0}(\tau^t(A))X,$$

which preserves the cone $\mathcal{C}$. The standard vector representative of the invariant state $\mu(A) = \operatorname{Tr}(\rho A)$ is $\rho^{1/2} \in \mathcal{C}$. One easily checks that the map $X \mapsto X|_{\mathcal{G}}$ is isometric from $\mathfrak{M}_{\omega_0}\rho^{1/2}$ into $\mathcal{L}^2(\mathcal{G}, \mathcal{H})$ and has a dense range. It extends to a unitary map U between $\mathcal{K} \equiv \overline{\mathfrak{M}_{\omega_0}\rho^{1/2}}$ and $\mathcal{H}_\mu = \mathcal{L}^2(\mathcal{G}, \mathcal{H})$ such that, according to Proposition 4.49

$$UL|_{\mathcal{K}}U^* = L_\mu.$$

4.6 Ergodic Properties of Quantum Dynamical Systems

My aim in this subsection is to extend to quantum dynamics the definitions and characterizations of the ergodic properties introduced in Subsection 3.1. Since there is no obvious way to formulate a quantum individual ergodic theorem (Theorem 3.3), I shall use the characterization (iv) of Proposition 3.6 to define ergodicity. Mixing will then be defined using Definition 3.8. To do so I only need to extend the notion of absolute continuity of two measures to two states on a C^*-algebra.

Example 4.52. Consider the classical C^*-dynamical system $(C(M), \tau^t)$ of Example 4.12 and let μ be a τ^t-invariant Baire probability measure on M. To construct the associated GNS representation denote by $\hat{\mu}$ the regular Borel extension of μ and set $\mathcal{H}_\mu \equiv L^2(M, \mathrm{d}\hat{\mu})$. The C^*-algebra $C(M)$ is represented on this Hilbert space as

$$\pi_\mu(f) : \psi \mapsto f\psi.$$

Since $C(M)$ is dense in $L^2(M, \mathrm{d}\mu)$ the vector $\Omega_\mu = 1$ is cyclic. Finally, one has

$$(\Omega_\mu, \pi_\mu(f)\Omega_\mu) = \int f(x)\,\mathrm{d}\hat{\mu}(x) = \mu(f).$$

We note in passing that the corresponding enveloping von Neumann algebra is $\pi_\mu(C(M))'' = L^\infty(M, \mathrm{d}\hat{\mu})$.

If ν is a μ-normal state on $C(M)$ then there exists a density matrix ρ on $\mathcal{H}_\mu$ such that $\nu(f) = \mathrm{Tr}(\rho\pi_\mu(f))$. Let

$$\rho = \sum_n p_n \psi_n(\psi_n, \cdot),$$

be the spectral representation of ρ. It follows that

$$F(x) \equiv \sum_n p_n|\psi_n(x)|^2 \in L^1(M, \mathrm{d}\hat{\mu}),$$

and hence

$$\nu(f) = \int F(x) f(x)\,\mathrm{d}\hat{\mu}(x).$$

Thus, if ν is μ-normal its regular Borel extension $\hat{\nu}$ is absolutely continuous with respect to $\hat{\mu}$.

Reciprocally, if $\hat{\nu} \ll \hat{\mu}$ and $G = (\frac{\mathrm{d}\hat{\nu}}{\mathrm{d}\hat{\mu}})^{1/2}$ one has

$$\nu(f) = \int f \frac{\mathrm{d}\hat{\nu}}{\mathrm{d}\hat{\mu}}\,\mathrm{d}\hat{\mu} = (G, \pi_\mu(f)G),$$

and we conclude that ν is μ-normal.

This example shows that for Abelian C^*- or W^*-algebras absolute continuity is equivalent to relative normality. If $A \subset M$ is a measurable set then its characteristic function χ_A, viewed as a multiplication operator on $\mathcal{H}_\mu$, is an orthogonal projection. In fact all orthogonal projections in $\pi_\mu(C(M))''$ are easily seen to be of this form. Since the absolute continuity of the Borel measure $\hat{\nu}$ with respect to $\hat{\mu}$ means that $\hat{\mu}(\chi_A) = 0$ implies $\hat{\nu}(\chi_A) = 0$ we see that if ν is μ-normal then $s_\nu \le s_\mu$. As the following example shows this is not necessarily true in the non-Abelian case.

Example 4.53. Let $\mathfrak{M} \equiv \mathfrak{B}(\mathcal{H})$ with $\dim \mathcal{H} > 1$ and $\mu \equiv (\psi, \cdot\,\psi)$ for some unit vector $\psi \in \mathcal{H}$. As already remarked in Example 4.32, any normal state on $\mathfrak{M}$ is μ-normal. In particular if ρ is a density matrix such that $\mathrm{Ker}\,\rho = \{0\}$ then $\nu \equiv \mathrm{Tr}(\rho\,\cdot)$ is a faithful μ-normal state and hence $I = s_\nu > s_\mu = \psi(\psi, \cdot)$.

Considering the special case of a W^*-dynamical system $(\mathfrak{M}, \tau^t)$ equipped with a normal invariant state μ, the following argument shows that the condition $s_\nu \leq s_\mu$ is necessary if we wish to define the ergodicity of the quantum dynamical system $(\mathfrak{M}, \tau^t, \mu)$ by the condition (iv) of Proposition 3.6, *i.e.,*

$$\lim_{T\to\infty} \frac{1}{T} \int_0^T \nu \circ \tau^t(A)\, \mathrm{d}t = \mu(A), \tag{32}$$

for all $A \in \mathfrak{M}$. Indeed, since μ is τ^t-invariant we have $\tau^t(s_\mu) = s_\mu$ and Equ. (32) leads to

$$\nu(s_\mu) = \lim_{T\to\infty} \frac{1}{T} \int_0^T \nu \circ \tau^t(s_\mu)\, \mathrm{d}t = \mu(s_\mu) = 1,$$

which means that $s_\nu \leq s_\mu$.

These considerations motivate the following definition.

Definition 4.54. *If $\nu, \mu \in E(\mathfrak{C})$ are such that $\nu \in N(\mathfrak{C}, \mu)$ and $s_\nu \leq s_\mu$ then we say that ν is absolutely continuous with respect to μ and we write $\nu \ll \mu$. We also set*

$$\mathcal{S}(\mathfrak{C}, \mu) \equiv \{\nu \in E(\mathfrak{C}) \,|\, \nu \ll \mu\}.$$

When dealing with the set $\mathcal{S}(\mathfrak{C}, \mu)$, the following lemma is often useful.

Lemma 4.55. *Let $\mathfrak{C}$ be a C^*-algebra, $\mu \in E(\mathfrak{C})$ and for $\lambda > 0$ set*

$$\mathcal{S}^\lambda(\mathfrak{C}, \mu) \equiv \{\nu \in E(\mathfrak{C}) \,|\, \nu \leq \lambda\mu\}.$$

(i) $\mathcal{S}^\lambda(\mathfrak{C}, \mu)$ is weak-$\star$ compact.
(ii) The set

$$\mathcal{S}_0(\mathfrak{C}, \mu) \equiv \bigcup_{n=1}^{\infty} \mathcal{S}^n(\mathfrak{C}, \mu), \tag{33}$$

is total in $\mathcal{S}(\mathfrak{C}, \mu)$, i.e., the set of finite convex linear combinations of elements of $\mathcal{S}_0(\mathfrak{C}, \mu)$ is a norm dense subset of $\mathcal{S}(\mathfrak{C}, \mu)$.
(iii) If $(\pi, \mathcal{H})$ is a representation of $\mathfrak{C}$ such that $\mu(A) = (\Psi, \pi(A)\Psi)$ for some $\Psi \in \mathcal{H}$ then one has

$$\mathcal{S}_0(\mathfrak{C}, \mu) = \{(\Phi, \pi(\,\cdot\,)\Phi) \,|\, \Phi \in \pi(\mathfrak{C})'\Psi, \|\Phi\| = 1\}.$$

Proof. We first prove (iii). Assume that $\nu \in \mathcal{S}_0(\mathfrak{C}, \mu)$ as defined in Equ. (33). Then there exists $\lambda > 0$ such that $\nu(A^*A) \leq \lambda\mu(A^*A) = \lambda\|\pi(A)\Psi\|^2$ for all $A \in \mathfrak{C}$. By the Cauchy-Schwarz inequality we further get

$$|\nu(A^*B)| \leq \lambda\|\pi(A)\Psi\|\|\pi(B)\Psi\|,$$

for all $A, B \in \mathfrak{C}$. It follows that the map

$$\begin{aligned} \pi(\mathfrak{C})\Psi \times \pi(\mathfrak{C})\Psi &\to \mathbb{C} \\ (\pi(A)\Psi, \pi(B)\Psi) &\mapsto \nu(A^*B), \end{aligned}$$

is well defined. As a densely defined, bounded, non-negative sesquilinear form on $\mathcal{K} \equiv \overline{\pi(\mathfrak{C})\Psi}$ it defines a bounded non-negative self-adjoint operator $M \in \mathfrak{B}(\mathcal{K})$, which we can extend by 0 on $\mathcal{K}^\perp$ and such that $\nu(A^*B) = (\pi(A)\Psi, M\pi(B)\Psi)$. Since $\nu(A^*BC) = \nu((B^*A)^*C)$ we get

$$(\pi(A)\Psi, M\pi(B)\pi(C)\Psi) = (\pi(A)\Psi, \pi(B)M\pi(C)\Psi),$$

for all $A, B, C \in \mathfrak{C}$. Since M vanishes on $\mathcal{K}^\perp$ we can conclude that $M \in \pi(\mathfrak{C})'$. Set $R \equiv M^{1/2}$, it follows that $R \in \pi(\mathfrak{C})'$,

$$\nu(A) = (R\Psi, \pi(A)R\Psi),$$

and $\|R\Psi\| = 1$. Reciprocally, if ν is given by the above formula one has $\nu(A^*A) = \|R\pi(A)\Psi\| \le \|R\|\|\pi(A)\Psi\|^2 = \|R\|\mu(A^*A)$ and thus $\nu \in S_0(\mathfrak{C}, \mu)$.

To prove (ii) let $(\mathcal{H}_\mu, \pi_\mu, \Omega_\mu)$ be the GNS representation of $\mathfrak{C}$ associated to μ. If $\nu \in \mathcal{S}(\mathfrak{C}, \mu)$ then there exists $\tilde{\nu} \in N(\mathfrak{C}_\mu)$ such that $\nu = \tilde{\nu} \circ \pi_\mu$ and $\tilde{\nu}(s_\mu) = 1$. Let ρ be a density matrix on $\mathcal{H}_\mu$ such that $\tilde{\nu}(A) = \mathrm{Tr}(\rho A)$. From the condition $\mathrm{Tr}(\rho s_\mu) = 1$ it follows that $\rho = s_\mu \rho s_\mu$. Thus, ρ is a density matrix in the subspace $\mathrm{Ran}\, s_\mu = \overline{\mathfrak{C}'_\mu \Omega_\mu}$ and Exercise 2.9 shows that $\tilde{\nu}$ can be approximated in norm by finite convex linear combinations of vector states $\tilde{\nu}_n(A) = (\psi_n, A\psi_n)$ with $\psi_n \in \mathfrak{C}'_\mu \Omega_\mu$. (ii) now follows from (iii) and the fact that the map $\pi_\mu^\star : \tilde{\nu} \mapsto \tilde{\nu} \circ \pi_\mu$ is norm continuous.

Since $E(\mathfrak{C})$ is weak-$\star$ compact in $\mathfrak{C}^\star$ (i) follows from the obvious fact that $\mathcal{S}^\lambda(\mathfrak{C}, \mu)$ is weak-$\star$ closed. □

Definition 4.56. *A quantum dynamical system $(\mathfrak{C}, \tau^t, \mu)$ is called:*

(i) Ergodic if, for any $\nu \ll \mu$ and any $A \in \mathfrak{C}$, one has

$$\lim_{T\to\infty} \frac{1}{T}\int_0^T \nu \circ \tau^t(A)\,\mathrm{d}t = \mu(A). \tag{34}$$

(ii) Mixing if, for any $\nu \ll \mu$ and any $A \in \mathfrak{C}$, one has

$$\lim_{t\to\infty} \nu \circ \tau^t(A) = \mu(A). \tag{35}$$

A state $\mu \in E(\mathfrak{C})$ is called ergodic (resp. mixing) if it belongs to $\mathcal{E}(\mathfrak{C}, \tau^t)$ and if the corresponding quantum dynamical system $(\mathfrak{C}, \tau^t, \mu)$ is ergodic (resp. mixing).

These definitions are consistent with the notion of normal form of a quantum dynamical system.

Proposition 4.57. *The quantum dynamical system $(\mathfrak{C}, \tau^t, \mu)$ is ergodic (resp. mixing) if and only if its normal form $(\pi_\mu, \mathfrak{C}_\mu, \mathcal{H}_\mu, L_\mu, \Omega_\mu)$ is ergodic (resp. mixing).*

Proof. We denote by $\hat{\mu}$ the vector state associated to Ω_μ and $\hat{\tau}^t_\mu$ the σ-weakly continuous group of $*$-automorphisms of $\mathfrak{C}_\mu$ generated by L_μ. Assume that $(\mathfrak{C}_\mu, \hat{\tau}^t_\mu, \hat{\mu})$ is ergodic (resp. mixing) and let $\nu \in \mathcal{S}(\mathfrak{C}, \mu)$ and $A \in \mathfrak{C}$. Then there exists

$\tilde{\nu} \in N(\mathfrak{C}_\mu)$ such that $\tilde{\nu} \ll \hat{\mu}$ and $\nu = \tilde{\nu} \circ \pi_\mu$. Since $\nu \circ \tau^t(A) = \tilde{\nu} \circ \hat{\tau}^t_\mu(\pi_\mu(A))$ and $\mu(A) = \hat{\mu}(\pi_\mu(A))$ the convergence in Equ. (34) (resp. Equ. (35)) follows directly from the corresponding statement for $(\mathfrak{C}_\mu, \hat{\tau}^t_\mu, \hat{\mu})$.

To prove the reverse statement assume that $(\mathfrak{C}, \tau^t, \mu)$ is ergodic (resp. mixing). Since $\hat{\tau}^t_\mu$ is isometric the map $\nu \mapsto \nu \circ \hat{\tau}^t_\mu$ is norm continuous uniformly in $t \in \mathbb{R}$. To prove convergence in Equ. (34) (resp. Equ. (35)) for all $\nu \in \mathcal{S}(\mathfrak{C}_\mu, \hat{\mu})$ it is therefore sufficient to prove it for all ν in a total subset of $\mathcal{S}(\mathfrak{C}_\mu, \hat{\mu})$. By Lemma 4.55, $\mathcal{S}_0(\mathfrak{C}_\mu, \hat{\mu})$ is total in $\mathcal{S}(\mathfrak{C}_\mu, \hat{\mu})$ and is the union of the sets $\mathcal{S}^n(\mathfrak{C}_\mu, \hat{\mu})$. Thus, it suffices to consider $\nu \in \mathcal{S}^n(\mathfrak{C}_\mu, \hat{\mu})$ for some $n > 0$. We set

$$\nu_t(A) \equiv \frac{1}{t} \int_0^t \nu \circ \hat{\tau}^s_\mu(A)\, \mathrm{d}s, \text{ (resp. } \nu_t(A) \equiv \nu \circ \hat{\tau}^t_\mu(A)) .$$

It follows from the fact that $\hat{\mu}$ is $\hat{\tau}^t_\mu$-invariant that $\nu_t \in \mathcal{S}^n(\mathfrak{C}_\mu, \hat{\mu})$ for all $t \in \mathbb{R}$. Since $\mathcal{S}^n(\mathfrak{C}_\mu, \hat{\mu})$ is weak-$\star$ compact the set of cluster point of the net ν_t is non-empty and contained in $\mathcal{S}^n(\mathfrak{C}_\mu, \hat{\mu}) \subset N(\mathfrak{C}_\mu)$. If $\bar{\nu}$ is such a cluster point then it follows from $\nu \circ \hat{\tau}^t_\mu(\pi_\mu(A)) = (\nu \circ \pi_\mu) \circ \tau^t(A)$ and $\nu \circ \pi_\mu \ll \mu$ that $\bar{\nu}(\pi_\mu(A)) = \hat{\mu}(\pi_\mu(A))$. Since $\bar{\nu}$ is normal we conclude that $\bar{\nu} = \hat{\mu}$ and hence that $\hat{\mu}$ is the weak-$\star$ limit of the net ν_t. □

Example 4.58. (Finite quantum systems, continuation of Example 4.32) If ψ is a normalized eigenvector of the Hamiltonian H then the state $\mu \equiv (\psi, (\,\cdot\,)\psi)$ is ergodic and even mixing. Indeed, since $s_\mu = \psi(\psi, \,\cdot\,)$, the only state ν satisfying $\nu \ll \mu$ is μ itself. Any normal invariant state which is a mixture of several such eigenstates is non-ergodic. In fact, the formula

$$\rho(A) = \sum_j p_j(\psi_j, A\psi_j),$$

expresses the decomposition of the state ρ into ergodic components.

Remark 4.59. Proposition 4.57 reduces the study of the ergodic properties of a quantum dynamical system to the special case of a W^*-dynamical systems $(\mathfrak{M}, \tau^t, \mu)$ with $\mu \in N(\mathfrak{M}, \tau^t)$. For such a quantum dynamical system one has

$$N(\mathfrak{M}, \mu) = \{\nu \in N(\mathfrak{M}) \,|\, s_\nu \le z_\mu\},$$

by Lemma 2.29. Since $s_\mu \le z_\mu$ we conclude that $\nu \ll \mu$ if and only if $\nu \in N(\mathfrak{M})$ and $\nu(s_\mu) = 1$, *i.e.*, $\nu = \nu(s_\mu(\,\cdot\,)s_\mu)$. For $\nu \in N(\mathfrak{M})$ one has either $\nu(s_\mu) = 0$ and thus $\nu(s_\mu(\,\cdot\,)s_\mu) = 0$ or $\nu(s_\mu(\,\cdot\,)s_\mu)/\nu(s_\mu) \ll \mu$. We conclude that

$$\{\nu \in E(\mathfrak{M}) \,|\, \nu \ll \mu\} = \{\nu(s_\mu(\,\cdot\,)s_\mu) \,|\, \nu \in \mathfrak{M}_\star, \nu \ge 0, \nu(s_\mu) = 1\}. \qquad (36)$$

In particular, the set of linear combinations of states $\nu \ll \mu$ is the set of linear functionals of the form $\nu(s_\mu(\,\cdot\,)s_\mu)$ where $\nu \in \mathfrak{M}_\star$.

As a quantum analogue of Proposition 3.6 we have the following characterizations of ergodicity which, using Proposition 4.57, can be applied to the normal form of a quantum dynamical system.

Proposition 4.60. *Let $(\mathfrak{M}, \tau^t)$ be a W^*-dynamical system and $\mu \in N(\mathfrak{M}, \tau^t)$. Denote by $\mathfrak{M}_{\tau,\mu} \equiv \{A \in s_\mu \mathfrak{M} s_\mu \,|\, \tau^t(A) = A \text{ for all } t \in \mathbb{R}\}$ the subalgebra of τ^t-invariant observables modulo μ. The following propositions are equivalent.*

(i) $(\mathfrak{M}, \tau^t, \mu)$ is ergodic.
(ii) If $P \in \mathfrak{M}_{\tau,\mu}$ is an orthogonal projection then $\mu(P) \in \{0, 1\}$.
(iii) If $A \in \mathfrak{M}_{\tau,\mu}$ then $A = \mu(A)s_\mu$.

Proof. (i)⇒(ii). By Remark 4.59, the system is ergodic if and only if

$$\lim_{T\to\infty} \frac{1}{T} \int_0^T \nu \circ \tau^t(s_\mu A s_\mu)\, \mathrm{d}t = \nu(s_\mu)\mu(A), \tag{37}$$

holds for all $A \in \mathfrak{M}$ and $\nu \in \mathfrak{M}_\star$. Let $P \in \mathfrak{M}_{\tau,\mu}$ be an orthogonal projection. Since $P = s_\mu P s_\mu$ is τ^t-invariant Equ. (37) yields $\nu(P) = \nu(s_\mu)\mu(P)$, *i.e.*, $\nu(P - \mu(P)s_\mu) = 0$ for any $\nu \in \mathfrak{M}_\star$. Thus, $P = \mu(P)s_\mu$ and from

$$0 = P(I - P) = \mu(P)s_\mu(I - \mu(P)s_\mu) = \mu(P)(1 - \mu(P))s_\mu,$$

we conclude that $\mu(P) \in \{0, 1\}$.

(ii)⇒(iii). It clearly suffices to prove that $A = \mu(A)s_\mu$ for all self-adjoint elements of $\mathfrak{M}_{\tau,\mu}$. Let $A \in \mathfrak{M}_{\tau,\mu}$ be self-adjoint and denote by P_a its spectral projection corresponding to the interval $]-\infty, a]$. Note that if $\mathfrak{M} \subset \mathfrak{B}(\mathcal{H})$ then we can interpret $\mathfrak{M}_{\tau,\mu}$ as a von Neumann algebra on $s_\mu\mathcal{H}$ and the functional calculus yields $P_a \in \mathfrak{M}_{\tau,\mu}$. It follows from (ii) that for any $a \in \mathbb{R}$ either $\mu(P_a) = 0$ or $\mu(P_a) = 1$. In the first case we get $P_a s_\mu = 0$ and hence $P_a = 0$. In the second case we have $P_a \geq s_\mu$ and hence $P_a = s_\mu$. By the spectral theorem A is a multiple of s_μ and (iii) follows.

(iii)⇒(i). Let $\nu \ll \mu$ and $\tilde{\nu} \in E(\mathfrak{M})$ be a weak-$\star$ cluster point of the net

$$\nu_T \equiv \frac{1}{T} \int_0^T \nu \circ \tau^t \, \mathrm{d}t.$$

Denote by T_α a net such that $\lim_\alpha \nu_{T_\alpha}(A) = \tilde{\nu}(A)$ for all $A \in \mathfrak{M}$. For $A \in \mathfrak{M}$ consider the corresponding net

$$A_{T_\alpha} \equiv \frac{1}{T_\alpha} \int_0^{T_\alpha} \tau^t(A)\, \mathrm{d}t \in \mathfrak{M}_{\|A\|} \equiv \{B \in \mathfrak{M} \,|\, \|B\| \leq \|A\|\}.$$

By the Banach-Alaoglu theorem $\mathfrak{M}_{\|A\|}$ is σ-weakly compact and there exists a subnet T_β of the net T_α such that $A_{T_\beta} \to \tilde{A}$ σ-weakly. Since μ is σ-weakly continuous and τ^t-invariant $\mu(\tilde{A}) = \mu(A)$. By the usual argument (see the proof of Theorem 4.20) we have $\tau^t(\tilde{A}) = \tilde{A}$ for all t. Since μ is τ^t-invariant we also have $\tau^t(s_\mu) = s_\mu$ and hence $s_\mu \tilde{A} s_\mu \in \mathfrak{M}_{\tau,\mu}$. It follows from (iii) that $s_\mu \tilde{A} s_\mu = \mu(s_\mu \tilde{A} s_\mu)s_\mu = \mu(\tilde{A})s_\mu = \mu(A)s_\mu$. On the other hand, since ν is σ-weakly continuous and $\nu(s_\mu) = 1$, we have

$$\tilde{\nu}(A) = \lim_{\alpha} \nu_{T_\alpha}(A) = \lim_{\beta} \nu_{T_\beta}(A) = \lim_{\beta} \nu(A_{T_\beta}) = \nu(\tilde{A}) = \nu(s_\mu \tilde{A} s_\mu),$$

and hence $\tilde{\nu}(A) = \nu(\mu(A)s_\mu) = \mu(A)$. Since this holds for any $A \in \mathfrak{M}$ and any cluster point of the net ν_T we conclude that ν_T converges to μ in the weak-$\star$ topology, *i.e.*, that Equ. (34) holds. □

Note that if $(\mathfrak{M}, \tau^t, \mu)$ is ergodic and μ is faithful ($s_\mu = I$) it follows immediately that μ is the only normal invariant state of the W^*-dynamical system $(\mathfrak{M}, \tau^t)$. More generally, one has

Proposition 4.61. *If the quantum dynamical systems $(\mathfrak{C}, \tau^t, \mu_1)$, $(\mathfrak{C}, \tau^t, \mu_2)$ are ergodic then either $\mu_1 = \mu_2$ or $\mu_1 \perp \mu_2$.*

Proof. Since $\mu_1, \mu_2 \in \mathcal{E}(\mathfrak{C}, \tau^t)$ we have $\omega \equiv (\mu_1 + \mu_2)/2 \in \mathcal{E}(\mathfrak{C}, \tau^t)$ and hence $(\mathfrak{C}, \tau^t, \omega)$ is a quantum dynamical system. Let $(\mathfrak{C}_\omega, \hat{\tau}^t_\omega, \hat{\omega})$ be its normal form. From the fact that $\mu_i \le 2\omega$ we conclude by Lemma 4.55 that μ_1 and μ_2 are ω-normal. Denote by $\tilde{\mu}_i$ the normal extension of μ_i to $\mathfrak{C}_\omega$, $s_i \equiv s_{\mu_i|\omega}$ its support, $z_i \equiv z_{\mu_i|\omega}$ its central support and set $\mathfrak{M}_i \equiv z_i \mathfrak{C}_\omega$. By Remark 4.31 the quantum dynamical system $(\mathfrak{M}_i, \hat{\tau}^t_\omega, \tilde{\mu}_i)$ is isomorphic to the normal form $(\mathfrak{C}_{\mu_i}, \hat{\tau}^t_{\mu_i}, \hat{\mu}_i)$ of $(\mathfrak{C}, \tau^t, \mu_i)$ and is therefore ergodic. Since the supports s_1, s_2 are invariant under $\hat{\tau}^t_\omega$ so is $P \equiv s_1 \wedge s_2$. From $s_i \le z_i$ we conclude that $P \le z_1 \wedge z_2 \le z_i$ and hence $P \in \mathfrak{M}_1^P \cap \mathfrak{M}_2^P$. Since $P = s_i P s_i$, Proposition 4.60 yields that $\tilde{\mu}_i(P) \in \{0, 1\}$. As in the proof of Proposition 4.60 this implies that $P = \tilde{\mu}_1(P)s_1 = \tilde{\mu}_2(P)s_2$ and hence either $P = 0$ or $P = s_1 = s_2$. In the later case one has $s_1 \le s_2 \le z_2$ and Corollary 2.24 yields $\mu_1 \ll \mu_2$. It immediately follows from Equ. (34) that $\mu_1 = \mu_2$. In the former case assume that for some $\lambda > 0$ and some state ν one has $\lambda\nu \le \mu_i$ for $i = 1, 2$. Then $\lambda\nu \le \omega$ and by Lemma 4.55, ν has a normal extension $\tilde{\nu}$ to $\mathfrak{C}_\omega$ which satisfies $\lambda\tilde{\nu} \le \tilde{\mu}_i$. It follows that $\tilde{\nu}(I - s_1) = \tilde{\nu}(I - s_2) = 0$ and hence $\tilde{\nu}(s_1) = \tilde{\nu}(s_2) = 1$. By Exercise 2.14 we get $\tilde{\nu}(P) = 1$, a contradiction which shows that $\mu_1 \perp \mu_2$. □

Proposition 4.62. *A quantum dynamical system $(\mathfrak{C}, \tau^t, \mu)$ is ergodic if and only if the state μ is an extremal point of the set $\mathcal{E}(\mathfrak{C}, \tau^t)$.*

Proof. Assume that $(\mathfrak{C}, \tau^t, \mu)$ is ergodic and that there exists $\mu_1, \mu_2 \in \mathcal{E}(\mathfrak{C}, \tau^t)$ and $\alpha_1, \alpha_2 \in]0, 1[$ such that $\alpha_1 + \alpha_2 = 1$ and $\mu = \alpha_1\mu_1 + \alpha_2\mu_2$. By Lemma 4.55 it follows from the fact that $\alpha_i\mu_i \le \mu$ that μ_i is μ-normal. Denote by $\tilde{\mu}_i$ its normal extension to $\mathfrak{C}_\mu$. By continuity one has $\alpha_i\tilde{\mu}_i \le \hat{\mu}$ and hence $\tilde{\mu}_i(I - s_\mu) = 0$, *i.e.*, $s_{\mu_i} \le s_\mu$. Equ. (34) then yields $\mu_i = \mu$ and we conclude that μ is extremal.

Assume now that μ is extremal in $\mathcal{E}(\mathfrak{C}, \tau^t)$ and let $(\mathfrak{C}_\mu, \hat{\tau}^t_\mu, \hat{\mu})$ be the corresponding normal form. Let $(\mathcal{H}, \pi)$ be a standard representation of $\mathfrak{C}_\mu$ and denote by $(\mathcal{H}, \mathfrak{M}, J, \mathcal{C})$ its standard form. Let L be the corresponding standard Liouvillean and Ψ_μ the standard vector representative of $\hat{\mu}$. The dynamics is implemented in $\mathfrak{M}$ by $\sigma^t(A) = \mathrm{e}^{itL} A \mathrm{e}^{-itL}$ and if we set $\bar{\mu}(A) \equiv (\Psi_\mu, A\Psi_\mu)$ then the normal form is isomorphic to the quantum dynamical system $(\mathfrak{M}, \sigma^t, \bar{\mu})$.

If $\bar{\nu} \in N(\mathfrak{M}, \sigma^t)$ then $\bar{\nu} \circ \pi \circ \pi_\mu$ is μ-normal and invariant, hence it belongs to $\mathcal{E}(\mathfrak{C}, \tau^t)$. In fact, since $\pi_\mu(\mathfrak{C})$ is σ-weakly dense in $\mathfrak{C}_\mu$ the map $\bar{\nu} \mapsto \bar{\nu} \circ \pi \circ \pi_\mu$ is an injection from $N(\mathfrak{M}, \sigma^t)$ into $\mathcal{E}(\mathfrak{C}, \tau^t)$. It follows that $\bar{\mu}$ is an extremal point of $N(\mathfrak{M}, \sigma^t)$. Indeed, let $\bar{\mu}_1, \bar{\mu}_2 \in N(\mathfrak{M}, \sigma^t)$ and $\alpha \in]0,1[$ be such that $\bar{\mu} = \alpha\bar{\mu}_1 + (1-\alpha)\bar{\mu}_2$. Then we have $\mu = \alpha\bar{\mu}_1 \circ \pi \circ \pi_\mu + (1-\alpha)\bar{\mu}_2 \circ \pi \circ \pi_\mu$ and since μ is extremal this yields $\bar{\mu}_1 = \bar{\mu}_2$.

Let $P \in \mathfrak{M}_{\sigma,\bar{\mu}}$ and assume that $\alpha \equiv \bar{\mu}(P) \in]0,1[$. Set $P' \equiv JPJ \in \mathfrak{M}'$ and define the states

$$\bar{\mu}_1(A) \equiv \frac{(P'\Psi_\mu, AP'\Psi_\mu)}{\|P'\Psi_\mu\|^2}, \qquad \bar{\mu}_2(A) \equiv \frac{((I-P')\Psi_\mu, A(I-P')\Psi_\mu)}{\|(I-P')\Psi_\mu\|^2},$$

on $\mathfrak{M}$. Since $\|P'\Psi_\mu\|^2 = (\Psi_\mu, JPJ\Psi_\mu) = (PJ\Psi_\mu, J\Psi_\mu) = \alpha$, we have

$$\begin{aligned} \alpha\bar{\mu}_1(A) + (1-\alpha)\bar{\mu}_2(A) &= (P'\Psi_\mu, A\Psi_\mu) + ((I-P')\Psi_\mu, A\Psi_\mu) \\ &= (\Psi_\mu, A\Psi_\mu) \\ &= \bar{\mu}(A), \end{aligned}$$

for any $A \in \mathfrak{M}$. Moreover, since

$$\begin{aligned} \mathrm{e}^{-itL}P'\mathrm{e}^{itL} &= \mathrm{e}^{-itL}JPJ\mathrm{e}^{itL} \\ &= J\mathrm{e}^{-itL}P\mathrm{e}^{itL}J \\ &= J\sigma^{-t}(P)J = JPJ = P', \end{aligned}$$

and $\mathrm{e}^{itL}\Psi_\mu = \Psi_\mu$, one easily checks that $\bar{\mu}_1$ and $\bar{\mu}_2$ are invariant and hence belong to $N(\mathfrak{M}, \tau^t)$. Let us assume that $\bar{\mu}_1 = \bar{\mu}_2$. Then $\bar{\mu} = \bar{\mu}_1$ and for any $A \in \mathfrak{M}$ we get $(\alpha\Psi_\mu, A\Psi_\mu) = (P'\Psi_\mu, A\Psi_\mu)$. It follows that

$$0 = [\mathfrak{M}\Psi_\mu](P'-\alpha)\Psi_\mu = [\mathfrak{M}\Psi_\mu]J(P-\alpha)J\Psi_\mu = J[\mathfrak{M}'\Psi_\mu](P-\alpha)\Psi_\mu,$$

from which we conclude that $s_\mu(P-\alpha)\Psi_\mu = 0$. Since $P \in \mathfrak{M}_{\sigma,\bar{\mu}}$ it satisfies $P = s_\mu P$ which yields $P\Psi_\mu = \alpha\Psi_\mu$. This is impossible since P is a projection and $\alpha \in]0,1[$. This contradiction shows that the states $\bar{\mu}_1 \neq \bar{\mu}_2$ provide a non-trivial decomposition of $\bar{\mu}$, a contradiction to its extremality. We conclude that $\bar{\mu}(P) \in \{0,1\}$ for all $P \in \mathfrak{M}_{\sigma,\bar{\mu}}$ and hence, by Proposition 4.60, that $(\mathfrak{M}, \sigma^t, \bar{\mu})$ is ergodic. Since this system is isomorphic to a normal form of $(\mathfrak{C}, \tau^t, \mu)$, the later is also ergodic by Proposition 4.57. □

The last result in this subsection is the following characterization of the mixing property.

Proposition 4.63. *Let $(\mathfrak{M}, \tau^t)$ be a W^*-dynamical system and $\mu \in N(\mathfrak{M}, \tau^t)$. The quantum dynamical system $(\mathfrak{M}, \tau^t, \mu)$ is mixing if and only if, for all $A, B \in \mathfrak{M}$, one has*

$$\lim_{t\to\infty} \mu(As_\mu\tau^t(B)) = \mu(A)\mu(B). \tag{38}$$

Proof. By Remark 4.59 the system is mixing if and only if

$$\lim_{t\to\infty} \nu \circ \tau^t(s_\mu B s_\mu) = \nu(s_\mu)\mu(B), \tag{39}$$

holds for any $\nu \in \mathfrak{M}_\star$ and $B \in \mathfrak{M}$. If the system is mixing then Equ. (38) follows from Equ. (39) with $\nu \equiv \mu(A(\cdot)) \in \mathfrak{M}_\star$.

To prove the reverse statement suppose that Equ. (38) holds. By Proposition 4.57 we may also assume that the system is in normal form, *i.e.*, that $\mathfrak{M} \subset \mathfrak{B}(\mathcal{H})$, $\mu(B) = (\Omega_\mu, B\Omega_\mu)$ for some unit vector $\Omega_\mu \in \mathcal{H}$, $\overline{\mathfrak{M}\Omega_\mu} = \mathcal{H}$ and $s_\mu = [\mathfrak{M}'\Omega_\mu]$.

Since τ^t is isometric it suffices to show Equ. (39) for a norm total set of ν in $\mathcal{S}(\mathfrak{M}, \mu)$. By Lemma 4.55, $\mathcal{S}_0(\mathfrak{M}, \mu)$ is total in $\mathcal{S}(\mathfrak{M}, \mu)$ and its elements are of the form $\nu(B) = (R^* R\Omega_\mu, B\Omega_\mu)$ with $R \in \mathfrak{M}'$. But $R^* R\Omega_\mu$ can be approximated in norm by vectors of the type $s_\mu A^* \Omega_\mu$ with $A \in \mathfrak{M}$ and hence the set of linear functionals of the form $\nu(B) = (s_\mu A^* \Omega_\mu, B\Omega_\mu)$ is also total in $\mathcal{S}(\mathfrak{M}, \mu)$. For such ν, it does indeed follow from Equ. (38) that

$$\nu \circ \tau^t(s_\mu B s_\mu) = \mu(A s_\mu \tau^t(B)) \to \mu(A)\mu(B) = \nu(s_\mu)\mu(B)$$

as $t \to \infty$. □

4.7 Quantum Koopmanism

To develop the spectral theory of quantum dynamical systems along the lines of Theorem 3.14 and 3.15 one may be tempted to use the μ-Liouvillean L_μ or the standard Liouvillean L. The next exercise shows that this is not possible.

Exercise 4.64. In Examples 4.32, 4.58, assume that E is a degenerate eigenvalue of H and ψ a corresponding normalized eigenvector. Show that the μ-Liouvillean L_μ corresponding to the state $\mu(A) = (\psi, A\psi)$ is unitarily equivalent to $H - E$. Thus, even though μ is ergodic, 0 is not a simple eigenvalue of L_μ. Show also that if H has N eigenvalues (counting multiplicities) then 0 is an N-fold degenerate eigenvalue of the standard Liouvillean L.

In fact a further reduction is necessary to obtain the analogue of the classical Liouvillean.

Theorem 4.65. *Let $(\mathfrak{C}, \tau^t, \mu)$ be a quantum dynamical system with normal form $(\pi_\mu, \mathfrak{C}_\mu, \mathcal{H}_\mu, L_\mu, \Omega_\mu)$. Then the support $s_\mu = [\mathfrak{C}'_\mu \Omega_\mu]$ reduces the μ-Liouvillean L_μ. We call reduced Liouvillean or simply "The Liouvillean" and denote by $\mathfrak{L}_\mu$ the restriction of L_μ to* Ran s_μ.

(i) $(\mathfrak{C}, \tau^t, \mu)$ is ergodic if and only if Ker $\mathfrak{L}_\mu$ *is one-dimensional.*
(ii) $(\mathfrak{C}, \tau^t, \mu)$ is mixing if and only if

$$\mathrm{w}-\lim_{t\to\infty} \mathrm{e}^{it\mathfrak{L}_\mu} = \Omega_\mu(\Omega_\mu, \cdot\,).$$

(iii) If the spectrum of $\mathfrak{L}_\mu$ on $\{\Omega_\mu\}^\perp$ is purely absolutely continuous then $(\mathfrak{C}, \tau^t, \mu)$ is mixing.

Proof. Let $\hat{\tau}^t(A) \equiv \mathrm{e}^{itL_\mu} A \mathrm{e}^{-itL_\mu}$. Since the vector state $\hat{\mu}(A) \equiv (\Omega_\mu, A\Omega_\mu)$ is $\hat{\tau}^t$-invariant its support s_μ satisfies

$$\mathrm{e}^{itL_\mu} s_\mu \mathrm{e}^{-itL_\mu} = s_\mu,$$

and hence L is reduced by $\mathrm{Ran}\, s_\mu$. As in the proof of Proposition 4.57 we can replace $\mathcal{S}(\mathfrak{C}, \mu)$ by $\mathcal{S}_0(\mathfrak{C}, \mu)$ in Definition 4.56. For $\nu \in \mathcal{S}_0(\mathfrak{C}, \mu)$ we have

$$\nu \circ \hat{\tau}^t(A) = (R\Omega_\mu, \mathrm{e}^{itL_\mu} A \mathrm{e}^{-itL_\mu} R\Omega_\mu) = (R^* R\Omega_\mu, \mathrm{e}^{itL_\mu} A\Omega_\mu).$$

Since $R^* R\Omega_\mu = s_\mu R^* R\Omega_\mu$ we can rewrite

$$\nu \circ \hat{\tau}^t(A) = (s_\mu R^* R\Omega_\mu, \mathrm{e}^{itL_\mu} s_\mu A\Omega_\mu) = (s_\mu R^* R\Omega_\mu, \mathrm{e}^{it\mathfrak{L}_\mu} s_\mu A\Omega_\mu).$$

From the mean ergodic theorem (Theorem 3.13) we get

$$\lim_{T\to\infty} \frac{1}{T} \int_0^T \nu \circ \hat{\tau}^t(A)\, \mathrm{d}t = (s_\mu R^* R\Omega_\mu, P s_\mu A\Omega_\mu),$$

where P is the orthogonal projection on $\mathrm{Ker}\, \mathfrak{L}_\mu$. Note that if $P = \Omega_\mu(\Omega_\mu, \cdot\,)$ then the right hand side of the last formula reduces to

$$(s_\mu R^* R\Omega_\mu, \Omega_\mu)(\Omega_\mu, s_\mu A\Omega_\mu) = (R\Omega_\mu, R\Omega_\mu)(\Omega_\mu, A\Omega_\mu) = \hat{\mu}(A).$$

Since the systems $\{s_\mu R^* R\Omega_\mu \,|\, R \in \mathfrak{C}_\mu'\}$ and $\{s_\mu A\Omega_\mu \,|\, A \in \mathfrak{C}_\mu\}$ are both total in $\mathrm{Ran}\, s_\mu$, we conclude that the system is ergodic if and only if $P = \Omega_\mu(\Omega_\mu, \cdot\,)$. Since Ω_μ always belongs to the kernel of $\mathfrak{L}_\mu$, this is equivalent to the condition (i). The proof of (ii) is completely similar. Finally (iii) is proved as in the classical case (Theorem 3.15). □

In many applications the invariant state $\hat{\mu}$ is faithful on $\mathfrak{C}_\mu$. In this case the μ-Liouvillean and the standard Liouvillean coincide, $s_\mu = I$ and the previous result reduces to the simpler

Corollary 4.66. *Let $(\mathfrak{C}, \tau^t, \mu)$ be a quantum dynamical system with normal form $(\pi_\mu, \mathfrak{C}_\mu, \mathcal{H}_\mu, L_\mu, \Omega_\mu)$ and assume that $s_\mu = [\mathfrak{C}_\mu' \Omega_\mu] = I$.*

(i) $(\mathfrak{C}, \tau^t, \mu)$ is ergodic if and only if $\mathrm{Ker}\, L_\mu$ *is one-dimensional.*
(ii) $(\mathfrak{C}, \tau^t, \mu)$ is mixing if and only if

$$\mathrm{w} - \lim_{t\to\infty} \mathrm{e}^{itL_\mu} = \Omega_\mu(\Omega_\mu, \cdot\,).$$

(iii) If the spectrum of L_μ on $\{\Omega_\mu\}^\perp$ is purely absolutely continuous then $(\mathfrak{C}, \tau^t, \mu)$ is mixing.

As in the classical case (see for example Theorem 9.12 in [5]), the spectrum of the Liouvillean of an ergodic quantum dynamical system is not arbitrary (see [24] and [31] for related results).

Theorem 4.67. *Let $\mathfrak{L}_\mu$ be the Liouvillean of the ergodic quantum dynamical system $(\mathfrak{C}, \tau^t, \mu)$.*

(i) The set of eigenvalues of $\mathfrak{L}_\mu$ is a subgroup Σ of the additive group $\mathbb{R}$.
(ii) The eigenvalues of $\mathfrak{L}_\mu$ are simple.
(iii) The spectrum of $\mathfrak{L}_\mu$ is invariant under translations in Σ, that is

$$\sigma(\mathfrak{L}_\mu) + \Sigma = \sigma(\mathfrak{L}_\mu).$$

(iv) If Φ is a normalized eigenvector of $\mathfrak{L}_\mu$ then the corresponding vector state is the normal extension of μ to $\mathfrak{C}_\mu$.
(v) If $(\mathfrak{C}, \tau^t, \mu)$ is mixing then 0 is the only eigenvalue of $\mathfrak{L}_\mu$.

Proof. Let $(\pi_\mu, \mathfrak{C}_\mu, \mathcal{H}_\mu, L_\mu, \Omega_\mu)$ be the normal form of $(\mathfrak{C}, \tau^t, \mu)$ and set $\hat{\mu}(A) = (\Omega_\mu, A\Omega_\mu)$, $\hat{\tau}^t(A) = \mathrm{e}^{itL_\mu} A \mathrm{e}^{-itL_\mu}$, $s_\mu = [\mathfrak{C}'_\mu \Omega_\mu]$ and $\mathcal{K}_\mu = s_\mu \mathcal{H}_\mu$. Note that $\hat{\tau}^t$ extends to $\mathfrak{B}(\mathcal{H}_\mu)$ and that

$$\hat{\tau}^t(\mathfrak{C}'_\mu)\mathfrak{C}_\mu = \hat{\tau}^t(\mathfrak{C}'_\mu \hat{\tau}^{-t}(\mathfrak{C}_\mu)) = \hat{\tau}^t(\hat{\tau}^{-t}(\mathfrak{C}_\mu)\mathfrak{C}'_\mu) = \mathfrak{C}_\mu \hat{\tau}^t(\mathfrak{C}'_\mu),$$

shows that $\hat{\tau}^t(\mathfrak{C}'_\mu) = \mathfrak{C}'_\mu$. By Proposition 4.57, $(\mathfrak{C}_\mu, \hat{\tau}^t, \hat{\mu})$ is ergodic. Denote by Σ the set of eigenvalues of $\mathfrak{L}_\mu$. If $\Sigma = \{0\}$ there is nothing to prove. Suppose that $\lambda \in \Sigma \setminus \{0\}$ and let $\Phi \in \mathcal{K}_\mu$ be a corresponding normalized eigenvector. Since $\mathrm{e}^{-itL_\mu}\Phi = \mathrm{e}^{-it\mathfrak{L}_\mu}\Phi = \mathrm{e}^{-it\lambda}\Phi$ it follows that the state $\nu(A) \equiv (\Phi, A\Phi)$ is invariant. Moreover since

$$\mathfrak{C}'_\mu \Phi \subset \mathfrak{C}'_\mu \mathcal{K}_\mu = \mathcal{K}_\mu, \tag{40}$$

one has $s_\nu \leq s_\mu$ and hence $\nu \ll \hat{\mu}$. Ergodicity yields that $\nu = \hat{\mu}$ and in particular $s_\nu = s_\mu$, *i.e.*, $\overline{\mathfrak{C}'_\mu \Phi} = \mathcal{K}_\mu$. This proves (iv).

For all $A \in \mathfrak{C}_\mu$ we have $\|A\Phi\| = \nu(A^*A) = \hat{\mu}(A^*A) = \|A\Omega_\mu\|$ which shows that the linear map

$$\begin{aligned} T_\lambda : \mathfrak{C}_\mu \Omega_\mu &\to \mathfrak{C}_\mu \Phi \\ A\Omega_\mu &\mapsto A\Phi, \end{aligned}$$

is well defined, densely defined and isometric on $\mathcal{H}_\mu$. We also denote by T_λ its unique isometric extension to $\mathcal{H}_\mu$. For $A, B \in \mathfrak{C}_\mu$ we obtain

$$(T_\lambda^* B\Phi, A\Omega_\mu) = (B\Phi, T_\lambda A\Omega_\mu) = (B\Phi, A\Phi) = \hat{\mu}(B^*A) = (B\Omega_\mu, A\Omega_\mu),$$

from which we conclude that

$$T_\lambda^* B\Phi = B\Omega_\mu. \tag{41}$$

It follows that $\mathrm{Ran}(T_\lambda^*)^\perp = \mathrm{Ker}\, T_\lambda = \{0\}$, *i.e.*, that T_λ is unitary. Since $T_\lambda BA\Omega_\mu = BA\Phi = BT_\lambda A\Omega_\mu$ for all $A, B \in \mathfrak{C}_\mu$ we further get

$$T_\lambda \in \mathfrak{C}'_\mu. \tag{42}$$

In particular, we have $s_\mu T_\lambda = T_\lambda s_\mu$, so that T_λ and T_λ^* map $\mathcal{K}_\mu$ into itself. Finally, for $A \in \mathfrak{C}_\mu$, we have

$$
\begin{aligned}
\hat{\tau}^t(T_\lambda)A\Omega_\mu &= A\hat{\tau}^t(T_\lambda)\Omega_\mu = A\mathrm{e}^{itL_\mu}T_\lambda\mathrm{e}^{-itL_\mu}\Omega_\mu \\
&= A\mathrm{e}^{itL_\mu}T_\lambda\Omega_\mu = A\mathrm{e}^{itL_\mu}\Phi \\
&= \mathrm{e}^{it\lambda}A\Phi = \mathrm{e}^{it\lambda}T_\lambda A\Omega_\mu,
\end{aligned}
$$

which yields

$$\hat{\tau}^t(T_\lambda) = \mathrm{e}^{it\lambda}T_\lambda. \tag{43}$$

The last relation can be rewritten as $\mathrm{e}^{itL_\mu} = T_\lambda\mathrm{e}^{it(L_\mu+\lambda)}T_\lambda^*$ which means that $L_\mu+\lambda$ is unitarily equivalent to L_μ. Property (iii) follows immediately.

To prove (ii) suppose that λ is not a simple eigenvalue of $\mathfrak{L}_\mu$. Then there exists a unit vector $\Psi \in \mathcal{K}_\mu$ such that $\Psi \perp \Phi$ and $\mathfrak{L}_\mu\Psi = \lambda\Psi$. Since (41) implies that $T_\lambda^*\Phi = \Omega_\mu$ it follows that $T_\lambda^*\Psi \in \mathcal{K}_\mu$ is a unit vector orthogonal to Ω_μ. Moreover, Equ. (43) yields

$$\mathrm{e}^{it\mathfrak{L}_\mu}T_\lambda^*\Psi = \mathrm{e}^{itL_\mu}T_\lambda^*\Psi = \hat{\tau}^t(T_\lambda)^*e^{itL_\mu}\Psi = (e^{it\lambda}T_\lambda)^*e^{it\lambda}\Psi = T_\lambda^*\Psi,$$

from which we conclude that $T_\lambda^*\Psi$ is an eigenvector of $\mathfrak{L}_\mu$ to the eigenvalue 0, a contradiction to the simplicity of this eigenvalue. Thus, λ is simple.

To prove (i) we first note that Equ. (43) yields that $\Phi' \equiv T_\lambda^*\Omega_\mu \in \mathcal{K}_\mu$ is an eigenvector of $\mathfrak{L}_\mu$ to the eigenvalue $-\lambda$. Second, suppose that $\lambda' \in \Sigma \setminus \{0\}$ and let $T_{\lambda'}$ be the corresponding unitary. We will then have $\hat{\tau}^t(T_\lambda T_{\lambda'}) = \hat{\tau}^t(T_\lambda)\hat{\tau}^t(T_{\lambda'}) = \mathrm{e}^{it(\lambda+\lambda')}T_\lambda T_{\lambda'}$ from which we can conclude again that $T_\lambda T_{\lambda'}\Omega_\mu$ is an eigenvector of $\mathfrak{L}_\mu$ to the eigenvalue $\lambda+\lambda'$.

To prove (v) suppose on the contrary that the system is mixing and that $\mathfrak{L}_\mu$ has a non-zero eigenvalue λ. Let $\Phi \in \mathcal{K}_\mu$ be a corresponding normalized eigenvector. It follows that $\Phi \perp \Omega_\mu$ and

$$(\Phi, \mathrm{e}^{it\mathfrak{L}_\mu}\Phi) = \mathrm{e}^{it\lambda},$$

does not converge to zero as $t \to \infty$, a contradiction to the mixing criterion (ii) of Theorem 4.65. □

Using the relation between μ-Liouvillean and standard Liouvillean described in Proposition 4.49 it is easy to rephrase Theorem 4.65 in terms of the standard Liouvillean.

Exercise 4.68. Let $(\mathfrak{C}, \tau^t, \mu)$ be a quantum dynamical system with standard form $(\eta, \mathfrak{M}, \mathcal{H}, J, \mathcal{C}, L, \Phi)$. Let $s_\mu \equiv [\mathfrak{M}'\Phi] \in \mathfrak{M}$ be the support of the vector state $\bar{\mu}(A) \equiv (\Phi, A\Phi)$ and set $s_\mu' \equiv Js_\mu J = [\mathfrak{M}\Phi] \in \mathfrak{M}'$.

(i) Show that $S_\mu \equiv s_\mu s_\mu'$ is an orthogonal projection which reduces the standard Liouvillean *i.e.*, $\mathrm{e}^{itL}S_\mu\mathrm{e}^{-itL} = S_\mu$.
(ii) Show that the Liouvillean $\mathfrak{L}_\mu$ of Theorem 4.49 is unitarily equivalent to the restriction of the standard Liouvillean L to Ran S_μ.

As we have seen, the μ-Liouvillean is good enough to study the basic ergodic properties of a quantum dynamical system. However, to get deeper results on the structure of the state space and in particular on the manifold of invariant states it is necessary to use the standard Liouvillean.

Theorem 4.69. *Let L be the standard Liouvillean of the W^*-dynamical system $(\mathfrak{M}, \tau^t)$. Then*

$$N(\mathfrak{M}, \tau^t) = \{\omega_\Phi \,|\, \Phi \in \operatorname{Ker} L, \|\Phi\| = 1\},$$

and in particular

(i) $\operatorname{Ker} L = \{0\}$ *if and only if* $N(\mathfrak{M}, \tau^t) = \varnothing$.
(ii) $\operatorname{Ker} L$ *is one-dimensional if and only if* $(\mathfrak{M}, \tau^t)$ *has a unique normal invariant state. In this case, the corresponding quantum dynamical system is ergodic.*

Proof. If $\mu \in N(\mathfrak{M}, \tau^t)$ then, by Proposition 4.41, μ has a standard vector representative Ψ_μ such that $\mathrm{e}^{-itL}\Psi_\mu = \Psi_{\mu\circ\tau^t} = \Psi_\mu$ and hence $\Psi_\mu \in \operatorname{Ker} L$. Reciprocally, any unit vector $\Phi \in \operatorname{Ker} L$ defines a normal τ^t-invariant state ω_Φ. □

4.8 Perturbation Theory

Let $(\mathfrak{A}, \tau_0^t)$ be a C^*-dynamical system and denote by δ_0 its generator. A *local perturbation* of the system is obtained by perturbing its generator with the bounded $*$-derivation associated with a self-adjoint element V of $\mathfrak{A}$:

$$\delta_V = \delta + i[V, \cdot\,],$$

with $\mathrm{D}(\delta_V) = \mathrm{D}(\delta)$. Using Theorem 4.3 it is easy to show that δ_V generates a strongly continuous group of $*$-automorphisms τ_V^t on $\mathfrak{A}$. In fact $\tau_{\lambda V}^t$ is an entire analytic function of λ. An expansion in powers of λ is obtained by solving iteratively the integral equation (Duhamel formula)

$$\tau_{\lambda V}^t(X) = \tau_0^t(X) + \lambda \int_0^t i[\tau_0^{t-s}(V), \tau_{\lambda V}^s(X)]\,\mathrm{d}s.$$

The result is the Dyson-Robinson expansion

$$\begin{aligned}\tau_{\lambda V}^t(X) &= \tau_0^t(X) + \\ &+ \sum_{N=1}^{\infty} \lambda^N \int_{0\leqslant t_1\leqslant\cdots\leqslant t_N\leqslant t} i[\tau_0^{t_1}(V), i[\cdots, i[\tau_0^{t_n}(V), \tau_0^t(X)]\cdots]]\mathrm{d}t_1\cdots\mathrm{d}t_N,\end{aligned} \tag{44}$$

which is norm convergent for any $\lambda \in \mathbb{C}$, $t \in \mathbb{R}$ and $X \in \mathfrak{A}$. Another useful representation of the locally perturbed dynamics is the *interaction picture* obtained through the Ansatz

$$\tau_V^t(X) = \Gamma_V^t \tau_0^t(X) \Gamma_V^{t*}. \tag{45}$$

It leads to the differential equation

$$\partial_t \Gamma_V^t = i\Gamma_V^t \tau_0^t(V),$$

with the initial condition $\Gamma_V^0 = I$. It follows that Γ_V^t is a unitary element of $\mathfrak{A}$ which has the norm convergent Araki-Dyson expansion

$$\Gamma_V^t \equiv I + \sum_{N=1}^{\infty} i^N \int_{0 \leqslant t_1 \leqslant \cdots \leqslant t_N \leqslant t} \tau_0^{t_1}(V) \cdots \tau_0^{t_N}(V) \mathrm{d}t_1 \cdots \mathrm{d}t_N. \tag{46}$$

Moreover, Γ_V^t satisfies the cocycle relations

$$\Gamma_V^{t+s} = \Gamma_V^t \tau_0^t(\Gamma_V^s) = \tau_V^t(\Gamma_V^s)\Gamma_V^t. \tag{47}$$

Note that the integrals in Equ. (44) and (46) are Riemann integrals of norm continuous $\mathfrak{A}$-valued functions.

The interaction picture allows to obtain unitary implementations of τ_V^t in an arbitrary representation $(\mathcal{H}, \pi)$ of $\mathfrak{A}$ carrying a unitary implementation U_0^t of the unperturbed dynamics τ_0^t. Indeed, one has

$$\begin{aligned}
\pi(\tau_V^t(X)) &= \pi(\Gamma_V^t \tau_0^t(X) \Gamma_V^{t*}) \\
&= \pi(\Gamma_V^t)\pi(\tau_0^t(X))\pi(\Gamma_V^t)^* \\
&= \pi(\Gamma_V^t) U_0^t \pi(X) U_0^{t*} \pi(\Gamma_V^t)^*,
\end{aligned}$$

from which we conclude that the unitary $U_V^t = \pi(\Gamma_V^t) U_0^t$ implements τ_V^t in $\mathcal{H}$. The cocycle property (47) shows that U_V^t has the group property. From the expansion (46) we get norm convergent expansion (the integral are in the strong Riemann sense)

$$\begin{aligned}
U_V^t &= U_0^t + \\
&+ \sum_{N=1}^{\infty} i^N \int_{0 \leqslant t_1 \leqslant \cdots \leqslant t_N \leqslant t} U_0^{t_1} \pi(V) U_0^{t_2 - t_1} \cdots U_0^{t_N - t_{N-1}} \pi(V) U_0^{t - t_N} \mathrm{d}t_1 \cdots \mathrm{d}t_N.
\end{aligned}$$

Let G_V be the self-adjoint generator of U_V^t. Applying the last formula to a vector $\Phi \in \mathrm{D}(G_0)$ and differentiating at $t = 0$ we obtain $\Phi \in \mathrm{D}(G_V)$ and

$$G_V = G_0 + \pi(V). \tag{48}$$

Note however that the unitary implementation of τ_V^t in $\mathcal{H}$ is by no means unique. Indeed, e^{itK} is another implementation if and only if

$$\mathrm{e}^{itG_V} \pi(X) \mathrm{e}^{-itG_V} = \mathrm{e}^{itK} \pi(X) \mathrm{e}^{-itK},$$

for all $X \in \mathfrak{A}$ and all t. Thus $\mathrm{e}^{-itK}\mathrm{e}^{itG_V}$ must be a unitary element of $\pi(\mathfrak{A})'$ for all t. Assuming that $\mathrm{D}(K) = \mathrm{D}(G_V)$, differentiation at $t = 0$ yields

$$K = G_V - W = G_0 + \pi(V) - W,$$

for some self-adjoint element $W \in \pi(\mathfrak{A})'$. Then $\Gamma_W'^t = \mathrm{e}^{itK}\mathrm{e}^{-itG_V}$ satisfies the differential equation

$$\partial_t \Gamma_W'^t = -i \Gamma_W'^t W^t,$$

with $W^t = \mathrm{e}^{itG_V} W \mathrm{e}^{-itG_V}$ and initial value $\Gamma_W'^0 = I$. Since $\mathrm{e}^{itG_V} = \pi(\Gamma_V^t)\mathrm{e}^{itG_0}$ and $\mathrm{e}^{itG_0} W \mathrm{e}^{-itG_0} \in \pi(\mathfrak{A})'$ for all t, we have $W^t = \mathrm{e}^{itG_0} W \mathrm{e}^{-itG_0}$ and $\Gamma_W'^t$ is given by the norm convergent expansion

$$\Gamma_W'^t = I + \sum_{N=1}^{\infty} (-i)^N \int_{0 \leqslant t_1 \leqslant \cdots \leqslant t_N \leqslant t} W^{t_1} \cdots W^{t_N} \mathrm{d}t_1 \cdots \mathrm{d}t_N.$$

Local perturbations of W^*-dynamical systems can be treated in a completely similar way, the only change is that the norm topology should be replaced with the σ-weak topology, *i.e.* the integrals in Equ. (44) and (46) have to be understood in the weak-$\star$ sense.

Unbounded perturbations of a W^*-dynamical systems $(\mathfrak{M}, \tau_0^t)$ are common in applications. They require a slightly more sophisticated treatment. I shall only consider the case of a unitarily implemented unperturbed dynamics[4]

$$\tau_0^t(X) = \mathrm{e}^{\mathrm{i}tG_0} X \mathrm{e}^{-\mathrm{i}tG_0}.$$

Let V be a self-adjoint operator affiliated to $\mathfrak{M}$, *i.e.*, $\mathrm{e}^{\mathrm{i}tV} \in \mathfrak{M}$ for all $t \in \mathbb{R}$, and such that $G_0 + V$ is essentially self-adjoint on $\mathrm{D}(G_0) \cap \mathrm{D}(V)$. Denote by G_V its self-adjoint extension and set

$$\tau_V^t(X) = \mathrm{e}^{\mathrm{i}tG_V} X \mathrm{e}^{-\mathrm{i}tG_V}.$$

Defining the unitary

$$\Gamma_V^t = \mathrm{e}^{\mathrm{i}tG_V} \mathrm{e}^{-\mathrm{i}tG_0},$$

the Trotter product formula (Theorem 15 in Lecture [32]) shows that

$$\Gamma_V^t = \mathrm{s} - \lim_{n\to\infty} \mathrm{e}^{\mathrm{i}tV/n} \tau_0^{t/n}(\mathrm{e}^{\mathrm{i}tV/n}) \cdots \tau_0^{(n-1)t/n}(\mathrm{e}^{\mathrm{i}tV/n}),$$

and since $\mathrm{e}^{\mathrm{i}tV/n} \in \mathfrak{M}$ we get $\Gamma_V^t \in \mathfrak{M}$. By construction Equ. (45) and (47) remain valid. In particular τ_V^t leaves $\mathfrak{M}$ invariant and $(\mathfrak{M}, \tau_V^t)$ is a W^*-dynamical system. Another application of the Trotter formula further gives, for $X' \in \mathfrak{M}'$,

$$\mathrm{e}^{\mathrm{i}tG_V} X' \mathrm{e}^{-\mathrm{i}tG_V} = \mathrm{s} - \lim_{n\to\infty} \left(\mathrm{e}^{\mathrm{i}tG_0/n} \mathrm{e}^{\mathrm{i}tV/n}\right)^n X' \left(\mathrm{e}^{-\mathrm{i}tV/n} \mathrm{e}^{-\mathrm{i}tG_0/n}\right)^n,$$

which allows to conclude that

$$\mathrm{e}^{\mathrm{i}tG_V} X' \mathrm{e}^{-\mathrm{i}tG_V} = \mathrm{e}^{\mathrm{i}tG_0} X' \mathrm{e}^{-\mathrm{i}tG_0}, \tag{49}$$

for all $X' \in \mathfrak{M}'$.

As in the C^*-case, other implementations of τ_V^t can be obtained by choosing a self-adjoint W affiliated to $\mathfrak{M}'$ and such that $G - W$ is essentially self-adjoint on $\mathrm{D}(G) \cap \mathrm{D}(W)$. Denoting by G_W' its self-adjoint extension we set

$$\Gamma_W'^t = \mathrm{e}^{\mathrm{i}tG_W'} \mathrm{e}^{-\mathrm{i}tG}.$$

By the Trotter formula argument we get $\Gamma_W'^t \in \mathfrak{M}'$ and Equ. (49) gives

[4] This is not a real restriction since it is always possible to go to a standard representation where such an implementation always exists.

$$\begin{aligned}(\Gamma'^t_W \Gamma^t_V \mathrm{e}^{itG})(\Gamma'^s_W \Gamma^s_V \mathrm{e}^{isG}) &= \Gamma'^t_W \mathrm{e}^{itG_V} \Gamma'^s_W \mathrm{e}^{-itG_V} \mathrm{e}^{i(t+s)G_V} \\ &= \Gamma'^t_W \mathrm{e}^{itG} \Gamma'^s_W \mathrm{e}^{-itG} \mathrm{e}^{i(t+s)G_V} \\ &= \Gamma'^{t+s}_W \mathrm{e}^{i(t+s)G_V} \\ &= \Gamma'^{t+s}_W \Gamma^{t+s}_V \mathrm{e}^{i(t+s)G}.\end{aligned}$$

Thus $\Gamma'^t_W \Gamma^t_V \mathrm{e}^{itG} = \mathrm{e}^{itG'_W} \mathrm{e}^{-itG} \mathrm{e}^{itG_V}$ is a unitary group, and differentiation shows that its generator is given on $\mathrm{D}(G) \cap \mathrm{D}(V) \cap \mathrm{D}(W)$ by $G + V - W$.

As an application let me now derive the perturbation formula for the standard Liouvillean.

Theorem 4.70. *Let L_0 be the standard Liouvillean of the W^*-dynamical system $(\mathfrak{M}, \tau^t)$ which we suppose to be in standard form $(\mathfrak{M}, \mathcal{H}, J, \mathcal{C})$. Let V be a self-adjoint operator affiliated to $\mathfrak{M}$ and such that*

(i) $L_0 + V$ is essentially self-adjoint on $\mathrm{D}(L_0) \cap \mathrm{D}(V)$.
(ii) $L_0 + V - JVJ$ essentially self-adjoint on $\mathrm{D}(L_0) \cap \mathrm{D}(V) \cap \mathrm{D}(JVJ)$.

Denote by L_V the self-adjoint extension of $L_0 + V - JVJ$. Then L_V is the standard Liouvillean of the perturbed dynamical system $(\mathfrak{M}, \tau^t_V)$.

Proof. $W = JVJ$ with $\mathrm{D}(W) = J\mathrm{D}(V)$ is affiliated to $\mathfrak{M}'$ since

$$\mathrm{e}^{itW} = J\mathrm{e}^{-itV} J \in \mathfrak{M}'.$$

Furthermore from Equ. (30) we get $\mathrm{D}(L_0) = J\mathrm{D}(L_0)$ and

$$L_0 - W = -J(L_0 + V)J,$$

which is essentially self-adjoint on $\mathrm{D}(L_0) \cap \mathrm{D}(W) = J(\mathrm{D}(L_0) \cap \mathrm{D}(V))$. Thus

$$\Gamma'^t_W = J\Gamma^t_V J,$$

and since L_V is essentially self-adjoint on $\mathrm{D}(L_0) \cap \mathrm{D}(V) \cap \mathrm{D}(W)$ we get from the above discussion

$$\mathrm{e}^{itL_V} = \Gamma^t_V J \Gamma^t_V J \mathrm{e}^{itL_0}.$$

The unitary group e^{itL_V} implements τ^t_V and leaves the cone $\mathcal{C}$ invariant by property (iii) of Definition 4.39. This show that L_V is the standard Liouvillean of the perturbed system. □

5 KMS States

5.1 Definition and Basic Properties

For a quantum system with an Hamiltonian H such that $\mathrm{Tr}\,\mathrm{e}^{-\beta H} < \infty$, the Gibbs-Boltzmann prescription for the canonical thermal equilibrium ensemble at inverse temperature $\beta = 1/k_B T$ is the density matrix

$$\rho_\beta = Z_\beta^{-1}\mathrm{e}^{-\beta H}, \tag{50}$$

where the normalization factor $Z_\beta = \mathrm{Tr}\,\mathrm{e}^{-\beta H}$ is the canonical partition function (see Section 3 in [33]). On the other hand the dynamics is given by the spatial automorphisms

$$\tau^t(A) = \mathrm{e}^{itH}A\mathrm{e}^{-itH}. \tag{51}$$

To avoid unnecessary technical problems, think of H as being bounded. The fact that the semi-group entering Equ. (50) reappears, after analytic continuation to the imaginary axis, in Equ. (51) expresses the very strong coupling that exists in quantum mechanics between dynamics and thermal equilibrium. To formalize this remark let us consider, for two observables A, B, the equilibrium correlation function

$$\omega(A\tau^t(B)) = Z_\beta^{-1}\,\mathrm{Tr}(\mathrm{e}^{-\beta H}A\mathrm{e}^{itH}B\mathrm{e}^{-itH}) = Z_\beta^{-1}\,\mathrm{Tr}(\mathrm{e}^{-i(t-i\beta)H}A\mathrm{e}^{itH}B).$$

Using the cyclicity of the trace, analytic continuation of this function to $t+i\beta$ gives

$$\omega(A\tau^{t+i\beta}(B)) = Z_\beta^{-1}\,\mathrm{Tr}(\mathrm{e}^{-itH}A\mathrm{e}^{i(t+i\beta)H}B) = \omega(\tau^t(B)A).$$

Thus there is a function $F_\beta(A,B;z)$, analytic in the strip

$$S_\beta \equiv \{z \in \mathbb{C} \,|\, 0 < \mathrm{Im}\, z < \beta\},$$

and taking boundary values

$$F_\beta(A,B;t) = \omega(A\tau^t(B)), \tag{52}$$
$$F_\beta(A,B;t+i\beta) = \omega(\tau^t(B)A), \tag{53}$$

on ∂S_β. These are the so called KMS boundary conditions. In some sense, the analytic function $F_\beta(A,B;z)$ encodes the non-commutativity of the product AB in the state ρ_β. A KMS state is a state for which such a function, satisfying the KMS boundary condition, exists for all observables A, B.

Definition 5.1. *Let $(\mathfrak{C}, \tau^t)$ be a C^*- or W^*-dynamical system. A state ω on $\mathfrak{C}$, assumed to be normal in the W^*-case, is said to be a (τ^t, β)-KMS state for $\beta > 0$ if for any $A, B \in \mathfrak{C}$ there exists a function $F_\beta(A,B;z)$ analytic in the strip S_β, continuous on its closure and satisfying the KMS conditions (52) and (53) on its boundary.*

Remark 5.2. i. KMS states for negative values of β have no physical meaning (except for very special systems). However, for historical reasons, they are widely used in the mathematical literature. These states have the same definition, the strip S_β being replaced by $S_\beta = \{z \in \mathbb{C} \,|\, \beta < \mathrm{Im}\, z < 0\}$.

ii. Our definition excludes the degenerate case $\beta = 0$. The KMS boundary condition then becomes $\omega(AB) = \omega(BA)$ for all $A, B \in \mathfrak{C}$. Such a state is called a trace.

iii. If ω is a (τ^t, β)-KMS state then it is also $(\tau^{\gamma t}, \beta/\gamma)$-KMS. Note however that there is no simple connection between KMS states at different temperatures for the same group τ^t.

If $(\mathfrak{C}, \tau^t)$ is a C^*- or W^*-dynamical system, an element $A \in \mathfrak{C}$ is analytic for τ^t if the function $t \mapsto \tau^t(A)$, defined for $t \in \mathbb{R}$, extends to an entire analytic function $z \mapsto \tau^z(A)$ of $z \in \mathbb{C}$. Let us denote by $\mathfrak{C}_\tau$ the set of analytic elements for τ^t. It is easy to see that $\mathfrak{C}_\tau$ is a $*$-subalgebra of $\mathfrak{C}$. Indeed, $\tau^z(A+B) = \tau^z(A) + \tau^z(B)$ and $\tau^z(AB) = \tau^z(A)\tau^z(B)$ hold for $z \in \mathbb{R}$ and hence extend to all $z \in \mathbb{C}$ by analytic continuation. Moreover, if $\tau^z(A)$ is entire analytic then so is $\tau^z(A^*) \equiv \tau^{\bar{z}}(A)^*$. For $A \in \mathfrak{C}$, set

$$A_n \equiv \sqrt{\frac{n}{\pi}} \int_{-\infty}^{\infty} \tau^t(A)\, \mathrm{e}^{-nt^2}\, \mathrm{d}t,$$

where the integral is understood in the Riemann sense in the C^*-case and in the weak-$\star$ sense in the W^*-case. One has $A_n \in \mathfrak{C}$ with $\|A_n\| \leq \|A\|$ and the formula

$$\tau^z(A_n) = \sqrt{\frac{n}{\pi}} \int_{-\infty}^{\infty} \tau^t(A)\, \mathrm{e}^{-n(t-z)^2}\, \mathrm{d}t,$$

clearly shows that $\tau^z(A_n)$ is an entire analytic function of z such that

$$\|\tau^z(A_n)\| \leq \mathrm{e}^{n\, \mathrm{Im}(z)^2} \|A\|,$$

and hence $A_n \in \mathfrak{C}_\tau$. Finally, since

$$A_n - A = \int_{-\infty}^{\infty} (\tau^{t/\sqrt{n}}(A) - A)\, \mathrm{e}^{-t^2}\, \frac{\mathrm{d}t}{\sqrt{\pi}},$$

it follows from Lebesgue dominated convergence theorem that the sequence A_n converges towards A in norm in the C^*-case and σ-weakly in the W^*-case. This shows that the $*$-subalgebra $\mathfrak{C}_\tau$ is dense in $\mathfrak{C}$ in the appropriate topology.

Let ω be a (τ^t, β)-KMS state on $\mathfrak{C}$, $A \in \mathfrak{C}$ and $B \in \mathfrak{C}_\tau$. Then the function $G(z) \equiv \omega(A\tau^z(B))$ is entire analytic and for $t \in \mathbb{R}$ one has $G(t) = F_\beta(A, B; t)$. Thus, the function $G(z) - F_\beta(A, B; z)$ is analytic on S_β, continuous on $S_\beta \cup \mathbb{R}$ and vanishes on $\mathbb{R}$. By the Schwarz reflection principle it extends to an analytic function on the strip $\{z \mid -\beta < \mathrm{Im}\, z < \beta\}$ which vanishes on $\mathbb{R}$ and therefore on the entire strip. By continuity

$$F_\beta(A, B; z) = \omega(A\tau^z(B)),$$

holds for $z \in \overline{S_\beta}$ and in particular one has

$$\omega(A\tau^{i\beta}(B)) = \omega(BA). \tag{54}$$

As a first consequence of this fact, let us prove the most important property of KMS states.

Theorem 5.3. *If ω is a (τ^t, β)-KMS state then it is τ^t-invariant.*

Proof. For $A \in \mathfrak{C}_\tau$, the function $f(z) = \omega(\tau^z(A))$ is entire analytic. Moreover, Equ. (54) shows that it is $i\beta$-periodic

$$f(z + i\beta) = \omega(I\tau^{i\beta}(\tau^z(A))) = \omega(\tau^z(A)I) = f(z).$$

On the closed strip $\overline{S_\beta}$ the estimate

$$|f(t+i\alpha)| \leqslant \|\tau^t(\tau^{i\alpha}(A))\| = \|\tau^{i\alpha}(A)\| \leqslant \sup_{0\leqslant\gamma\leqslant\beta} \|\tau^{i\gamma}(A)\| < \infty,$$

holds and therefore f is bounded on whole complex plane. By Liouville Theorem, f is constant. By continuity, this property extends to all observables. □

As shown by the next result, Property (54) characterizes KMS states in a way that is often more convenient than Definition 5.1 (See Proposition 5.3.7 in [12]).

Theorem 5.4. *A state ω on a C^*- (resp. W^*-) algebra $\mathfrak{C}$ is (τ^t, β)-KMS if and only it is normal in the W^*-case and there exists a norm dense (resp. σ-weakly dense), τ^t-invariant $*$-subalgebra $\mathfrak{D}$ of analytic elements for τ^t such that*

$$\omega(A\tau^{i\beta}(B)) = \omega(BA),$$

holds for all $A, B \in \mathfrak{D}$.

Proof. It remains to prove sufficiency. We will only consider the C^*-case and refer the reader to the proof of Proposition 5.3.7 in [12] for the W^*-case. For $A, B \in \mathfrak{D}$, the function defined by

$$F_\beta(A, B; z) \equiv \omega(A\tau^z(B)),$$

is analytic on the strip S_β and continuous on its closure. Since $\mathfrak{D}$ is invariant under τ^t one has $\tau^t(B) \in \mathfrak{D}$ and hence

$$F_\beta(A, B; t) = \omega(A\tau^t(B)),$$
$$F_\beta(A, B; t+i\beta) = \omega(A\tau^{i\beta}(\tau^t(B))) = \omega(\tau^t(B)A).$$

From the bound $|\omega(A\tau^z(B))| \leq \|A\|\,\|\tau^{i\,\mathrm{Im}\,z}(B)\|$ we deduce that $F_\beta(A, B; z)$ is bounded on $\overline{S_\beta}$ and Hadamard tree line theorem yields

$$\sup_{z\in\overline{S_\beta}} |F_\beta(A, B; z)| \leq \|A\|\,\|B\|. \tag{55}$$

Since $\mathfrak{D} \subset \mathfrak{C}_\tau$ is norm dense, any $A, B \in \mathfrak{C}$ can be approximated by sequences $A_n, B_n \in \mathfrak{D}$. From the identity

$$F_\beta(A_n, B_n; z) - F_\beta(A_m, B_m; z) = F_\beta(A_n - A_m, B_n; z) + F_\beta(A_m, B_n - B_m; z),$$

and the bound (55) we conclude that the sequence $F_\beta(A_n, B_n; z)$ is uniformly Cauchy in $\overline{S_\beta}$. Its limit, which we denote by $F_\beta(A, B; z)$, is therefore analytic on S_β and continuous on its closure where it satisfies (55). Finally, for $t \in \mathbb{R}$ one has

$$F_\beta(A, B; t) = \lim_n \omega(A_n\tau^t(B_n)) = \omega(A\tau^t(B)),$$
$$F_\beta(A, B; t+i\beta) = \lim_n \omega(\tau^t(B_n)A_n) = \omega(\tau^t(B)A),$$

which concludes the proof. □

Example 5.5. (Finite quantum systems, continuation of Example 4.16) Let H be a self-adjoint operator on the Hilbert space $\mathcal{H}$ and consider the induced W^*-dynamical system $(\mathfrak{B}(\mathcal{H}), \tau^t)$. If ω is a (τ, β)-KMS state then there exists a density matrix ρ on $\mathcal{H}$ such that $\omega(A) = \operatorname{Tr}(\rho A)$ for all $A \in \mathfrak{B}(\mathcal{H})$. Relation (54) with $A \equiv \phi\,(\psi, \cdot)$ yields

$$(\psi, \tau^{i\beta}(B)\rho\,\phi) = (\psi, \rho B\,\phi),$$

and hence $\tau^{i\beta}(B)\rho = \rho B$ for all τ^t-analytic elements B. If ϕ and ψ belong to the dense subspace of entire analytic vectors for e^{itH} then $B \equiv \phi(\psi, \cdot)$ is analytic for τ^t and $\tau^{i\beta}(B) = \mathrm{e}^{-\beta H}\phi\,(\mathrm{e}^{\beta H}\psi, \cdot)$. For $\psi \neq 0$ we can rewrite the relation $\tau^{i\beta}(B)\rho\psi = \rho B\psi$ as

$$\rho\phi = \frac{(\mathrm{e}^{\beta H}\psi, \rho\psi)}{(\psi, \psi)}\,\mathrm{e}^{-\beta H}\phi,$$

from which we can conclude that

$$\rho = \frac{\mathrm{e}^{-\beta H}}{\operatorname{Tr}\mathrm{e}^{-\beta H}}.$$

Thus, the W^*-dynamical system $(\mathfrak{B}(\mathcal{H}), \tau^t)$ admits a (τ^t, β)-KMS state if and only if $\mathrm{e}^{-\beta H}$ is trace class. Moreover, if such a state exists then it is unique.

Example 5.6. (Ideal Fermi gas) Let ω_β be a (τ^t, β)-KMS state for the C^*-dynamical system of Example 4.6. Then $\omega_\beta(a^*(g)a(f))$ is a sesquilinear form on $\mathfrak{h}$. Since

$$|\omega_\beta(a^*(g)a(f)| \leqslant \|a^*(g)\|\|a(f)\| \leqslant \|g\|\|f\|,$$

there exists a bounded self-adjoint T on $\mathfrak{h}$ such that

$$\omega_\beta(a^*(g)a(f)) = (f, Tg). \tag{56}$$

Moreover, since $\omega_\beta(a^*(f)a(f)) \geqslant 0$, T satisfies the inequalities $0 \leqslant T \leqslant I$. For $t \in \mathbb{R}$ one has $\omega_\beta(a^*(g)\tau^t(a(f))) = (\mathrm{e}^{ith}f, Tg)$ and for f in the dense subspace of entire analytic vectors for h the analytic continuation of this function to $t + i\beta$ is given by $(\mathrm{e}^{\beta h}\mathrm{e}^{ith}f, Tg) = (T\mathrm{e}^{\beta h}\mathrm{e}^{ith}f, g)$. The CAR gives $\tau^t(a(f))a^*(g) = -a^*(g)\tau^t(a(f)) + (\mathrm{e}^{ith}f, g)$ and hence

$$\omega_\beta(\tau^t(a(f))a^*(g)) = (\mathrm{e}^{ith}f, g) - \omega_\beta(a^*(g)a(\mathrm{e}^{ith}f)) = ((I - T)\mathrm{e}^{ith}f, g),$$

and the KMS boundary condition implies $T\mathrm{e}^{\beta h} = I - T$ from which we conclude that

$$T = \frac{1}{1 + \mathrm{e}^{\beta h}}.$$

Consider now the expectation of an arbitrary even monomial ($m + n$ even)

$$W_{m,n}(g_1, \cdots g_m; f_1, \cdots f_n) = \omega_\beta(a^*(g_m)\cdots a^*(g_1)a(f_1)\cdots a(f_n)).$$

We first show that $W_{m,0} = W_{0,n} = 0$. For any f in the dense subspace of entire analytic vectors for h, the KMS condition and the CAR lead to

$$W_{0,n}(f, f_2, \cdots f_n) = W_{0,n}(f_2, \cdots f_n, e^{\beta h} f) = (-1)^{n-1} W_{0,n}(e^{\beta h} f, f_2, \cdots f_n),$$

from which we get $W_{0,n}((1+e^{\beta h})f, f_2, \cdots f_n) = 0$. If g is an entire analytic vector for h, so is $f = (1+e^{\beta h})^{-1} g$ and we can conclude that $W_{0,n}(g, f_2, \cdots, f_n) = 0$ for all g in a dense subspace and all $f_2, \cdots f_n \in \mathfrak{h}$ and hence for all $g, f_2, \cdots f_n \in \mathfrak{h}$. By conjugation, this implies that for all $g_1, \cdots g_m \in \mathfrak{h}$ one also has $W_{m,0}(g_1, \cdots g_m) = 0$.

We consider now the general case. Using the KMS boundary condition, we can write

$$W_{m,n}(g_1, \ldots; f_1, \ldots) = \omega_\beta(a(e^{-\beta h} f_n) a^*(g_m) \cdots a^*(g_1) a(f_1) \cdots a(f_{n-1})).$$

Using the CAR we can commute back the first factor through the others to bring it again in the last position. This leads to the formula

$$W_{m,n}(g_1, \cdots g_m; f_1, \cdots f_n) = -W_{m,n}(g_1, \cdots g_m; f_1, \cdots e^{-\beta h} f_n) +$$

$$+ \sum_{j=1}^{m} (-1)^{m-j} (g_j, e^{-\beta h} f_n) W_{m-1,n-1}(g_1, \cdots g_{j-1}, g_{j+1}, \cdots g_m; f_1, \cdots f_{n-1}).$$

Replacing f_n by $(1+e^{-\beta h})^{-1} f_n$ and using Equ. (56), this can be rewritten as

$$W_{m,n}(g_1, \cdots g_m; f_1, \cdots f_n) =$$

$$\sum_{j=1}^{m} (-1)^{m-j} W_{1,1}(g_j; f_n)\, W_{m-1,n-1}\, (g_1, \cdots g_{j-1}, g_{j+1}, \cdots g_m; f_1, \cdots f_{n-1}).$$

Iteration of this formula shows that $W_{m,n} = 0$ if $m \neq n$, and $W_{n,n}$ can be expressed as sum of products of $W_{1,1}(g_j; f_k) = (g_j, T f_k)$. In fact, a closer look at this formula shows that it is nothing but the usual formula for the expansion of the determinant of the $n \times n$ matrix $(W_{1,1}(g_j; f_k))_{jk}$, *i.e.*,

$$\omega_\beta(a^*(g_m) \cdots a^*(g_1) a(f_1) \cdots a(f_n)) = \delta_{nm} \det\{(f_j, T g_k)\}.$$

Definition 5.7. *A state* ω *on* $\mathrm{CAR}(\mathfrak{h})$ *or* $\mathrm{CAR}^+(\mathfrak{h})$ *is called gauge-invariant if it is invariant under the gauge group* ϑ^t *(recall Equ. (13)). It is called gauge-invariant quasi-free if it satisfies*

$$\omega(a^*(g_m) \cdots a^*(g_1) a(f_1) \cdots a(f_n)) = \delta_{nm} \det\{(f_j, T g_k)\}, \qquad (57)$$

for some self-adjoint operator T *on* $\mathfrak{h}$ *such that* $0 \leq T \leq I$*, all integers* n*,* m *and all* $f_1, \ldots, f_n, g_1, \ldots, g_m \in \mathfrak{h}$.

Remark 5.8. It is easy to see that a gauge-invariant quasi-free state is indeed gauge-invariant. One can show that, given an operator T on $\mathfrak{h}$ such that $0 \leq T \leq I$, there exists a unique gauge-invariant quasi-free state such that Equ. (57) holds (see the Notes to Section 2.5.3 in [12] for references).

Thus, we have shown:

Theorem 5.9. *Let* $\tau^t(a(f)) = a(e^{ith} f)$*, then the unique* (τ^t, β)*-KMS state of the* C^**-dynamical system* $(\mathrm{CAR}^+(\mathfrak{h}), \tau^t)$ *is the quasi-free gauge invariant state generated by* $T = (1 + e^{\beta h})^{-1}$.

Example 5.10. (Quantum spin system, continuation of Example 4.11) The reader should consult Chapter 6.2 in [12] as well as [38] for more complete discussions.

For any finite subset $\Lambda \subset \Gamma$ and any local observable $A \in \mathfrak{A}_{\mathrm{loc}}$ we set

$$\omega_\Lambda(A) \equiv \frac{\mathrm{Tr}_{\mathfrak{h}_{\Lambda \cup X}}(e^{-\beta H_\Lambda} A)}{\mathrm{Tr}_{\mathfrak{h}_{\Lambda \cup X}}(e^{-\beta H_\Lambda})},$$

whenever $A \in \mathfrak{A}_X$. It is easy to see that $\omega_\Lambda(A)$ does not depend on the choice of the finite subset X. By continuity, ω_Λ extends to a state on $\mathfrak{A}$. We say that a state ω on $\mathfrak{A}$ is a thermodynamic limit of the net $\omega_\Lambda, \Lambda \uparrow \Gamma$, if there exists a subnet Λ_α such that

$$\omega(A) = \lim_\alpha \omega_{\Lambda_\alpha}(A),$$

for all $A \in \mathfrak{A}$. Since $E(\mathfrak{A})$ is weak-$\star$ compact the set of thermodynamic limits of the net ω_Λ, is not empty. Let us now prove that any thermodynamic limit of ω_Λ is a (τ, β)-KMS state.

Following the discussion at the beginning of this subsection we remark that for any finite subset $\Lambda \subset \Gamma$ and any $A, B \in \mathfrak{A}_{\mathrm{loc}}$ the function

$$F_{\beta,\Lambda}(A, B; z) \equiv \omega_\Lambda(A \tau_\Lambda^z(B)),$$

is entire analytic and satisfies the (τ_Λ, β)-KMS conditions (52), (53). Since for $t, \eta \in \mathbb{R}$ one has $|\omega_\Lambda(A \tau_\Lambda^{t+i\eta}(B))| \le \|A\| \, \|\tau^{i\eta}(B)\|$, it is also bounded on any horizontal strip $\{t + i\eta \,|\, t \in \mathbb{R}, a \le \eta \le b\}$. The boundary conditions (52), (53) and Hadamard three line theorem further yields the bound

$$\sup_{z \in \overline{S_\beta}} |F_{\beta,\Lambda}(A, B; z)| \le \|A\| \, \|B\|. \tag{58}$$

Assume now that ω is the weak-$\star$ limit of the net ω_{Λ_α} and let $A, B \in \mathfrak{A}_{\mathrm{loc}}$. Since for $t \in \mathbb{R}$ one has

$$\lim_\alpha \|\tau_{\Lambda_\alpha}^t(B) - \tau^t(B)\| = 0,$$

it follows that

$$\lim_\alpha F_{\beta,\Lambda_\alpha}(A, B; t) = \omega(A \tau^t(B)),$$
$$\lim_\alpha F_{\beta,\Lambda_\alpha}(A, B; t + i\beta) = \omega(\tau^t(B) A).$$

From the bound (58) and Montel theorem we conclude that some subsequence $F_{\beta,\Lambda_{\alpha_n}}(A, B; z)$ is locally uniformly convergent in the open strip S_β. Let us denote its limit, which is analytic in S_β, by $F_\beta(A, B; z)$.

The estimate (16) and the expansion (18) show that, for $B \in \mathfrak{A}_{\rm loc}$,

$$\lim_{\epsilon\to 0} \|\tau_\Lambda^{i\epsilon}(B) - B\| = 0,$$

holds uniformly in Λ. From the identity

$$F_\beta(A, B; t + i\epsilon) - \omega(A\tau^t(B)) = \lim_n \omega_{\Lambda_{\alpha_n}}(A(\tau_{\Lambda_{\alpha_n}}^{t+i\epsilon}(B) - \tau_{\Lambda_{\alpha_n}}^t(B))),$$

and the bound

$$\begin{aligned}|\omega_{\Lambda_{\alpha_n}}(A(\tau_{\Lambda_{\alpha_n}}^{t+i\epsilon}(B) - \tau_{\Lambda_{\alpha_n}}^t(B)))| &= |\omega_{\Lambda_{\alpha_n}}(\tau_{\Lambda_{\alpha_n}}^{-t}(A)(\tau_{\Lambda_{\alpha_n}}^{i\epsilon}(B) - B))|\\ &\leq \|A\|\,\|\tau_{\Lambda_{\alpha_n}}^{i\epsilon}(B) - B\|,\end{aligned}$$

we thus conclude that

$$\lim_{\epsilon\downarrow 0} F_\beta(A, B; t + i\epsilon) = \omega(A\tau^t(B)).$$

Using the (τ_Λ, β)-KMS conditions (53) we prove in a completely similar way that

$$\lim_{\epsilon\downarrow 0} F_\beta(A, B; t + i\beta - i\epsilon) = \omega(\tau^t(B)A),$$

and we conclude that $F_\beta(A, B; z)$ extends to a continuous function on the closed strip $\overline{S_\beta}$. Moreover, this function clearly satisfies the bound

$$\sup_{z\in\overline{S_\beta}} |F_\beta(A, B; z)| \leq \|A\|\,\|B\|, \tag{59}$$

as well as the (τ, β)-KMS conditions (52) and (53).

Finally, we note that any $A, B \in \mathfrak{A}$ can be approximated by sequences $A_n, B_n \in \mathfrak{A}_{\rm loc}$. As in the proof of Theorem 5.4 the bound (59) shows that the sequence $F_\beta(A_n, B_n; z)$ converges uniformly on $\overline{S_\beta}$. Its limit, which we denote by $F_\beta(A, B; z)$ then satisfies all the requirements of Definition 5.1.

I conclude this brief introduction to KMS states with the following complement to Proposition 4.48 which is very useful in many applications to open quantum systems.

Proposition 5.11. *Let $(\mathfrak{C}, \tau^t, \omega)$ be a quantum dynamical system with normal form $(\pi_\omega, \mathfrak{C}_\omega, \mathcal{H}_\omega, L_\omega, \Omega_\omega)$. Assume that there exists a $*$-subalgebra $\mathfrak{D} \subset \mathfrak{C}$ with the following properties.*

(i) $\pi_\omega(\mathfrak{D})$ is σ-weakly dense in $\mathfrak{C}_\omega$.
(ii) For each $A, B \in \mathfrak{D}$ there exists a function $F_\beta(A, B; z)$ analytic in the strip S_β, continuous and bounded on its closure and satisfying the KMS boundary conditions (52), (53).

Denote by $\hat\omega = (\Omega_\omega, (\,\cdot\,)\Omega_\omega)$ the normal extension of ω to the enveloping von Neumann algebra $\mathfrak{C}_\omega$ and by $\hat\tau^t(A) = \mathrm{e}^{itL_\omega}A\mathrm{e}^{-itL_\omega}$ the W^*-dynamics induced by τ^t on this algebra. Then the following hold.

(i) $\hat{\omega}$ *is* $(\hat{\tau}^t, \beta)$*-KMS and faithful.*
(ii) $(\mathfrak{C}_\omega, \hat{\tau}^t, \hat{\omega})$ *is in standard form. In particular the* ω*-Liouvillean* L_ω *coincide with the standard Liouvillean* L*.*
(iii) The modular operator of the state $\hat{\omega}$ *is given by* $\Delta_{\hat{\omega}} = \mathrm{e}^{-\beta L}$*.*

Proof. Let $A, B \in \mathfrak{C}_\omega$ be such that $\|A\| \leq 1$ and $\|B\| \leq 1$. By the Kaplansky density theorem there exists nets A_α, B_α in $\pi_\omega(\mathfrak{D})$ such that $\|A_\alpha\| \leq 1$, $\|B_\alpha\| \leq 1$ and $A_\alpha \to A$, $B_\alpha \to B$ in the σ-strong* topology. In particular, if we set

$$d_\alpha \equiv \max(\|(A_\alpha - A)\Omega_\omega\|, \|(A_\alpha^* - A^*)\Omega_\omega\|, \|(B_\alpha - B)\Omega_\omega\|, \|(B_\alpha^* - B^*)\Omega_\omega\|),$$

then $\lim_\alpha d_\alpha = 0$. By Hypothesis (ii) for each α there exists a function $F_\alpha(z)$ analytic in S_β, continuous and bounded on its closure and such that $F_\alpha(t) = \hat{\omega}(A_\alpha \hat{\tau}^t(B_\alpha))$ and $F_\alpha(t+i\beta) = \hat{\omega}(\hat{\tau}^t(B_\alpha)A_\alpha)$ for $t \in \mathbb{R}$. Hence, $F_\alpha(z) - F_{\alpha'}(z)$ is also analytic in S_β, continuous and bounded on its closure. By the Cauchy-Schwarz inequality we have the bounds

$$|F_\alpha(t) - F_{\alpha'}(t)| \leq 2(d_\alpha + d_{\alpha'}),$$
$$|F_\alpha(t+i\beta) - F_{\alpha'}(t+i\beta)| \leq 2(d_\alpha + d_{\alpha'}),$$

and the Hadamard three line theorem yields

$$\sup_{z \in \overline{S_\beta}} |F_\alpha(z) - F_{\alpha'}(z)| \leq 2(d_\alpha + d_{\alpha'}).$$

Thus, F_α is a Cauchy net for the uniform topology. It follows that $\lim_\alpha F_\alpha = F$ exists, is analytic in S_β and continuous on its closure. Moreover, we have

$$\begin{aligned} F(t) &= \lim_\alpha \hat{\omega}(A_\alpha \hat{\tau}^t(B_\alpha)) = \lim_\alpha (A_\alpha^* \Omega_\alpha, \mathrm{e}^{itL_\omega} B_\alpha \Omega_\omega) \\ &= (A^*\Omega_\omega, \mathrm{e}^{itL_\omega} B\Omega_\omega) = \hat{\omega}(A\hat{\tau}^t(B)) \end{aligned}$$

and similarly $F(t+i\beta) = \hat{\omega}(\hat{\tau}^t(B)A)$. This shows that $\hat{\omega}$ is $(\hat{\tau}^t, \beta)$-KMS.

Let now $A \in \mathfrak{C}_\omega$ be such that $A\Omega_\omega = 0$. Then the function $F_\beta(A^*, A; z)$ of Definition 5.1 satisfies

$$F_\beta(A^*, A; t) = \hat{\omega}(A^* \hat{\tau}^t(A)) = (A\Omega_\omega, \mathrm{e}^{itL_\omega} A\Omega_\omega) = 0.$$

By the Schwarz reflection principle, this function extends to an analytic function on the strip $\{z \in \mathbb{C} \mid -\beta < \operatorname{Im} z < \beta\}$ which vanishes on $\mathbb{R}$. It is therefore identically zero. In particular $F_\beta(A^*, A; i\beta) = \hat{\omega}(AA^*) = \|A^*\Omega_\omega\|^2 = 0$. For any $B \in \mathfrak{C}_\omega$, the same argument shows that $(BA)^*\Omega_\omega = A^*B^*\Omega_\omega = 0$ and since Ω_ω is cyclic we can conclude that $A^* = 0$ and $A = 0$. Thus, $\hat{\omega}$ is faithful and (ii) follows from Proposition 4.48.

Note that by Remark 5.2 *iii* the state $\hat{\omega}$ is $(\hat{\tau}^{-\beta t}, -1)$-KMS. (iii) follows from the fact that the modular group $\sigma_{\hat{\omega}}^t(A) = \Delta_{\hat{\omega}}^{it} A \Delta_{\hat{\omega}}^{-it}$ of a faithful normal state is the unique W^*-dynamics for which ω is a $\beta = -1$ KMS state (Takesaki's theorem, see Lecture [7], Theorem 18). □

Example 5.12. (Ideal Bose gas, continuation of Example 4.35) The characteristic function of the thermal equilibrium state of an ideal Bose gas can be obtained from the explicit calculation of the thermodynamic limit of the unique Gibbs state of a finite Bose gas (see Chapter 1 in [12], see also Subsection 4.4 in [36]). We shall get it by assuming that, as in the Fermionic case, the KMS state is quasi-free. Thus, we are looking for a non-negative operator ρ on $\mathfrak{h}$ such that

$$\omega(W(f)) = \mathrm{e}^{-(f,(I+2\rho)f)/4}, \tag{60}$$

satisfies the (τ, β)-KMS condition. Using the CCR relation (19) we can compute

$$\omega(W(f)\tau^t(W(f))) = \omega(W(f+\mathrm{e}^{ith}f))\mathrm{e}^{-i\,\mathrm{Im}(f,\mathrm{e}^{ith}f)} = \mathrm{e}^{-(f,\gamma(t)f)/4},$$

where

$$\gamma(t) = (I+\mathrm{e}^{-ith})(I+2\rho)(I+\mathrm{e}^{ith}) + 2i\sin th.$$

Assuming that f is an entire analytic vector for the group e^{ith}, analytic continuation to $t = i\beta$ yields

$$\gamma(i\beta) = (I+\mathrm{e}^{\beta h})(I+2\rho)(I+\mathrm{e}^{-\beta h}) - 2\,\mathrm{sh}\,\beta h,$$

and the KMS condition (54) requires that $\gamma(i\beta) = \gamma(0)$. This equation is easily solved for ρ and its solution is given by Planck's black-body radiation law

$$\rho = \frac{1}{\mathrm{e}^{\beta h} - 1}. \tag{61}$$

It is clear from the singularity at $h = 0$ in Equ. (61) that if $0 \in \sigma(h)$ one can not hope to get a state on $\mathrm{CCR}(\mathcal{D})$ without further assumption on $\mathcal{D}$.

Assume that $h > 0$, *i.e.*, that 0 is not an eigenvalue of h and that $\mathcal{D} = \mathrm{D}(h^{-1/2})$. Then according to Definition 4.37, Equ. (60), (61) define a quasi-free, τ^t-invariant state ω on $\mathrm{CCR}(\mathcal{D})$. Let $(\mathrm{CCR}(\mathcal{D}), \tau^t, \omega)$ be the associated quantum dynamical system and $(\pi_\omega, \mathfrak{M}_\omega, \mathcal{H}_\omega, L_\omega, \Omega_\omega)$ its normal form as described in Example 4.35. Note that since finite linear combinations of elements of $W(\mathcal{D})$ are dense in $\mathrm{CCR}(\mathcal{D})$ one has $\mathfrak{M}_\omega' = \pi_\omega(W(\mathcal{D}))'$ and $\mathfrak{M}_\omega = \pi_\omega(W(\mathcal{D}))''$.

$\mathcal{D}$ equipped with the scalar product $(f,g)_\mathcal{D} \equiv (f,(I+2\rho)g)$ is a Hilbert space. Since e^{ith} is a strongly continuous unitary group on $\mathcal{D}$ the subspace $\mathcal{A} \subset \mathcal{D}$ of entire analytic vectors for e^{ith} is dense. It follows that any $f \in \mathcal{D}$ can be approximated in the norm of $\mathcal{D}$ by a sequence $f_n \in \mathcal{A}$.

A simple calculation using the CCR shows that for $f, f', g \in \mathcal{D}$,

$$\|(\pi_\omega(W(f)) - \pi_\omega(W(f')))\pi_\omega(W(g))\Omega_\omega\|^2 = 2(1 - \mathrm{e}^{-\|f-f'\|_\mathcal{D}^2/4}\cos\theta),$$

where $\theta \equiv \mathrm{Im}(f-f',g) + \mathrm{Im}(f,f')/2$. Since the elements of $\pi_\omega(W(\mathcal{D}))$ are unitary and Ω_ω is cyclic it follows that

$$\mathrm{s}-\lim_n \pi_\omega(W(f_n)) = \pi_\omega(W(f)),$$

whenever f_n converges to f in $\mathcal{D}$. We conclude that

$$\pi_\omega(W(\mathcal{A}))' = \pi_\omega(W(\mathcal{D}))' = \mathfrak{M}_\omega',$$

and hence $\pi_\omega(W(\mathcal{A}))'' = \mathfrak{M}_\omega$.

Denote by $\mathfrak{D} \subset \mathrm{CCR}(\mathcal{D})$ the linear span of $W(\mathcal{A})$. By the von Neumann density theorem $\pi_\omega(\mathfrak{D})$ is σ-weakly dense in $\mathfrak{M}_\omega$. Moreover, an explicit calculation shows that $\omega(W(f)\tau^t(W(g))) = \mathrm{e}^{-\phi_t(f,g)/4}$ where

$$\phi_t(f,g) = \|f\|_{\mathcal{D}}^2 + \|g\|_{\mathcal{D}}^2 + 2(g, \frac{\mathrm{e}^{-ith}}{I + \mathrm{e}^{\beta h}} f)_{\mathcal{D}} + 2(f, \frac{\mathrm{e}^{ith}}{I + \mathrm{e}^{-\beta h}} g)_{\mathcal{D}}.$$

If $f, g \in \mathcal{A}$ then $\phi_t(f,g)$ is an entire analytic function of t and $\phi_{t+i\beta}(f,g) = \phi_t(g,f)$. Thus, $\mathfrak{D}$ fulfills the requirements of Proposition 5.11. It follows that $\hat{\omega}(A) = (\Omega_\omega, A\Omega_\omega)$ is a β-KMS state for the W^*-dynamics generated by the Liouvillean $L = L_\omega$. Note that since $\mathrm{Ker}\,\rho = \{0\}$ the GNS representation associated to ω is given by

$$\mathcal{H}_\omega = \mathcal{L}^2(\Gamma_+(\mathfrak{h})),$$
$$\pi_\omega(W(f))X = W((I+\rho)^{1/2} f) X W(\rho^{1/2} f)^*,$$
$$\Omega_\omega = \Omega(\Omega, \cdot),$$

where Ω is the Fock vacuum in $\Gamma_+(\mathfrak{h})$. The Liouvillean L is given by

$$\mathrm{e}^{itL} X = \Gamma(\mathrm{e}^{ith}) X \Gamma(\mathrm{e}^{-ith}),$$

Moreover, since $\hat{\omega}$ is faithful we have $[\mathfrak{M}_\omega' \Omega_\omega] = I$ and Corollary 4.66 applies.

Exercise 5.13. Show that the standard form of $\mathfrak{M}_\omega$ on $\mathcal{H}_\omega$ is specified by conjugation $J : X \mapsto X^*$ and the cone $\mathcal{C} = \{X \in \mathcal{L}^2(\Gamma_+(\mathfrak{h})) \,|\, X \geq 0\}$.

Exercise 5.14. Show that if h has purely absolutely continuous spectrum then 0 is a simple eigenvalue of L with eigenvector Ω_ω. Show that the spectrum of L on $\{\Omega_\omega\}^\perp$ is purely absolutely continuous. Conclude that the quantum dynamical system $(\mathrm{CCR}(\mathcal{D}), \tau^t, \omega)$ is mixing. What happens if h has non-empty singular spectrum ?

The thermodynamics of the ideal Bose gas is more complex than the above picture. In fact, due to the well known phenomenon of Bose-Einstein condensation, non-unique KMS states are possible. I refer to Lecture [36] and [12] for a detailed discussion.

5.2 Perturbation Theory of KMS States

Consider the finite dimensional C^*-dynamical system defined by $\mathfrak{A} = \mathfrak{B}(\mathbb{C}^n)$ and $\tau^t(X) = \mathrm{e}^{itH} X \mathrm{e}^{-itH}$ for some self-adjoint matrix H. Then

$$\omega(X) = \frac{\mathrm{Tr}(\mathrm{e}^{-\beta H}X)}{\mathrm{Tr}(\mathrm{e}^{-\beta H})} = \frac{\mathrm{Tr}(\mathrm{e}^{-\beta H/2}X\mathrm{e}^{-\beta H/2})}{\mathrm{Tr}(\mathrm{e}^{-\beta H})},$$

is the unique (τ^t, β)-KMS state. If V is another self-adjoint matrix then the perturbed dynamics τ_V^t as well as the perturbed KMS state ω_V are obtained by replacing H by $H+V$. Note that, in the present situation, the definition (45) of the unitary cocycle Γ_V^t reads

$$\Gamma_V^t = \mathrm{e}^{it(H+V)}\mathrm{e}^{-itH},$$

which is obviously an entire function of t. Thus, we can express ω_V in terms of ω as

$$\omega_V(X) = \frac{\omega(X\Gamma_V^{i\beta})}{\omega(\Gamma_V^{i\beta})} = \frac{\omega(\Gamma_V^{i\beta/2*}X\Gamma_V^{i\beta/2})}{\omega(\Gamma_V^{i\beta/2*}\Gamma_V^{i\beta/2})}. \tag{62}$$

Let $(\mathcal{H}_\omega, \pi_\omega, \Omega_\omega)$ be the cyclic representation of ω and L the Liouvillean (ω-Liouvillean and standard Liouvillean coincide since ω is faithful). Then on has

$$\pi_\omega(\Gamma_V^{i\beta/2})\Omega_\omega = \pi_\omega(\Gamma_V^{i\beta/2})\mathrm{e}^{-\beta L/2}\Omega_\omega = \mathrm{e}^{-\beta(L+\pi_\omega(V))/2}\Omega_\omega, \tag{63}$$

by Equ. (48). Thus we can write Equ. (62) as

$$\omega_V(X) = \frac{(\Omega_{\omega_V}, \pi_\omega(X)\Omega_{\omega_V})}{(\Omega_{\omega_V}, \Omega_{\omega_V})},$$

where $\Omega_{\omega_V} = \mathrm{e}^{-\beta(L+\pi_\omega(V))/2}\Omega_\omega$. The cocycle property (47) further gives

$$\Gamma_V^{i\beta/2} = \Gamma_V^{i\beta/4}\tau^{i\beta/4}(\Gamma_V^{i\beta/4}) = \Gamma_V^{i\beta/4}\tau^{i\beta/2}(\tau^{-i\beta/4}(\Gamma_V^{i\beta/4})),$$

and $\tau^{-i\beta/4}(\Gamma_V^{i\beta/4}) = (\Gamma_V^{-i\beta/4})^{-1}$. Since $\Gamma_V^{\bar{z}*}$ is analytic and equals $(\Gamma_V^z)^{-1}$ for real z, they are equal for all z and

$$\Gamma_V^{i\beta/2} = \Gamma_V^{i\beta/4}\tau^{i\beta/2}(\Gamma_V^{i\beta/4*}).$$

We can rewrite the perturbed vector Ω_{ω_V} as

$$\Omega_{\omega_V} = \pi_\omega(\Gamma_V^{i\beta/4})\mathrm{e}^{-\beta L/2}\pi_\omega(\Gamma_V^{i\beta/4})^*\Omega_\omega = \pi_\omega(\Gamma_V^{i\beta/4})\mathrm{e}^{-\beta L/2}J\Delta_\omega^{1/2}\pi_\omega(\Gamma_V^{i\beta/4})\Omega_\omega,$$

and since $J\Delta_\omega^{1/2} = J\mathrm{e}^{-\beta L/2} = \mathrm{e}^{\beta L/2}J$ we conclude that

$$\Omega_{\omega_V} = \pi_\omega(\Gamma_V^{i\beta/4})J\pi_\omega(\Gamma_V^{i\beta/4})\Omega_\omega \in \mathcal{C}.$$

Thus Ω_{ω_V} is, up to normalization, the unique standard vector representative of the perturbed KMS-state ω_V.

The main difficulty in extending this formula to more general situations is to show that $\Omega_\omega \in \mathrm{D}(\mathrm{e}^{-\beta(L+\pi_\omega(V))/2})$. Indeed, even if V is bounded, the Liouvillean L is usually unbounded below and ordinary perturbation theory of quasi-bounded semi-groups fails. If V is such that $\tau^t(V)$ is entire analytic, this can be done using (63) since the cocycle Γ_V^t is then analytic, as the solution of a linear differential equation with analytic coefficients. It is possible to extend the result to general bounded perturbations using an approximation argument.

Theorem 5.15. *Let $(\mathfrak{C}, \tau^t)$ be a C^*- or W^*-dynamical system and $V \in \mathfrak{C}$ a local perturbation. There exists a bijective map $\omega \mapsto \omega_V$ between the set of (τ^t, β)-KMS states and the set of (τ_V^t, β)-KMS states on $\mathfrak{C}$ such that $\omega_V \in N(\mathfrak{C}, \omega)$ and $(\omega_{V_1})_{V_2} = \omega_{V_1+V_2}$.*

Let ω be a (τ^t, β)-KMS state on $\mathfrak{C}$. Denote by L the (standard) Liouvillean of the quantum dynamical system $(\mathfrak{C}_\omega, \hat{\tau}^t, \hat{\omega})$ and by $\Omega \in \mathcal{H}_\omega$ the standard vector representative of $\hat{\omega}$. For any local perturbation $V \in \mathfrak{C}_\omega$ one has:

(i) $\Omega \in \mathrm{D}(\mathrm{e}^{-\beta(L+V)/2})$.
(ii) Up to normalization, $\Omega_V = \mathrm{e}^{-\beta(L+V)/2}\Omega$ *is the standard vector representative of* $\hat{\omega}_V$.
(iii) Ω_V *is cyclic and separating for* $\mathfrak{C}_\omega$.
(iv) For any $V_1, V_2 \in \mathfrak{C}_\omega$ *one has* $(\Omega_{V_1})_{V_2} = \Omega_{V_1+V_2}$.
(v) The Peierls-Bogoliubov inequality holds: $\mathrm{e}^{-\beta\hat{\omega}(V)} \leqslant \|\Omega_V\|^2$.
(vi) The Golden-Thompson inequality holds: $\|\Omega_V\|^2 \leqslant \hat{\omega}(\mathrm{e}^{-\beta V})$.
(vii) If $V_n \in \mathfrak{C}_\omega$ *strongly converges to* $V \in \mathfrak{C}_\omega$ *then* Ω_{V_n} *converges in norm to* Ω_V *and* ω_{V_n} *converges in norm to* ω_V.

In the case of unbounded perturbations, one can use an approximation by bounded perturbations to obtain

Theorem 5.16. *Let $(\mathfrak{M}, \tau^t, \omega)$ be a quantum dynamical system and suppose that ω is a (τ^t, β)-KMS state. Assume that the system is in standard form $(\mathfrak{M}, \mathcal{H}, J, \mathcal{C})$. Denote by L its standard Liouvillean and by Ω the standard vector representative of ω. Let V be a self-adjoint operator affiliated to $\mathfrak{M}$ and such that the following conditions are satisfied:*

(i) $L + V$ *is essentially self-adjoint on* $\mathrm{D}(L) \cap \mathrm{D}(V)$.
(ii) $L + V - JVJ$ *is essentially self-adjoint on* $\mathrm{D}(L) \cap \mathrm{D}(V) \cap \mathrm{D}(JVJ)$.
(iii) $\Omega \in \mathrm{D}(\mathrm{e}^{-\beta V/2})$.

Then the following conclusions hold:

(i) $\Omega \in \mathrm{D}(\mathrm{e}^{-\beta(L+V)/2})$.
(ii) Up to normalization, $\Omega_V = \mathrm{e}^{-\beta(L+V)/2}\Omega$ *is the standard vector representative of a* (τ_V^t, β)*-KMS state* ω_V.
(iii) Ω_V *is cyclic and separating for* $\mathfrak{M}$.
(iv) The Peierls-Bogoliubov inequality holds: $\mathrm{e}^{-\beta\hat{\omega}(V)} \leqslant \|\Omega_V\|^2$.
(v) The Golden-Thompson inequality holds: $\|\Omega_V\|^2 \leqslant \hat{\omega}(\mathrm{e}^{-\beta V})$.
(vi) For any $\lambda \in [0, 1]$ *the operator* λV *satisfies the hypotheses (i), (ii) and (iii) and one has* $\lim_{\lambda\downarrow 0} \|\Omega_{\lambda V} - \Omega\| = 0$ *and* $\lim_{\lambda\downarrow 0} \|\omega_{\lambda V} - \omega\| = 0$.

References

1. Araki, H., Woods, E.J.: Representations of the canonical commutation relations describing a nonrelativistic infinite free Bose gas. J. Math. Phys. **4** (1963), 637.

2. Araki, H.: Relative entropy of states on von Neumann algebras. Publ. Res. Inst. Math. Sci. Kyoto Univ. **11** (1975), 809.
3. Araki, H.: Positive cone, Radon-Nikodym theorems, relative Hamiltonian and the Gibbs condition in statistical mechanics. In *C^*-Algebras and their Applications to Statistical Mechanics and Quantum Field Theory.* D. Kastler editor. North-Holand, Amsterdam, 1976.
4. Araki, H.: Relative entropy of states on von Neumann algebras II. Publ. Res. Inst. Math. Sci. Kyoto Univ. **13** (1977), 173.
5. Arnold, V.I., Avez, A.: *Ergodic Problems of Classical Mechanics.* Benjamin, New York, 1968.
6. Aschbacher, W., Jakšić, V., Pautrat, Y., Pillet, C.-A.: Topics in nonequilibrium quantum statistical mechanics. Volume 3 in this series.
7. Attal, S.: Elements of operator algebras. Part 3 in this volume.
8. Bach, V., Fröhlich, J., Sigal, I.M.: Return to equilibrium. J. Math. Phys. **41** (2000), 3985.
9. Benatti, F.: *Deterministic Chaos in Infinite Quantum Systems.* Springer, Berlin, 1993.
10. Birkhoff, G.D.: Proof of the ergodic theorem. Proc. Nat. Acad. Sci. (U.S.A.) **17** (1931), 656.
11. Bratteli, O., Robinson, D.W.: *Operator Algebras and Quantum Statistical Mechanics I.* Springer, Berlin, 1979.
12. Bratteli, O., Robinson, D.W.: *Operator Algebras and Quantum Statistical Mechanics II.* Springer, Berlin, 1996.
13. Cohen-Tanoudji, C., Diu, B., Laloe, F.: *Quantum Mechanics.* Wiley, New York, 1977.
14. Connes, A.: Une classification des facteurs de type III. Ann. Sci. Ecole Norm. Sup. **6** (1973), 133.
15. Cornfeld, I.P., Fomin, S.V., Sinai, Ya G.: *Ergodic Theory.* Springer, Berlin, 1982.
16. Dereziński, J., Jakšić, V.: Return to equilibrium for Pauli-Fierz systems. Ann. H. Poincaré **4** (2003), 739.
17. Dereziński, J., Jakšić, V., Pillet, C.-A.: Perturbation theory of W^*-dynamics, Liouvilleans and KMS-states. Rev. Math. Phys. **15** (2003), 447.
18. Dereziński, J.: Introduction to representations of canonical commutation and anticommutation relations. In the lecture notes of the summer school "Large Coulomb Systems—QED", held in Nordfjordeid, August 11—18 2004. To be published in *Lecture Notes in Mathematics.*
19. Donald, M.J.: Relative Hamiltonians which are not bounded from above. J. Funct. Anal. **91** (1990), 143.
20. Ford, G.W., Kac, M., Mazur, P.: Statistical mechanics of assemblies of coupled oscillators. J. Math. Phys. **6** (1965), 504.
21. Fröhlich, J., Merkli, M.: Thermal ionization. Math. Phys., Analysis and Geometry **7** (2004), 239.
22. Fröhlich, J., Merkli, M.: Another return of "Return to Equilibrium". Commun. Math. Phys. **251** (2004), 235.
23. Haagerup, U.: The standard form of von Neumann algebras. Math. Scand. **37** (1975), 271.
24. Jadczyk, A.Z.: On some groups of automorphisms of von Neumann algebras with cyclic and separating vector. Commun. Math. Phys. **13** (1969), 142.
25. Jakšić, V.: Topics in spectral theory. Part 6 in this volume.
26. Jakšić, V., Kritchevski, E., Pillet, C.-A.: Mathematical theory of the Wigner-Weisskopf atom. In the lecture notes of the summer school "Large Coulomb

Systems—QED", held in Nordfjordeid, August 11—18 2004. To be published in *Lecture Notes in Mathematics.*
27. Jakšić, V., Pillet, C.-A.: On a Model for Quantum Friction III. Ergodic Properties of the Spin-Boson System. Commun. Math. Phys. **178** (1996), 627.
28. Jakšić, V., Pillet, C.-A.: Ergodic properties of classical dissipative systems I. Acta Math. **181** (1998), 245.
29. Jakšić, V., Pillet, C.-A.: Non-equilibrium steady states of finite quantum systems coupled to thermal reservoirs. Commun. Math. Phys. **226** (2002), 131.
30. Jakšić, V., Pillet, C.-A.: Mathematical theory of non-equilibrium quantum statistical mechanics. J. Stat. Phys. **108** (2002), 787.
31. Jakšić, V., Pillet, C.-A.: A note on eigenvalues of Liouvilleans. J. Stat. Phys. **105** (2001), 937.
32. Joye, A.: Introduction to the theory of linear operators. Part 1 in this volume.
33. Joye, A.: Introduction to quantum statistical mechanics. Part 2 in this volume.
34. Koopman, B.O.: Hamiltonian systems and transformations in Hilbert spaces. Proc. Nat. Acad. Sci. (U.S.A.) **17** (1931), 315.
35. Mané, R.: *Ergodic Theory and Differentiable Dynamics.* Springer, Berlin, 1987.
36. Merkli, M.: The ideal quantum gas. Part 5 in this volume.
37. Rey-Bellet, L.: Classical open systems. Volume 2 in this series.
38. Ruelle, D.:*Statistical Mechanics. Rigorous Results.* Benjamin, New York, 1969.
39. Ruelle, D.: Smooth dynamics and new theoretical ideas in nonequilibrium statistical mechanics. J. Stat. Phys. **95** (1999), 393.
40. Ruelle, D.: Natural nonequilibrium states in quantum statistical mechanics. J. Stat. Phys. **98** (2000), 57.
41. Reed, M., Simon, B.: *Methods of Modern Mathematical Physics I: Functional Analysis.* Academic Press, New York, 1972.
42. Sakai, S.: *C^*-Algebras and W^*-Algebras.* Springer, Berlin, 1971.
43. Sakai, S.: Perturbations of KMS states in C^*-dynamical systems. Contemp. Math. **62** (1987), 187.
44. Sakai, S.: *Operator Algebras in Dynamical Systems.* Cambridge University Press, Cambridge, 1991.
45. Stratila, S., Zsido, L.: *Lectures on von Neumann Algebras.* Abacus Press, Tunbridge Wells, 1979.
46. Takesaki, M.: *Theory of Operator Algebras II.* Springer, Berlin, 2003.
47. Thirring, W.: *Quantum Mechanics of Large Systems.* Springer, Berlin, 1980.
48. von Neumann, J: Proof of the quasiergodic hypothesis. Proc. Nat. Acad. Sci. (U.S.A.) **17** (1932), 70.
49. Walters, P.: *An Introduction to Ergodic Theory.* Springer, Berlin, 1981.

The Ideal Quantum Gas

Marco Merkli

Institute of Theoretical Physics, ETH Zürich,
CH-8093 Zurich, Switzerland**
e-mail: merkli@itp.phyx.ethz.ch

** present address: Department of Mathematics and Statistics, McGill University, 805 Sherbrooke Street West, Montreal, QC, H3A 2K6, Canada
e-mail: merkli@math.mcgill.ca

1 Introduction

The goal of these lecture notes is to give an introduction to the mathematical description of a system of identical non-interacting quantum particles. An important characteristic of the systems considered is their "size", which may refer to *spatial extension* or to the *number of particles*, or to a combination of both. Certain physical phenomena occur only for very large systems, say for systems which occupy an immense region of the universe or for a system the size of a laboratory, if the observed phenomenon takes place on a microscopic level. For the mathematical analysis it is often convenient to make an abstraction and to consider systems which are spatially infinitely extended (and which contain infinitely many particles). From a physical point of view, such a description can only be an approximation which is, however, justified by the fact that the mathematical models lead to correct answers to physical questions. An important part of these lectures is concerned with the description of infinite systems, or the passage of a finite system (a confined one, or one with only finitely many particles) to an infinite one. In some instances, this passage is called the *thermodynamic limit*.

It is natural to consider first a system of finitely many (identical) quantum particles. States of such a system are described by vectors in Fock space, a Hilbert space that has a direct sum decomposition into subspaces, each of which describes a system with a fixed number $n = 0, 1, 2, \ldots$ of particles. The action of operators which are not reduced by this direct sum decomposition is interpreted as creation or annihilation of particles. So Fock space provides us already with a nice toolbox enabling the modeling of many physical processes. However, not all physical situations can be described by Fock space. Given any state in Fock space the probability of finding at least n particles in it decreases to zero, as $n \to \infty$. Imagine a gas of particles which has a uniform nonzero density (say one particle per unit volume) and which is spatially infinitely extended. Such a state cannot be described by a vector in Fock space!

How can we thus describe the infinitely extended system at positive density? The observable algebra (the one generated by the creation and annihilation operators on Fock space) has a certain structure determined by algebraic relations. Those are called the canonical commutation relations (CCR) or the canonical anticommutation relations (CAR) depending on whether one considers Bosons or Fermions. It can be viewed as an *abstract* algebra, merely determined by its relations, and not a priori represented as an operator algebra on a Hilbert space. Fock space emerges then just as one possible representation Hilbert space of the abstract algebra (called the CCR or the CAR algebra). A fundamental theorem regarding this setting is the von Neumann uniqueness theorem. It says that if we consider only *finitely* many particles then all the representations of the corresponding algebra are (spatially) equivalent. However, in the case of a system with infinitely many particles there are non-equivalent representations of the algebra! This is what happens in the case of the infinitely extended system with nonzero density; it is described by a vector in some Hilbert space which is not compatible with Fock space (the corresponding representations of the algebras are not spatially equivalent).

It is one of the goals of these notes to calculate the representation Hilbert space of the infinitely extended gas for arbitrary densities.

It may have become clear from this short introduction what kind of mathematics is involved in these notes. In the first chapter we will mainly deal with operators on Fock space (bounded and unbounded ones) and, in the second chapter, we move on to some aspects of the theory of C^*algebras in relation with the CCR and CAR algebras. The last chapter is devoted to the Araki-Woods representation, which gives the above mentioned representation of the infinitely extended free Bose gas for arbitrary momentum density distributions.

These notes represent a composition of mostly well known concepts and results relevant to this collection of lecture notes, and they have, in the author's view, an interest on their own. An effort has been made to render the material easy to understand for anybody with basic knowledge in functional analysis.

2 Fock space

Fock space is the Hilbert space suitable to describe a system of arbitrarily many (identical) quantum particles. We start this section by introducing the Bosonic and Fermionic Fock spaces and the corresponding creation and annihilation operators. We will see that in the case of Bosons those operators are unbounded and it is thus convenient to "replace" them by Weyl operators. This leads us to the definition of the C^*algebras CCR_F and CAR_F, for Bosons and Fermions, respectively. We discuss the "shortcomings" of these algebras in the last section, motivating the definition of the abstract CCR and CAR algebras.

2.1 Bosons and Fermions

An ideal quantum gas is a system of identical (meaning indistinguishable), non-interacting quantum particles.

A single particle is described by a complex Hilbert space $\mathfrak{H}$, i.e., a normalized $\psi \in \mathfrak{H}$ is a (pure) state of the particle (ψ is also called the state vector). It is often useful to consider states which are determined by a linear (not necessarily closed) subspace

$$\mathfrak{D} \subseteq \mathfrak{H}. \tag{1}$$

Typically, one may think of $\mathfrak{H} = L^2(\mathbb{R}^3, d^3x)$, then a normalized vector $\psi \in \mathfrak{H}$ is called the *wave function* of the particle and has the following physical interpretation: $|\psi(x)|^2$ is the probability density of finding the particle at location $x \in \mathbb{R}^3$. An example for $\mathfrak{D}$ is the set $\{f \in C_0^\infty(\mathbb{R}^3) \,|\, \mathrm{supp} f \subset V\}$ of all smooth functions with support in some compact set $V \subset \mathbb{R}^3$; $\mathfrak{D}$ is called the *test function space*. We will see that the choice of the test function space often reflects physical properties of the system at hand, e.g., we may want to look only at particles confined to a region V in space.

The Hilbert space of n *distinguishable* particles is given by the n-fold tensor product

$$\mathfrak{H}^n = \mathfrak{H} \otimes \cdots \otimes \mathfrak{H}. \tag{2}$$

If we restrict our attention to one-particle states in $\mathfrak{D}$ then of course only the subspace $\mathfrak{D} \otimes \cdots \otimes \mathfrak{D}$ of $\mathfrak{H}^n$ is relevant. To be able to describe processes involving creation and annihilation of particles, we build the direct sum Hilbert space

$$\mathcal{F}(\mathfrak{H}) = \bigoplus_{n \geq 0} \mathfrak{H}^n, \tag{3}$$

where $\mathfrak{H}^0 = \mathbb{C}$. $\mathcal{F}(\mathfrak{H})$ is called the *Fock space over the Hilbert space* $\mathfrak{H}$. The Hilbert space $\mathfrak{H}^n$ identified as a subspace of Fock space is called the *n-sector* (or the n-th chaos, in quantum probability). The zero-sector is also called the *vacuum sector*. An element ψ of $\mathcal{F}(\mathfrak{H})$ is a sequence $\psi = \{\psi_n\}_{n \geq 0}$ with $\psi_n \in \mathfrak{H}^n$. We write sometimes the n-particle component ψ_n of ψ as $[\psi]_n$. The scalar product on $\mathcal{F}(\mathfrak{H})$ is given by

$$\langle \psi, \phi \rangle = \sum_{n \geq 0} \langle \psi_n, \phi_n \rangle_{\mathfrak{H}^n}, \tag{4}$$

where $\langle \cdot, \cdot \rangle_{\mathfrak{H}^n}$ is the scalar product of $\mathfrak{H}^n$, which we take to be antilinear in the first argument and linear in the second one. The direct sum in (3) is the decomposition of Fock space into spectral subspaces (eigenspaces) of the selfadjoint number operator, N, defined as follows. The domain of N is

$$\mathcal{D}(N) = \left\{ \psi \in \mathcal{F}(\mathfrak{H}) \,\middle|\, \sum_{n \geq 0} n^2 \|\psi_n\|^2_{\mathfrak{H}^n} < \infty \right\}, \tag{5}$$

and the action of N is given, for $\psi \in \mathcal{D}(N)$, by

$$[N\psi]_n = n[\psi]_n. \tag{6}$$

Clearly, the spectrum of N is discrete and consists of all integers $n \in \mathbb{N}$. The vector $\Omega \in \mathcal{F}(\mathfrak{H})$ given by

$$[\Omega]_0 = 1 \in \mathbb{C}, \ \ [\Omega]_n = 0 \in \mathfrak{H}^n, \text{if } n > 0, \tag{7}$$

is called the vacuum (vector). It spans the one-dimensional kernel of N. The degree of degeneracy of the eigenvalue n of N equals $\dim(\mathfrak{H}^n) = (\dim \mathfrak{H})^n$.

Let us now consider a system of *indistinguishable* particles. The indistinguishability is reflected in the symmetry of the state vector (wave function) under the exchange of particle labels. We are adopting in these notes the view that all quantum particles fall into two categories: either the state vectors are *symmetric* under permutations of indices, in which case the particles are called *Bosons*, or the state vectors are *anti-symmetric* under permutations of indices, in which case the particles are called *Fermions*. For example, let $\{f_k\}_{k=1}^n \subset \mathfrak{H}$ be n state vectors of a single particle. The vector $f_1 \otimes \cdots \otimes f_n \in \mathfrak{H}^n$ is the state of an n-particle system where the particle labelled by k is in the state f_k. The state describing n Bosons, one of which (but we cannot say which one, because they are indistinguishable) is in the state f_1, one of which is in the state f_2, and so on, is given by the symmetric state vector

$$\frac{1}{n!}\sum_{\pi\in S_n} f_{\pi(1)}\otimes\cdots\otimes f_{\pi(n)}\in\mathfrak{H}^n, \tag{8}$$

where S_n is the group of all permutations π of n objects. The corresponding vector describing n Fermions is given by

$$\frac{1}{n!}\sum_{\pi\in S_n}\epsilon(\pi)f_{\pi(1)}\otimes\cdots\otimes f_{\pi(n)}\in\mathfrak{H}^n, \tag{9}$$

where $\epsilon(\pi)$ is the *signature* of the permutation π. [3]

Let us introduce the symmetrization operator P_+ and the anti-symmetrization operator P_- on $\mathcal{F}(\mathfrak{H})$. Set $P_\pm\Omega=\Omega$ and for $\{f_k\}_{k=1}^n\subset\mathfrak{H}$, $n\geq 1$, set

$$P_+f_1\otimes\cdots\otimes f_k=\frac{1}{n!}\sum_{\pi\in S_n} f_{\pi(1)}\otimes\cdots\otimes f_{\pi(n)}, \tag{10}$$

$$P_-f_1\otimes\cdots\otimes f_k=\frac{1}{n!}\sum_{\pi\in S_n}\epsilon(\pi)f_{\pi(1)}\otimes\cdots\otimes f_{\pi(n)}, \tag{11}$$

and extend the action of $P_\pm$ by linearity to the sets

$$D^n=\left\{\sum_{k=1}^{K} f_1^{(k)}\otimes\cdots\otimes f_n^{(k)}\;\middle|\; K\in\mathbb{N}, f_l^{(k)}\in\mathfrak{H}\right\}\subset\mathfrak{H}^n,\quad n\geq 1. \tag{12}$$

It is clear that $\|P_\pm f_1\otimes\cdots\otimes f_n\|\leq\frac{1}{n!}\sum_{n\in S_n}\|f_1\|\cdots\|f_n\|=\|f_1\otimes\cdots\otimes f_n\|$, so $P_\pm$ is a contraction on D^n, $\|P_\pm\psi\|\leq\|\psi\|$ for $\psi\in D^n$. Consequently the operators $P_\pm$ extend to all of $\mathfrak{H}^n$, for all n, and to $\mathcal{F}(\mathfrak{H})$ by sector-wise action. [4] Of course $P_\pm$ are actually selfadjoint projections; i.e., $P_\pm^2=P_\pm={P_\pm}^*$ and they satisfy $\|P_\pm\|=1$. We define the *Bosonic Fock space*, $\mathcal{F}_+(\mathfrak{H})$, and the *Fermionic Fock space*, $\mathcal{F}_-(\mathfrak{H})$, to be the symmetric and anti-symmetric parts of $\mathcal{F}(\mathfrak{H})$:

$$\mathcal{F}_\pm(\mathfrak{H})=P_\pm\mathcal{F}(\mathfrak{H})=\bigoplus_{n\geq 0}P_\pm\mathfrak{H}^n. \tag{13}$$

The number operator (6) leaves $\mathcal{F}_\pm(\mathfrak{H})$ invariant. We will not distinguish in our notation between N and its restriction to those invariant subspaces.

[3] Let us recall that every permutation $\pi\in S_n$ is uniquely decomposed into a (commutative) product of cycles and that every cycle is a (non commutative, non unique) product of transpositions (a cycle of length two). The number of transpositions in the decomposition of each cycle is a constant modulo 2. One defines the signature of π to be $\epsilon(\pi)=(-1)^{\#(\mathrm{transp})}$, where $\#(\mathrm{transp})$ is the number of transpositions in any decomposition of π. The permutation π is called even if $\epsilon(\pi)=1$ and odd if $\epsilon(\pi)=-1$. Each cycle of length $l(\mathrm{cycle})\geq 2$ is the product of $l(\mathrm{cycle})-1$ transpositions, so we have the relations $\epsilon(\pi)=(-1)^{\sum_{\mathrm{c:cycles}}\#(\mathrm{transp\ in\ }c)}=(-1)^{\sum_{\mathrm{c:cycles}}(l(\mathrm{cycle})-1)}=(-1)^{n-\#(\mathrm{cycles})}=(-1)^{n+\#(\mathrm{cycles})}$, where we use $\sum_{c:\mathrm{cycles}}l(\mathrm{cycle})=n$.

[4] Formally this means that we consider $\bigoplus_{n\geq 0}P_\pm$ on $\mathcal{F}(\mathfrak{H})$, which we denote simply again by $P_\pm$.

2.2 Creation and annihilation operators

Given $f \in \mathfrak{H}$, we define the annihilation operator $a(f)$ in the following way: $a(f) : \mathfrak{H}^0 \to 0 \in \mathcal{F}(\mathfrak{H})$, $a(f) : \mathfrak{H}^n \to \mathfrak{H}^{n-1}$, $n \geq 1$, and for $\{f_k\}_{k=1}^n \subset \mathfrak{H}$,

$$a(f) f_1 \otimes \cdots \otimes f_n \mapsto \sqrt{n}\, \langle f, f_1 \rangle\, f_2 \otimes \cdots \otimes f_n, \tag{14}$$

where $\langle \cdot, \cdot \rangle$ is the scalar product in $\mathfrak{H}$. Similarly, we define the creation operator $a^*(f) : \mathfrak{H}^n \to \mathfrak{H}^{n+1}$ by

$$a^*(f) f_1 \otimes \cdots \otimes f_n \mapsto \sqrt{n+1}\, f \otimes f_1 \otimes \cdots \otimes f_n. \tag{15}$$

The map $f \mapsto a(f)$ is antilinear, while $f \mapsto a^*(f)$ is linear. We extend the action of the creation and annihilation operators by linearity to D^n, see (12), for all n. We have the following relations, for $\psi_n \in D^n$ and $f \in \mathfrak{H}$:

$$\|a(f)\psi_n\| \leq \sqrt{n}\, \|f\|\, \|\psi_n\|, \tag{16}$$

$$\|a^*(f)\psi_n\| = \sqrt{n+1}\, \|f\|\, \|\psi_n\|, \tag{17}$$

where the symbol $\|\cdot\|$ denotes the norm in the obvious spaces. The bound (16) follows from

$$\begin{aligned}
\|a(f)\psi_n\| &= \sup_{\phi \in \mathfrak{H}^{n-1},\, \|\phi\|=1} |\langle \phi, a(f)\psi_n \rangle| \\
&= \sup_{\phi \in \mathfrak{H}^{n-1},\, \|\phi\|=1} \left| \sqrt{n} \sum_{k=1}^K \left\langle f, f_1^{(k)} \right\rangle \left\langle \phi, f_2^{(k)} \otimes \ldots \otimes f_n^{(k)} \right\rangle \right| \\
&= \sup_{\phi \in \mathfrak{H}^{n-1},\, \|\phi\|=1} \left| \sqrt{n} \sum_{k=1}^K \left\langle f \otimes \phi, f_1^{(k)} \otimes \cdots \otimes f_n^{(k)} \right\rangle \right| \\
&\leq \sqrt{n}\, \|f\| \sup_{\Phi \in \mathfrak{H}^n,\, \|\Phi\|=1} \left| \sum_{k=1}^K \left\langle \Phi, f_1^{(k)} \otimes \cdots \otimes f_n^{(k)} \right\rangle \right| \\
&= \sqrt{n}\, \|f\|\, \|\psi_n\|.
\end{aligned}$$

Equality (17) is shown as follows

$$\begin{aligned}
\|a^*(f)\psi_n\| &= \sqrt{n+1} \left\| \sum_{k=1}^K f \otimes f_1^{(k)} \otimes \cdots \otimes f_n^{(k)} \right\| \\
&= \sqrt{n+1} \left\| f \otimes \left(\sum_{k=1}^K f_1^{(k)} \otimes \cdots \otimes f_n^{(k)} \right) \right\| \\
&= \sqrt{n+1}\, \|f\|\, \|\psi_n\|.
\end{aligned}$$

By continuity, the action of $a(f)$ and $a^*(f)$ extends to $\mathfrak{H}^n$, for all n, and hence by component-wise action to the domain $\mathcal{D}(N^{1/2}) \subset \mathcal{F}(\mathfrak{H})$. We have

$$\|a^{\#}(f)\psi\| \le \|f\| \, \|(N+1)^{1/2}\psi\|, \tag{18}$$

for $\psi \in \mathcal{D}(N^{1/2})$, where $a^{\#}$ stands for either a or a^*. The bound (18) is easily obtained from $\|a^{\#}(f)\psi\|^2 = \sum_{n\ge 0} \|a^{\#}(f)\psi_n\|^2$, (16), (17) and the definition of the number operator N, (6). The appearance of the star in $a^*(f)$ is not an arbitrary piece of notation, it signifies that $a^*(f)$ is the adjoint operator $a(f)^*$ of $a(f)$. We show this now. For all $\psi, \phi \in \mathcal{D}(N^{1/2})$, $f \in \mathfrak{H}$, we have

$$\langle \psi, a(f)\phi\rangle = \langle a^*(f)\psi, \phi\rangle \, . \tag{19}$$

Relation (19) follows easily from

$$\langle f_1 \otimes \cdots \otimes f_{n-1}, a(f) g_1 \otimes \cdots \otimes g_n\rangle = \langle a^*(f) f_1 \otimes \cdots \otimes f_{n-1}, g_1 \otimes \cdots \otimes g_n\rangle \, ,$$

for any n, $f, f_j, g_j \in \mathfrak{H}$, which in turn follows directly from the definitions of $a^{\#}(f)$, see (14), (15). Equality (19) shows that $a^*(f) \subseteq a(f)^*$ (the adjoint of $a(f)$ is an extension of $a^*(f)$), so $a(f)^*$ is densely defined and consequently $a(f)$ is closable (a closed extension of $a(f)$ is $a(f)^{**}$). Similarly, one sees that $a^*(f)$ is a closable operator. We denote from now on by $a^{\#}(f)$ the closed creation and annihilation operators. To show that $a^*(f) = a(f)^*$ it remains to prove that $a^*(f) \supseteq a(f)^*$. Let $\psi \in \mathcal{D}(a(f)^*)$ then

$$\varphi \mapsto \langle \psi, a(f)\varphi\rangle \tag{20}$$

is a bounded linear map on $\mathcal{D}(a(f))$. Given $\varphi \in \mathcal{D}(a(f))$ we choose $\varphi^{(n)}$ to be the vector in Fock space obtained by setting all components φ_k of φ equal to zero, for $k > n$. Due to the boundedness of the map (20) we have

$$\langle \psi, a(f)\varphi\rangle = \lim_{n\to\infty} \left\langle \psi, a(f)\varphi^{(n)}\right\rangle = \lim_{n\to\infty} \sum_{k=0}^{n-1} \langle \psi_k, a(f)\varphi_{k+1}\rangle \, . \tag{21}$$

Equality (19) shows that for each fixed n we can move $a(f)$ to the left factor in the inner product, so

$$\langle \psi, a(f)\varphi\rangle = \lim_{n\to\infty} \sum_{k=0}^{n-1} \langle a^*(f)\psi_k, \varphi_{k+1}\rangle \, . \tag{22}$$

By the density of $\mathcal{D}(a(f))$ the last equality extends to all vectors $\varphi \in \mathcal{F}(\mathfrak{H})$ and choosing $\varphi_{k+1} = a^*(f)\psi_k$ shows that $\sum_{k=0}^{\infty} \|a^*(f)\psi_k\|^2 < \infty$, so that $\psi \in \mathcal{D}(a^*(f))$. We conclude that $\mathcal{D}(a(f)^*) = \mathcal{D}(a^*(f))$. Since $a^*(f)$ is closed we have $a^*(f)\psi = \lim_n a^*(f)\psi^{(n)}$, where $\psi^{(n)}$ is the truncation of ψ as explained above in the case of φ. Using this in (22) gives

$$\langle \psi, a(f)\varphi\rangle = \langle a^*(f)\psi, \varphi\rangle \, , \tag{23}$$

for any φ in the dense set $\mathcal{D}(a(f))$. Consequently, we have $a(f)^*\psi = a^*(f)\psi$ which shows that $a^*(f) \supseteq a(f)^*$. This finishes the proof of the statement $a^*(f) = a(f)^*$.

Notice that $a(f)\Omega = 0$ for all $f \in \mathfrak{H}$ and conversely, if $\psi \in \mathcal{F}(\mathfrak{H})$ is s.t. $a(f)\psi = 0$ for all $f \in \mathfrak{H}$ then $\psi = z\Omega$, for some $z \in \mathbb{C}$.

The annihilation operators $a(f)$ leave the subspaces $\mathcal{F}_\pm(\mathfrak{H})$ invariant. This can be seen as follows. Let $\tau_{i,j}$ be the bounded linear operator on $\mathcal{F}(\mathfrak{H})$ which interchanges indices i and j in the tensor product, e.g. $\tau_{1,2}$ is determined by $\tau_{1,2} f_1 \otimes f_2 \otimes f_3 \otimes \cdots \otimes f_n = f_2 \otimes f_1 \otimes f_3 \otimes \cdots \otimes f_n$. An element $\psi_n \in \mathfrak{H}^n$ is in the range of $P_\pm$ if and only if $\tau_{i,j}\psi_n = \pm\psi_n$, for all $1 \leq i < j \leq n$. From the definition (14) of $a(f)$ we have for instance

$$\begin{aligned}\tau_{1,2} a(f) f_1 \otimes \cdots \otimes f_n &= \sqrt{n}\,\langle f, f_1\rangle\, f_3 \otimes f_2 \otimes \cdots \otimes f_n \\ &= a(f) f_1 \otimes f_3 \otimes f_2 \otimes \cdots \otimes f_n \\ &= a(f)\tau_{2,3} f_1 \otimes \cdots \otimes f_n,\end{aligned}$$

and in a similar fashion one sees that $\tau_{i,j} a(f) = a(f)\tau_{i+1,j+1}$. Consequently, if ψ_n is in the range of $P_\pm$, then we have $\tau_{i,j} a(f)\psi_n = a(f)\tau_{i+1,j+1}\psi_n = \pm a(f)\psi_n$, so $a(f)\psi_n$ is in the range of $P_\pm$. We may write this also as $P_\pm a(f) P_\pm = a(f) P_\pm$.

The Bosonic $(+)$ and Fermionic $(-)$ creation and annihilation operators are defined to be the restrictions

$$a^\#_\pm(f) = P_\pm a^\#(f) P_\pm. \tag{24}$$

One then has $a_\pm(f) = a(f)P_\pm$ and $a^*_\pm(f) = P_\pm a^*(f)$. Using (14) and (15), it is not difficult to verify that

$$a_+(g)a^*_+(f) f_1 \otimes \cdots \otimes f_n = \sum_{k=1}^{n+1} \langle g, f_k\rangle\, P_+ f_1 \otimes \cdots \otimes \widehat{f_k} \otimes \cdots \otimes f_{n+1}, \tag{25}$$

where the hat $\widehat{\ }$ means that the corresponding symbol is omitted, and where we have set $f_{n+1} = f$. Similarly,

$$a^*_+(f)a_+(g) f_1 \otimes \cdots \otimes f_n = \sum_{k=1}^{n} \langle g, f_k\rangle\, P_+ f_1 \otimes \cdots \otimes \widehat{f_k} \otimes \cdots \otimes f_{n+1}. \tag{26}$$

Bosonic creation and annihilation operators satisfy the *canonical commutation relations (CCR)*:

$$[a_+(g), a^*_+(f)] = \langle g, f\rangle\, \mathbb{1}_{\mathcal{F}_+(\mathfrak{H})}, \tag{27}$$

$$[a_+(f), a_+(g)] = [a^*_+(f), a^*_+(g)] = 0, \tag{28}$$

for any $f, g \in \mathfrak{H}$, and where $[x, y] = xy - yx$ is the commutator. Equations (27), (28) are understood in the strong sense on $\mathcal{D}(N)$, on which products of two creation and annihilation operators are defined. Relation (27) follows directly from (25) and (26), and (28) can be established similarly.

Fermionic creation and annihilation operators satisfy the *canonical anti-commutation relations (CAR)*:

$$\{a_-(g), a_-^*(f)\} = \langle g, f \rangle \, \mathbb{1}_{\mathcal{F}_-(\mathfrak{H})}, \tag{29}$$

$$\{a_-(f), a_-(g)\} = \{a_-^*(f), a_-^*(g)\} = 0, \tag{30}$$

for any $f, g \in \mathfrak{H}$, and where $\{x, y\} = xy + yx$ is the anti-commutator (a priori again understood in the strong sense on $\mathcal{D}(N)$. However, it turns out that this relation extends to an equality of bounded operators, as we show now).

Although the CCR and the CAR have a similar structure (just interchange commutators with anti-commutators), they impose very different properties on the respective creation and annihilation operators. For instance, it turns out that the Fermionic creation and annihilation operators extend to bounded operators, while this is not true in the Bosonic case. We see this by using the CAR to obtain

$$\|a_-^*(f)\psi\|^2 = \langle \psi, a_-(f)a_-^*(f)\psi \rangle = -\|a_-(f)\psi\|^2 + \|f\|^2 \, \|\psi\|^2, \tag{31}$$

for all $\psi \in \mathcal{D}(N)$, from which it follows that $\|a_-^{\#}(f)\| \leq \|f\|$. On the other hand, $\|a_-^*(f)\Omega\| = \|f\| = \|f\| \, \|\Omega\|$, so $\|a_-(f)a_-^*(f)\Omega\| = \|f\|^2 = \|f\| \, \|a_-^*(f)\Omega\|$, hence

$$\|a_-^{\#}(f)\| = \|f\|. \tag{32}$$

Notice that this reasoning does not work for Bosons, because the minus sign on the r.h.s. of (31) would have to be replaced by a plus sign.

The fact that $a_+^*(f)$ is an unbounded operator can be seen as follows. Let $\psi_n \in \mathcal{F}_+(\mathfrak{H})$ be the normalized vector whose components are all zero except the n-particle component, which is $f \otimes f \otimes \cdots \otimes f$, for some $f \in \mathfrak{H}$, $\|f\| = 1$. Then we have $a_+^*(f)\psi_n = \sqrt{n+1}\psi_{n+1}$, hence $\|a_+^*(f)\psi_n\| = \sqrt{n+1} \to \infty$, as $n \to \infty$. This reasoning does not work for Fermions, because the vector ψ_n is not in the Fermionic Fock space. More generally, the *Pauli principle* says that it is impossible to have a state of several Fermions in which two among them are in the same one-particle state. This is expressed as

$$a_-^*(f)a_-^*(f) = 0, \tag{33}$$

for all $f \in \mathfrak{H}$, which follows immediately from (30).

2.3 Weyl operators

On a mathematical level, dealing with unbounded operators is a delicate affair so from this point of view Fermionic creation and annihilation operators are more easily handled than the Bosonic ones. It is desirable to replace the set of Bosonic creation and annihilation operators by a set of bounded operators which are in a certain sense equivalent to the set of creation and annihilation operators. These bounded operators are called Weyl operators.

We first form the (normalized) real and imaginary parts of $a_+(f)$

$$\Phi(f) = \frac{a_+(f) + a_+^*(f)}{\sqrt{2}}, \quad \Pi(f) = \frac{a_+(f) - a_+^*(f)}{\sqrt{2}\, i}, \tag{34}$$

defined as operators on $\mathcal{D}(N^{1/2})$. We do not equip Φ and Π with an index $+$ since we are going to use them only for Bosons (although one can do the same procedure for Fermions as well). We have $\Pi(f) = \Phi(if)$, so it suffices to consider the operators $\Phi(f)$. Notice though that $f \mapsto \Phi(f)$ is not a linear nor an antilinear map; it is only a real-linear map. Define the *finite particle subspace* of Fock space by

$$\mathcal{F}_+^0(\mathfrak{H}) = \{\psi = \{\psi_n\}_{n\geq 0} \in \mathcal{F}_+(\mathfrak{H}) \mid \text{all but finitely many } \psi_n \text{ are zero}\}. \quad (35)$$

Clearly, $\mathcal{F}_+^0(\mathfrak{H}) \subset \mathcal{D}(N^\nu)$ for any $\nu > 0$. In particular, any polynomial in creation and annihilation operators is well defined as an operator on $\mathcal{F}_+^0(\mathfrak{H})$.

Proposition 2.1. *1. For any $f \in \mathfrak{H}$, $\Phi(f)$ is essentially selfadjoint on $\mathcal{F}_+^0(\mathfrak{H})$. If $\{f_n\}$ is a sequence in $\mathfrak{H}$ converging to $f \in \mathfrak{H}$, i.e. $\|f_n - f\| \to 0$, then $\Phi(f_n) \to \Phi(f)$ in the strong sense on $\mathcal{D}(N^{1/2})$, i.e. $\|(\Phi(f_n) - \Phi(f))\psi\| \to 0$, for all $\psi \in \mathcal{D}(N^{1/2})$.*

2. On $\mathcal{F}_+^0(\mathfrak{H})$, we have

$$e^{itN}\Phi(f)e^{-itN} = \Phi(e^{it}f), \quad (36)$$

for any $t \in \mathbb{R}$, $f \in \mathfrak{H}$.

3. For $f, g \in \mathfrak{H}$, we have the CCR

$$[\Phi(f), \Phi(g)] = i\mathrm{Im}\,\langle f, g\rangle\,, \quad (37)$$

understood in the strong sense on $\mathcal{D}(N)$.

Proof. An elegant proof of essential selfadjointness can be given using the Glimm-Jaffe-Nelson commutator theorem, c.f. [13]. We opt for a more pedestrian proof involving analytic vectors, [5] because these are useful for concrete calculations. Nelson's analytic vector theorem says that if the domain of a symmetric operator contains an invariant subspace C which itself contains a dense set (in Hilbert space) of analytic vectors, then the symmetric operator is essentially selfadjoint on C. (See e.g. [13], Theorem X.39).

Let $f \in \mathfrak{H}$ be fixed. The dense set $\mathcal{F}_+^0$ is invariant under $\Phi(f)$. We show that each vector $\psi \in \mathcal{F}_+^0$ is analytic for $\Phi(f)$. Because ψ is a finite sum of vectors $\psi_n \in P_+\mathfrak{H}^n$ (for possibly varying n), it is enough to show that ψ_n is an analytic vector for $\Phi(f)$, for any n. It is clear that $\psi_n \in \mathcal{D}(\Phi(f)^k)$, for all $k \geq 0$ and from

$$\|\Phi(f)^k\psi_n\| \leq \sqrt{2}\|f\|\,\|(N+1)^{1/2}\Phi(f)^{k-1}\Psi_n\| \leq \sqrt{2}\sqrt{n+k}\,\|f\|\,\|\Phi(f)^{k-1}\psi_n\|$$

it follows that

$$\|\Phi(f)^k\psi_n\| \leq 2^{k/2}\sqrt{(n+k)!}\,\|f\|^k\,\|\psi_n\|.$$

This means that the series

$$\sum_{k\geq 0}\frac{t^k}{k!}\|\Phi(f)^k\psi_n\|$$

[5] Let A be a linear operator on a Hilbert space $\mathcal{H}$. A vector $\psi \in \mathcal{H}$ is called *analytic* for A if $\psi \in \cap_{k\geq 0}\mathcal{D}(A^k)$ and the complex power series $\sum_{k\geq 0} t^k\|A^k\psi\|/k!$ has a nonzero radius of convergence. If the radius of convergence is infinite then ψ is said to be *entire* for A.

converges for any $t \in \mathbb{C}$, hence ψ_n is an analytic (even an entire) vector for $\Phi(f)$.

We now show the strong continuity property. Let $\psi \in \mathcal{D}(N^{1/2}) \cap \mathcal{F}_+(\mathfrak{H})$. Then

$$\begin{aligned}\|(\Phi(f_n) - \Phi(f))\psi\| &\leq 2^{-1/2}\|a^*(f_n - f)\psi\| + 2^{-1/2}\|a(f_n - f)\psi\| \\ &\leq \sqrt{2}\|f_n - f\| \, \|(N+1)^{1/2}\psi\|,\end{aligned}$$

and the result follows.

To see 2., simply use the definition of the creation operator to obtain

$$\begin{aligned}e^{itN}a_+^*(f)e^{-itN}P_+f_1 \otimes \cdots f_n &= \sqrt{n+1}e^{it}P_+f \otimes f_1 \otimes \cdots \otimes f_n \\ &= a_+^*(e^{it}f)P_+f_1 \otimes \cdots \otimes f_n,\end{aligned}$$

and similarly for annihilation operators.

The proof of 3. is immediate from (27), (28). □

From now on we denote by $\Phi(f)$ the selfadjoint closure of (34). It generates a strongly continuous one-parameter group of unitaries on the Hilbert space $\mathcal{F}_+(\mathfrak{H})$,

$$\mathbb{R} \ni t \mapsto e^{it\Phi(f)}. \tag{38}$$

We define the *Weyl operator* $W(f)$, for $f \in \mathfrak{H}$, to be the unitary operator

$$W(f) = e^{i\Phi(f)}. \tag{39}$$

We have encountered the CCR expressed in terms of creation and annihilation operators (see (27), (28)) and in terms of the operators $\Phi(f)$ (see (37)). How are they expressed in terms of the Weyl operators? Taking into account (37), the Baker-Campbell-Hausdorff formula gives (formally)

$$W(f)W(g) = e^{-\frac{i}{2}\mathrm{Im}\langle f,g\rangle}W(f+g) = e^{-i\mathrm{Im}\langle f,g\rangle}W(g)W(f).\,[6] \tag{40}$$

Relation (40) is called the Weyl form of the CCR. The following result is sometimes useful.

Proposition 2.2. *On the domain $\mathcal{D}(N)$ of the number operator we have*

$$NW(f) = W(f)N + W(f)(\Phi(if) + \|f\|^2/2), \tag{41}$$

for any $f \in \mathfrak{H}$. This means in particular that the Weyl operators leave $\mathcal{D}(N)$ invariant. It follows thus from (40) that any finite sum of products of Weyl operators leave $\mathcal{D}(N)$ invariant.

[6] The BCH formula is the non-commutative analogue of the formula $e^a e^b = e^{a+b}$. Let A, B be bounded operators on a Hilbert space $\mathfrak{H}$. Then $e^A e^B = \exp\{A + B + \frac{1}{2}[A,B] + \frac{1}{12}([A,[A,B]] - [B,[A,B]]) + \cdots\}$ (these are the first explicit terms in the BCH formula). In case the commutator $[A, B]$ is proportional to the identity the BCH formula simply reduces to $e^A e^B = e^{A+B+\frac{1}{2}[A,B]} = e^{A+B}e^{\frac{1}{2}[A,B]}$. Formally (40) follows thus from (37). Recall though that the $\Phi(f)$, $\Phi(g)$ are unbounded operators. It is correct to say that (40) implies (37); this can be seen by noticing that, on $\mathcal{F}_+^0(\mathfrak{H})$, one has $[\Phi(f), \Phi(g)] = \frac{1}{i^2}\partial_{st}^2|_{s=t=0}(W(tf)W(sg) - W(sg)W(tf))$, and then calculating the r.h.s. using (40).

Proof. To show (41) we notice first that $e^{itN}W(f) = W(e^{it}f)e^{itN}$ (which follows from (36)). Using

$$\partial_t|_{t=0}\,\Phi(e^{it}f) = \Phi(if),$$
$$\partial_t|_{t=0}\,\Phi(e^{it}f)^n = n\Phi(f)^{n-1}\Phi(if) - i\|f\|^2\frac{n(n-1)}{2}\Phi(f)^{n-2}, \quad \text{for } n \geq 2,$$

we obtain, in the strong sense on $\mathcal{F}_+^0$,

$$\frac{1}{i}\partial_t|_{t=0}\,W(e^{it}f)e^{itN} = W(f)N + \frac{1}{i}\partial_t|_{t=0}\sum_{n\geq 0}\frac{i^n\Phi(e^{it}f)^n}{n!}$$
$$= W(f)N + W(f)\left(\Phi(if) + \|f\|^2/2\right),$$

which extends to $\mathcal{D}(N)$, giving (41). □

We finish this section by examining the continuity properties of the map $f \mapsto W(f)$. Recall that for Fermionic creation and annihilation operators, $f \mapsto a_-^{\#}(f)$ is a continuous map from $\mathfrak{H}$ into the bounded operators (equipped with the operator-norm topology), see (32). As we show now only a weaker form of continuity holds for the map $f \mapsto W(f)$. This is a source of considerable trouble in many applications.

Theorem 2.3. *If $f_n \to f$ in $\mathfrak{H}$, then $W(f_n) \to W(f)$ in the strong sense on $\mathcal{F}_+(\mathfrak{H})$, i.e., for any $\psi \in \mathcal{F}_+(\mathfrak{H})$, $\|(W(f_n) - W(f))\psi\| \to 0$. However, for any $f \in \mathfrak{H}$, $f \neq 0$, we have $\|W(f) - \mathbb{1}\| = 2$.*

Let e^{ith} be a strongly continuous unitary group on $\mathfrak{H}$ (h being its selfadjoint generator). Due to the theorem we have $\|W(e^{ith}f) - \mathbb{1}\| = 2$ (for $f \neq 0$), which implies that $t \mapsto W(e^{ith}f)$ is not norm continuous (the dynamics defined by e^{ith} is not continuous in the C^*algebra topology).

Proof of Theorem 2.3. The previous proposition tells us that $\Phi(f_n) \to \Phi(f)$, in the strong sense on $\mathcal{F}_+^0(\mathfrak{H})$, which is a joint core for all the operators $\Phi(f_n)$ and $\Phi(f)$. Therefore, $\Phi(f_n)$ converges to $\Phi(f)$ in the strong resolvent sense (see e.g. [13], Theorem VII.25]), from which it follows that $e^{it\Phi(f_n)}$ converges to $e^{it\Phi(f)}$ in the strong sense, for all t ([13], Theorem VII.21).

Let us show $\|W(f) - \mathbb{1}\| = 2$, for any $f \neq 0$. The CCR (40) give

$$W(g)^*W(f)W(g) = e^{-i\mathrm{Im}\langle f,g\rangle}W(f),$$

for any $g \in \mathfrak{H}$. Since $W(g)$ is unitary, this tells us that the spectrum of $W(f)$ is invariant under rotations, hence it must be the whole unit circle. The assertion $\|W(f) - \mathbb{1}\| = 2$ follows now from the spectral theorem. □

2.4 The C^*-algebras $\mathrm{CAR}_\mathrm{F}(\mathfrak{H})$, $\mathrm{CCR}_\mathrm{F}(\mathfrak{H})$

The set of all Fermionic creation and annihilation operators generates a C^*-algebra of operators on $\mathcal{F}_-(\mathfrak{H})$, which we call $\mathrm{CAR}_\mathrm{F}(\mathfrak{H})$. The index $_F$ reminds us that the elements of this C^*-algebra are viewed as operators on Fock space $\mathcal{F}_-(\mathfrak{H})$. Similarly, the set of all Weyl operators generates a C^*-algebra of operators on $\mathcal{F}_+(\mathfrak{H})$, which we shall call $\mathrm{CCR_F}(\mathfrak{H})$. Both algebras are unital C^*-algebras. For $\mathrm{CAR_F}(\mathfrak{H})$ this follows from (29), and for $\mathrm{CCR_F}(\mathfrak{H})$ it follows from $W(0) = \mathbb{1}$.

Theorem 2.4. *Let $a^*(f)$ and $a(f)$ denote the Fermionic creation and annihilation operators, acting on $\mathcal{F}_-(\mathfrak{H})$. The linear span of vectors of the form*

$$a^*(f_1)\cdots a^*(f_n)\Omega,$$

with $f_k \in \mathfrak{H}$, $n \geq 0$, is dense in $\mathcal{F}_-(\mathfrak{H})$. In particular, Ω is cyclic for $\mathrm{CAR_F}(\mathfrak{H})$ in $\mathcal{F}_-(\mathfrak{H})$. [7] *Moreover, $\mathrm{CAR_F}(\mathfrak{H})$ acts irreducibly on $\mathcal{F}_-(\mathfrak{H})$.* [8]

Proof. The first statement follows from

$$a^*(f_1)a^*(f_2)\cdots a^*(f_n)\Omega = \sqrt{n!}\, P_- f_1 \otimes f_2 \otimes \cdots \otimes f_n.$$

To see irreducibility, we suppose that T is a bounded operator on $\mathcal{F}_-(\mathfrak{H})$ that commutes with all operators $a^\#(f)$, $f \in \mathfrak{H}$, and show that $T = z\mathbb{1}$, for some $z \in \mathbb{C}$. We have $a(f)T\Omega = Ta(f)\Omega = 0$, for all $f \in \mathfrak{H}$, so $T\Omega = z\Omega$, for some complex number z (see after (19)). It follows that

$$Ta^*(f_1)\cdots a^*(f_n)\Omega = a^*(f_1)\cdots a^*(f_n)T\Omega = za^*(f_1)\cdots a^*(f_n)\Omega,$$

so by cyclicity of Ω, $T\psi = z\psi$, for all $\psi \in \mathcal{F}_-(\mathfrak{H})$. □

Theorem 2.5. *The vacuum vector $\Omega \in \mathcal{F}_+(\mathfrak{H})$ is cyclic for $\mathrm{CCR_F}(\mathfrak{H})$ in $\mathcal{F}_+(\mathfrak{H})$, and $\mathrm{CCR_F}(\mathfrak{H})$ acts irreducibly on $\mathcal{F}_+(\mathfrak{H})$.*

Proof. As in the case of Fermions, it is clear that the span of

$$\{a^*(f_1)\cdots a^*(f_n)\Omega \mid f_k \in \mathfrak{H}, n \geq 0\}$$

is dense in $\mathcal{F}_+(\mathfrak{H})$. But this is the same as the span of $\{\Phi(f_1)\cdots\Phi(f_n)\Omega \mid f_k \in \mathfrak{H}, n \geq 0\}$, so it is enough to prove that $\mathrm{CCR_F}(\mathfrak{H})$ is dense in that latter span.

We show first that

$$(N+1)^k W(f)(N+1)^{-k-1} \tag{42}$$

is a bounded operator for all $f \in \mathfrak{H}$ and all $k \geq 0$. We proceed by induction in k. The statement is obvious for $k = 0$. Using (41) of Proposition 2.2 we get

---

[7] Let ψ be a vector in a Hilbert space $\mathcal{H}$ and let $\mathfrak{M}$ be a set of bounded operators on $\mathcal{H}$, $\mathfrak{M} \subseteq \mathcal{B}(\mathcal{H})$. We say that ψ is cyclic for $\mathfrak{M}$ in $\mathcal{H}$ if $\mathfrak{M}\psi = \{M\psi \mid M \in \mathfrak{M}\}$ is dense in $\mathcal{H}$.

[8] Let $\mathfrak{M}$ be a set of bounded operators acting on a Hilbert space $\mathcal{H}$. We say that $\mathfrak{M}$ acts irreducibly if the only closed subspaces of $\mathcal{H}$ which are invariant under the action of $\mathfrak{M}$ are the trivial subspaces $\{0\}$ and $\mathcal{H}$. $\mathfrak{M}$ acts irreducibly on $\mathcal{H}$ if and only if its commutant is trivial, $\mathfrak{M}' = \{T \in \mathcal{B}(\mathcal{H}) \mid TM = MT,\ \forall M \in \mathfrak{M}\} = \mathbb{C}\mathbb{1}$.

$$
\begin{aligned}
&(N+1)^k(N+1)W(f)(N+1)^{-1}(N+1)^{-k-1}\\
&= (N+1)^k W(f)(N+1)^{-k-1} \qquad (43)\\
&\quad +(N+1)^k\left\{W(f)(\Phi(if)+\|f\|/2)\right\}(N+1)^{-k-2}, \qquad (44)
\end{aligned}
$$

where we commuted $N+1$ through $W(f)$ in the r.h.s. By the induction assumption, (43) is a bounded operator. The term with the field operator in (44) can be written as

$$
(N+1)^k W(f)(N+1)^{-k-1}(N+1)^{k+1}\Phi(if)(N+1)^{-k-2},
$$

where the product of the first three operators is again bounded. It suffices thus to show that

$$
(N+1)^k\Phi(f)(N+1)^{-k-1} \qquad (45)
$$

is bounded, for all $f \in \mathfrak{H}$ and $k \geq 0$. Clearly we have $a(f)N = (N+1)a(f)$, $a^*(f)N = (N-1)a^*(f)$, and (45) follows easily. This finishes the proof of (42).

Since the product of Weyl operators is again a Weyl operator (modulo a phase) we get a bounded operator also if we replace $W(f)$ in (42) by any sum of products of Weyl operators. Given any $\epsilon > 0$ there exists a $T_n(\epsilon) < \infty$ such that

$$
\Phi(f_1)\cdots\Phi(f_n)\Omega = \Phi(f_1)\cdots\Phi(f_{n-1})\frac{W(t_n f_n)-\mathbb{1}}{it_n}\Omega + O(\epsilon),
$$

provided $t_n \leq T_n(\epsilon)$, and where $O(\epsilon)$ denotes a vector with norm less than ϵ. There exists a $T_{n-1}(\epsilon, t_n)$ such that

$$
\begin{aligned}
&\Phi(f_1)\cdots\Phi(f_n)\Omega\\
&= \Phi(f_1)\cdots\Phi(f_{n-2})\frac{W(t_{n-1}f_{n-1})-\mathbb{1}}{it_{n-1}}\,\frac{W(t_n f_n)-\mathbb{1}}{it_n}\Omega + O(\epsilon),
\end{aligned}
$$

provided $t_{n-1} \leq T_{n-1}(\epsilon, t_n)$. Continuing this process we see that there are numbers $T_n(\epsilon), T_k(\epsilon, t_n, \ldots, t_{k+1})$, $1 \leq k \leq n-1$, such that if $t_k \leq T_k(\epsilon, t_n, \ldots, t_{k+1})$ and $t_n \leq T_n(\epsilon)$ then

$$
\begin{aligned}
&\Phi(f_1)\cdots\Phi(f_n)\Omega\\
&= \frac{W(t_1 f_1)-\mathbb{1}}{it_1}\cdots\frac{W(t_{n-1}f_{n-1})-\mathbb{1}}{it_{n-1}}\,\frac{W(t_n f_n)-\mathbb{1}}{it_n}\Omega + O(\epsilon).
\end{aligned}
$$

Since the operator acting on Ω in the above r.h.s. is an element of $\mathrm{CCR}_F(\mathfrak{H})$ cyclicity of Ω is shown.

We finish the proof by showing irreducibility. Suppose T is a bounded operator on $\mathcal{F}_+(\mathfrak{H})$ that commutes with all $W(f)$, $f \in \mathfrak{H}$. It follows that for any $\psi \in \mathcal{D}(\Phi(f))$,

$$
\frac{e^{it\Phi(f)}-\mathbb{1}}{it}T\psi = T\frac{e^{it\Phi(f)}-\mathbb{1}}{it}\psi \longrightarrow T\Phi(f)\psi,
$$

as $t \to 0$. This shows that $T\psi \in \mathcal{D}(\Phi(f))$ and that $\Phi(f)T\psi = T\Phi(f)\psi$, i.e. T leaves the domain of every $\Phi(f)$ invariant and T commutes strongly with every $\Phi(f)$. Since $a(f) = 2^{-1/2}(\Phi(f)+i\Phi(if))$, this means that T commutes with $a(f)$, in the strong sense, for all $f \in \mathfrak{H}$. Irreducibility is now shown exactly as in Theorem 2.4. □

2.5 Leaving Fock space

We explain in this section why Fock space is not always the right Hilbert space to describe a physical system.

As we have pointed out in Section 1.1, the very definition of Fock space gives the existence of a number operator, N, which is the operator of multiplication by n on the n-sector. Let $\psi \in \mathcal{F}(\mathfrak{H})$ be a (pure) state of the quantum gas (the following reasoning applies equally well to mixed states given by density matrices, i.e., convex combinations of pure states). The probability of finding more than a fixed number n of particles in the state ψ is given by

$$\langle \psi, P(N \geq n)\psi \rangle = \sum_{k \geq n} \| [\psi]_k \|^2, \tag{46}$$

where $P(N \geq n)$ is the spectral projection of N onto the set $\{n, n+1, \ldots\}$. The probability (46) vanishes in the limit $n \to \infty$, simply because ψ is in Fock space (the series converges). This shows that, a priori, any state described by a vector (or a density matrix) in Fock space has only finitely many particles in the sense that the probability of finding n particles approaches zero as n increases to infinity.

We will be interested in describing an ideal quantum gas which is extended in all of physical space $\mathbb{R}^3$, and which has a *nonzero density*, say one particle per unit volume. Such a state cannot be described by a vector (or density matrix) in Fock space! We may describe such a state as a limit of states "living" in Fock space (i.e., given by a density matrix on Fock space), e.g. by saying that the system should first be confined to a finite box $\Lambda_0 \subset \mathbb{R}^3$, in which case it is described by a vector $\psi_{\Lambda_0} \in \mathcal{F}(L^2(\Lambda_0))$ (of course, since the box is finite, and we specify a fixed density, there are only finitely many particles and Fock space can describe such a state). One then takes a sequence of nested boxes, $\Lambda_0 \subset \Lambda_1 \subset \cdots$ which increase to all of $\mathbb{R}^3$, $\cup_{k \geq 0}\Lambda_k = \mathbb{R}^3$, hence obtaining a sequence of states $\psi_{\Lambda_k} \in \mathcal{F}(L^2(\Lambda_k))$. If one can show that ψ_{Λ_k} has a limit ψ_∞, in a suitable sense, and where the density or particles is fixed, as $k \to \infty$, then ψ_∞ can be regarded as being the infinitely extended state with nonzero density. This limit is called the thermodynamic limit.

The limit state ψ_∞ is naturally not a vector in Fock space any more. What kind of object is it? To answer this, we have to say in what sense we take the thermodynamic limit. To be specific, we carry out the following discussion for Bosons. It can be repeated for Fermions. For any finite box Λ, the vector $\psi_\Lambda \in \mathcal{F}_+(L^2(\Lambda))$ gives rise to a positive, linear, normalized map on the von Neumann algebra of all bounded operators on $\mathcal{F}_+(L^2(\Lambda))$ by the assignment

$$\mathcal{B}(\mathcal{F}_+(L^2(\Lambda))) \ni A \mapsto \omega_\Lambda(A) = \langle \psi_\Lambda, A\psi_\Lambda \rangle \tag{47}$$

(for a mixed state determined by the density matrix ρ_Λ, we set $\omega_\Lambda(A) = \mathrm{tr}(\rho_\Lambda A)$). Since $\mathrm{CCR_F}(L^2(\Lambda))$ is irreducible (see Theorem 2.5), its weak closure is the set of all bounded operators (indeed, irreducibility implies that $\mathrm{CCR_F}(\mathfrak{H})' = \mathbb{C}\mathbb{1}$, so $\mathrm{CCR_F}(\mathfrak{H})'' = \mathcal{B}(\mathfrak{H})$). Without loss of generality, we may therefore consider (47) only for $A \in \mathrm{CCR_F}(L^2(\Lambda))$, i.e. we view ω_Λ as a state on $\mathrm{CCR_F}(L^2(\Lambda))$, in the sense of the theory of C^*-algebras. [9]

Consider the (so-called *quasi-local*) C^*-algebra

$$\mathfrak{A}_0 = \overline{\bigcup_{n\geq 0} \mathrm{CCR_F}(L^2(\Lambda_n))}^{\mathrm{norm}} \subset \mathcal{B}\left(\mathcal{F}_+(L^2(\mathbb{R}^3))\right)$$

where $-^{\mathrm{norm}}$ means that we take the norm closure in the operator norm. Assume that the limit

$$\omega_\infty(A) = \lim_{k\to\infty} \omega_{\Lambda_k}(A) \tag{48}$$

exists, for any $A \in \mathrm{CCR_F}(L^2(\Lambda_n))$, any n. Then ω_∞ defines a state on $\mathfrak{A}_0$. We point out once more that in general, ω_∞ cannot be represented by a density matrix on Fock space $\mathcal{F}_+(L^2(\mathbb{R}^3, d^3x))$. One says that ω_∞ is not *normal* with respect to the states ω_{Λ_k}. [10] In the GNS representation $(\mathcal{H}_\infty, \pi_\infty, \psi_\infty)$ of $(\mathfrak{A}_0, \omega_\infty)$, the state ω_∞ is represented as

$$\omega_\infty(A) = \langle \psi_\infty, \pi_\infty(A)\psi_\infty\rangle.$$

In Section 4 we will discuss in detail the construction of the infinite-volume limit of a state describing a Bose gas with a given momentum density distribution and we will explicitly construct the corresponding GNS representation (the Araki-Woods representation).

One may wonder about the dependence of the C^*-algebra $\mathrm{CCR_F}(\mathfrak{H})$ on its underlying Hilbert space, $\mathcal{F}_+(\mathfrak{H})$. After all, we have just seen that density matrices on $\mathcal{F}_+(\mathfrak{H})$ cannot describe certain states of physical interest. Therefore Fock space should not play a central role in the definition of a physical system. In an attempt to detach ourselves from Fock space we may define the CCR and CAR algebras as *abstract C^*-algebras*, without referring to a Hilbert space. Fock space is then just the GNS representation space of a certain state on the abstract algebras, represented by the Fock vacuum vector (recall that the Fock vacuum vector is cyclic for $\mathrm{CCR_F}(\mathfrak{H})$ and $\mathrm{CAR_F}(\mathfrak{H})$, as we have shown in Theorems 2.4 and 2.5 above).

3 The CCR and CAR algebras

In this section we introduce abstract CAR and CCR algebras and review some of their properties. Useful references are [4], [5] [14].

We remind the reader of the notion of the "test function space" $\mathfrak{D} \subseteq \mathfrak{H}$, introduced at the beginning of Section 2.1, see (1).

---

[9] Let $\mathfrak{A}$ be a (unital) C^*-algebra. A state ω on $\mathfrak{A}$ is a positive linear functional $\omega : \mathfrak{A} \to \mathbb{C}$ which is normalized as $\omega(\mathbb{1}) = 1$.

[10] Let ω_1 and ω_2 be two states on a C^*-algebra $\mathfrak{A}$. Then ω_1 is called normal with respect to ω_2 iff $\omega_1(A) = \mathrm{tr}(\rho\pi_2(A))$, where ρ is a trace class operator (density matrix) on $\mathcal{H}_2$, and where $(\mathcal{H}_2, \pi_2, \Omega_2)$ is the GNS representation of $(\mathfrak{A}, \omega_2)$.

3.1 The algebra CAR($\mathfrak{D}$)

An (abstract) CAR algebra CAR($\mathfrak{D}$) over $\mathfrak{D} \subseteq \mathfrak{H}$ (where $\mathfrak{H}$ is a Hilbert space) is defined to be a unital C^*-algebra generated by elements written as $a(f)$, $f \in \mathfrak{D}$, where the assignment $f \mapsto a(f)$ is an antilinear map, and where the following relations hold

$$\{a(f), a(g)\} = 0, \quad \{a(f), a^*(g)\} = \langle f, g \rangle \, \mathbb{1}. \tag{49}$$

Here $a^*(f)$ is the element in the C^*algebra obtained by applying the $*$operation to $a(f)$, and $\{a, b\} = ab + ba$ is the anticommutator. We have already seen in the previous section that a C^*-algebra with these properties exists. Let us mention that the CAR (49) imply that

$$\|a(f)\| = \|f\|, \tag{50}$$

where $\|\cdot\|$ on the left hand side is the C^* norm and on the right hand side it is the norm of $\mathfrak{D}$ induced by $\mathfrak{H}$. This follow since $(a(f)a(f) = 0$ by the Pauli principle, see (33))

$$\left(a^*(f)a(f)\right)^2 = a^*(f)\{a(f), a^*(f)\}a(f) = \|f\|^2 a^*(f)a(f),$$

so that by the C^*norm property ($\|A^*A\| = \|A\|^2$), we have

$$\|a(f)\|^4 = \|f\|^2 \, \|a(f)\|^2.$$

Alternatively, boundedness of the Fermionic creation and annihilation operators follows from the fact that

$$\|\pi(a(f))\| = \|f\| \tag{51}$$

in any representation π of the CAR, which is shown as in (32). Let f_α be a net in $\mathfrak{D}$ converging to $f \in \overline{\mathfrak{D}}$ (the closure of $\mathfrak{D} \subseteq \mathfrak{H}$). Then $\|a(f) - a(f_\alpha)\| = \|f - f_\alpha\| \to 0$, so $a(f) \in \mathrm{CAR}(\overline{\mathfrak{D}})$ because CAR($\mathfrak{D}$), being a C^*-algebra, is uniformly closed. This shows that

$$\mathrm{CAR}(\mathfrak{D}) = \mathrm{CAR}(\overline{\mathfrak{D}}). \tag{52}$$

The next result tells us that given $\mathfrak{D}$, the corresponding CAR algebra CAR($\mathfrak{D}$) is unique.

Theorem 3.1. (Uniqueness of the CAR algebra). *Let $\mathfrak{D} \subseteq \mathfrak{H}$ be a given test function space (see (1)), and let $\mathfrak{A}_1$, $\mathfrak{A}_2$ be two CAR algebras over $\mathfrak{D}$ (generated by $a_1(f)$ and $a_2(f)$, respectively, with $f \in \mathfrak{D}$). There is a unique $*$isomorphism $\alpha : \mathfrak{A}_1 \to \mathfrak{A}_2$ such that $\alpha(a_1(f)) = a_2(f)$, for all $f \in \mathfrak{D}$.*

A proof can be found for instance in [5]. Once uniqueness is known in the sense above, one can easily prove the following result.

Theorem 3.2. *The C^*algebra* CAR($\mathfrak{D}$) *is simple.* [11]

---

[11] A C^*algebra $\mathfrak{A}$ is called simple if it has no nontrivial closed two-sided ideals, i.e., if the only closed two-sided ideals are $\{0\}$ and $\mathfrak{A}$. A subspace $\mathfrak{I} \subseteq \mathfrak{A}$ is a two-sided ideal if $A \in \mathfrak{A}$ and $I \in \mathfrak{I}$ implies that IA and AI are in $\mathfrak{I}$.

Proof. Let $\mathfrak{I} \neq \mathfrak{A}_1 = \mathrm{CAR}(\mathfrak{D})$ be a closed two-sided ideal of $\mathrm{CAR}(\mathfrak{D})$. Define $\mathfrak{A}_2 = \mathrm{CAR}(\mathfrak{D})/\mathfrak{I}$ to be the C^*algebra generated by the equivalence classes $a_2(f) = [a(f)]$. Theorem 3.1 tells us that the projection $P : \mathrm{CAR}(\mathfrak{D}) \mapsto \mathrm{CAR}(\mathfrak{D})/\mathfrak{I}$ is an isomorphism. Therefore the kernel of P, which is the span of $\mathfrak{I}$, must be zero: $\mathfrak{I} = \{0\}$. □

An interesting consequence of the simplicity is that every representation of $\mathrm{CAR}(\mathfrak{D})$ is faithful (has trivial kernel). Indeed, let π be a (nonzero) representation of $\mathrm{CAR}(\mathfrak{D})$. It is readily verified that $\ker\pi$ is a two-sided, closed ideal of $\mathrm{CAR}(\mathfrak{D})$. Hence by Theorem 3.2, $\ker\pi = \{0\}$.

3.2 The algebra $\mathrm{CCR}(\mathfrak{D})$

An (abstract) Weyl algebra, or CCR algebra $\mathrm{CCR}(\mathfrak{D})$ over a test function space $\mathfrak{D} \subseteq \mathfrak{H}$ is defined to be the unital C^*algebra generated by elements $W(f)$, $f \in \mathfrak{D}$, satisfying the relations

$$W(-f) = W(f)^*, \quad W(f)W(g) = e^{-\frac{i}{2}\mathrm{Im}\langle f,g\rangle} W(f+g). \tag{53}$$

We have seen in the previous section that an algebra with these properties exists. The CCR (53) imply that $f \mapsto W(f)$ is not continuous (in the C^* norm topology). Indeed, the proof of Theorem 2.3 shows that we have $\|W(f) - \mathbb{1}\| = 2$, for any $f \neq 0$. Similarly to the CAR case, the Weyl algebra is unique.

Theorem 3.3. (Uniqueness of the Weyl algebra). *Let $\mathfrak{D} \subseteq \mathfrak{H}$ be given and let $\mathfrak{W}_1$ and $\mathfrak{W}_2$ be two Weyl algebras over $\mathfrak{H}$ (generated by $W_1(f)$ and $W_2(f)$, $f \in \mathfrak{D}$). There is a unique $*$isomorphism $\alpha : \mathfrak{W}_1 \to \mathfrak{W}_2$ such that $\alpha(W_1(f)) = W_2(f)$, for all $f \in \mathfrak{D}$.*

A proof can be found in [5],[12]. As for the CAR algebra, simplicity of the CCR algebra follows from uniqueness.

Theorem 3.4. *The C^*algebra $\mathrm{CCR}(\mathfrak{D})$ is simple.*

Due to the lack of continuity of the map $f \mapsto W(f)$ it is not true that the Weyl algebra over $\mathfrak{D}$ is the same as the one over $\overline{\mathfrak{D}}$ if $\mathfrak{D} \neq \overline{\mathfrak{D}}$. One can show that if $\mathfrak{D}_1$ and $\mathfrak{D}_2$ are two linear (not necessarily closed) subspaces of $\mathfrak{H}$ then

$$\mathrm{CCR}(\mathfrak{D}_1) = \mathrm{CCR}(\mathfrak{D}_1) \Longleftrightarrow \mathfrak{D}_1 = \mathfrak{D}_2,$$

see e.g. [5], Proposition 5.2.9. In particular, $\mathrm{CCR}(\overline{\mathfrak{D}}) = \mathrm{CCR}(\mathfrak{D})$ if and only if $\mathfrak{D}$ is closed. Another difficulty is generated by the lack of continuity of the map

$$t \mapsto W(e^{ith} f), \tag{54}$$

where $t \in \mathbb{R}$ and h is some selfadjoint operator on $\mathfrak{H}$ (leaving $\mathfrak{D}$ invariant). The assignment (54) is called a *Bogoliubov transformation*. It represents a *dynamics* of the system, where h is interpreted as the one-particle Hamiltonian. The lack of continuity prevents us from treating the dynamics with ease on an algebraic level; for instance, one cannot take the derivative (nor the integral) of the r.h.s. of (54) w.r.t. t – and these operations are important e.g. to define a perturbed dynamics. There are representations of the CCR for which weaker continuity properties hold; we look at them now.

By a *regular representation* π of $\mathrm{CCR}(\mathfrak{D})$ we understand one with the property that $t \mapsto \pi(W(tf))$ is continuous in the strong operator topology on the representation Hilbert space $\mathcal{H}$, for all $f \in \mathfrak{D}$. A state ω on $\mathrm{CCR}(\mathfrak{D})$ is called a *regular state* if its GNS representation is regular (see also Theorem 4.1). For a regular representation the map $t \mapsto \pi(W(tf))$ is a strongly continuous one-parameter group of unitaries on $\mathcal{H}$. [12] The Stone-von Neumann theorem tells us that this group has a selfadjoint generator on $\mathcal{H}$, which we denote by $\Phi_\pi(f)$,

$$\pi(W(tf)) = e^{it\Phi_\pi(f)}.$$

It is convenient to introduce annihilation and creation operators in the regular representation π by setting

$$a_\pi(f) = \frac{\Phi_\pi(f) + i\Phi_\pi(if)}{\sqrt{2}}, \quad a_\pi^*(f) = \frac{\Phi_\pi(f) - i\Phi_\pi(if)}{\sqrt{2}}. \tag{55}$$

Compare this with (34)! Definition (55) needs some explanation because $\Phi_\pi(f)$ and $\Phi_\pi(if)$ are both unbounded operators on $\mathcal{H}$.

Proposition 3.5. *Let $F = \{f_1, \dots, f_n\}$ be a finite collection of elements in $\mathfrak{D}$. The operators $\{\Phi_\pi(f_j), \Phi_\pi(if_j)\}_{j=1}^N$ have a common set of analytic vectors which is dense in the representation Hilbert space $\mathcal{H}$. This means that, for $f \in \mathfrak{D}$ fixed, the domain*

$$\mathcal{D}_{\pi,f} := \mathcal{D}(a_\pi(f)) := \mathcal{D}(a_\pi^*(f)) := \mathcal{D}(\Phi_\pi(f)) \cap \mathcal{D}(\Phi_\pi(if)) \tag{56}$$

is dense in $\mathcal{H}$. We understand the equalities (55) in the sense of operators on $\mathcal{D}_{\pi,f}$. Both $a_\pi(f)$ and $a_\pi^*(f)$ are closed operators on $\mathcal{D}_{\pi,f}$.

We have proved after equation (19) above that, for $a^\#(f)$ defined as in Section 2.2, the adjoint operator of $a(f)$ is $a^*(f)$. This can be shown for any regular representation, i.e., we have

$$a_\pi^*(f) = a_\pi(f)^*. \tag{57}$$

A proof of (57) can be found in [5].

Proof of Proposition 3.5. The following "smoothing" is useful: let $f \in \mathfrak{D}$ and consider the integral (understood in the strong sense on $\mathcal{H}$)

[12] The group properties follow from $\pi(W(tf))\pi(W(sf)) = \pi(W(tf)W(sf))$ $= e^{\frac{i}{2}st\mathrm{Im}\langle f,f\rangle}\pi(W((s+t)f)) = \pi(W((s+t)f))$ and $\pi(W(f))^* = \pi(W(f)^*)$ $= \pi(W(-f)) = \pi(W(f)^{-1}) = \pi(W(f))^{-1}$.

$$\sqrt{\frac{n}{\pi}} \int_{\mathbb{R}} ds\, e^{-ns^2} W_\pi(sf), \tag{58}$$

where $n > 0$ and where we set $\pi(W(f)) = W_\pi(f)$. The strong limit of (58), as $n \to \infty$, is just the identity operator on $\mathcal{H}$. We apply the operator $W_\pi(tf)$ to the integral in (58) and obtain, after a change of variable,

$$W_\pi(tf) \int_{\mathbb{R}} ds\, e^{-ns^2} W_\pi(sf) = \int_{\mathbb{R}} ds\, e^{-n(s-t)^2} W_\pi(sf). \tag{59}$$

The r.h.s. of (59) has an analytic extension in t to the whole complex plane. Similarly, if f_k is any element in F then the map

$$t \mapsto W_\pi(tf_k) \left(\frac{n}{\pi}\right)^{N/2} \int_{\mathbb{R}} ds_1 \cdots \int_{\mathbb{R}} ds_N\, e^{-n(s_1^2+\cdots+s_N^2)} W_\pi\left(\sum_{j=1}^N s_j f_j\right) \tag{60}$$

is easily seen to have an analytic extension in t to all of $\mathbb{C}$, and the r.h.s. of (60) converges in the strong sense to $W_\pi(tf_k)$, as $n \to \infty$. This means that any vector of the form

$$\left(\frac{n}{\pi}\right)^{N/2} \int_{\mathbb{R}} ds_1 \cdots \int_{\mathbb{R}} ds_N\, e^{-n(s_1^2+\cdots+s_N^2)} W_\pi\left(\sum_{j=1}^N s_j f_j\right) \psi, \tag{61}$$

where $\psi \in \mathcal{H}$ is arbitrary, is an analytic (entire) vector for all operators in the set $\{\Phi_\pi(f_j), \Phi_\pi(if_j)\}_{j=1}^N$. The set (61), where ψ varies over all of $\mathcal{H}$, is dense in $\mathcal{H}$, because (61) converges to ψ, as $n \to \infty$. This shows the first part of the proposition.

Let us now prove that the $a_\pi^\#(f)$ are closed operators on $\mathcal{D}_{\pi,f}$, where $f \in \mathfrak{D}$ is fixed. For any $\psi \in \mathcal{D}_{\pi,f}$ we have, by (55),

$$\|\Phi_\pi(f)\psi\|^2 + \|\Phi_\pi(if)\psi\|^2 = \|a_\pi(f)\psi\|^2 + \|a_\pi^*(f)\psi\|^2. \tag{62}$$

We use $W_\pi(sf)W_\pi(itf) = e^{-ist\|f\|^2} W_\pi(itf)W_\pi(sf)$ to get

$$\begin{aligned} \frac{1}{i^2}\partial_{st}^2|_{s=t=0}\, \langle\psi, W_\pi(sf)W_\pi(itf)\psi\rangle &= \langle\Phi_\pi(f)\psi, \Phi_\pi(if)\psi\rangle \\ &= \langle\Phi_\pi(if)\psi, \Phi_\pi(f)\psi\rangle + i\|f\|^2\,\|\psi\|^2, \end{aligned}$$

which implies that

$$\|a_\pi^*(f)\psi\|^2 = \|a_\pi(f)\psi\|^2 + \|f\|^2\,\|\psi\|^2. \tag{63}$$

Combining (62) and (63) yields the identity

$$\|\Phi_\pi(f)\psi\|^2 + \|\Phi_\pi(if)\psi\|^2 = 2\|a_\pi(f)\psi\|^2 + \|f\|^2\|\psi\|^2. \tag{64}$$

To show that $a_\pi(f)$ is a closed operator on $\mathcal{D}_{\pi,f}$ assume that $\psi_n \in \mathcal{D}_{\pi,f}$ is a sequence of vectors converging to some $\psi \in \mathcal{H}$, such that $a_\pi(f)\psi_n$ converges as $n \to \infty$, i.e.,

$\|a_\pi(f)(\psi_n - \psi_m)\| \to 0$, as $n, m \to \infty$. It follows from (64) that both $\|\Phi_\pi(f)(\psi_n - \psi_m)\|$ and $\|\Phi_\pi(if)(\psi_n - \psi_m)\|$ converge to zero as $n \to \infty$. Since $\Phi_\pi(f)$ and $\Phi_\pi(if)$ are closed operators (they are selfadjoint) we conclude that $\psi \in \mathcal{D}(\Phi_\pi(f))$ and $\psi \in \mathcal{D}(\Phi_\pi(if))$, i.e., $\psi \in \mathcal{D}_{\pi,f}$, and that $\Phi_\pi(f)\psi_n \to \Phi_\pi(f)\psi$, $\Phi_\pi(if)\psi_n \to \Phi_\pi(if)\psi$. Another application of (64) (with ψ replaced by $\psi_n - \psi$) shows that $\|a_\pi(f)(\psi_n - \psi)\|^2 \to 0$ as $n \to \infty$. Consequently $a_\pi(f)$ is a closed operator. In the same way one sees that $a_\pi^*(f)$ is a closed operator. □

The *Fock representation* of $\mathrm{CCR}(\mathfrak{D})$ is the regular representation defined by $\pi_{\mathrm{F}} : \mathrm{CCR}(\mathfrak{D}) \to \mathcal{B}(\mathcal{F}_+(\mathfrak{H}))$,

$$\pi_{\mathrm{F}}(W(f)) = W_{\mathrm{F}}(f), \tag{65}$$

where the operator on the r.h.s. is given by (39), and where the Bosonic Fock space $\mathcal{F}_+(\mathfrak{H})$ was defined in (13).

We mention another structural property of the Weyl algebra. Let $\mathfrak{D}_1 \subseteq \mathfrak{H}_1$ and $\mathfrak{D}_2 \subseteq \mathfrak{H}_2$ be two linear subspaces and let $\mathfrak{D}_1 \oplus \mathfrak{D}_2 \subseteq \mathfrak{H}_1 \oplus \mathfrak{H}_2$ be their direct sum (i.e., the not necessarily closed set of all $f \oplus g$, $f \in \mathfrak{D}_1, g \in \mathfrak{D}_2$ equipped with the usual direct sum operations). We have the relation

$$\mathrm{CCR}(\mathfrak{D}_1 \oplus \mathfrak{D}_2) = \mathrm{CCR}(\mathfrak{D}_1) \otimes \mathrm{CCR}(\mathfrak{D}_2). \tag{66}$$

This follows simply from the CCR (53),

$$W(f_1 \oplus f_2) = W(f_1 \oplus 0 + 0 \oplus f_2) = e^{\frac{i}{2}\mathrm{Im}\langle f_1 \oplus 0, 0 \oplus f_2\rangle_{\mathfrak{H}\oplus\mathfrak{H}}} W(f_1 \oplus 0) W(0 \oplus f_2),$$

$\langle f_1 \oplus 0, 0 \oplus f_2\rangle_{\mathfrak{H}\oplus\mathfrak{H}} = \langle f_1, 0\rangle + \langle 0, f_2\rangle = 0$ and the identifications

$$W(f_1 \oplus 0) \mapsto W(f_1) \otimes W(0), \quad W(0 \oplus f_2) \mapsto W(0) \otimes W(f_2).$$

3.3 Schrödinger representation and Stone – von Neumann uniqueness theorem

Let us consider the easiest Weyl algebra $\mathrm{CCR}(\mathbb{C})$, where $\mathfrak{H} = \mathbb{C}$ is a one-dimensional Hilbert space. The Weyl operators are given by $W(z)$, with $z = s + it \in \mathbb{C}$, $s, t \in \mathbb{R}$. They satisfy

$$W(z) = W(s + it) = e^{\frac{i}{2}\mathrm{Im}\langle s, it\rangle} W(s)W(it) = e^{\frac{i}{2}st} W(s)W(it).$$

Let us assume that we are in a regular representation of the CCR, i.e., $\tau \mapsto W(\tau z)$ is a strongly continuous one parameter group ($\tau \in \mathbb{R}$) of unitaries on a (representation) Hilbert space. In particular, there are selfadjoint operators Φ, Π such that

$$W(\tau) = e^{i\tau\Phi}, \quad W(i\tau) = e^{i\tau\Pi}.$$

It is suggestive to write $\Phi = \Phi(1)$ and $\Pi = \Phi(i)$, compare with (39). The generators satisfy the commutation relations $[\Phi, \Pi] = [\Phi(1), \Phi(i)] = i\mathrm{Im}\,\langle 1, i\rangle = i\mathbb{1}$ which can be seen by noticing that

$$W(s)W(it)W(-s) = e^{-i\mathrm{Im}\langle s,it\rangle}W(it) = e^{-ist}W(it), \tag{67}$$

which yields (by applying $-i\partial_s|_{s=0}$) $[\Phi, W(it)] = -tW(it)$, and hence (by applying $-i\partial_t|_{t=0}$) $[\Phi, \Pi] = i\mathbb{1}$.

These commutation relations remind us of $[x, -i\partial_x] = i$, where x and $-i\partial_x$ are selfadjoint operators on $L^2(\mathbb{R}, dx)$. We can define a regular representation π_{S} of CCR($\mathbb{C}$) on $L^2(\mathbb{R}, dx)$ by

$$\pi_{\mathrm{S}}(W(z)) = e^{\frac{i}{2}st}U(s)V(t), \tag{68}$$

where $z = s + it \in \mathbb{C}$, and $U(s)$ and $V(t)$ are the one-parameter ($s, t \in \mathbb{R}$) unitary groups on $L^2(\mathbb{R}, dx)$ given by

$$(U(s)\psi)(x) = e^{isx}\psi(x),$$
$$(V(t)\psi)(x) = \psi(x+t),$$

with selfadjoint generators $\Phi_{\mathrm{S}} = x$ and $\Pi_{\mathrm{S}} = -i\partial_x$. The representation (68) is called the *Schrödinger representation* of the CCR.

Since this representation is regular we can introduce creation and annihilation operators (c.f. (34)) by

$$a_{\mathrm{S}} = \frac{\Phi_{\mathrm{S}} + i\Pi_{\mathrm{S}}}{\sqrt{2}} = \frac{1}{\sqrt{2}}(x + \partial_x) \tag{69}$$

$$a_{\mathrm{S}}^* = \frac{\Phi_{\mathrm{S}} - i\Pi_{\mathrm{S}}}{\sqrt{2}} = \frac{1}{\sqrt{2}}(x - \partial_x). \tag{70}$$

Since both Φ_{S} and Π_{S} are unbounded operators one has to take care in the exact definition of the unbounded (non-selfadjoint) operators $a_{\mathrm{S}}^{\#}$ in (69), (70). This can be done by proceeding as in Proposition 3.5.

These considerations show that $L^2(\mathbb{R}, dx)$ carries a Fock space structure, i.e., there are two (densely defined, closed, unbounded, non symmetric) operators a_{S} and a_{S}^* acting on $L^2(\mathbb{R}, dx)$ and satisfying the commutation relation $[a_{\mathrm{S}}, a_{\mathrm{S}}^*] = \mathrm{id}$. The commutator is understood in the strong sense on some dense set of vectors (e.g. the functions in C_0^∞).

The vacuum vector $\Omega_{\mathrm{S}} \in L^2(\mathbb{R}, dx)$ is given by the normalized solution of $(x + \partial_x)\Omega_{\mathrm{S}}(x) = 0$ (i.e. $a_{\mathrm{S}}\Omega_{\mathrm{S}} = 0$),

$$\Omega_{\mathrm{S}}(x) = \pi^{-1/4}e^{-x^2/2}.$$

We introduce a sequence of one-dimensional subspaces $\mathcal{H}_n \subset L^2(\mathbb{R}, dx)$ spanned by $(a_{\mathrm{S}}^*)^n\Omega_{\mathrm{S}}$. Using the commutation relations for the creation and annihilation operator, one easily sees that the operator

$$N_{\mathrm{S}} = a_{\mathrm{S}}^* a_{\mathrm{S}} = \frac{1}{2}\left(-\partial_x^2 + x^2 - 1\right) \tag{71}$$

leaves each $\mathcal{H}_n$ invariant, and that $N_{\mathrm{S}} \restriction \mathcal{H}_n = n\,\mathrm{id} \restriction \mathcal{H}_n$. Notice that N_{S} is just the Schrödinger operator (Hamiltonian) corresponding to a one-dimensional quantum harmonic oscillator (modulo the constant term $-1/2$). There are various ways to see that we have

$$L^2(\mathbb{R}, dx) = \bigoplus_{n \geq 0} \mathcal{H}_n. \tag{72}$$

For instance, one knows that the eigenvalues of N_{S} are $0, 1, 2, 3, \ldots$ and they are simple (harmonic oscillator!), so (72) is a consequence of the fact that the eigenvectors of N_{S} span the entire space. The eigenvector ψ_n of N_{S} with eigenvalue $n \in \mathbb{N}$ satisfies the equation $N_{\mathrm{S}}\psi_n = n\psi_n$, which is equivalent to (c.f. (71))

$$(-\partial_x^2 + x^2)\psi_n = (2n+1)\psi_n, \tag{73}$$

i.e. ψ_n is a harmonic oscillator eigenvector. The ψ_n are the Hermite functions, they have the form

$$\psi_n(x) = \frac{1}{\sqrt{n!}}(a_{\mathrm{S}}^*)^n \Omega_{\mathrm{S}}(x) = \frac{1}{\sqrt{2^n n!}}(-1)^n \pi^{-1/4} e^{\frac{1}{2}x^2}(\partial_x)^n e^{-x^2}, \tag{74}$$

where $\frac{1}{\sqrt{n!}}$ is a normalization factor.

The Schrödinger representation of $\mathrm{CCR}(\mathbb{C}^n)$ is defined as the n-fold tensor product representation of $\mathrm{CCR}(\mathbb{C})$,

$$\pi_{\mathrm{S}}(W(z_1, \ldots, z_n)) = \prod_{j=1}^{n} e^{\frac{i}{2}s_j t_j} U_j(s_j) V_j(t_j), \tag{75}$$

acting on $L^2(\mathbb{R}^n, d^n x)$ and where $U_j(s)$, $V_j(t)$ act on the variable x_j in the obvious way. We may view $\mathbb{C}^n$ as $\mathbb{C} \oplus \cdots \oplus \mathbb{C}$ and compare (75) with (66).

The above discussion shows that the Schrödinger and the Fock representations of $\mathrm{CCR}(\mathbb{C})$ are unitarily equivalent, the correspondence being

$$(a_{\mathrm{S}}^*)^n \Omega_{\mathrm{S}} \mapsto (a_{\mathrm{F}}^*)^n \Omega_{\mathrm{F}}, \tag{76}$$

where a_{F}^* is the creation operator in the Fock representation, (65). This is not a coincidence, it can be viewed as a consequence of the following result.

Theorem 3.6. (Stone – von Neumann uniqueness theorem). *Let $\mathfrak{H}$ be a finite dimensional Hilbert space. Any irreducible regular representation of* $\mathrm{CCR}(\mathfrak{H})$ *is unitarily equivalent to the Fock representation of* $\mathrm{CCR}(\mathfrak{H})$.

It is instructive to have a look at the mechanism behind the proof of the Stone – von Neumann uniqueness theorem.

Outline of the proof of Theorem 3.6. Let $\{f_1, \ldots, f_n\}$ be an orthonormal basis of $\mathfrak{H}$ and define the non-negative operator (the number operator)

$$N_\pi = \sum_{j=1}^{n} a_\pi^*(f_j) a_\pi(f_j), \tag{77}$$

where the creation and annihilation operators are defined as in Proposition 3.5. One can show that N_π is a non-negative selfadjoint operator on the representation Hilbert space which we shall call $\mathcal{H}$. Using the CCR we find that, for any $f \in \mathfrak{H}$,

$$N_\pi a_\pi(f) = a_\pi(f)(N_\pi - 1). \tag{78}$$

Let $n_0 > -\infty$ be the infimum of the spectrum of N_π and let $P(N_\pi \leq n_0 + 1/2)$ denote the spectral projection of N_π associated with the interval $[n_0, n_0 + 1/2]$. Take any normalized $\Omega_\pi \in \operatorname{Ran} P(N_\pi \leq n_0 + 1/2)$. Relation (78) tells us that $P(N_\pi \leq x) a_\pi(f) = a_\pi(f) P(N_\pi \leq x+1)$ for any x, so we have $a_\pi(f)\Omega_\pi = 0$, for any $f \in \mathfrak{H}$.

Since π is irreducible the set $\mathcal{H}_\pi = \{W_\pi(f)\Omega_\pi \mid f \in \mathfrak{H}\}$, where $W_\pi(f) = \pi(W(f))$, is dense in $\mathcal{H}$ (the closure of $\mathcal{H}_\pi$ is a closed subspace of $\mathcal{H}$ which is invariant under $\pi(\mathrm{CCR}(\mathfrak{H}))$, and $\mathcal{H}_\pi \neq \{0\}$ since $\mathbb{1} \in \mathcal{H}_\pi$). Proceeding as in the proof of Theorem 2.5 one shows that the closure of $\mathcal{H}_\pi$ is the same as the closure of the set of vectors of the form $a^*_\pi(f_1)\cdots a^*_\pi(f_n)\Omega_\pi$,

$$\mathcal{H} = \mathrm{closure}\{a^*_\pi(f_1)\cdots a^*_\pi(f_n)\Omega_\pi \mid n \in \mathbb{N},\ f_1, \ldots, f_n \in \mathfrak{H}\}. \tag{79}$$

Now we define the linear map $U : \mathcal{H} \to \mathcal{F}_+(\mathfrak{H})$ by

$$U a^*_\pi(f_1)\cdots a^*_\pi(f_n)\Omega_\pi = a^*_{\mathrm{F}}(f_1)\cdots a^*_{\mathrm{F}}(f_n)\Omega_{\mathrm{F}}. \tag{80}$$

It is easy to verify that U extends to a unitary map because the norms of

$$a^*_\#(f_1)\cdots a^*_\#(f_n)\Omega_\#,$$

$\# = \pi, \mathrm{F}$, can be calculated purely by using the fact that $a_\#(f_j)\Omega_\# = 0$ and the canonical commutation relations. This finishes the outline of the proof of the Stone – von Neumann uniqueness theorem. □

Since every representation can be decomposed into a direct sum of irreducible representations, Theorem 3.6 says that every regular representation of $\mathrm{CCR}(\mathfrak{H})$, $\dim \mathfrak{H} < \infty$, is a direct sum of Fock representations (in which case we say that the representation is *quasi-equivalent* to the Fock representation). If $\dim \mathfrak{H} = \infty$ this is no longer true. In particular, the GNS representation corresponding to states of the infinitely extended free Bose gas with nonzero density which we will construct in Section 4 are not quasi-equivalent to the Fock representation.

There is however a characterization of representations of $\mathrm{CCR}(\mathfrak{H})$, where $\dim \mathfrak{H} = \infty$, which are quasi-equivalent to the Fock representation. In view of the outline of the proof of the Stone – von Neumann uniqueness theorem this characterization is very natural, although its exact formulation is somewhat technical. The central object in the above proof of Theorem 3.6 is the number operator (77). It can be generalized by putting

$$N_\pi = \sup_F \sum_j a^*_\pi(f_j) a_\pi(f_j), \tag{81}$$

where the supremum is over all finite-dimensional subspaces F of $\mathfrak{H}$, and the sum extends over an orthonormal basis $\{f_j\}$ of F. It is clear that a rigorous definition of (81) is not trivial. It can be given using quadratic forms rather than operators, see e.g. [5] Section 5.2.3. By proceeding as in the above outline of the proof of the Stone – von Neumann theorem one can show that *a representation* π *of* CCR($\mathfrak{H}$) *is a direct sum of Fock representations of* CCR($\mathfrak{H}$) *if and only if the number operator (81) can be defined as a densely defined selfadjoint operator.* This may be phrased as "π is quasi-equivalent to the Fock representation if and only if there is a number operator in the representation space of π". A precise statement of this result can be found in [5], Theorem 5.2.14.

3.4 *Q*–space representation

Our goal is to examine the unitary equivalence obtained from (76) when $\mathbb{C}$ is first replaced by $\mathbb{C}^n$, and then n is taken to infinity. This will provide us with another representation of CCR($\mathfrak{H}$), where $\mathfrak{H}$ is a separable Hilbert space. The representation Hilbert space we construct is $L^2(Q, d\mu)$, where μ is a probability measure on Q, $\mu(Q) = 1$. We give an explicit unitary equivalence between $L^2(Q, d\mu)$ and the bosonic Fock space $\mathcal{F}(\mathfrak{H})$ (we write $\mathcal{F}$ instead of $\mathcal{F}_+$). The Q-space representation is particularly useful in the analysis of interacting fields, see e.g. [13].

The assignment

$$
\begin{aligned}
&a_1^*(f_1)\cdots a_1^*(f_m)\Omega_1 \otimes a_2^*(g_1)\cdots a_2^*(g_n)\Omega_2 \\
&\mapsto a^*(f_1 \oplus 0)\cdots a^*(f_m \oplus 0)a^*(0 \oplus g_1)\cdots a^*(0 \oplus g_n)\Omega,
\end{aligned}
$$

where $f_j \in \mathfrak{H}_1$, $g_j \in \mathfrak{H}_2$, establishes a unitary map between the Fock spaces $\mathcal{F}(\mathfrak{H}_1) \otimes \mathcal{F}(\mathfrak{H}_2)$ and $\mathcal{F}(\mathfrak{H}_1 \oplus \mathfrak{H}_2)$ (compare with (66)). This means that

$$
\mathcal{F}(\mathbb{C}^n) = \mathcal{F}(\mathbb{C} \oplus \cdots \oplus \mathbb{C}) \cong \mathcal{F}(\mathbb{C}) \otimes \cdots \otimes \mathcal{F}(\mathbb{C}), \tag{82}
$$

and taking into account the identification (76) we obtain

$$
\mathcal{F}(\mathbb{C}^n) \cong L^2(\mathbb{R}, dx) \otimes \cdots \otimes L^2(\mathbb{R}, dx) \cong L^2(\mathbb{R}^n, d^n x). \tag{83}
$$

Let C be a conjugation on $\mathfrak{H}$, i.e., C is an antilinear isometry satisfying $C^2 = 1$. One may think of C as the operation of taking the complex conjugate of coordinates in a given basis of $\mathfrak{H}$. [13] Let $\{e_j\}_{j=1}^{\infty}$ be an orthonormal basis of $\mathfrak{H}$ such that each e_j is invariant under C, $Ce_j = e_j$. A consequence of introducing a basis of C invariant vectors is that if $f = Cf$ and $g = Cg$ then $\langle f, g\rangle = \langle Cf, Cg\rangle = \overline{\langle f, g\rangle}$, so the corresponding Weyl operators commute, $W(f)W(g) = W(g)W(f)$; c.f. (53) (similarly the field operators in a regular representation commute in the strong sense on a dense set of vectors).

[13] If $\{f_j\}$ is any basis of $\mathfrak{H}$ then define $f_j' = f_j + Cf_j$. The f_j' are invariant under C, $Cf_j' = f_j'$, and they span $\mathfrak{H}$. A Gram-Schmidt procedure yields an orthonormal basis $\{e_j\}$ of vectors satisfying $Ce_j = e_j$. The action of C on coordinates w.r.t. the basis $\{e_j\}$ is complex conjugation.

Let $\{f_1, \dots f_n\}$ be a finite collection of elements in $\{e_j\}$ and define

$$\mathcal{F}_n = \text{closure}\{P(a^*(f_1), \dots, a^*(f_n))\Omega \mid P \text{ a polynomial}\} \subset \mathcal{F}(\mathfrak{H}), \quad (84)$$

where Ω is the Fock vacuum and the a^* are the creation operators in Fock representation, defined by (34) (we write a^* instead of a_+^*). Clearly, the map

$$a^*(f_1)^{k_1} \cdots a^*(f_n)^{k_n} \Omega \mapsto a^*(\zeta_1)^{k_1} \cdots a^*(\zeta_n)^{k_n} \Omega_{\mathcal{F}(\mathbb{C}^n)}, \quad (85)$$

where $\Omega_{\mathcal{F}(\mathbb{C}^n)}$ is the vacuum vector in $\mathcal{F}(\mathbb{C}^n)$ and $\zeta_j \in \mathbb{C}^n$ has zero components except for the j-th which equals one, extends to a unitary map between $\mathcal{F}_n$ and $\mathcal{F}(\mathbb{C}^n)$. The r.h.s. of (85) can be identified, via (83), (70), with the vector

$$\zeta_1^{k_1} \cdots \zeta_n^{k_n} \left(\frac{x_1 - \partial_{x_1}}{\sqrt{2}}\right)^{k_1} \cdots \left(\frac{x_n - \partial_{x_n}}{\sqrt{2}}\right)^{k_n} \Omega_n \in L^2(\mathbb{R}^n, d^n x), \quad (86)$$

where

$$\Omega_n = \pi^{-n/4} \exp\left(-\frac{1}{2}\sum_{j=1}^n x_j^2\right). \quad (87)$$

We normalize Ω_n to be the constant function by introducing the unitary map

$$(Tf)(x) = \pi^{n/4} \exp\left(\frac{1}{2}\sum_{j=1}^n x_j^2\right) f(x)$$

between $L^2(\mathbb{R}^n, d^n x)$ and $L^2(\mathbb{R}^n, d\mu_1 \times \cdots \times d\mu_n)$, where

$$d\mu_j = \pi^{-1/2} e^{-x_j^2} dx_j.$$

Thus (85), (86) give a unitary map U_n between $\mathcal{F}_n$ and $L^2(\mathbb{R}^n, d\mu_1 \times \cdots \times d\mu_n)$ such that $U_n \Omega = 1$ (the constant function) and $U_n a^*(f_j) U_n^{-1} = \frac{2x_j - \partial_{x_j}}{\sqrt{2}}$, $U_n a(f_j) U_n^{-1} = \frac{1}{\sqrt{2}} \partial_{x_j}$ [14] so that

$$U_n \Phi(f_j) U_n^{-1} = U_n \frac{a^*(f_j) + a(f_j)}{\sqrt{2}} U_n^{-1} = x_j.$$

Let P_j, $j = 1, \dots, n$ be n polynomials in one variable. The unitarity of U_n gives

$$\begin{aligned} \langle \Omega, P_1(\Phi(f_1)) \cdots P_n(\Phi(f_n))\Omega \rangle &= \int_{\mathbb{R}^n} P_1(x_1) \cdots P_n(x_n) d\mu_1 \cdots d\mu_n \\ &= \prod_{j=1}^n \langle \Omega, P_j(\Phi(f_j))\Omega \rangle . \end{aligned} \quad (88)$$

[14] We have $T\partial_{x_j} T^{-1} = \partial_{x_j} - x_j$

Let $Q = \times_{j=1}^{\infty}\mathbb{R}$ be the set of sequences $q = (q_1, q_2, \ldots)$ equipped with the $\sigma-$algebra generated by countable products of measurable sets in $\mathbb{R}$, and let $\mu = \otimes_{j=1}^{\infty}\mu_j$. The pair (Q, μ) is a measure space (see e.g. Chapter VI of [9]), and the set of all polynomials $P(q_1, \ldots, q_n)$, $n \in \mathbb{N}$, is dense in $L^2(Q, d\mu)$.

The space $\mathcal{F}_n$, (84), equals the closure of $\{P(\Phi(f_1), \ldots, \Phi(f_n))\Omega\}$, where P ranges over all polynomials in n variables (see also the proof of Theorem 2.5). For any $n \in \mathbb{N}$ and any polynomial P in n variables,

$$P(x_{j_1}, \ldots, x_{j_n}) = \sum_{p_1,\ldots,p_n} c(p_1, \ldots, p_n) x_{j_1}^{p_1} \cdots x_{j_n}^{p_n},$$

set

$$UP(\Phi(f_{j_1}), \ldots, \Phi(f_{j_n}))\Omega = P(q_{j_1}, \ldots, q_{j_n}) \in L^2(Q, d\mu). \tag{89}$$

Let us verify that U is norm preserving:

$$\begin{aligned}
&\|P(\Phi(f_{j_1}), \ldots, \Phi(f_{j_n}))\Omega\|^2 \\
&= \sum_{p_1,\ldots,p_n} \sum_{p'_1,\ldots,p'_n} \overline{c(p_1, \ldots, p_n)} c(p'_1, \ldots, p'_n) \left\langle \Omega, \Phi(f_{j_1})^{p_1+p'_1} \cdots \Phi(f_{j_n})^{p_n+p'_n}\Omega \right\rangle \\
&= \sum_{p_1,\ldots,p_n} \sum_{p'_1,\ldots,p'_n} \overline{c(p_1, \ldots, p_n)} c(p'_1, \ldots, p'_n) \int_{\mathbb{R}^n} q_{j_1}^{p_1+p'_1} \cdots q_{j_n}^{p_n+p'_n} d\mu_{j_1} \cdots d\mu_{j_n} \\
&= \int_Q |P(q_{j_1}, \ldots, q_{j_n})|^2 d\mu.
\end{aligned} \tag{90}$$

We use in the first step that the Φ's commute, which is due to the fact that $Cf_j = f_j$ and in the second step we make use of (88). Since the set of vectors

$$P(\Phi(f_{j_1}), \ldots, \Phi(f_{j_n}))\Omega,$$

is dense in $\mathcal{F}(\mathfrak{H})$ formula (90) shows that U extends to a unitary map from $\mathcal{F}(\mathfrak{H})$ to $L^2(Q, d\mu)$, s.t. $U\Omega = 1$ and $U\Phi(f_j)U^{-1} = q_j$.

3.5 Equilibrium state and thermodynamic limit

We focus in this subsection on Bosons and refer for more detail, as well as for the Fermionic case, to [5], Section 5.2.5.

Let H be the one-particle Hamiltonian, acting on the one-particle Hilbert space $\mathfrak{H}$, and denote by $\mathrm{d}\Gamma(H)$ its second quantization acting on Bosonic Fock space $\mathcal{F}_+(\mathfrak{H})$. $\mathrm{d}\Gamma(H)$ acts on the n sector as

$$H \otimes \cdots \otimes \mathbb{1} + \mathbb{1} \otimes H \otimes \cdots \otimes \mathbb{1} + \cdots + \mathbb{1} \otimes \cdots \otimes H.$$

We set $N = \mathrm{d}\Gamma(\mathbb{1})$, put

$$K_\mu = \mathrm{d}\Gamma(H) - \mu N = \mathrm{d}\Gamma(H - \mu\mathbb{1}), \tag{91}$$

where $\mu \in \mathbb{R}$ is called the chemical potential, and assume that

$$Z_{\beta,\mu} = \mathrm{tr} e^{-\beta K_\mu} \tag{92}$$

exists, for some inverse temperature $\beta > 0$. Here, tr denotes the trace on the Hilbert space $\mathcal{F}_+(\mathfrak{H})$. It is not hard to show that (92) is finite if and only if

$$\mathrm{tr} e^{-\beta H} < \infty \text{ and } H - \mu \mathbb{1} > 0, \tag{93}$$

see [5], Proposition 5.2.27; the trace here is of course over $\mathfrak{H}$. From the latter inequality it follows ($H - \mu\mathbb{1}$ has purely discrete spectrum) that there is a number $\eta > 0$ s.t.

$$\mathrm{d}\Gamma(H - \mu\mathbb{1}) = K_\mu \geq \eta \mathrm{d}\Gamma(\mathbb{1}) = \eta N. \tag{94}$$

The *Gibbs (equilibrium) state* on $\mathrm{CCR}(\mathfrak{H})$ is defined by

$$\omega_{\beta,\mu}(A) = Z_{\beta,\mu}^{-1} \,\mathrm{tr}\left(e^{-\beta K_\mu} A\right). \tag{95}$$

It depends on the inverse temperature β and the chemical potential μ. The Gibbs state satisfies the KMS relation

$$\omega_{\beta,\mu}(A\alpha_t(B)) = \omega_{\beta,\mu}\left(\alpha_{t-i\beta}\left(e^{-\beta\mu N} B e^{\beta\mu N}\right) A\right), \tag{96}$$

where $\alpha_t(A) = e^{it\mathrm{d}\Gamma(H)} A e^{-it\mathrm{d}\Gamma(H)}$ is the Heisenberg dynamics generated by the Hamiltonian H. Identity (96) makes sense for operators B s.t.

$$e^{\beta(\mathrm{d}\Gamma(H) - \mu N)} B e^{-\beta(\mathrm{d}\Gamma(H) - \mu N)}$$

exists. If $\mu = 0$, (96) reduces to the usual KMS relation

$$\omega_\beta(A\alpha_t(B)) = \omega_\beta(\alpha_{t-i\beta}(B)A).$$

In order to calculate (95) explicitly it is useful to extend the domain of definition of $\omega_{\beta,\mu}$ to arbitrary (finite) products of creation and annihilation operators, i.e., to the polynomial $*$algebra $\mathfrak{P}$ of unbounded operators on $\mathcal{F}_+(\mathfrak{H})$, generated by $\{a^\#(f) \mid f \in \mathfrak{H}\}$. This can be done in the following way.

From (94) we see that

$$\|N^k e^{-tK_\mu}\| < \infty, \tag{97}$$

for any $t > 0$ and for any $k \geq 0$ [15]. The operator $e^{-\beta K_\mu/2}$ leaves the finite particle subspace $\mathcal{F}_+^0$ invariant (see (35)). If $Q \in \mathfrak{P}$ is any polynomial in creation and annihilation operators then Qe^{-tK_μ} is well defined on $\mathcal{F}_+^0$ and, by (97), extends to a bounded operator on $\mathcal{F}_+$, satisfying

$$\|a^\#(f_1) \cdots a^\#(f_k) e^{-tK_\mu}\| \leq C\|f_1\| \cdots \|f_k\|. \tag{98}$$

Let $\mu > 0$. For $\psi \in \mathcal{F}_+(\mathfrak{H})$ we have

[15] This follows from $\|N^k e^{-tK_\mu}\psi\| \leq \left\langle N^k\psi, e^{-2t\eta N} N^k \psi\right\rangle^{1/2} = \|N^k e^{-t\eta N}\psi\|$.

$$
\begin{aligned}
&\left|\left\langle \psi, e^{-\beta K_\mu/2} Q e^{-\beta K_\mu/2}\psi\right\rangle\right| \\
&\quad = \left|\left\langle \psi, e^{-\beta K_\mu/2} e^{\beta\mu N/4} e^{-\beta\mu N/4} Q e^{-\beta\mu N/4} e^{\beta\mu N/4} e^{-\beta K_\mu/2}\psi\right\rangle\right| \\
&\quad \leq \|Q e^{-\beta\mu N/4}\| \left|\left\langle \psi, e^{-\beta K_\mu} e^{\beta\mu N/2}\psi\right\rangle\right| \\
&\quad = \|Q e^{-\beta\mu N/4}\| \left|\left\langle \psi, e^{-\beta K_{\mu/2}}\psi\right\rangle\right|. \qquad (99)
\end{aligned}
$$

Since $e^{-\beta K_\mu}$ is trace class $e^{-\beta K_{\mu/2}}$ is too (see (93)), so (99) shows that for any $Q \in \mathfrak{P}$,

$$
\operatorname{tr}\left(e^{-\beta K_\mu/2} Q e^{-\beta K_\mu/2}\right) \leq C, \qquad (100)
$$

where $C < \infty$ depends on Q, β and $\mu > 0$. Therefore $\omega_{\beta,\mu}$ can be extended to $\mathfrak{P}$, and we have

$$
\omega_{\beta,\mu}\left(a^{\#}(f_1)\cdots a^{\#}(f_k)\right) \leq C\|f_1\|\cdots\|f_k\|. \qquad (101)
$$

Note that since $e^{-\beta K_\mu}$ commutes with the number operator, the l.h.s. of (101) is actually zero unless k is even and $k/2$ of the operators $a^{\#}$ are creation operators.

We have in the strong sense on $\mathcal{F}_+^0$

$$
e^{-\beta K_\mu/2} a^*(f) = a^*\left(e^{-\beta(H-\mu)/2} f\right) e^{-\beta K_\mu/2}, \qquad (102)
$$

and hence, using the cyclicity of the trace and the CCR, we obtain

$$
\begin{aligned}
\omega_{\beta,\mu}(a^*(f)a(g)) &= Z_{\beta,\mu}^{-1} \operatorname{tr}\left(a^*(e^{-\beta(H-\mu)/2} f)\, e^{-\beta K_\mu}\, a(e^{-\beta(H-\mu)/2} g)\right) \\
&= \omega_{\beta,\mu}\left(a(e^{-\beta(H-\mu)/2} g) a^*(e^{-\beta(H-\mu)/2} f)\right) \\
&= \left\langle g, e^{-\beta(H-\mu)} f\right\rangle \\
&\quad + \omega_{\beta,\mu}\left(a^*(e^{-\beta(H-\mu)/2} f) a(e^{-\beta(H-\mu)/2} g)\right).
\end{aligned}
$$

Iterating this m times gives

$$
\begin{aligned}
\omega_{\beta,\mu}(a^*(f)a(g)) &= \sum_{j=1}^{m} \left\langle g, e^{-j\beta(H-\mu)} f\right\rangle \\
&\quad + \omega_{\beta,\mu}\left(a^*(e^{-m\beta(H-\mu)/2} f) a(e^{-m\beta(H-\mu)/2} g)\right). \qquad (103)
\end{aligned}
$$

In the limit $m \to \infty$, the last term on the r.h.s. of (103) tends to zero, which follows from $\lim_m \|e^{-m\beta(H-\mu)/2} f\| = 0$ ($H - \mu > 0$!) and the continuity of $\omega_{\beta,\mu}$, (101). The first term on the r.h.s. of (103) can be summed explicitly and we obtain

$$
\omega_{\beta,\mu}(a^*(f)a(g)) = \left\langle g, \frac{1}{e^{\beta(H-\mu)} - 1} f\right\rangle. \qquad (104)
$$

Viewed as a function on $\mathfrak{H} \times \mathfrak{H}$, (104) is called the two-point function of the state $\omega_{\beta,\mu}$. Similarly, one defines n-point functions for all $n \geq 1$ by

$$\omega_{\beta,\mu}(a^*(f_1)\cdots a^*(f_n)a(g_1)\cdots a(g_n)). \tag{105}$$

Notice that the average of a product of m creation operators and n annihilation operators in the state $\omega_{\beta,\mu}$ vanishes unless $m = n$. A state with this property is called *gauge invariant*. The average of an arbitrary polynomial $Q \in \mathfrak{P}$ is expressed in terms of the n-point functions by first normal ordering Q. This means that the CCR are used repeatedly to write Q as a sum of polynomials in $a^\#$, where in each polynomial all creation operators stand to the left of all annihilation operators.

Proceeding in the same way as above, one can show that the n-point function (105) can be expressed as a sum of products of two-point functions. Consequently, (104) determines the state uniquely. Any state which is determined uniquely by its one- and two-point functions is called *quasi-free*.

Using the quasi-free structure one can show that

$$\begin{aligned}\omega_{\beta,\mu}(W(f)) &= \exp\left\{-\frac{1}{4}\left\langle f, \frac{e^{\beta(H-\mu)}+1}{e^{\beta(H-\mu)}-1} f\right\rangle\right\} \\ &= \exp\left\{-\frac{1}{4}\left\langle f, \coth\left(\frac{\beta(H-\mu)}{2}\right) f\right\rangle\right\}.\end{aligned}$$

So far, we have treated a general Hilbert space $\mathfrak{H}$ and a Hamiltonian H with the property that $H - \mu\mathbb{1} > 0$ is trace class. We consider now the case of the free Bose gas. The following discussion of the thermodynamic limit of the free Bose gas is summarized in [5], Proposition 5.2.29, see also [12].

Let $\{\Lambda_k\}_{k\geq 0} \subset \mathbb{R}^3$ be an increasing sequence of bounded regions in $\mathbb{R}^3$, s.t. $\bigcup_k \Lambda_k = \mathbb{R}^3$. Denote by $-H_k$ the selfadjoint Laplace operator on $L^2(\Lambda_k, d^3x)$ corresponding to a classical boundary condition. We choose μ s.t. there is a $C > 0$ satisfying

$$H_k - \mu\mathbb{1} \geq C\mathbb{1}, \tag{106}$$

uniformly in k. Let $\omega_{\beta,\mu}^{\Lambda_k}$ denote the Gibbs state on $\mathrm{CCR}(L^2(\Lambda_k, d^3x))$, see (95). The following results hold.

1. For any k and any $A \in \mathrm{CCR}(L^2(\Lambda_k, d^3x))$, the limit

$$\lim_{k'\to\infty} \omega_{\beta,\mu}^{\Lambda_{k'}}(A) = \omega_{\beta,\mu}(A) \tag{107}$$

exists and defines a state $\omega_{\beta,\mu}$ on $\mathrm{CCR}(\mathfrak{D})$, where $\mathfrak{D}$ is the dense subspace of $L^2(\mathbb{R}^3, d^3x)$ given by

$$\mathfrak{D} = \bigcup_{k\geq 0} L^2(\Lambda_k, d^3x). \tag{108}$$

The generating functional of $\omega_{\beta,\mu}$ is given by

$$\omega_{\beta,\mu}(W(f)) = \exp\left\{-\frac{1}{4}\left\langle f, \coth\left(\frac{\beta(H-\mu)}{2}\right) f\right\rangle\right\}, \tag{109}$$

for $f \in \mathfrak{D}$ and where $-H$ is the selfadjoint Laplace operator on $L^2(\mathbb{R}^3, d^3x)$. Note that due to (106) we can extend (109) to all $f \in L^2(\mathbb{R}^3, d^3x)$.

2. The GNS representation $(\mathcal{H}_{\beta,\mu}, \pi_{\beta,\mu}, \Omega_{\beta,\mu})$ of $(\mathrm{CCR}(\mathfrak{D}), \omega_{\beta,\mu})$ is regular. Let $a^{\#}_{\beta,\mu}(f)$, $f \in \mathfrak{D}$, denote the creation and annihilation operators in this representation. The state $\omega_{\beta,\mu}$ can be extended to the polynomial algebra $\mathfrak{P}_{\beta,\mu}$ generated by $\{a^{\#}_{\beta,\mu}(f) \mid f \in \mathfrak{D}\}$. The extension is the gauge-invariant quasi-free state with two-point function

$$\omega_{\beta,\mu}\left(a^*_{\beta,\mu}(f) a_{\beta,\mu}(g)\right) = \left\langle g, \frac{1}{e^{\beta(H-\mu)} - 1} f \right\rangle. \tag{110}$$

3. Let $f \in L^2(\mathbb{R}^3, d^3x)$ and let $\{f_n\} \subset \mathfrak{D}$ be a sequence approximating f, i.e., $\|f - f_n\| \to 0$. The strong limit

$$\lim_n \pi_{\beta,\mu}(W(f_n)) = W_{\beta,\mu}(f) \tag{111}$$

exists and defines a unitary operator $W_{\beta,\mu}(f)$ in the von Neumann algebra $\pi_{\beta,\mu}(\mathfrak{A})'' \subset \mathcal{B}(\mathcal{H}_{\beta,\mu})$. The operators $W_{\beta,\mu}(f)$, for $f \in L^2(\mathbb{R}^3, d^3x)$, satisfy the Weyl CCR, (40). In other words, they define a representation of the algebra $\mathrm{CCR}(L^2(\mathbb{R}^3, d^3x))$.
4. The state $\omega_{\beta,\mu}$, viewed as a state on the von Neumann algebra $\pi_{\beta,\mu}(\mathfrak{A})''$ determined by the vector $\Omega_{\beta,\mu}$, is a (β, α_t)-KMS state, where α_t is the $*$automorphism group given by $\alpha_t(W_{\beta,\mu}(f)) = W_{\beta,\mu}(e^{-iHt} f)$, for $f \in L^2(\mathbb{R}^3, d^3x)$.

We point out that condition (106) gives a restriction on the possible values of μ. We must require $\mu < \mu_0$, where μ_0 depends on the choice of the boundary condition. On the other hand, μ is related to the particle density of the system. It turns out that under condition (106), one cannot describe high particle densities – e.g. the situation where most particles are in the state of lowest energy. In order to describe this phenomenon, called Bose-Einstein condensation, one needs a more careful analysis of the thermodynamic limit. We refer for more detail to [5], Section 5.2.5.

4 Araki-Woods representation of the infinite free Boson gas

The goal of this section is to find the GNS representation of states ω on $\mathrm{CCR}(\mathfrak{D})$ which represent the infinitely extended ideal Bose gas in which the momentum density distribution of the particles is prescribed. Our approach is based on the original paper [3]. In a first step we show that the states of $\mathrm{CCR}(\mathfrak{D})$ are in one-to-one correspondence with so-called *generating* functionals on $\mathfrak{D}$. Then we calculate explicitly the generating functional corresponding to the Bose gas in a box and with a prescribed momentum density distribution. We take the thermodynamic limit of the finite-volume generating functionals, where the box size tends to infinity and the momentum density distribution approaches a given limit. The infinite-limit generating functional corresponds to a unique state on $\mathrm{CCR}(\mathfrak{D})$. We construct explicitly its GNS representation, which is commonly called the Araki-Woods representation.

4.1 Generating functionals

We consider in the remaining part of the notes the C^*algebra CCR($\mathfrak{D}$), where $\mathfrak{D} \subseteq \mathfrak{H}$. Given a state ω on CCR($\mathfrak{D}$), we may consider the (nonlinear) generating functional defined by

$$\begin{aligned} E : \mathfrak{D} &\to \mathbb{C} \\ f &\mapsto E(f) = \omega(W(f)). \end{aligned} \tag{112}$$

The generating functional satisfies the following properties:

1. (normalization) $E(0) = 1$
2. (unitarity) $\overline{E(f)} = E(-f)$, $f \in \mathfrak{D}$
3. (positivity) for any $K \geq 1$, $z_k \in \mathbb{C}$, $f_k \in \mathfrak{D}$, $k = 1 \ldots K$,

$$\sum_{k,k'=1}^{K} z_k \, \overline{z_{k'}} \, e^{-\frac{i}{2}\mathrm{Im}\langle f_k, f_{k'}\rangle} E(f_k - f_{k'}) \geq 0. \tag{113}$$

Properties 1. and 2. are obvious, and (113) is a consequence of the positivity of the state ω. Any positive element in a C^*algebra can be written as A^*A, so in CCR($\mathfrak{D}$), any positive element is approximated in C^*algebra norm by elements of the form

$$\left(\sum_{k=1}^{K} z_k W(f_k)\right)^* \left(\sum_{k=1}^{K} z_k W(f_k)\right) = \sum_{k,k'=1}^{K} z_k \, \overline{z_{k'}} \, e^{-\frac{i}{2}\mathrm{Im}\langle -f_{k'}, f_k\rangle} W(f_k - f_{k'}),$$

for some $z_k \in \mathbb{C}$ and $f_k \in \mathfrak{D}$. Hence (113) is equivalent to $\omega(A^*A) \geq 0$, for any $A \in$ CCR($\mathfrak{D}$).

We now show that conversely, if a functional E with properties 1.-3. is given, then it determines uniquely a state on CCR($\mathfrak{D}$), with respect to which it is the generating functional.

Theorem 4.1. *Suppose a map $E : \mathfrak{D} \to \mathbb{C}$ satisfies 1.-3. above. For $f \in \mathfrak{D}$, set $\omega(W(f)) = E(f)$ and extend ω by linearity to the linear span of the Weyl operators,*

$$\omega\left(\sum_{k=1}^{K} z_k W(f_k)\right) = \sum_{k=1}^{K} z_k \, E(f_k).$$

Then ω extends uniquely to a state on CCR($\mathfrak{D}$). *Moreover, E is continuous in the topology of $\mathfrak{D} \subseteq \mathfrak{H}$ if and only if $f \mapsto \pi_\omega(W(f))$ is a strongly continuous map from $\mathfrak{D}$ into the bounded operators on the GNS Hilbert space associated to* (CCR($\mathfrak{D}$), ω).

Remark. The statement "$f \mapsto E(f)$ is continuous in the topology of $\mathfrak{D}$" is equivalent to the statement "$f \mapsto E(f + f_0)$ is continuous in the topology of $\mathfrak{D}$, for any fixed $f_0 \in \mathfrak{D}$". We incorporate the proof of this remark in the proof of Theorem 4.1 given below.

This theorem can be viewed as a *non-commutative analog of the Bochner-Minlos theorem* (see e.g. [8], Section 3.2 or [7], Sections 3.4 and A.6). Let $\mathcal{S}$ be the Schwartz space on $\mathbb{R}^n$, [16] $\mathcal{S}'$ its dual (the set of continuous linear functionals on $\mathcal{S}$), and let $\mathcal{S}'_{\mathbb{R}}$ be the set of real Schwartz distributions, i.e. the set of $\chi \in \mathcal{S}'$ satisfying $\overline{\langle \chi; f\rangle} = \langle \chi; \overline{f}\rangle$, for all $f \in \mathcal{S}$, where $\langle \cdot\,;\cdot\rangle$ is the dual pairing. Let ν be a positive regular Borel measure on $\mathcal{S}'_{\mathbb{R}}$, [17] s.t. $\int_{\mathcal{S}'_{\mathbb{R}}} d\nu(\chi) = 1$. We define the Fourier transform of ν by

$$E(f) = \int_{\mathcal{S}'_{\mathbb{R}}} e^{-i\langle \chi; f\rangle} d\nu(\chi),$$

$f \in \mathcal{S}$. Then E satisfies

1'. $E(0) = 1$,
2'. $f \mapsto E(f)$ is continuous,
3'. for any $K \geq 1$, $z_k \in \mathbb{C}$, $f_k \in \mathcal{S}$, $k = 1 \ldots K$, we have

$$\sum_{k,k'=1}^{K} z_k\, \overline{z_{k'}}\, E(f_k - \overline{f_{k'}}) \geq 0. \tag{114}$$

Inequality (114) holds because the l.h.s. is just $\int_{\mathcal{S}'_{\mathbb{R}}} \left|\sum_{k=1}^{K} z_k e^{-i\langle \chi; f_k\rangle}\right|^2 d\nu(\chi)$. Here is the Bochner–Minlos theorem:

Theorem 4.2. *Suppose a map $E : \mathcal{S} \to \mathbb{C}$ satisfies 1'.-3'. above. Then there exists a unique normalized positive regular Borel measure ν on the real Schwartz distribution space $\mathcal{S}'_{\mathbb{R}}$ such that E is the Fourier transform of ν.*

Proof of Theorem 4.1. Parts of our proof are inspired by [1]. Let $\mathcal{S}$ be the $*$algebra generated by the Weyl operators $W(f)$, i.e., $\mathcal{S}$ is the $*$algebra of finite linear combinations of products of elements $W(f) \in \mathrm{CCR}(\mathfrak{D})$. $\mathcal{S}$ is a subalgebra of $\mathrm{CCR}(\mathfrak{D})$ and inherits the notion of positivity induced by $\mathrm{CCR}(\mathfrak{D})$. Inequality (113) implies that ω is a positive linear map on $\mathcal{S}$. Positivity implies boundedness as follows: let first $A \in \mathcal{S}$ be a *selfadjoint* element satisfying $\|A\| < 1$. Then we have $\mathcal{S} \ni \mathbb{1} - A > 0$, so $\omega(\mathbb{1}) - \omega(A) = \omega(\mathbb{1} - A) \geq 0$, and consequently $\omega(A) \leq \omega(\mathbb{1}) = E(0) = 1$. Next consider any $A \in \mathcal{S}$ s.t. $\|A\| < 1$. From $A^*A \leq \|A^*A\|\mathbb{1} \leq \|A\|^2\mathbb{1} < \mathbb{1}$ and the Cauchy-Schwarz inequality, $|\omega(A^*B)|^2 \leq \omega(A^*A)\omega(B^*B)$, which is valid for $A, B \in \mathcal{S}$ (note that we only need $\mathcal{S}$ to be a $*$algebra here, it does not have to be a Banach algebra), we obtain the estimate

[16] the set of $f \in C^\infty(\mathbb{R}^n)$ s.t. all seminorms $\|f\|_{\mathbf{k},\mathbf{l}} = \sup_{x\in\mathbb{R}^n} \|x^{\mathbf{k}}\partial^{\mathbf{l}} f(x)\|$ are finite, for any multi-indices $\mathbf{k}, \mathbf{l} \in \mathbb{N}^n$. The topology of $\mathcal{S}$ is the one induced by these seminorms.

[17] A Borel measure μ on Q is called regular if for any Borel subset E of Q we have $\mu(E) = \inf\{\mu(U) \mid U \supset E, U \text{ open }\}$ and $\mu(E) = \sup\{\mu(K) \mid K \subset E, K \text{ compact }\}$.

The Borel σ-algebra of $\mathcal{S}'$ is generated by the open sets of the weak$*$ topology on $\mathcal{S}'$. A base for this topology is given by the collection of all *cylinder sets*. Cylinder sets are of the form $\{\chi \in \mathcal{S}' \mid (\langle\chi; f_1\rangle, \ldots, \langle\chi; f_n\rangle) \in B \subset \mathbb{C}^n\}$, where $f_1, \ldots, f_n \in \mathcal{S}$ and B is open. An open set in the weak$*$ topology is a union of cylinder sets.

$$|\omega(A)|^2 = |\omega(\mathbb{1}A)|^2 \le \omega(\mathbb{1})\omega(A^*A) \le \|A\|^2\omega(\mathbb{1})^2 = \|A\|^2.$$

This shows that $|\omega(A)| \le \|A\|$, for $A \in \mathcal{S}$. Thus ω extends to a state on $\mathrm{CCR}(\mathfrak{D})$.

Next we show that if $f \mapsto E(f)$ is continuous then ω is a regular state. Let $(\mathcal{H}_\omega, \pi_\omega, \Omega_\omega)$ be the GNS representation of $(\mathrm{CCR}(\mathfrak{D}), \omega)$. Suppose $f_n \to f$ is a convergent sequence in $\mathfrak{D}$. Define a family of unitary operators $U_n = \pi_\omega(W(f_n))$ and $U = \pi_\omega(W(f))$ on the Hilbert space $\mathcal{H}_\omega$. We show that

$$\lim_{n\to\infty} \|(U_n - U)\psi\| = 0, \tag{115}$$

for any $\psi \in \mathcal{H}_\omega$. Due to unitarity it suffices to show weak convergence, i.e. (115) is equivalent to

$$\lim_{n\to\infty} \langle \phi, (U_n - U)\psi\rangle = 0, \tag{116}$$

for all $\phi, \psi \in \mathcal{H}_\omega$. Because Ω_ω is cyclic, we have that for any $\epsilon > 0$, there are vectors $\phi_\epsilon = \sum_{k=1}^K z_k \pi_\omega(W(g_k))\Omega_\omega$ and $\psi_\epsilon = \sum_{l=1}^L \zeta_l \pi_\omega(W(h_l))\Omega_\omega$, s.t. $\|\phi - \phi_\epsilon\| < \epsilon$ and $\|\psi - \psi_\epsilon\| < \epsilon$. Now

$$\begin{aligned}|\langle \phi, (U_n - U)\psi\rangle - \langle \phi_\epsilon, (U_n - U)\psi_\epsilon\rangle| &\le \|\phi - \phi_\epsilon\|\,\|(U_n - U)\|\,\|\psi\| \\ &\quad + 2\|\phi\|\,\|(U_n - U)\|\,\|\psi - \psi_\epsilon\| \\ &\le 4\epsilon(\|\phi\| + \|\psi\|),\end{aligned}$$

uniformly in n (we use here that $\|\phi_\epsilon\| < 2\|\phi\|$, for small ϵ). Thus it is enough to prove that

$$\lim_{n\to\infty} \langle \pi_\omega(W(g))\Omega_\omega, (U_n - U)\pi_\omega(W(h))\Omega_\omega\rangle = 0, \tag{117}$$

for any $g, h \in \mathfrak{D}$. This scalar product is just

$$\begin{aligned}&\langle \Omega_\omega, \pi_\omega(W(-g)(W(f_n) - W(f))W(h))\Omega_\omega\rangle \\ &\quad = e^{-\frac{i}{2}\mathrm{Im}(\langle -g, f_n\rangle + \langle -g+f_n, h\rangle)} E(-g + f_n + h) \\ &\quad\quad - e^{-\frac{i}{2}\mathrm{Im}(\langle -g, f\rangle + \langle -g+f, h\rangle)} E(-g + f + h),\end{aligned}$$

which converges to zero as $n \to \infty$.

Next we show that if ω is a regular state then $f \mapsto E(f)$ is continuous. Let f_n be a sequence in $\mathfrak{D}$ converging to $f \in \mathfrak{D}$. We have

$$\begin{aligned}E(f_n) - E(f) &= \omega(W(f_n) - W(f)) = \omega\big((W(f_n)W(-f) - \mathbb{1})W(f)\big) \\ &= \omega\left(\left(e^{\frac{i}{2}\mathrm{Im}\langle f_n, f\rangle} W(f_n - f) - \mathbb{1}\right)W(f)\right) \\ &= \left\langle \Omega_\omega, \left(e^{\frac{i}{2}\mathrm{Im}\langle f_n, f\rangle}\pi_\omega(W(f_n - f)) - \mathbb{1}\right)\pi_\omega(W(f))\Omega_\omega\right\rangle.\end{aligned} \tag{118}$$

Since $\pi_\omega(W(f_n - f))$ converges strongly to zero as $n \to \infty$ we have the desired continuity of E.

Finally we prove the assertion of the remark after Theorem 4.1. Suppose that $f_n \to 0$ and that $E(f_n) \to 1$. Our goal is to show that $E(f_n + f_0) \to E(f_0)$, where

$f_0 \in \mathfrak{H}$ is fixed. The above considerations leading to (115) show that $\pi_\omega(W(f_n)) \to \mathbb{1}$ in the strong sense on $\mathcal{H}_\omega$. Then we write, as in (118),

$$\begin{aligned} &E(f_n + f_0) - E(f_0) \\ &= \left\langle \Omega_\omega, \left(e^{\frac{i}{2}\mathrm{Im}\langle f_n, f_0\rangle} \pi_\omega(W(f_n)) - \mathbb{1} \right) \pi_\omega(W(f_0))\Omega_\omega \right\rangle . \end{aligned}$$

The r.h.s. converges to zero as $n \to \infty$. □

Suppose that E_k, $k = 1, 2, \ldots$ is a sequence of generating functionals, each satisfying conditions 1.-3. above. If E_k has a limit in the sense that there is a map $E : \mathfrak{D} \to \mathbb{C}$ s.t. $E(f) = \lim_{k\to\infty} E_k(f)$, for any $f \in \mathfrak{D}$, then it is clear that E satisfies conditions 1.-3. as well. In the next section we use this fact to construct the generating functional and the GNS representation of the free Bose gas extended to all of physical space in a state determined by a given momentum density distribution.

We close this section with the calculation of $E_{\mathrm{F}}(f) = \langle \Omega, W(f)\Omega \rangle$, the Fock generating functional, corresponding to the vacuum state on $\mathrm{CCR}_{\mathrm{F}}(\mathfrak{H})$. Using the series expansion of the Weyl operator in Fock space, we can write

$$E_{\mathrm{F}}(f) = \sum_{n\geq 0} \frac{i^{2n}}{(2n)!} \langle \Omega, \Phi(f)^{2n} \Omega \rangle , \tag{119}$$

where we have used that all odd powers of $\Phi(f)$ have a vanishing vacuum expectation value. We use the commutation relations (27), (28), and the fact that $a(f)\Omega = 0$ to get

$$\langle \Omega, \Phi(f)^{2n} \Omega \rangle = \frac{1}{\sqrt{2}} \langle \Omega, a(f)\Phi(f)^{2n-1}\Omega \rangle = \frac{2n-1}{2} \|f\|^2 \langle \Omega, \Phi(f)^{2n-2}\Omega \rangle .$$

By induction, we arrive at $\langle \Omega, \Phi(f)^{2n}\Omega \rangle = \left(\frac{\|f\|^2}{2} \right)^n \frac{(2n)!}{2^n n!}$, which we use in (119) to obtain

$$E_{\mathrm{F}}(f) = e^{-\|f\|^2/4}. \tag{120}$$

4.2 Ground state (condensate)

We construct in this section the representation of the CCR describing the infinitely extended Bose gas in its ground state where all particles are in the same state. The ground state is an example of a *condensate* (macroscopic occupation of a particular one-particle state of an infinitely extended system), it is parametrized by the particle density $\rho \geq 0$.

Consider first the free non-relativistic Bose gas confined to a finite box $V = \frac{1}{8}[-|V|^{1/3}, |V|^{1/3}]^3 \subset \mathbb{R}^3$ of volume $|V|$. We will let the volume and the number of particles, n, tend to infinity while keeping the density $\rho = n/|V|$ fixed. For any finite n, the Bose gas is described using Fock space $\mathcal{F}(L^2(V, d^3x))$, it is just a system of n non-interacting particles whose symmetric wave function $\psi \in L^2(V^n, d^{3n}x)$ evolves according to the Schrödinger equation

$$i\partial_t \psi(x_1, \dots, x_n) = (H\psi)(x_1, \dots, x_n), \tag{121}$$

where

$$H = \sum_{j=1}^{n} -\Delta_j \tag{122}$$

is the selfadjoint Hamiltonian operator on $L^2(V^n, d^{3n}x)$ with periodic boundary conditions (of course, Δ_j is the Laplacian with respect to the variable x_j). The system is in its ground state Ψ_V (the one having the lowest energy) if each of the n particles is in the state f_V of minimal energy (relative to $-\Delta_j$), given by

$$f_V(x) = |V|^{-1/2}, \ \ x \in V, \tag{123}$$

which we have normalized as $\|f_V\|_{L^2(V,d^3x)} = 1$. Consequently,

$$\Psi_V = \frac{1}{\sqrt{n!}} a_{\mathrm{F}}^*(f_V)^n \Omega_{\mathrm{F}}, \tag{124}$$

where $\frac{1}{\sqrt{n!}}$ is a normalization factor, a_{F}^* is the Bosonic creation operator on Fock space, and Ω_{F} is the Fock vacuum. The generating functional corresponding to Ψ_V is

$$E_V(f) = \langle \Psi_V, W(f)\Psi_V \rangle = \frac{1}{n!} \langle \Omega_{\mathrm{F}}, a_{\mathrm{F}}(f_V)^n W_{\mathrm{F}}(f) a_{\mathrm{F}}^*(f_V)^n \Omega_{\mathrm{F}} \rangle , \tag{125}$$

where $W_{\mathrm{F}} = e^{i\Phi_{\mathrm{F}}(f)}$ is a Weyl operator in the Fock representation,

$$\Phi_{\mathrm{F}} = \frac{1}{\sqrt{2}}(a_{\mathrm{F}}^*(f) + a_{\mathrm{F}}(f)).$$

Our plan is to calculate the right hand side of (125) explicitly and take the limit $n \to \infty$, keeping ρ fixed. This provides us with a generating functional E_{GS} (depending on the number ρ) which we interpret as the generating functional of the ground state of the infinite system. Knowing E_{GS}, we explicitly construct the GNS representation of the ground state of the infinite system (it will not be the Fock representation – i.e., there is no vector (or density matrix) on Fock space representing the ground state of the infinite system – we have already discussed this in Section 1.5.).

In order to calculate the r.h.s. of (125) we "pull" (or commute) the annihilation operators to the right, through $W_{\mathrm{F}}(f)$ and through the creation operators, by using the canonical commutation relations. Whenever an annihilation operator is completely pulled through, it hits the vacuum Ω_{F} yielding zero. The value of the r.h.s. of (125) is given by all extra terms (contractions) one generates, using the CCR, in this procedure. Let us first show how to pull the annihilation operators through the Weyl operator. Using the series expansion of $W_{\mathrm{F}}(f) = e^{i\Phi_{\mathrm{F}}(f)}$ and that

$$[a_{\mathrm{F}}(f), \Phi_{\mathrm{F}}(g)^k] = 2^{-1/2} k \langle f, g \rangle \Phi_{\mathrm{F}}(g)^{k-1}, \tag{126}$$

which follows easily from the CCR (27), (28), one verifies without difficulty that

$$[a_{\mathrm{F}}(f), W_{\mathrm{F}}(g)] = 2^{-1/2} i \langle f, g \rangle W_{\mathrm{F}}(g). \tag{127}$$

All these relations can be understood in the strong sense on the finite particle subspace. We view the pulling through procedure as follows. Consider $a_{\mathrm{F}}(f_V)^n W_{\mathrm{F}}(f)$. Among the n annihilation operators, k $(= 0, 1, \dots, n)$ are commuted through $W_{\mathrm{F}}(f)$ to the right while $n - k$ have undergone a contraction of the form (127). For each fixed value of k, there are $\binom{n}{k}$ ways of choosing which annihilation operators are safely pulled through the Weyl operator. We obtain

$$a_{\mathrm{F}}(f_V)^n W_{\mathrm{F}}(f) = \sum_{k=0}^{n} \binom{n}{k} \left(\frac{i \langle f_V, f \rangle}{\sqrt{2}} \right)^{n-k} W_{\mathrm{F}}(f) a_{\mathrm{F}}(f_V)^k. \tag{128}$$

One can of course prove (128) as well by induction, which is an easy task. The generating functional can thus be written as

$$E_V(f) = \frac{1}{n!} \sum_{k=0}^{n} \binom{n}{k} \left(\frac{i \langle f_V, f \rangle}{\sqrt{2}} \right)^{n-k} \langle \Omega_{\mathrm{F}}, W_{\mathrm{F}}(f) a_{\mathrm{F}}(f_V)^k a_{\mathrm{F}}^*(f_V)^n \Omega_{\mathrm{F}} \rangle .$$

A similar pull through argument as above, plus the facts that $a_{\mathrm{F}}(f_V)\Omega_{\mathrm{F}} = 0$ and $\|f_V\| = 1$ yields

$$\begin{aligned} a_{\mathrm{F}}(f_V)^k a_{\mathrm{F}}^*(f_V)^n \Omega_{\mathrm{F}} &= n(n-1) \cdots (n-k+1) \langle f_V, f_V \rangle^k a_{\mathrm{F}}^*(f_V)^{n-k} \Omega_{\mathrm{F}} \\ &= \frac{n!}{(n-k)!} a_{\mathrm{F}}^*(f_V)^{n-k} \Omega_{\mathrm{F}}, \end{aligned}$$

from which we get

$$E_V(f) = \frac{1}{n!} \sum_{k=0}^{n} \binom{n}{k} \frac{n!}{(n-k)!} \left(\frac{i \langle f_V, f \rangle}{\sqrt{2}} \right)^{n-k} \langle \Omega_{\mathrm{F}}, W_{\mathrm{F}}(f) a_{\mathrm{F}}^*(f_V)^{n-k} \Omega_{\mathrm{F}} \rangle . \tag{129}$$

We pull the $n - k$ creation operators to the left through the Weyl operator by using the adjoint relation to (128) (recall also that $W_{\mathrm{F}}(f)^* = W_{\mathrm{F}}(-f)$)

$$W_{\mathrm{F}}(f) a_{\mathrm{F}}^*(f_V)^{n-k} = \sum_{l=0}^{n-k} \binom{n-k}{l} \left(\frac{-i \overline{\langle f_V, -f \rangle}}{\sqrt{2}} \right)^{n-k-l} a_{\mathrm{F}}^*(f_V)^l W_{\mathrm{F}}(f).$$

Clearly, only the term $l = 0$, where no creation annihilator arrives safely to the left of $W_{\mathrm{F}}(f)$ will give a non-zero contribution to expression (129) (because $a_{\mathrm{F}}(f_V)\Omega_{\mathrm{F}} = 0$, once again), so

$$E_V(f) = \frac{1}{n!} \sum_{k=0}^{n} \binom{n}{k} \frac{n!}{(n-k)!} \left(-\frac{1}{2} |\langle f_V, f \rangle|^2 \right)^{n-k} \langle \Omega_{\mathrm{F}}, W_{\mathrm{F}}(f) \Omega_{\mathrm{F}} \rangle .$$

We denote the Fock vacuum generating functional by $E_{\rm F}(f) = \langle \Omega, W_{\rm F}(f)\Omega\rangle$ (see (119)) and observe that for f with compact support and large enough $|V|$, we have

$$\langle f_V, f\rangle = |V|^{-1/2}\int_V f(x)d^3x = \left((2\pi)^3\frac{\rho}{n}\right)^{1/2}\widehat{f}(0),$$

where

$$\widehat{f}(k) = \frac{1}{(2\pi)^{3/2}}\int_{\mathbb{R}^3} d^3k\, e^{-ikx}f(x) \tag{130}$$

is the Fourier transform. Consequently we have

$$E_V(f) = E_{\rm F}(f)\,\frac{1}{n!}\sum_{k=0}^{n}\binom{n}{k}\frac{n!}{(n-k)!}\left(-(2\pi)^3\frac{\rho}{2n}|\widehat{f}(0)|^2\right)^{n-k}. \tag{131}$$

We recall that the Laguerre polynomials are defined by

$$L_n(z) = \frac{1}{n!}e^z\frac{d^n}{dz^n}(e^{-z}z^n) = \frac{1}{n!}\sum_{k=0}^{n}\binom{n}{k}\frac{n!}{(n-k)!}(-z)^{n-k} \tag{132}$$

for $n = 0, 1, \ldots$ and $z \in \mathbb{C}$. Next, it is known that (see e.g. [2], formula 22.15.2)

$$\lim_{n\to\infty} L_n(z/n) = J_0(2\sqrt{z}), \tag{133}$$

where the Bessel function J_0 satisfies

$$\int_{-\pi}^{\pi}\frac{d\theta}{2\pi}e^{-i(\alpha\cos\theta+\beta\sin\theta)} = J_0\left(\sqrt{\alpha^2+\beta^2}\right), \tag{134}$$

for any $\alpha, \beta \in \mathbb{R}$ (see e.g. [11], formula (5.3.66)). In conclusion, we have calculated the infinite volume generating functional to be

$$E_{\rm GS}(f) = E_{\rm F}(f)J_0\left((2\pi)^{3/2}\sqrt{2\rho}\,|\widehat{f}(0)|\right). \tag{135}$$

This generating functional defines a state on $\mathrm{CCR}(\mathfrak{D})$ which we view as being the ground state of the infinite Bose gas. The test function space is given by

$$\mathfrak{D} = \left\{ f \in L^2(\mathbb{R}^3, d^3x) \mid \widehat{f}(0) \text{ exists} \right\}. \tag{136}$$

Any function in Schwarz space is contained in $\mathfrak{D}$, so $\mathfrak{D}$ is dense in $L^2(\mathbb{R}^3, d^3x)$.

Let us now construct the GNS representation of the infinite Bose gas. Consider the Hilbert space

$$\mathcal{H}_{\rm GS} = \mathcal{F}(L^2(\mathbb{R}^3, d^3x)) \otimes L^2\left(S^1, d\sigma\right), \tag{137}$$

where the left factor is the Bosonic Fock space over $L^2(\mathbb{R}^3, d^3x)$ and the right one is the Hilbert space of all square integrable functions on the unit circle with uniform measure. It is convenient to parametrize the circle by the angle $\theta \in [-\pi, \pi]$. Set

$$\Omega_{\rm GS} = \Omega_{\rm F} \otimes 1, \tag{138}$$

where $1 \in L^2\left(S^1, d\sigma\right)$ is the constant function. We define the representation map $\pi_{\rm GS} : {\rm CCR}(\mathfrak{D}) \to \mathcal{B}(\mathcal{H}_{\rm GS})$ as

$$\pi_{\rm GS} : W(f) \mapsto W_{\rm F}(f) \otimes e^{-i(2\pi)^{3/2}\sqrt{2\rho}\,\left({\rm Re}\widehat{f}(0)\cos\theta + {\rm Im}\widehat{f}(0)\sin\theta\right)}. \tag{139}$$

Using relation (134) it is easily seen that for any $f \in \mathfrak{D}$ we have

$$\langle \Omega_{\rm GS}, \pi_{\rm GS}(W(f))\Omega_{\rm GS}\rangle = E_{\rm GS}(f) \tag{140}$$

so the representation gives the correct generating functional. To show that the GNS Hilbert space of $({\rm CCR}(\mathfrak{D}), \omega_{\rm GS})$, where the state $\omega_{\rm GS}$ is represented by $\Omega_{\rm GS}$ in $\mathcal{H}_{\rm GS}$, is actually the entire $\mathcal{H}_{\rm GS}$, we need to verify that $\Omega_{\rm GS}$ is cyclic for $\pi_{\rm GS}$. To show this, define the family of functions

$$f_{z,s}(x) = \sqrt{\frac{\pi}{2}}\, sz \frac{e^{-s|x|}}{x^2+1},$$

where $x \in \mathbb{R}^3$, $z \in \mathbb{C}$ and $s > 0$. Clearly, $\|f_{z,s}\|_{L^2(\mathbb{R}^3, d^3x)} \to 0$ as $s \to 0_+$, while $\widehat{f}_{z,s}(0) \to z$ for $s \to 0_+$. Due to the strong continuity of the Weyl operators in Fock space, we have for any $z \in \mathbb{C}$

$$\pi_{\rm GS}(W(f_{z,s})) \to \mathbb{1} \otimes e^{-i(2\pi)^{3/2}\sqrt{2\rho}\,({\rm Re}z\cos\theta + {\rm Im}z\sin\theta)},$$

in the strong sense on $\mathcal{H}_{\rm GS}$, as $s \to 0_+$. Since the constant function 1 is cyclic in $L^2\left(S^1, d\sigma\right)$ for the set of multiplication operators $\{e^{i(a\cos\theta + b\sin\theta)} \mid a, b \in \mathbb{R}\}$ [18], we see that

$$\Omega_{\rm F} \otimes L^2\left(S^1, d\sigma\right)$$

is contained in the closure of

$$\pi_{\rm GS}\left({\rm CCR}(\mathfrak{D})\right)\Omega_{\rm GS}.$$

Because $\Omega_{\rm F}$ is cyclic for $\{W_{\rm F}(f) \mid f \in \mathfrak{D}\}$ in $\mathcal{F}(L^2(\mathbb{R}^3, d^3x))$ (we use here that $\mathfrak{D}$ is dense in $L^2(\mathbb{R}^3, d^3x)$) we can approximate arbitrarily well any given $\psi \in \mathcal{F}(L^2(\mathbb{R}^3, d^3x))$ by some $W_{\rm F}(f)\Omega_{\rm F}$, $f \in \mathfrak{D}$. It follows that for an appropriate choice of $z \in \mathbb{C}$ and for s small enough the vector $\psi \otimes 1 \in \mathcal{H}_{\rm GS}$ is approximated arbitrarily well by $\pi_{\rm GS}(W(f)W(f_{z,s}))\Omega_{\rm GS}$. This shows that $\Omega_{\rm GS}$ is cyclic for $\pi_{\rm GS}$ in $\mathcal{H}_{\rm GS}$.

The representation $(\mathcal{H}_{\rm GS}, \pi_{\rm GS}, \Omega_{\rm GS})$ is a regular. The creation and annihilation operators are given by

$$a^*_{\rm GS}(f) = a^*_{\rm F}(f) \otimes 1 - (2\pi)^{3/2}\sqrt{\rho}\,\widehat{f}(0)\,\mathbb{1} \otimes e^{-i\theta},$$
$$a_{\rm GS}(f) = a_{\rm F}(f) \otimes 1 - (2\pi)^{3/2}\sqrt{\rho}\,\overline{\widehat{f}(0)}\,\mathbb{1} \otimes e^{i\theta}.$$

At zero density, $\rho = 0$, the ground state representation of the infinite Bose gas coincides with (is isomorphic to) the Fock representation.

[18] Any function $f \in L^2(S^1, d\sigma)$ has a Fourier series expansion.

4.3 Excited states

Our goal for this section is to extend the above method to construct the generating functional and the GNS representation corresponding to an (infinite volume) state of the CCR with a continuous momentum distribution $\rho(k)$.

Consider first the situation where, in our box V (as in the last section), we have n_j particles with momentum k_j, i.e., with wave function $f_V^j(x) = |V|^{-1/2}e^{ik_j x}$, where $j = 1, \ldots, p$, and where $|k_j|^2$ are (discrete) eigenvalues of the Laplacian (in the box $V \subset \mathbb{R}^3$ with periodic boundary conditions). The f_V^j are eigenfunctions of the Laplacian and satisfy the orthonormality condition $\left\langle f_V^j, f_V^l \right\rangle = \delta_{jl}$ (Kronecker symbol). We will let the box tend to all of $\mathbb{R}^3$ with the result that in the limit, the values of k can range continuously throughout $\mathbb{R}^3$ (this reflects the fact that $-\Delta$ on $L^2(\mathbb{R}^3, d^3x)$ has purely absolutely continuous spectrum).

The state of the gas in the box with densities $\rho_j = n_j/|V|$ of particles with momenta k_j, $j = 1, \ldots, p$, is given by

$$\Psi_V = \frac{1}{\sqrt{n_1! \cdots n_p!}} a_{\mathrm{F}}^*(f_V^1)^{n_1} \cdots a_{\mathrm{F}}^*(f_V^p)^{n_p} \Omega_{\mathrm{F}},$$

and the corresponding generating functional

$$E_V(f) = \langle \Psi_V, W_{\mathrm{F}}(f)\Psi_V \rangle$$

can be calculated just as in the previous section. It is an easy exercise to obtain the expression

$$E_V(f) = E_{\mathrm{F}}(f) L_{n_1}\left((2\pi)^3 \frac{\rho_1}{2n_1} |\widehat{f}(k_1)|^2 \right) \cdots L_{n_p}\left((2\pi)^3 \frac{\rho_p}{2n_p} |\widehat{f}(k_p)|^2 \right),$$

where $L_n(z)$ are the Laguerre polynomials defined in (132). We have used that for any f with compact support,

$$\langle f_j, f \rangle = |V|^{-1/2} \int_V e^{-ik_j x} f(x) d^3x = \left(\frac{\rho_j}{n_j} \right)^{1/2} (2\pi)^{3/2} \widehat{f}(k_j),$$

for $|V|$ big enough and where $\widehat{f}$ is the Fourier transform (130). Using (133), we take the limits $n_j \to \infty$, $j = 1, \ldots, p$, while leaving ρ_j fixed. The infinite volume generating functional is

$$E(f) = E_{\mathrm{F}}(f) J_0\left((2\pi)^{3/2} \sqrt{2\rho_1}\, |\widehat{f}(k_1)| \right) \cdots J_0\left((2\pi)^{3/2} \sqrt{2\rho_p}\, |\widehat{f}(k_p)| \right). \quad (141)$$

Our next task is to let the discrete distribution ρ_j, $j = 1, \ldots, p$, tend to a continuous distribution $\rho(k)$. We do this for simplicity first in the one-dimensional case, $k \in \mathbb{R}$, and we will deduce the general formula afterwards. Let thus $\mathbb{R} \ni k \mapsto \rho(k) \in \mathbb{R}_+$ be a given momentum density and consider an interval $[-L, L]$. We partition $[-L, L]$

into small intervals with endpoints $k_j = -L + 2Lj/p, j = 0, \ldots, p$, each of length $2L/p$, and we will let $p \to \infty$. The density ρ_j of particles having momenta in the interval with left endpoint k_j is given by $\rho_j = 2L\rho(k_j)/p$. Our goal is to calculate

$$\lim_{p\to\infty} \log\left(\frac{E(f)}{E_{\mathrm{F}}(f)}\right) = \lim_{p\to\infty} \sum_{j=0}^{p} \log J_0\left((2\pi)^{1/2}\sqrt{\frac{4L\rho(k_j)}{p}}\,|\widehat{f}(k_j)|\right). \quad (142)$$

Notice that the power of 2π is now $1/2$, in one dimension. A simple Taylor expansion shows that the leading term of $\log J_0(\epsilon)$ for small ϵ is $-\frac{\epsilon^2}{4}$, so that we have

$$\lim_{p\to\infty} \log\left(\frac{E(f)}{E_{\mathrm{F}}(f)}\right)$$
$$= -\frac{1}{2}2\pi \lim_{p\to\infty} \sum_{j=0}^{p} \frac{2L\rho(k_j)}{p} |\widehat{f}(k_j)|^2 = -\frac{1}{2}2\pi \int_{-L}^{L} dk\ \rho(k)|\widehat{f}(k)|^2.$$

If we take f such that $\sqrt{\rho}\widehat{f}$ is square integrable then we can take $L \to \infty$ and obtain for the generating functional of the infinite Bose gas *in three dimensions* and with momentum density ρ

$$E_\rho(f) = E_{\mathrm{F}}(f) \exp\left\{-\frac{(2\pi)^3}{2} \int_{\mathbb{R}^3} d^3k\ \rho(k)|\widehat{f}(k)|^2\right\}. \quad (143)$$

The test function space consists of functions s.t. the integral in (143) exists,

$$\mathfrak{D}' = \left\{f \in L^2(\mathbb{R}^3, d^3x) \mid \sqrt{\rho}\widehat{f} \in L^2(\mathbb{R}^3, d^3k)\right\}, \quad (144)$$

it depends on the function ρ. It is convenient to pass to a representation of the one-particle Hilbert space where the energy operator is diagonal; in the case of the Laplacian this means that we consider $L^2(\mathbb{R}^3, d^3k)$, the Fourier-transformed position space $L^2(\mathbb{R}^3, d^3x)$. The Fourier transform is an isometric isomorphism between $L^2(\mathbb{R}^3, d^3x)$ and $L^2(\mathbb{R}^3, d^3k)$ which induces a C^*algebra isomorphism between $\mathrm{CCR}(\mathfrak{D}')$ and $\mathrm{CCR}(\mathfrak{D})$, where

$$\mathfrak{D} = \{f \in L^2(\mathbb{R}^3, d^3k) \mid \sqrt{\rho}f \in L^2(\mathbb{R}^3, d^3k)\}, \quad (145)$$

and the corresponding generating functional is given by

$$E_\rho(f) = E_{\mathrm{F}}(f) \exp\left\{-\frac{(2\pi)^3}{2} \int_{\mathbb{R}^3} d^3k\ \rho(k)|f(k)|^2\right\}, \quad f \in \mathfrak{D}, \quad (146)$$

where $E_{\mathrm{F}}(f) = e^{-\|f\|^2/4}$, for $f \in \mathfrak{D}$ (see (120)). Hence

$$E_\rho(f) = \exp\left\{-\frac{1}{4}\langle f, (1 + 16\pi^3\rho)f\rangle\right\}, \quad f \in \mathfrak{D}. \quad (147)$$

One can carry out the construction for a general selfadjoint Hamiltonian H (not necessarily of the form (122)) and one arrives at (143) where $\widehat{f}$ stands for the eigenfunction expansion of f corresponding to H.

Formula (147) gives a generating functional which defines a state ω_ρ on $\mathrm{CCR}(\mathfrak{D})$, according to Theorem 4.1. We give now the GNS representation of $(\mathrm{CCR}(\mathfrak{D}), \omega_\rho)$ for densities $\rho(k)$ such that

$$k \mapsto \rho(k) \text{ is continuous}, \quad \rho(k) > 0 \text{ a.e.}, \quad \int_{\mathbb{R}^3} d^3k\, \rho(k) < \infty. \tag{148}$$

The representation Hilbert space $\mathcal{H}_\rho$ and the cyclic vector Ω_ρ are

$$\mathcal{H}_\rho = \mathcal{F}(L^2(\mathbb{R}^3, d^3k)) \otimes \mathcal{F}(L^2(\mathbb{R}^3, d^3k)) \tag{149}$$

$$\Omega_\rho = \Omega_{\mathrm{F}} \otimes \Omega_{\mathrm{F}}, \tag{150}$$

where $\mathcal{F}(L^2(\mathbb{R}^3, d^3k))$ is the Bosonic Fock space over $L^2(\mathbb{R}^3, d^3k)$ and Ω_{F} is the vacuum therein. The representation map $\pi_\rho : \mathrm{CCR}(\mathfrak{D}) \to \mathcal{B}(\mathcal{H}_\rho)$ is given by

$$\pi_\rho(W(f)) = W_{\mathrm{F}}\left(\sqrt{1+\mu}\, f\right) \otimes W_{\mathrm{F}}\left(\sqrt{\mu}\, \overline{f}\right), \tag{151}$$

$$\mu(k) = 8\pi^3 \rho(k). \tag{152}$$

Notice that in the Weyl operator on the right factor there appears the complex conjugate of f. Using expression (120) it is an easy matter to verify that

$$\langle \Omega_\rho, \pi_\rho(W(f))\Omega_\rho \rangle = E_\rho(f), \tag{153}$$

where $E_\rho(f)$ is given by (147). π_ρ is a regular representation and the creation and annihilation operators are given by

$$a^*_\rho(f) = a^*_{\mathrm{F}}\left(\sqrt{1+\mu}\, f\right) \otimes \mathbb{1} + \mathbb{1} \otimes a_{\mathrm{F}}\left(\sqrt{\mu}\, \overline{f}\right), \tag{154}$$

$$a_\rho(f) = a_{\mathrm{F}}\left(\sqrt{1+\mu} f\right) \otimes \mathbb{1} + \mathbb{1} \otimes a^*_{\mathrm{F}}\left(\sqrt{\mu}\, \overline{f}\right). \tag{155}$$

Since the $a^\#_\rho(f)$ are obtained from the represented Weyl operators by strong differentiation it follows that Ω_ρ is cyclic for π_ρ if Ω_ρ is cyclic for the polynomial algebra $\mathfrak{P}$ generated by all creation and annihilation operators $a^\#_\rho(f)$, $f \in \mathfrak{D}$. The set $\{\sqrt{\mu} f \mid f \in \mathfrak{D}\}$ is dense in $L^2(\mathbb{R}^3, d^3k)$ due to condition (148). Since Ω_{F} is cyclic for the Fock creation operators it follows from (155) that $\Omega_{\mathrm{F}} \otimes \mathcal{F}(L^2(\mathbb{R}^3, d^3k))$ lies in the closure of $\mathfrak{P}\Omega_\rho$. Similarly (154) shows that $\mathcal{F}(L^2(\mathbb{R}^3, d^3k)) \otimes \Omega_{\mathrm{F}}$ is in that closure. Hence Ω_ρ is cyclic for π_ρ.

If $\rho(k) = 0$ then $\mu(k) = 0$ and the representation (151) reduces to the Fock representation.

4.4 Equilibrium states

The results of the previous two sections can be combined to describe the infinitely extended free Bose gas with a momentum density distribution which has some condensate part characterized by the density $\rho_0 \in \mathbb{R}_+$ and some continuous part given by $\rho(k)$. The corresponding generating functional is

$$E_{\rho_0,\rho}(f) \tag{156}$$
$$= E_{\mathrm{F}}(f)\exp\left\{-\frac{(2\pi)^3}{2}\int_{\mathbb{R}^3} d^3k\,\rho(k)|f(k)|^2\right\} J_0\left((2\pi)^{3/2}\sqrt{2\rho_0}\,|f(0)|\right),$$

compare with (135) and (146), for functions f in the test function space

$$\mathfrak{D} = \left\{f \in L^2(\mathbb{R}^3, d^3k) \mid \sqrt{\rho} f \in L^2(\mathbb{R}^3, d^3k),\ |f(0)| < \infty\right\}. \tag{157}$$

The GNS representation Hilbert space $\mathcal{H}_{\rho_0,\rho}$ and the cyclic vector $\Omega_{\rho_0,\rho}$ associated to $(\mathrm{CCR}(\mathfrak{D}), \omega_{\rho_0,\rho})$, where $\omega_{\rho_0,\rho}$ is the state defined by (156), are given by

$$\begin{aligned}\mathcal{H}_{\rho_0,\rho} &= \mathcal{F}(L^2(\mathbb{R}^3, d^3k)) \otimes \mathcal{F}(L^2(\mathbb{R}^3, d^3k)) \otimes L^2\left(S^1, d\sigma\right), \\ \Omega_{\rho_0,\rho} &= \Omega_{\mathrm{F}} \otimes \Omega_{\mathrm{F}} \otimes 1,\end{aligned} \tag{158}$$

and the representation map is

$$\pi_{\rho_0,\rho}(W(f)) = W_{\mathrm{F}}(\sqrt{1+\mu}\, f) \otimes W_{\mathrm{F}}(\sqrt{\mu}\,\overline{f}\,) \otimes e^{-i\Phi(f,\theta)}, \tag{159}$$

with $\mu(k) = 8\pi^3\rho(k)$, and where we introduce the phase

$$\Phi(f,\theta) = (2\pi)^{3/2}\sqrt{2\rho_0}\ (\mathrm{Re} f(0)\cos\theta + \mathrm{Im} f(0)\sin\theta)\,. \tag{160}$$

Note that Φ is real linear in the first argument. The creation and annihilation operators in this regular representation are not difficult to calculate:

$$\begin{aligned}a^*_{\rho_0,\rho}(f) &= a^*_{\mathrm{F}}\left(\sqrt{1+\mu}\, f\right) \otimes \mathbb{1} \otimes 1 + \mathbb{1} \otimes a_{\mathrm{F}}\left(\sqrt{\mu}\,\overline{f}\,\right) \otimes 1 \\ &\quad -(2\pi)^{3/2}\sqrt{\rho_0}\, f(0)\ \mathbb{1} \otimes \mathbb{1} \otimes e^{-i\theta}, \end{aligned}\tag{161}$$
$$\begin{aligned}a_{\rho_0,\rho}(f) &= a_{\mathrm{F}}\left(\sqrt{1+\mu} f\right) \otimes \mathbb{1} \otimes 1 + \mathbb{1} \otimes a^*_{\mathrm{F}}\left(\sqrt{\mu}\,\overline{f}\,\right) \otimes 1 \\ &\quad -(2\pi)^{3/2}\sqrt{\rho_0}\, \overline{f(0)}\ \mathbb{1} \otimes \mathbb{1} \otimes e^{i\theta}. \end{aligned}\tag{162}$$

The dynamics on $\mathrm{CCR}(\mathfrak{D})$ generated by the Hamiltonian (122) is given by

$$W(f) \mapsto \alpha_t(W(f)) = W(e^{it\omega} f), \tag{163}$$

where $\omega(k) = |k|^2$. It is clear from (156) that $E_{\rho_0,\rho}(e^{it\omega} f) = E_{\rho_0,\rho}(f)$, for all $t \in \mathbb{R}$. Consequently $\omega_{\rho_0,\rho}$ is a time translation invariant i.e. stationary state, for any choice of $\rho_0, \rho(k)$. We wish to examine which particular momentum density distributions correspond to *equilibrium states* of the system.

We have

$$W_{\mathrm{F}}(e^{it\omega} f) = e^{itH} W_{\mathrm{F}}(f) e^{-itH}, \tag{164}$$

where H is the free field Hamiltonian (in Fock space) given by the second quantization of the multiplication by $\omega(k)$. It is easy to see from (159) and (164) that the dynamics (163) is unitarily implemented as

$$\pi_{\rho_0,\rho}(\alpha_t(W(f)) = e^{itL}\pi_{\rho_0,\rho}(W(f))e^{-itL}, \tag{165}$$

where L is the so-called Liouvillian, given by

$$L = H \otimes \mathbb{1} \otimes 1 - \mathbb{1} \otimes H \otimes 1. \tag{166}$$

An equilibrium state ω is a state that satisfies the KMS condition

$$\omega(A\alpha_t(B)) = \omega(\alpha_{t-i\beta}(B)A), \tag{167}$$

see also (96). We assume here that B is such that $\alpha_z(B)$ exists for values of z in a strip around the real axis. Since an equilibrium state is necessarily α_t–invariant, (167) is equivalent to $\omega(AB) = \omega(\alpha_{-i\beta}(B)A)$. It is evident from the explicit form of $\Omega_{\rho_0,\rho}$ that $\omega_{\rho_0,\rho}$ can be extended to the polynomial algebra generated by the creation and annihilation operators (161), (162), giving a gauge-invariant quasifree state (see after (105)). To see which densities give an equilibrium state it is thus necessary and sufficient to solve the equation

$$\left\langle \Omega_{\rho_0,\rho}, a^*_{\rho_0,\rho}(f) a_{\rho_0,\rho}(g) \Omega_{\rho_0,\rho} \right\rangle = \left\langle \Omega_{\rho_0,\rho}, e^{\beta L} a_{\rho_0,\rho}(g) e^{-\beta L} a^*_{\rho_0,\rho}(f) \Omega_{\rho_0,\rho} \right\rangle, \tag{168}$$

which should hold for all f, g, for ρ_0 and $\rho(k)$. We calculate

$$\left\langle \Omega_{\rho_0,\rho}, a^*_{\rho_0,\rho}(f) a_{\rho_0,\rho}(g) \Omega_{\rho_0,\rho} \right\rangle = \langle g, \mu f \rangle + (2\pi)^3 \rho_0\, f(0) \overline{g(0)}$$

and

$$\left\langle \Omega_{\rho_0,\rho}, e^{\beta L} a_{\rho_0,\rho}(g) e^{-\beta L} a^*_{\rho_0,\rho}(f) \Omega_{\rho_0,\rho} \right\rangle = \langle g, e^{-\beta\omega}(1+\mu) f \rangle + (2\pi)^3 \rho_0\, f(0) \overline{g(0)}.$$

Consequently $\rho_0 \geq 0$ can be arbitrary and μ must satisfy $\mu = e^{-\beta\omega}(1+\mu)$, i.e., $\mu(k) = \frac{1}{e^{\beta\omega(k)}-1}$. The density is thus given by (see (152))

$$\rho(k) = (2\pi)^{-3} \frac{1}{e^{\beta\omega(k)} - 1}, \tag{169}$$

which is the Planck distribution of black body radiation.

Let us focus on massless relativistic Bosons, where $\omega(k) = |k|$. Other dispersion relations are discussed in an analogous way. The total density of particles in the state of equilibrium is

$$\rho_{\mathrm{tot}} = \rho_0 + \int_{\mathbb{R}^3} d^3k\, \rho(k) = \rho_0 + \frac{1}{8\pi^3} \int_{\mathbb{R}^3} \frac{d^3k}{e^{\beta|k|} - 1} = \rho_0 + \frac{c}{\beta^3}, \tag{170}$$

where $c = \frac{1}{2\pi^2} \int_0^\infty \frac{s^2}{e^s - 1} ds$ is a fixed constant. We can deduce from (170) the following qualitative behavior of the system. Suppose ρ_{tot} is fixed and suppose we decrease the temperature of the system ($\beta \to \infty$). Then ρ_0 tends to ρ_{tot} which means that for low temperatures the system likes to form a condensate. If we fix an inverse temperature β and increase the total density ρ_{tot} of the system then ρ_0 increases as well. These considerations show that we are likely to observe a condensate if either the temperature is low or the density is high (this, of course, is also an experimental fact).

We close this section with a result about the thermodynamic limit of Gibbs states which is due to Cannon, [6]. Fix an inverse temperature $0 < \beta < \infty$ and define the *critical density* by

$$\rho_{\mathrm{crit}}(\beta) = (2\pi)^{-3} \int_{\mathbb{R}^3} \frac{d^3k}{e^{\beta\omega} - 1}, \tag{171}$$

which coincides with the total density (170) in the equilibrium state for $\rho_0 = 0$. Let V be the box defined by $-L/2 \leq x_j \leq L/2$ $(j = 1, 2, 3)$ and define the canonical state at inverse temperature β and density ρ_{tot} by

$$\langle A \rangle^{\mathrm{c}}_{\beta,\rho_{\mathrm{tot}},V} = \frac{\mathrm{tr} A P_{\rho_{\mathrm{tot}}V} e^{-\beta H_V}}{\mathrm{tr} P_{\rho_{\mathrm{tot}}V} e^{-\beta H_V}}, \tag{172}$$

where the trace is over Fock space over $L^2(V, d^3x)$, $P_{\rho_{\mathrm{tot}}V}$ is the projection onto the subspace of Fock space with $\rho_{\mathrm{tot}}V$ particles (if $\rho_{\mathrm{tot}}V$ is not an integer take a convex combination of canonical states with integer values $\rho_1 V$ and $\rho_2 V$ extrapolating $\rho_{\mathrm{tot}}V$). The Hamiltonian H_V is negative the Laplacian with periodic boundary conditions. The observable A in (172) belongs to the Weyl algebra over the test function space C_0^∞, realized as a C^*algebra acting on Fock space. Cannon shows that for any $\beta, \rho_{\mathrm{tot}} > 0$ and $f \in C_0^\infty$,

$$\langle W(f) \rangle^{\mathrm{c}}_{\beta,\rho_{\mathrm{tot}},V} \longrightarrow \begin{cases} e^{-\frac{1}{4}\|f\|^2} e^{-\frac{1}{2}\left\langle f, \frac{z_\infty}{e^{\beta\omega} - z_\infty} f \right\rangle}, & \rho_{\mathrm{tot}} \leq \rho_{\mathrm{crit}}(\beta) \\ E_{\rho_0,\rho}(f), & \rho_{\mathrm{tot}} \geq \rho_{\mathrm{crit}}(\beta) \end{cases} \tag{173}$$

for any sequence $L \to \infty$. Here, $z_\infty \in [0, 1]$ is such that for subcritical density, the momentum density distribution of the gas is given by

$$\rho(k) = (2\pi)^{-3} \frac{z_\infty}{e^{\beta\omega} - z_\infty}, \tag{174}$$

so that z_∞ is the solution of

$$\rho_{\mathrm{tot}} = (2\pi)^{-3} \int \frac{z}{e^{\beta\omega} - z} d^3k. \tag{175}$$

The generating functional $E_{\rho_0,\rho}$ in (173) is the one obtained by Araki and Woods, (156), where ρ is the continuous momentum density distribution prescribed by Planck's law of black body radiation (169), and where

$$\rho_0 = \rho_{\mathrm{tot}} - \rho_{\mathrm{crit}}. \tag{176}$$

This gives the following picture for equilibrium states: if the system has density $\rho_{\mathrm{tot}} \leq \rho_{\mathrm{crit}}$ then the particle momentum distribution of the equilibrium state is purely continuous, meaning that below critical density there is no condensate. As ρ_{tot} increases and surpasses the critical value, $\rho_{\mathrm{tot}} > \rho_{\mathrm{crit}}$, the "excess" particles form a condensate which is immersed in a gas of particles radiating according to Planck's law.

Finally we mention the work [10] which treats the thermodynamic limit for the grand-canonical ensemble.

4.5 Dynamical stability of equilibria

Take the infinitely extended Bose gas initially in a state which differs from the equilibrium state at a given temperature only inside a bounded region of space. As time goes on we expect the local perturbation to spread out and propagate off to spatial infinity. This property, sometimes called the property of return to equilibrium, is a priori not built into the definition of equilibrium states, i.e., the KMS condition, but it has to be verified "by hand". In this section we investigate the large time limit of initial states which are local perturbations of an equilibrium state.

Let us first describe sates which are *local perturbations* of a given state ω of the infinitely extended Bose gas. Let $f \in \mathfrak{D} \subset L^2(\mathbb{R}^3, d^3x)$ be a test function which is supported in a compact region $\Lambda_0 \subset \mathbb{R}^3$. If g is supported in the complement $\mathbb{R}^3 \backslash \Lambda_0$ then we have

$$W(f)W(g) = e^{-i\mathrm{Im}\langle f,g\rangle} W(g)W(f) = W(g)W(f). \tag{177}$$

Consequently the state $A \mapsto \omega'(A) := \omega(W(f)^* A W(f))$ does not differ from the state ω on observables supported away from Λ_0 (i.e., on observables $A = \sum_{j=1}^n z_j W(f_j)$, where the f_j are supported away from Λ_0). The state ω' is a local perturbation of ω. More generally, if B is an observable (an element of the Weyl algebra) we say the state

$$\omega_B(\cdot) := \frac{\omega(B^* \cdot B)}{\omega(B^* B)} \tag{178}$$

is a local perturbation of ω. The set of all local perturbations of ω is defined to be the set of all convex combinations of states of the form (178). The dynamical stability of an equilibrium state ω_β (w.r.t. the dynamics α_t) is expressed as

$$\lim_{t\to\infty} \omega_B(\alpha_t(A)) = \omega_\beta(A), \tag{179}$$

for all observables A, B.

We start our investigation of return to equilibrium by some purely algebraic considerations. Let A be an element in the Weyl algebra $\mathrm{CCR}(\mathfrak{D})$. Given any ϵ there are complex numbers z_j and test functions $f_j \in \mathfrak{D}$ s.t.

$$\left\| A - \sum_{j=1}^n z_j W(f_j) \right\| = \left\| \alpha_t(A) - \sum_{j=1}^n z_j W(e^{i\omega t} f_j) \right\| < \epsilon, \tag{180}$$

where we use the fact that α_t is an isometry. Let g be fixed. We have

$$\left\| W(g)^* \left(\alpha_t(A) - \sum_{j=1}^n z_j W(e^{i\omega t} f_j) \right) W(g) \right\|$$
$$= \left\| \alpha_t(A) - \sum_{j=1}^n z_j W(e^{i\omega t} f_j) \right\| < \epsilon,$$

and the l.h.s. equals

$$\left\| W(g)^*\alpha_t(A)W(g) - \sum_{j=1}^{n} z_j e^{-\frac{i}{2}\mathrm{Im}[\langle -g, e^{i\omega t} f_j\rangle + \langle e^{i\omega t} f_j, g\rangle]} W(e^{i\omega t} f_j) \right\| < \epsilon. \tag{181}$$

Since $\lim_{t\to\infty} \langle -g, e^{i\omega t} f_j\rangle + \langle e^{i\omega t} f_j, g\rangle = 0$ (this follows from the Riemann-Lebesgue Lemma) there is a number $T_0(\epsilon) < \infty$ s.t. if $t > T_0$ then

$$\left\| \sum_{j=1}^{n} z_j e^{-\frac{i}{2}\mathrm{Im}[\langle -g, e^{i\omega t} f_j\rangle + \langle e^{i\omega t} f_j, g\rangle]} W(e^{i\omega t} f_j) - \sum_{j=1}^{n} z_j \alpha_t(W(f_j)) \right\| < \epsilon. \tag{182}$$

It follows from (180), (181) and (182) that $\|W(g)^*\alpha_t(A)W(g) - \alpha_t(A)\| < 2\epsilon$, for $t > T_0(\epsilon)$, and consequently

$$\lim_{t\to\infty} \|W(g)^*\alpha_t(A)W(g) - \alpha_t(A)\| = 0, \tag{183}$$

for all observables A and all test functions g. Relation (183), which merely involves observables and the dynamics (and no reference to any state is made), is a form of *asymptotic abelianness* w.r.t. α_t. In fact, it follows easily from (183) and (180) that

$$\lim_{t\to\infty} \|B\alpha_t(A) - \alpha_t(A)B\| = 0, \tag{184}$$

for any observables $A, B \in \mathrm{CCR}(\mathfrak{D})$.

If ω is any α_t–invariant state then (183) shows that

$$\lim_{t\to\infty} \omega(W(g)^*\alpha_t(A)W(g)) = \omega(A). \tag{185}$$

To prove that (185) holds if ω is an equilibrium state, and for $W(g)$ replaced by *any* observable B, i.e. to show return to equilibrium as in (179), we need to use special properties of equilibrium states. The property of asymptotic abelianness, (184), does not suffice.

Let ω_β denote an equilibrium state of the free Bose gas with a continuous density (169) and with a fixed condensate density $\rho_0 \geq 0$. The expectation functional is given by (156). We have

$$\omega_\beta(W(g)W(e^{i\omega t} f)W(h)) = e^{-\frac{i}{2}\mathrm{Im}[\langle g, e^{i\omega t} f\rangle + \langle g + e^{i\omega t} f, h\rangle]}\omega_\beta(W(g + e^{i\omega t} f + h))$$

and (using again the Riemann-Lebesgue Lemma)

$$\begin{aligned}
&\lim_{t\to\infty} \omega_\beta(W(g)W(e^{i\omega t} f)W(h)) \\
&= e^{-\frac{i}{2}\mathrm{Im}\langle g,h\rangle} E_{\mathrm{F}}(g+h) \exp\left\{-\frac{(2\pi)^3}{2}\|\sqrt{\rho}(g+h)\|^2\right\} \\
&\quad \times E_{\mathrm{F}}(f) \exp\left\{-\frac{(2\pi)^3}{2}\|\sqrt{\rho} f\|^2\right\} \\
&\quad \times J_0\left((2\pi)^{3/2}\sqrt{2\rho_0}\,|g(0) + f(0) + h(0)|\right).
\end{aligned} \tag{186}$$

In the absence of a condensate, $\rho_0 = 0$, $J_0(0) = 1$, equation (186) is just

$$\lim_{t\to\infty} \omega_\beta(W(g)W(e^{i\omega t}f)W(h)) = \omega_\beta(W(g)W(h))\,\omega_\beta(W(f)). \tag{187}$$

Using an easy approximation argument (as in (180)), this yields the property of return to equilibrium ($\rho_0 = 0$),

$$\lim_{t\to\infty} \omega_\beta(B\alpha_t(A)C) = \omega_\beta(BC)\omega_\beta(A), \tag{188}$$

for any $A, B, C \in \mathrm{CCR}(\mathfrak{D})$.

What happens in presence of a condensate, $\rho_0 > 0$? Formula (188) is *not* valid in this case, because the Bessel function in (186) does not split into a product. However, the integrand in the representation (134) of J_0 does split into a product according to

$$\begin{aligned} &J_0\left((2\pi)^{3/2}\sqrt{2\rho_0}\,|g(0)+f(0)+h(0)|\right) \\ &= \int_{-\pi}^{\pi} \frac{d\theta}{2\pi} \exp -i(2\pi)^{3/2}\sqrt{2\rho_0}[\mathrm{Re}(g(0)+h(0))\cos\theta + \mathrm{Im}(g(0)+h(0))\sin\theta] \\ &\quad \times \exp -i(2\pi)^{3/2}\sqrt{2\rho_0}[\mathrm{Re}(f(0))\cos\theta + \mathrm{Im}(f(0))\sin\theta]. \end{aligned} \tag{189}$$

This suggests that for an equilibrium state with a condensate and a fixed value of θ, the property of return to equilibrium holds. To cast this into a precise form we write

$$\omega_\beta(W(f)) = \int_{-\pi}^{\pi} \frac{d\theta}{2\pi}\, \omega_\beta^\theta(W(f)), \tag{190}$$

where

$$\omega_\beta^\theta(W(f)) = e^{-i\Phi(f,\theta)} \langle \Omega_\rho, \pi_\rho(W(f))\Omega_\rho\rangle, \tag{191}$$

with Ω_ρ given in (150) and π_ρ defined in (151), (152). This decomposition is in accordance with the decomposition of the Hilbert space into a direct integral,

$$\mathcal{H}_{\rho_0,\rho} = \int_{S^1}^{\oplus} d\sigma\, \mathcal{H}_\rho, \tag{192}$$

see (149), (158). The GNS representation of $(\mathrm{CCR}(\mathfrak{D}), \omega_\beta^\theta)$ is given by

$$(\mathcal{H}_\rho, \pi_\beta^\theta, \Omega_\rho),$$

where

$$\pi_\beta^\theta(W(f)) = e^{-i\Phi(f,\theta)}\pi_\rho(W(f)). \tag{193}$$

The representation map π_β associated to the state ω_β is the direct integral of the fibers π_β^θ, and the representation vector Ω_β of ω_β is the direct integral with constant fiber Ω_ρ:

$$\pi_\beta = \int_{[-\pi,\pi]}^{\oplus} \frac{d\theta}{2\pi}\, \pi_\beta^\theta,\quad \Omega_\beta = \int_{[-\pi,\pi]}^{\oplus} \frac{d\theta}{2\pi}\, \Omega_\rho. \tag{194}$$

It is clear from (191) that for each θ fixed, ω_β^θ is a (α_t, β)-KMS state. The (α_t, β)-KMS state ω_β is a uniform superposition of the ω_β^θ, $\theta \in S^1$. We can form other equilibrium states by taking different superpositions of the ω_β^θ: Given any probability measure $d\mu$ on $[-\pi, \pi]$,

$$\omega_\mu(W(f)) := \int_{-\pi}^{\pi} d\mu(\theta)\, \omega_\beta^\theta(W(f)) \tag{195}$$

is an (α_t, β)-KMS state. As follows directly from (188) and (191), for each fixed θ we have $\lim_{t\to\infty} \omega_\beta^\theta(B\alpha_t(A)C) = \omega_\beta^\theta(BC)\omega_\beta^\theta(A)$, so

$$\lim_{t\to\infty} \omega_\mu(B\alpha_t(A)C) = \int_{-\pi}^{\pi} d\mu(\theta)\, \omega_\beta^\theta(BC)\, \omega_\beta^\theta(A). \tag{196}$$

In general, the r.h.s. of (196) does *not* equal ω_μ, so the time-asymptotic state depends on the initial state. If the perturbation of the state ω_μ is given by B, C s.t. $\omega_\beta^\theta(BC) = 1$ for all θ then we have return to ω_μ in the usual sense.

What is special about the equilibrium states ω_β^θ? They are *factor*[19] equilibrium states. The fact that each ω_β^θ is a factor state follows from $\mathfrak{M} := \pi_{\omega_\beta^\theta}(\mathcal{A})'' = \mathcal{B}(\mathcal{F}(L^2(\mathbb{R}^3, d^3k))) \otimes \mathbb{1}$, $\mathfrak{M}' = \mathbb{1} \otimes \mathcal{B}(\mathcal{F}(L^2(\mathbb{R}^3, d^3k)))$, hence $\mathfrak{M} \cap \mathfrak{M}' = \mathbb{C}\mathbb{1} \otimes \mathbb{1}$. On the other hand, it is clear that ω_β, (190), is not a factor state since $\mathbb{1} \otimes \mathbb{1} \otimes \mathcal{M}$ (denoting the multiplication operators on $L^2(S^1, d\sigma)$) belongs to the center of its von Neumann algebra, see (159). The decomposition (192)–(194) is called a *factor decomposition* of the state ω_β, or a decomposition into extremal states.

Let us now see how the emergence of a multitude of equilibrium states for a fixed inverse temperature β can be viewed as a consequence of *spontaneous symmetry breaking* – here, the gauge group symmetry is broken. The general scheme is this: suppose a dynamics α_t on a C^*-algebra $\mathfrak{A}$ has a symmetry group σ_s, i.e. σ_s is a group of automorphisms of $\mathfrak{A}$ satisfying $\sigma_s \circ \alpha_t = \alpha_t \circ \sigma_s$, for all s, t (s may belong to a discrete or continuous set, $t \in \mathbb{R}$). If ω is any (β, α_t)-KMS state then $\omega_s := \omega \circ \sigma_s$ is a (β, α_t)-KMS state as well:

$$\begin{aligned}\omega_s(A\alpha_t(B)) &= \omega(\sigma_s(A)\alpha_t(\sigma_s(B)))\\ &= \omega(\alpha_{t-i\beta}(\sigma_s(B))\sigma_s(A)) = \omega_s(\alpha_{t-i\beta}(B)A).\end{aligned} \tag{197}$$

(We implicitly assume that $\alpha_{t-i\beta}(\sigma_s(B))$ is well defined.) If there is a (β, α_t)-KMS state which is not invariant under σ_s for some s, i.e., $\omega \circ \sigma_s \neq \omega$, then we say the symmetry σ_s is spontaneously broken, because there are equilibrium states which

[19] A state ω on a C^*–algebra $\mathfrak{A}$ is a factor state iff the von Neumann algebra $\pi_\omega(\mathfrak{A})''$ is a factor. (Here, π_ω is the GNS representation map associated to $(\mathfrak{A}, \omega)$.) A von Neumann algebra $\mathfrak{M} \subset \mathcal{B}(\mathcal{H})$ is a factor iff its center is trivial, $\mathfrak{Z} := \mathfrak{M} \cap \mathfrak{M}' = \mathbb{C}\mathbb{1}$.

We point out that it follows from general considerations that an equilibrium state is a factor state iff it is *extremal* (see [5], Theorem 5.3.30). A state ω is called extremal iff the relation $\{\omega = \lambda\omega_1 + (1-\lambda)\omega_2$, for some $0 < \lambda < 1$ and some states $\omega_1, \omega_2\}$ implies that $\omega_1 = \omega_2 = \omega$.

"have less symmetry" than the dynamics. This gives rise to the existence of several equilibrium states at the same temperature.

Consider the continuous symmetry group σ_s on CCR($\mathfrak{D}$) given by

$$\sigma_s(W(f)) = W(e^{is}f), \;\; s \in \mathbb{R}, f \in \mathfrak{D}, \tag{198}$$

called the gauge group. Clearly we have $\alpha_t \circ \sigma_s = \sigma_s \circ \alpha_t$ (where α_t is given in (163)). Using (191) we obtain

$$\omega_\beta^\theta(\sigma_s(W(f))) = e^{-i\Phi(e^{is}f,\theta)} \left\langle \Omega_\rho, \pi_\rho(W(e^{is}f))\Omega_\rho \right\rangle, \tag{199}$$

and (147), (153) show that $\left\langle \Omega_\rho, \pi_\rho(W(e^{is}f))\Omega_\rho \right\rangle = \langle \Omega_\rho, \pi_\rho(W(f))\Omega_\rho \rangle$, while (160) gives

$$\begin{aligned} \Phi(e^{is}f,\theta) &= (2\pi)^{3/2}\sqrt{2\rho_0}\,\left(\mathrm{Re}(e^{is}f(0))\cos\theta + \mathrm{Im}(e^{is}f(0))\sin\theta\right) \\ &= (2\pi)^{3/2}\sqrt{2\rho_0}\,\left(\mathrm{Re}(f(0))\cos(\theta - s) + \mathrm{Im}(f(0))\sin(\theta - s)\right) \\ &= \Phi(f, \theta - s). \end{aligned} \tag{200}$$

This shows that the equilibrium states ω_β^θ break the gauge group symmetry, hence giving rise to an S^1-multitude of equilibrium states ((200) shows that we get the whole family ω_β^θ by varying s in any interval of length 2π).

Let us finally examine the *mixing properties* of the equilibrium states with respect to space translations. Given a vector $a \in \mathbb{R}^3$ we define

$$\tau_a(W(f)) := W(f_a), \tag{201}$$

where $f_a(x) := f(x-a)$ is the translate of f by a. τ_a defines a (three parameter) group of automorphisms on CCR($\mathfrak{D}$). We say that a state ω on CCR($\mathfrak{D}$) has the property of *strong mixing* w.r.t. space translations if

$$\lim_{|a|\to\infty} \omega(W(f)\tau_a(W(g))) = \omega(W(f))\omega(W(g)), \tag{202}$$

for any $f, g \in \mathfrak{D}$. This means that if two observables ($W(f)$ and $W(g)$) are spatially separated far from each other then the expectation of the product of the observables is close to the product of the expectations (independence of random variables). Intuitively, this means that the state ω has a certain property of locality in space: what happens far out in space does not influence events taking place, say, around the origin. Condition (202) is also called a *cluster property*. It is easy to calculate explicitly the l.h.s. of (202) for $\omega = \omega_\beta$, the equilibrium state of the free Bose gas with a continuous density (169) and with a fixed condensate density $\rho_0 \geq 0$ (whose expectation functional is given by (156)):

$$\begin{aligned} &\lim_{|a|\to\infty} \omega_\beta(W(f)\tau_a(W(g))) \\ &\quad = \omega_\beta(W(f))\omega_\beta(W(g)) \exp\left[-8\pi^3\rho_0\,\mathrm{Re}(\overline{f(0)}g(0))\right]. \end{aligned} \tag{203}$$

Consequently, the equilibrium state is strongly mixing w.r.t. space translations if and only if $\rho_0 = 0$, i.e., if and only if there is no condensation. In presence of a condensate, the system exhibits *long range correlations* (what happens far out does influence what happens say at the origin). On the other hand, it is easily verified that each state ω_β^θ is strongly mixing. We may understand that limit states ($\lim_{t\to\infty}$ of states of the form (195)) depend on the initial state as a consequence of the long-range correlations in presence of a condensate. Even as time reaches its asymptotic value the system "remembers" the initial state.

References

1. Araki, H.: Hamiltonian Formalism and the Canonical Commutation Relations in Quantum Field Theory. J. Math. Phys., **1**, 492-504 (1960)
2. Abramowitz, A., Stegun, I. A.: *Handbook of mathematical functions.* Dover Publications, Inc., New York, Ninth Printing, 1970
3. Araki, H., Woods, E.: Representations of the canonical commutation relations describing a non-relativistic infinite free Bose gas. J. Math. Phys. **4**, 637-662 (1963)
4. Bratteli, O., Robinson, D.W.: *Operator Algebras and Quantum Statistical Mechanics I.* Springer, Berlin, second edition, 1987
5. Bratteli, O., Robinson, D.: *Operator Algebras and Quantum Statistical Mechanics II.* Springer, Berlin, second edition, 1997
6. Cannon, J.T.: Infinite volume limits of the canonical free Bose gas states on the Weyl algebra. Commun. Math. Phys., **29**, 89-104 (1973)
7. Glimm, J., Jaffe, A.: *Quantum Physics. A functional integral point of view.* Springer, New York, second edition, 1981
8. Hida, T.: *Brownian Motion.* Springer-Verlag, Berlin, 1980
9. Jacobs, K.: *Measure and Integral.* Academic Press, New York, 1978
10. Lewis, J.T., Pulè, J.V.: The equilibrium states of the free Boson gas. Commun. Math. Phys. **36**, 1-18 (1974)
11. Morse, P. M., Feshbach, H.: *Methods of Theoretical Physics, Part I.* McGraw-Hill Book Company, Inc., New York, Toronto, London, 1953
12. Petz, D.: *An invitation to the Algebra of Canonical Commutation Relations.* Leuven Notes in Mathematical and Theoretical Physics, Volume 2, Series A: Mathematical Physics, Leuven University Press, 1990
13. Reed, M., Simon, B.: *Methods of Modern Mathematical Physics, II. Fourier Analysis, Self-Adjointness.* Academic Press, London, 1975.
14. Takesaki, M. :*Theory of Operator Algebra I.* Springer, Berlin, second edition, 2002

Topics in Spectral Theory

Vojkan Jakšić

Department of Mathematics and Statistics, McGill University,
805 Sherbrooke Street West, Montreal, QC, H3A 2K6, Canada
e-mail: jaksic@math.mcgill.ca

1 Introduction

These lecture notes are an expanded version of the lectures I gave in the Summer School on Open Quantum Systems, held in Grenoble June 16—July 4, 2003. Shortly afterwards, I also lectured in the Summer School on Large Coulomb Systems—QED, held in Nordfjordeid August 11—18, 2003 [13]. The Nordfjordeid lectures were a natural continuation of the material covered in Grenoble, and [13] can be read as Section 6 of these lecture notes.

The subject of these lecture notes is spectral theory of self-adjoint operators and some of its applications to mathematical physics. This topic has been covered in many places in the literature, and in particular in [4, 23, 24, 25, 26, 17]. Given the clarity and precision of these references, there appears to be little need for additional lecture notes on the subject. On the other hand, the point of view adopted in these lecture notes, which has its roots in the developments in mathematical physics which primarily happened over the last decade, makes the notes different from most expositions and I hope that the reader will learn something new from them.

The main theme of the lecture notes is the interplay between spectral theory of self-adjoint operators and classical harmonic analysis. In a nutshell, this interplay can be described as follows. Consider a self-adjoint operator A on a Hilbert space $\mathcal{H}$ and a vector $\varphi \in \mathcal{H}$. The function

$$F(z) = (\varphi|(A - z)^{-1}\varphi)$$

is analytic in the upper half-plane $\mathrm{Im}\, z > 0$ and satisfies the bound

$$|F(z)| \leq \|\varphi\|^2 / \mathrm{Im}\, z.$$

By a well-known result in harmonic analysis (see Theorem 3.11) there exists a positive Borel measure μ_φ on $\mathbb{R}$ such that for $\mathrm{Im}\, z > 0$,

$$F(z) = \int_{\mathbb{R}} \frac{\mathrm{d}\mu_\varphi(x)}{x - z}.$$

The measure μ_φ is *the spectral measure* for A and φ. Starting with this definition we will develop the spectral theory of A. In particular, we will see that many properties of the spectral measure can be characterized by the boundary values $\lim_{y \downarrow 0} F(x+\mathrm{i}y)$ of the corresponding function F. The resulting theory is mathematically beautiful and has found many important applications in mathematical physics. In Section 5 we will discuss a simple and important application to the spectral theory of rank one perturbations. A related application concerns the spectral theory of the Wigner-Weisskopf atom and is discussed in the lecture notes [13].

Although we are mainly interested in applications of harmonic analysis to spectral theory, it is sometimes possible to turn things around and use the spectral theory to prove results in harmonic analysis. To illustrate this point, in Section 5 we will prove Boole's equality and the celebrated Poltoratskii theorem using spectral theory of rank one perturbations.

The lecture notes are organized as follows. In Section 2 we will review the results of the measure theory we will need. The proofs of less standard results are given in detail. In particular, we present detailed discussion of the differentiation of measures based on the Besicovitch covering lemma. The results of harmonic analysis we will need are discussed in Section 3. They primarily concern Poisson and Borel transforms of measures and most of them can be found in the classical references [18, 19]. However, these references are not concerned with applications of harmonic analysis to spectral theory, and the reader would often need to go through a substantial body of material to extract the needed results. To aid the reader, we have provided proofs of all results discussed in Section 3. Spectral theory of self-adjoint operators is discussed in Section 4. Although this section is essentially self-contained, many proofs are omitted and the reader with no previous exposition to spectral theory would benefit by reading it in parallel with Chapters VII-VIII of [23] and Chapters I-II of [4]. Spectral theory of rank one perturbations is discussed in Section 5.

Concerning the prerequisites, it is assumed that the reader is familiar with basic notions of real, functional and complex analysis. Familiarity with [23] or the first ten Chapters of [27] should suffice.

Acknowledgment. I wish to thank Stéphane Attal, Alain Joye and Claude-Alain Pillet for the invitation to lecture in the Grenoble summer school. I am also grateful to Jonathan Breuer, Eugene Kritchevski, and in particular Philippe Poulin for many useful comments about the lecture notes. The material in the lecture notes is based upon work supported by NSERC.

2 Preliminaries: measure theory

2.1 Basic notions

Let M be a set and $\mathcal{F}$ a σ-algebra in M. The pair $(M, \mathcal{F})$ is called a measure space. We denote by χ_A the characteristic function of a subset $A \subset M$.

Let μ be a measure on $(M, \mathcal{F})$. We say that μ is concentrated on the set $A \in \mathcal{F}$ if $\mu(E) = \mu(E \cap A)$ for all $E \in \mathcal{F}$. If $\mu(M) = 1$, then μ is called a probability measure.

Assume that M is a metric space. The minimal σ-algebra in M that contains all open sets is called the Borel σ-algebra. A measure on the Borel σ-algebra is called a Borel measure. If μ is Borel, the complement of the largest open set $\mathcal{O}$ such that $\mu(\mathcal{O}) = 0$ is called the support of μ and is denoted by $\operatorname{supp}\mu$.

Assume that M is a locally compact metric space. A Borel measure μ is called *regular* if for every $E \in \mathcal{F}$,

$$\mu(E) = \inf\{\mu(V) : E \subset V, V \text{ open}\}$$
$$= \sup\{\mu(K) : K \subset E, K \text{ compact}\}.$$

Theorem 2.1. *Let M be a locally compact metric space in which every open set is σ-compact (that is, a countably union of compact sets). Let μ be a Borel measure finite on compact sets. Then μ is regular.*

The measure space we will most often encounter is $\mathbb{R}$ with the usual Borel σ-algebra. Throughout the lecture notes we will denote by $\mathbb{1}$ the constant function $\mathbb{1}(x) = 1 \; \forall x \in \mathbb{R}$.

2.2 Complex measures

Let $(M, \mathcal{F})$ be a measure space. Let $E \in \mathcal{F}$. A countable collection of sets $\{E_i\}$ in $\mathcal{F}$ is called a partition of E if $E_i \cap E_j = \emptyset$ for $i \neq j$ and $E = \cup_j E_j$. A *complex measure* on $(M, \mathcal{F})$ is a function $\mu : \mathcal{F} \to \mathbb{C}$ such that

$$\mu(E) = \sum_{i=1}^{\infty} \mu(E_i), \tag{1}$$

for every $E \in \mathcal{F}$ and *every* partition $\{E_i\}$ of E. In particular, the series (1) is absolutely convergent.

Note that complex measures take only finite values. The usual positive measures, however, are allowed to take the value ∞. In the sequel, the term *positive measure* will refer to the standard definition of a measure on a σ-algebra which takes values in $[0, \infty]$.

The set function $|\mu|$ on $\mathcal{F}$ defined by

$$|\mu|(E) = \sup \sum_i |\mu(E_i)|,$$

where the supremum is taken over all partitions $\{E_i\}$ of E, is called the total variation of the measure μ.

Theorem 2.2. *Let μ be a complex measure. Then:*
(1) *The total variation $|\mu|$ is a positive measure on $(M, \mathcal{F})$.*
(2) $|\mu|(M) < \infty$.
(3) *There exists a measurable function $h : M \to \mathbb{C}$ such that $|h(x)| = 1$ for all $x \in M$ and*

$$\mu(E) = \int_E h(x)\mathrm{d}|\mu|(x),$$

for all $E \in \mathcal{F}$. The last relation is abbreviated $\mathrm{d}\mu = h\mathrm{d}|\mu|$.

A complex measure μ is called regular if $|\mu|$ is a regular measure. Note that if ν is a positive measure, $f \in L^1(M, \mathrm{d}\nu)$ and

$$\mu(E) = \int_E f\mathrm{d}\nu,$$

then

$$|\mu|(E) = \int_E |f|\mathrm{d}\nu.$$

The integral with respect to a complex measure is defined in the obvious way, $\int f\mathrm{d}\mu = \int fh\mathrm{d}|\mu|$.

Notation. Let μ be a complex or positive measure and $f \in L^1(M, \mathrm{d}|\mu|)$. In the sequel we will often denote by $f\mu$ the complex measure

$$(f\mu)(E) = \int_E f\mathrm{d}\mu.$$

Note that $|f\mu| = |f||\mu|$.

Every complex measure can be written as a linear combination of four finite positive measures. Let $h_1(x) = \mathrm{Re}\, h(x)$, $h_2(x) = \mathrm{Im}\, h(x)$, $h_i^+(x) = \max(h_i(x), 0)$, $h_i^-(x) = -\min(h_i(x), 0)$, and $\mu_i^\pm = h_i^\pm|\mu|$, $i = 1, 2$. Then

$$\mu = (\mu_1^+ - \mu_1^-) + \mathrm{i}(\mu_2^+ - \mu_2^-).$$

A complex measure μ which takes values in $\mathbb{R}$ is called a *signed* measure. Such a measure can be decomposed as

$$\mu = \mu^+ - \mu^-,$$

where $\mu^+ = (|\mu| + \mu)/2$, $\mu^- = (|\mu| - \mu)/2$. If $A = \{x \in M : h(x) = 1\}$, $B = \{x \in M : h(x) = -1\}$, then for $E \in \mathcal{F}$,

$$\mu^+(E) = \mu(E \cap A), \qquad \mu^-(E) = -\mu(E \cap B).$$

This fact is known as the Hahn decomposition theorem.

2.3 Riesz representation theorem

In this subsection we assume that M is a locally compact metric space.

A continuous function $f : M \to \mathbb{C}$ *vanishes at infinity* if $\forall \epsilon > 0$ there exists a compact set K_ϵ such that $|f(x)| < \epsilon$ for $x \notin K_\epsilon$. Let $C_0(M)$ be the vector space of all continuous functions that vanish at infinity, endowed with the supremum norm $\|f\| = \sup_{x\in M} |f(x)|$. $C_0(M)$ is a Banach space and we denote by $C_0(M)^*$ its dual. The following result is known as the Riesz representation theorem.

Theorem 2.3. *Let $\phi \in C_0(M)^*$. Then there exists a unique regular complex Borel measure μ such that*

$$\phi(f) = \int_M f \mathrm{d}\mu,$$

for all $f \in C_0(M)$. Moreover, $\|\phi\| = |\mu|(M)$.

2.4 Lebesgue-Radon-Nikodym theorem

Let $(M, \mathcal{F})$ be a measure space. Let ν_1 and ν_2 be complex measures concentrated on disjoint sets. Then we say that ν_1 and ν_2 are mutually singular (or orthogonal), and write $\nu_1 \perp \nu_2$. If $\nu_1 \perp \nu_2$, then $|\nu_1| \perp |\nu_2|$.

Let ν be a complex measure and μ a positive measure. We say that ν is absolutely continuous w.r.t. μ, and write $\nu \ll \mu$, if $\mu(E) = 0 \Rightarrow \nu(E) = 0$. The following result is known as the Lebesgue-Radon-Nikodym theorem.

Theorem 2.4. *Let ν be a complex measure and μ a positive σ-finite measure on $(M, \mathcal{F})$. Then there exists a unique pair of complex measures ν_{a} and ν_{s} such that $\nu_{\mathrm{a}} \perp \nu_{\mathrm{s}}$, $\nu_{\mathrm{a}} \ll \mu$, $\nu_{\mathrm{s}} \perp \mu$, and*

$$\nu = \nu_{\mathrm{a}} + \nu_{\mathrm{s}}.$$

Moreover, there exists a unique $f \in L^1(\mathbb{R}, \mathrm{d}\mu)$ such that $\forall E \in \mathcal{F}$,

$$\nu_{\mathrm{a}}(E) = \int_E f \mathrm{d}\mu.$$

The Radon-Nikodym decomposition is abbreviated as $\nu = f\mu + \nu_s$ (or $d\nu = f d\mu + d\nu_s$).

If $M = \mathbb{R}$ and μ is the Lebesgue measure, we will use special symbols for the Radon-Nikodym decomposition. We will denote by ν_{ac} the part of ν which is absolutely continuous (abbreviated ac) w.r.t. the Lebesgue measure and by ν_{sing} the part which is singular with respect to the Lebesgue measure. A point $x \in \mathbb{R}$ is called an atom of ν if $\nu(\{x\}) \neq 0$. Let A_ν be the set of all atoms of ν. The set A_ν is countable and $\sum_{x \in A_\nu} |\nu(\{x\})| < \infty$. The pure point part of ν is defined by

$$\nu_{pp}(E) = \sum_{x \in E \cap A_\nu} \nu(\{x\}).$$

The measure $\nu_{sc} = \nu_{sing} - \nu_{pp}$ is called the singular continuous part of ν.

2.5 Fourier transform of measures

Let μ be a complex Borel measure on $\mathbb{R}$. Its Fourier transform is defined by

$$\hat{\mu}(t) = \int_{\mathbb{R}} e^{-itx} d\mu(x).$$

$\hat{\mu}(t)$ is also called the characteristic function of the measure μ. Note that

$$|\hat{\mu}(t+h) - \hat{\mu}(t)| \leq \int_{\mathbb{R}} \left| e^{-ihx} - 1 \right| d|\mu|,$$

and so the function $\mathbb{R} \ni t \mapsto \hat{\mu}(t) \in \mathbb{C}$ is uniformly continuous.

The following result is known as the Riemann-Lebesgue lemma.

Theorem 2.5. *Assume that μ is absolutely continuous w.r.t. the Lebesgue measure. Then*

$$\lim_{|t| \to \infty} |\hat{\mu}(t)| = 0. \tag{2}$$

The relation (2) may hold even if μ is singular w.r.t. the Lebesgue measure. The measures for which (2) holds are called Rajchman measures. A geometric characterization of such measures can be found in [20].

Recall that A_μ denotes the set of atoms of μ. In this subsection we will prove the Wiener theorem:

Theorem 2.6. *Let μ be a signed Borel measure. Then*

$$\lim_{T \to \infty} \frac{1}{T} \int_0^T |\hat{\mu}(t)|^2 dt = \sum_{x \in A_\mu} \mu(\{x\})^2.$$

Proof: Note first that

$$|\hat{\mu}(t)|^2 = \hat{\mu}(t)\overline{\hat{\mu}(t)} = \int_{\mathbb{R}^2} \mathrm{e}^{-\mathrm{i}t(x-y)}\mathrm{d}\mu(x)\mathrm{d}\mu(y).$$

Let

$$K_T(x,y) = \frac{1}{T}\int_0^T \mathrm{e}^{-\mathrm{i}t(x-y)}\mathrm{d}t = \begin{cases} (1-\mathrm{e}^{-\mathrm{i}T(x-y)})/(\mathrm{i}T(x-y)) & \text{if } x \neq y, \\ 1 & \text{if } x = y. \end{cases}$$

Then

$$\frac{1}{T}\int_0^T |\hat{\mu}(t)|^2\mathrm{d}t = \int_{\mathbb{R}^2} K_T(x,y)\mathrm{d}\mu(x)\mathrm{d}\mu(y).$$

Obviously,

$$\lim_{T\to\infty} K_T(x,y) = \begin{cases} 0 & \text{if } x \neq y, \\ 1 & \text{if } x = y. \end{cases}$$

Since $|K_T(x,y)| \leq 1$, by the dominated convergence theorem we have that for all x,

$$\lim_{T\to\infty} \int_{\mathbb{R}} K_T(x,y)\mathrm{d}\mu(y) = \mu(\{x\}).$$

By Fubini's theorem,

$$\int_{\mathbb{R}^2} K_T(x,y)\mathrm{d}\mu(x)\mathrm{d}\mu(y) = \int_{\mathbb{R}}\left[\int_{\mathbb{R}} K_T(x,y)\mathrm{d}\mu(y)\right]\mathrm{d}\mu(x),$$

and by the dominated convergence theorem,

$$\lim_{T\to\infty}\frac{1}{T}\int_0^T |\hat{\mu}(t)|^2\mathrm{d}t = \int_{\mathbb{R}} \mu(\{x\})\mathrm{d}\mu(x) = \sum_{x\in A_\nu} \mu(\{x\})^2.$$

□

2.6 Differentiation of measures

We will discuss only the differentiation of Borel measures on $\mathbb{R}$. The differentiation of Borel measures on $\mathbb{R}^n$ is discussed in the problem set.

We start by collecting some preliminary results. The first result we will need is the Besicovitch covering lemma.

Theorem 2.7. *Let A be a bounded set in $\mathbb{R}$ and, for each $x \in A$, let I_x be an open interval with center at x.*

(1) *There is a countable subcollection* $\{I_j\}$ *of* $\{I_x\}_{x\in A}$ *such that* $A \subset \cup I_j$ *and that each point in* $\mathbb{R}$ *belongs to at most two intervals in* $\{I_j\}$, *i.e.* $\forall y \in \mathbb{R}$,

$$\sum_j \chi_{I_j}(y) \leq 2.$$

(2) *There is a countable subcollection* $\{I_{i,j}\}$, $i = 1, 2$, *of* $\{I_x\}_{x\in A}$ *such that* $A \subset \cup I_{i,j}$ *and* $I_{i,j} \cap I_{i,k} = \emptyset$ *if* $j \neq k$.

In the sequel we will refer to $\{I_i\}$ and $\{I_{i,j}\}$ as the *Besicovitch subcollections.*

Proof. $|I|$ denotes the length of the interval I. We will use the shorthand

$$I(x,r) = (x - r, x + r).$$

Setting $r_x = |I_x|/2$, we have $I_x = I(x, r_x)$.

Let $d_1 = \sup\{r_x : x \in A\}$. Choose $I_1 = I(x_1, r_1)$ from the family $\{I_x\}_{x\in A}$ such that $r_1 > 3d_1/4$. Assume that $I_1, \ldots, I_{j-1}$ are chosen for $j \geq 1$ and that $A_j = A \setminus \cup_{i=1}^{j-1} I_i$ is non-empty. Let $d_j = \sup\{r_x : x \in A_j\}$. Then choose $I_j = I(x_j, r_j)$ from the family $\{I_x\}_{x\in A_j}$ such that $r_j > 3d_j/4$. In this way be obtain a countable (possibly finite) subcollection $I_j = I(x_j, r_j)$ of $\{I_x\}_{x\in A}$.

Suppose that $j > i$. Then $x_j \in A_i$ and

$$r_i \geq \frac{3}{4}\sup\{r_x : x \in A_i\} \geq \frac{3r_j}{4}. \tag{3}$$

This observation yields that the intervals $I(x_j, r_j/3)$ are disjoint. Indeed, if $j > i$, then $x_j \notin I(x_i, r_i)$, and (3) yields

$$|x_i - x_j| > r_i = \frac{r_i}{3} + \frac{2r_i}{3} > \frac{r_i}{3} + \frac{r_j}{3}.$$

Since A is a bounded set and $x_j \in A$, the disjointness of $I_j = I(x_j, r_j/3)$ implies that if the family $\{I_j\}$ is infinite, then

$$\lim_{j\to\infty} r_j = 0. \tag{4}$$

The relation (4) yields that $A \subset \cup_j I(x_j, r_j)$. Indeed, this is obvious if there are only finitely many I_j's. Assume that there are infinitely many I_j's and let $x \in A$. By (4), there is j such that $r_j < 3r_x/4$, and by the definition of r_j, $x \in \cup_{i=1}^{j-1} I_i$.

Notice that if three intervals in $\mathbb{R}$ have a common point, then one of the intervals is contained in the union of the other two. Hence, by dropping superfluous intervals from the collection $\{I_j\}$, we derive that $A \subset \cup_j I_j$ and that each point in $\mathbb{R}$ belongs to no more than two intervals I_j. This proves (1).

To prove (2), we enumerate I_j's as follows. To I_1 is associated the number 0. The intervals to right of I_1 are enumerated in succession by positive integers, and the intervals to the left by negative integers. (The "succession" is well-defined, since no point belongs simultaneously to three intervals). The intervals associated to even integers are mutually disjoint, and so are the intervals associated to odd integers. Finally, denote the interval associated to $2n$ by $I_{1,n}$, and the interval associated to $2n+1$ by $I_{2,n}$. This construction yields (2). □

Let μ be a positive Borel measure on $\mathbb{R}$ finite on compact sets and let ν be a complex measure. The corresponding maximal function is defined by

$$M_{\nu,\mu}(x) = \sup_{r>0} \frac{|\nu|(I(x,r))}{\mu(I(x,r))}, \qquad x \in \operatorname{supp}\mu.$$

If $x \notin \operatorname{supp}\mu$ we set $M_{\nu,\mu}(x) = \infty$. It is not hard (Problem 1) to show that the function $\mathbb{R} \ni x \mapsto M_{\nu,\mu}(x) \in [0,\infty]$ is Borel measurable.

Theorem 2.8. *For any* $t > 0$,

$$\mu\{x : M_{\nu,\mu}(x) > t\} \le \frac{2}{t}|\nu|(\mathbb{R}).$$

Proof. Let $[a,b]$ be a bounded interval and

$$E_t = [a,b] \cap \{x : M_{\nu,\mu}(x) > t\}.$$

Every point x in E_t is the center of an open interval I_x such that

$$|\nu|(I_x) \ge t\mu(I_x).$$

Let $I_{i,j}$ be the Besicovitch subcollection of $\{I_x\}$. Then,

$$E_t \subset \cup I_{i,j},$$

and

$$\mu(E_t) \le \sum_{i,j} \mu(I_{i,j}) \le \frac{1}{t}\sum_{i,j} |\nu|(I_{i,j})$$

$$= \frac{1}{t}\sum_{i=1}^{2} |\nu|(\cup_j I_{i,j}) \le \frac{2}{t}|\nu|(\mathbb{R}).$$

The statement follows by taking $a \to -\infty$ and $b \to +\infty$. □

In Problem 3 the reader is asked to prove:

Proposition 2.9. *Let* A *be a bounded Borel set. Then for any* $0 < p < 1$,

$$\int_A M_{\nu,\mu}(x)^p \mathrm{d}\mu(x) < \infty.$$

We will also need:

Proposition 2.10. *Let ν_j be a sequence of Borel complex measures such that*

$$\lim_{j\to\infty} |\nu_j|(\mathbb{R}) = 0.$$

Then there is a subsequence ν_{j_k} such that

$$\lim_{k\to\infty} M_{\nu_{j_k},\mu}(x) = 0 \qquad \text{for } \mu - a.e.\ x.$$

Proof. By Theorem 2.8, for each $k = 1, 2, \ldots$, we can find j_k so that

$$\mu\left\{x : M_{\nu_{j_k},\mu}(x) > 2^{-k}\right\} \leq 2^{-k}.$$

Hence,

$$\sum_{k=1}^{\infty} \mu\left\{x : M_{\nu_{j_k},\mu}(x) > 2^{-k}\right\} < \infty,$$

and so for μ-a.e. x, there is k_x such that for $k > k_x$, $M_{\nu_{j_k},\mu}(x) \leq 2^{-k}$. This yields the statement. □

We are now ready to prove the main theorem of this subsection.

Theorem 2.11. *Let ν be a complex Borel measure and μ a positive Borel measure finite on compact sets. Let $\nu = f\mu + \nu_s$ be the Radon-Nikodym decomposition. Then:*
(1)

$$\lim_{r\downarrow 0} \frac{\nu(I(x,r))}{\mu(I(x,r))} = f(x), \qquad \text{for } \mu - a.e.\ x.$$

In particular, $\nu \perp \mu$ iff

$$\lim_{r\downarrow 0} \frac{\nu(I(x,r))}{\mu(I(x,r))} = 0, \qquad \text{for } \mu - a.e.\ x.$$

(2) *Let in addition ν be positive. Then*

$$\lim_{r\downarrow 0} \frac{\nu(I(x,r))}{\mu(I(x,r))} = \infty, \qquad \text{for } \nu_s - a.e.\ x.$$

Proof. (1) We will split the proof into two steps.
Step 1. Assume that $\nu \ll \mu$, namely that $\nu = f\mu$. Let g_n be a continuous function with compact support such that $\int_{\mathbb{R}} |f - g_n| \mathrm{d}\mu < 1/n$. Set $h_n = f - g_n$. Then, for $x \in \operatorname{supp} \mu$ and $r > 0$,

$$\left|\frac{f\mu(I(x,r))}{\mu(I(x,r))} - f(x)\right| \leq \frac{|h_n|\mu(I(x,r))}{\mu(I(x,r))} + \left|\frac{g_n\mu(I(x,r))}{\mu(I(x,r))} - g_n(x)\right| + |g_n(x) - f(x)|.$$

Since g_n is continuous, we obviously have

$$\lim_{r\downarrow 0}\left|\frac{g_n\mu(I(x,r))}{\mu(I(x,r))} - g_n(x)\right| = 0,$$

and so for all n and $x \in \operatorname{supp}\mu$,

$$\limsup_{r\downarrow 0}\left|\frac{f\mu(I(x,r))}{\mu(I(x,r))} - f(x)\right| \leq M_{h_n\mu,\mu}(x) + |g_n(x) - f(x)|.$$

Let n_j be a subsequence such that $g_{n_j} \to f(x)$ for μ-a.e. x. Since $\int |h_{n_j}|\mathrm{d}\mu \to 0$ as $j \to \infty$, Proposition 2.10 yields that there is a subsequence of n_j (which we denote by the same letter) such that $M_{h_{n_j}\mu,\mu}(x) \to 0$ for μ-a.e. x. Hence, for μ-a.e. x,

$$\limsup_{r\downarrow 0}\left|\frac{f\mu(I(x,r))}{\mu(I(x,r))} - f(x)\right| = 0,$$

and (1) holds if $\nu \ll \mu$.

Step 2. To finish the proof of (1), it suffices to show that if ν is a complex measure such that $\nu \perp \mu$, then

$$\lim_{r\downarrow 0}\frac{|\nu|(I(x,r))}{\mu(I(x,r))} = 0, \qquad \text{for } \mu - a.e.\ x. \tag{5}$$

Let S be a Borel set such that $\mu(S) = 0$ and $|\nu|(\mathbb{R}\setminus S) = 0$. Then

$$\frac{|\nu|(I(x,r))}{(\mu+|\nu|)(I(x,r))} = \frac{\chi_S(|\nu|+\mu)(I(x,r))}{(\mu+|\nu|)(I(x,r))}. \tag{6}$$

By Step 1,

$$\lim_{r\downarrow 0}\frac{\chi_S(|\nu|+\mu)(I(x,r))}{(|\nu|+\mu)(I(x,r))} = \chi_S(x), \qquad \text{for } |\nu|+\mu - a.e.\ x. \tag{7}$$

Since $\chi_S(x) = 0$ for μ-a.e. x, (6) and (7) yield (5).

(2) Since ν is positive, $\nu(I(x,r)) \geq \nu_{\mathrm{s}}(I(x,r))$, and we may assume that $\nu \perp \mu$. By (6) and (7),

$$\lim_{r\downarrow 0}\frac{\nu(I(x,r))}{(\nu+\mu)(I(x,r))} = 1, \qquad \text{for } \nu - a.e.\ x,$$

and so

$$\lim_{r\downarrow 0}\frac{\mu(I(x,r))}{\nu(I(x,r))} = 0, \qquad \text{for } \nu - a.e.\ x.$$

This yields (2). □

We finish this subsection with several remarks. If μ is the Lebesgue measure, then the results of this section reduce to the standard differentiation results discussed, for example, in Chapter 7 of [27]. The arguments in [27] are based on the Vitali covering lemma which is specific to the Lebesgue measure. The proofs of this subsection are based on the Besicovitch covering lemma and they apply to an arbitrary positive measure μ. In fact, the proofs directly extend to $\mathbb{R}^n$ (one only needs to replace the intervals $I(x,r)$ with the balls $B(x,r)$ centered at x and of radius r) if one uses the following version of the Besicovitch covering lemma.

Theorem 2.12. *Let A be a bounded set in $\mathbb{R}^n$ and, for each $x \in A$, let B_x be an open ball with center at x. Then there is an integer N, which depends only on n, such that:*
(1) There is a countable subcollection $\{B_j\}$ of $\{B_x\}_{x\in A}$ such that $A \subset \cup B_j$ and each point in $\mathbb{R}^n$ belongs to at most N balls in $\{B_j\}$, i.e. $\forall y \in \mathbb{R}$,

$$\sum_j \chi_{B_j}(y) \le N.$$

(2) There is a countable subcollection $\{B_{i,j}\}$, $i = 1, \cdots, N$, of $\{B_x\}_{x\in A}$ such that $A \subset \cup B_{i,j}$ and $B_{i,j} \cap B_{i,k} = \emptyset$ if $j \neq k$.

Unfortunately, unlike the proof of the Vitali covering lemma, the proof of Theorem 2.12 is somewhat long and complicated.

2.7 Problems

[1] *Prove that the maximal function $M_{\nu,\mu}(x)$ is Borel measurable.*

[2] *Let μ be a positive σ-finite measure on $(M, \mathcal{F})$ and let f be a measurable function. Let*

$$m_f(t) = \mu\{x : |f(x)| > t\}.$$

Prove that for $p \ge 1$,

$$\int_M |f(x)|^p \mathrm{d}\mu(x) = p \int_0^\infty t^{p-1} m_f(t)\mathrm{d}t.$$

This result can be generalized as follows. Let $\alpha : [0,\infty] \mapsto [0,\infty]$ be monotonic and absolutely continuous on $[0,T]$ for every $T < \infty$. Assume that $\alpha(0) = 0$ and $\alpha(\infty) = \infty$. Prove that

$$\int_M (\alpha \circ f)(x)\mathrm{d}\mu(x) = \int_0^\infty \alpha'(t) m_f(t)\mathrm{d}t.$$

Hint: See Theorem 8.16 in [27].

[3] *Prove Proposition 2.9. Hint: Use Problem 2.*

[4] *Prove the Riemann-Lebesgue lemma (Theorem 2.5).*

[5] *Let μ be a complex Borel measure on $\mathbb{R}$. Prove that $|\mu_{\rm sing}| = |\mu|_{\rm sing}$.*

[6] *Let μ be a positive measure on $(M, \mathcal{F})$. A sequence of measurable functions f_n converges in measure to zero if*

$$\lim_{n\to\infty} \mu(\{x : |f_n(x)| > \epsilon\}) = 0,$$

for all $\epsilon > 0$. The sequence f_n converges almost uniformly to zero if for all $\epsilon > 0$ there is a set $M_\epsilon \in \mathcal{F}$, such that $\mu(M_\epsilon) < \epsilon$ and f_n converges uniformly to zero on $M \setminus M_\epsilon$.

Prove that if f_n converges to zero in measure, then there is a subsequence f_{n_j} which converges to zero almost uniformly.

[7] *Prove Theorem 2.12. (The proof can be found in* [9]*).*

[8] *State and prove the analog of Theorem 2.11 in $\mathbb{R}^n$.*

[9] *Let μ be a positive Borel measure on $\mathbb{R}$. Assume that μ is finite on compact sets and let $f \in L^1(\mathbb{R}, \mathrm{d}\mu)$. Prove that*

$$\lim_{r\downarrow 0} \frac{1}{\mu(I(x,r))} \int_{I(x,r)} |f(t) - f(x)| \mathrm{d}\mu(t) = 0, \qquad \text{for } \mu - a.e.\ x.$$

Hint: You may follow the proof of Theorem 7.7 in [27].

[10] *Let $p \geq 1$ and $f \in L^p(\mathbb{R}, \mathrm{d}x)$. The maximal function of f, M_f, is defined by*

$$M_f(x) = \sup_{r>0} \frac{1}{2r} \int_{I(x,r)} |f(t)| \mathrm{d}t.$$

(1) *If $p > 1$, prove that $M_f \in L^p(\mathbb{R}, \mathrm{d}x)$. Hint: See Theorem 8.18 in* [27].
(2) *Prove that if f and M_f are in $L^1(\mathbb{R}, \mathrm{d}x)$, then $f = 0$.*

[11] *Denote by $B_{\rm b}(\mathbb{R})$ the algebra of the bounded Borel functions on $\mathbb{R}$. Prove that $B_{\rm b}(\mathbb{R})$ is the smallest algebra of functions which includes $C_0(\mathbb{R})$ and is closed under pointwise limits of uniformly bounded sequences.*

3 Preliminaries: harmonic analysis

In this section we will deal only with Borel measures on $\mathbb{R}$. We will use the shorthand $\mathbb{C}_+ = \{z : \mathrm{Im}\, z > 0\}$. We denote the Lebesgue measure by m and write $\mathrm{d}m = \mathrm{d}x$.

Let μ be a complex measure or a positive measure such that

$$\int_{\mathbb{R}} \frac{\mathrm{d}\mu(t)}{1 + |t|} < \infty.$$

The Borel transform of μ is defined by

$$F_\mu(z) = \int_{\mathbb{R}} \frac{\mathrm{d}\mu(t)}{t-z}, \qquad z \in \mathbb{C}_+. \tag{8}$$

The function $F_\mu(z)$ is analytic in $\mathbb{C}_+$.

Let μ be a complex measure or positive measure such that

$$\int_{\mathbb{R}} \frac{\mathrm{d}\mu(t)}{1+t^2} < \infty. \tag{9}$$

The Poisson transform of μ is defined by

$$P_\mu(x+\mathrm{i}y) = y \int_{\mathbb{R}} \frac{\mathrm{d}\mu(t)}{(x-t)^2+y^2}, \qquad y > 0.$$

The function $P_\mu(z)$ is harmonic in $\mathbb{C}_+$. If μ is the Lebesgue measure, then $P_\mu(z) = \pi$ for all $z \in \mathbb{C}_+$. Note that F_μ and P_μ are linear functions of μ, i.e. for $\lambda_1, \lambda_2 \in \mathbb{C}$, $F_{\lambda_1\mu_1+\lambda_2\mu_2} = \lambda_1 F_{\mu_1} + \lambda_2 F_{\mu_2}$, $P_{\lambda_1\mu_1+\lambda_2\mu_2} = \lambda_1 P_{\mu_1} + \lambda_2 P_{\mu_2}$. If μ is a positive or signed measure, then $\operatorname{Im} F_\mu = P_\mu$.

Our goal in this section is to study the boundary values of $P_\mu(x+\mathrm{i}y)$ and $F_\mu(x+\mathrm{i}y)$ as $y \downarrow 0$. More precisely, we wish to study how these boundary values reflect the properties of the measure μ.

Although we will restrict ourselves to the radial limits, all the results discussed in this section hold for the non-tangential limits (see the problem set). The non-tangential limits will not be needed for our applications.

3.1 Poisson transforms and Radon-Nikodym derivatives

This subsection is based on [14]. Recall that $I(x,r) = (x-r, x+r)$.

Lemma 3.1. *Let μ be a positive measure. Then for all $x \in \mathbb{R}$ and $y > 0$,*

$$\frac{1}{y} P_\mu(x+\mathrm{i}y) = \int_0^{1/y^2} \mu(I(x, \sqrt{u^{-1}-y^2}))\mathrm{d}u.$$

Proof. Note that

$$\begin{aligned}\int_0^{1/y^2} \mu(I(x, \sqrt{u^{-1}-y^2}))\mathrm{d}u &= \int_0^{1/y^2} \left[\int_{\mathbb{R}} \chi_{I(x,\sqrt{u^{-1}-y^2})}(t)\mathrm{d}\mu(t)\right] \mathrm{d}u \\ &= \int_{\mathbb{R}} \left[\int_0^{1/y^2} \chi_{I(x,\sqrt{u^{-1}-y^2})}(t)\mathrm{d}u\right] \mathrm{d}\mu(t).\end{aligned} \tag{10}$$

Since

$$|x-t| < \sqrt{u^{-1}-y^2} \quad \Longleftrightarrow \quad 0 \le u < ((x-t)^2+y^2)^{-1},$$

we have

$$\chi_{I(x,\sqrt{u^{-1}-y^2})}(t) = \chi_{[0,((x-t)^2+y^2)^{-1})}(u),$$

and

$$\int_0^{1/y^2} \chi_{I(x,\sqrt{u^{-1}-y^2})}(t)\mathrm{d}u = ((x-t)^2+y^2))^{-1}.$$

Hence, the result follows from (10). □

Lemma 3.2. *Let ν be a complex and μ a positive measure. Then for all $x \in \mathbb{R}$ and $y > 0$,*

$$\frac{|P_\nu(x+\mathrm{i}y)|}{P_\mu(x+\mathrm{i}y)} \le M_{\nu,\mu}(x).$$

Proof. Since $|P_\nu| \le P_{|\nu|}$, w.l.o.g. we may assume that ν is positive. Also, we may assume that $x \in \operatorname{supp}\mu$ (otherwise $M_{\nu,\mu}(x) = \infty$ and there is nothing to prove). Set

$$I_{x,y}(u) = I(x, \sqrt{u^{-1}-y^2}).$$

Since

$$\int_0^{1/y^2} \nu(I_{x,y}(u))\mathrm{d}u = \int_0^{1/y^2} \frac{\nu(I_{x,y}(u))}{\mu(I_{x,y}(u))}\mu(I_{x,y}(u))\mathrm{d}u$$

$$\le M_{\nu,\mu}(x)\int_0^{1/y^2} \mu(I_{x,y}(u))\mathrm{d}u,$$

the result follows from Lemma 3.1. □

Lemma 3.3. *Let μ be a positive measure. Then for μ-a.e. x,*

$$\int_{\mathbb{R}} \frac{\mathrm{d}\mu(t)}{(x-t)^2} = \infty. \tag{11}$$

The proof of this lemma is left for the problem set.

Lemma 3.4. *Let μ be a positive measure and $f \in C_0(\mathbb{R})$. Then for μ-a.e. x,*

$$\lim_{y\downarrow 0} \frac{P_{f\mu}(x+\mathrm{i}y)}{P_\mu(x+\mathrm{i}y)} = f(x). \tag{12}$$

Remark. The relation (12) holds for all x for which (11) holds. For example, if μ is the Lebesgue measure, then (12) holds for all x.
Proof. Note that

$$\left|\frac{P_{f\mu}(x+iy)}{P_\mu(x+iy)} - f(x)\right| \leq \frac{P_{|f-f(x)|\mu}(x+iy)}{P_\mu(x+iy)}.$$

Fix $\epsilon > 0$ and let $\delta > 0$ be such that $|x-t| < \delta \Rightarrow |f(x) - f(t)| < \epsilon$. Let $M = \sup|f(t)|$ and

$$C = 2M \int_{|x-t|\geq\delta} \frac{\mathrm{d}\mu(t)}{(x-t)^2}.$$

Then

$$P_{|f-f(x)|\mu}(x+iy) \leq \epsilon P_\mu(x+iy) + Cy,$$

and

$$\left|\frac{P_{f\mu}(x+iy)}{P_\mu(x+iy)} - f(x)\right| \leq \epsilon + \frac{Cy}{P_\mu(x+iy)}.$$

Let x be such that (11) holds. The monotone convergence theorem yields that

$$\lim_{y\downarrow 0} \frac{y}{P_\mu(x+iy)} = \left(\int \frac{\mathrm{d}\mu(t)}{(x-t)^2}\right)^{-1} = 0$$

and so for all $\epsilon > 0$,

$$\limsup_{y\downarrow 0} \left|\frac{P_{f\mu}(x+iy)}{P_\mu(x+iy)} - f(x)\right| \leq \epsilon.$$

This yields the statement. □

The main result of this subsection is:

Theorem 3.5. *Let ν be a complex measure and μ a positive measure. Let $\nu = f\mu + \nu_{\mathrm{s}}$ be the Radon-Nikodym decomposition. Then:*
(1)

$$\lim_{y\downarrow 0} \frac{P_\nu(x+iy)}{P_\mu(x+iy)} = f(x), \qquad \text{for } \mu - a.e.\ x.$$

In particular, $\nu \perp \mu$ iff

$$\lim_{y\downarrow 0} \frac{P_\nu(x+iy)}{P_\mu(x+iy)} = 0, \qquad \text{for } \mu - a.e.\ x.$$

(2) *Assume in addition that ν is positive. Then*

$$\lim_{y\downarrow 0} \frac{P_\nu(x+iy)}{P_\mu(x+iy)} = \infty, \qquad \text{for } \nu_{\mathrm{s}} - a.e.\ x.$$

Proof. The proof is very similar to the proof of Theorem 2.11 in Section 2.

(1) We will split the proof into two steps.

Step 1. Assume that $\nu \ll \mu$, namely that $\nu = f\mu$. Let g_n be a continuous function with compact support such that $\int_{\mathbb{R}} |f - g_n| \mathrm{d}\mu < 1/n$. Set $h_n = f - g_n$. Then,

$$\left| \frac{P_{f\mu}(x+\mathrm{i}y)}{P_\mu(x+\mathrm{i}y)} - f(x) \right| \le \frac{P_{|h_n|\mu}(x+\mathrm{i}y)}{P_\mu(x+\mathrm{i}y)} + \left| \frac{P_{g_n\mu}(x+\mathrm{i}y)}{P_\mu(x+\mathrm{i}y)} - g_n(x) \right| + |g_n(x) - f(x)|.$$

It follows from Lemmas 3.2 and 3.4 that for μ-a.e. x,

$$\limsup_{y\downarrow 0} \left| \frac{P_{f\mu}(x+\mathrm{i}y)}{P_\mu(x+\mathrm{i}y))} - f(x) \right| \le M_{|h_n|\mu,\mu}(x) + |g_n(x) - f(x)|.$$

As in the proof of Theorem 2.11, there is a subsequence $n_j \to \infty$ such that $g_{n_j}(x) \to f(x)$ and $M_{|h_n|\mu,\mu}(x) \to 0$ for μ-a.e. x, and (1) holds if $\nu \ll \mu$.

Step 2. To finish the proof of (1), it suffices to show that if ν is a finite positive measure such that $\nu \perp \mu$, then

$$\lim_{y\downarrow 0} \frac{P_\nu(x+\mathrm{i}y)}{P_\mu(x+\mathrm{i}y)} = 0, \qquad \text{for } \mu - a.e.\ x. \tag{13}$$

Let S be a Borel set such that $\mu(S) = 0$ and $\nu(\mathbb{R} \setminus S) = 0$. Then

$$\frac{P_\nu(x+\mathrm{i}y)}{P_\nu(x+\mathrm{i}y) + P_\mu(x+\mathrm{i}y)} = \frac{P_{\chi_S(\nu+\mu)}(x+\mathrm{i}y)}{P_{\nu+\mu}(x+\mathrm{i}y)}. \tag{14}$$

By Step 1,

$$\lim_{y\downarrow 0} \frac{P_{\chi_S(\nu+\mu)}(x+\mathrm{i}y)}{P_{\nu+\mu}(x+\mathrm{i}y)} = \chi_S(x), \qquad \text{for } \nu+\mu - a.e.\ x. \tag{15}$$

Since $\chi_S(x) = 0$ for μ-a.e. x,

$$\lim_{y\downarrow 0} \frac{P_\nu(x+\mathrm{i}y)}{P_\nu(x+\mathrm{i}y) + P_\mu(x+\mathrm{i}y)} = 0, \qquad \text{for } \mu - a.e.\ x,$$

and (13) follows.

(2) Since ν is positive, $\nu(I(x,r)) \ge \nu_{\mathrm{s}}(I(x,r))$, and we may assume that $\nu \perp \mu$. By (14) and (15),

$$\lim_{y\downarrow 0} \frac{P_\nu(x+\mathrm{i}y)}{P_\nu(x+\mathrm{i}y) + P_\mu(x+\mathrm{i}y)} = 1, \qquad \text{for } \nu - a.e.\ x,$$

and so

$$\lim_{y\downarrow 0} \frac{P_\mu(x+\mathrm{i}y)}{P_\nu(x+\mathrm{i}y)} = 0, \qquad \text{for } \nu - a.e.\ x.$$

This yields part (2). □

3.2 Local L^p norms, $0 < p < 1$.

In this subsection we prove Theorem 3.1 of [28]. ν is a complex measure, μ is a positive measure and $\nu = f\mu + \nu_s$ is the Radon-Nikodym decomposition.

Theorem 3.6. *Let A be a bounded Borel set and $0 < p < 1$. Then*

$$\lim_{y \downarrow 0} \int_A \left| \frac{P_\nu(x+\mathrm{i}y)}{P_\mu(x+\mathrm{i}y)} \right|^p \mathrm{d}\mu(x) = \int_A |f(x)|^p \mathrm{d}\mu(x).$$

(Both sides are allowed to be ∞). In particular, $\nu \restriction A \perp \mu \restriction A$ iff for some $p \in (0,1)$,

$$\lim_{y \downarrow 0} \int_A \left| \frac{P_\nu(x+\mathrm{i}y)}{P_\mu(x+\mathrm{i}y)} \right|^p \mathrm{d}\mu(x) = 0.$$

Proof. By Theorem 3.5,

$$\lim_{y \downarrow 0} \left| \frac{P_\nu(x+\mathrm{i}y)}{P_\mu(x+\mathrm{i}y)} \right|^p = |f(x)|^p \qquad \text{for } \mu - \text{a.e. } x.$$

By Lemma 3.2,

$$\left| \frac{P_\nu(x+\mathrm{i}y)}{P_\mu(x+\mathrm{i}y)} \right|^p \leq M_{\nu,\mu}(x)^p.$$

Hence, Proposition 2.9 and the dominated convergence theorem yield the statement.
□

3.3 Weak convergence

Let ν be a complex or positive measure and

$$\mathrm{d}\nu_y(x) = \frac{1}{\pi} P_\nu(x+\mathrm{i}y)\mathrm{d}x. \tag{16}$$

Theorem 3.7. *For any $f \in C_c(\mathbb{R})$ (continuous functions of compact support),*

$$\lim_{y \downarrow 0} \int_{\mathbb{R}} f(x)\mathrm{d}\nu_y(x) = \int_{\mathbb{R}} f(x)\mathrm{d}\nu(x). \tag{17}$$

In particular, $P_{\nu_1} = P_{\nu_2} \Rightarrow \nu_1 = \nu_2$.

Proof. Note that

$$\int_{\mathbb{R}} f(x)\mathrm{d}\nu_y(x) = \int_{\mathbb{R}} \left[\frac{y}{\pi} \int_{\mathbb{R}} \frac{f(x)\mathrm{d}x}{(x-t)^2+y^2} \right] \mathrm{d}\nu(t),$$

and so

$$\left|\int_{\mathbb{R}} f(x)\mathrm{d}\nu_y(x) - \int_{\mathbb{R}} f(x)\mathrm{d}\nu(x)\right| \le \int_{\mathbb{R}} \frac{|L_y(t)|\mathrm{d}|\nu|(t)}{1+t^2}, \tag{18}$$

where

$$L_y(t) = (1+t^2)\left(f(t) - \frac{y}{\pi}\int_{\mathbb{R}} \frac{f(x)\mathrm{d}x}{(x-t)^2+y^2}\right).$$

Clearly, $\sup_{y>0,t\in\mathbb{R}} |L_y(t)| < \infty$. By Lemma 3.4 and Remark after it,

$$\lim_{y\downarrow 0} L_y(t) = 0,$$

for all $t \in \mathbb{R}$ (see also Problem 2). Hence, the statement follows from the estimate (18) and the dominated convergence theorem. □

3.4 Local L^p-norms, $p > 1$

In this subsection we will prove Theorem 2.1 of [28].

Let ν be a complex or positive measure and let $\nu = fm + \nu_{\mathrm{sing}}$ be its Radon-Nikodym decomposition w.r.t. the Lebesgue measure.

Theorem 3.8. *Let $A \subset \mathbb{R}$ be open, $p > 1$, and assume that*

$$\sup_{0<y<1} \int_A |P_\nu(x+\mathrm{i}y)|^p \mathrm{d}x < \infty.$$

Then:
(1) $\nu_{\mathrm{sing}} \restriction A = 0$.
(2) $\int_A |f(x)|^p \mathrm{d}x < \infty$.
(3) *For any $[a,b] \subset A$, $\pi^{-1}P_\nu(x+\mathrm{i}y) \to f(x)$ in $L^p([a,b],\mathrm{d}x)$ as $y \downarrow 0$.*

Proof. We will prove (1) and (2); (3) is left to the problems.

Let g be a continuous function with compact support contained in A and let q be the index dual to p, $p^{-1} + q^{-1} = 1$. Then, by Theorem 3.7,

$$\int_A g\mathrm{d}\nu = \lim_{y\downarrow 0} \pi^{-1} \int_A g(x)P_\nu(x+\mathrm{i}y)\mathrm{d}x,$$

and

$$\left|\int_A g(x)P_\nu(x+\mathrm{i}y)\mathrm{d}x\right| \le \left(\int_A |g(x)|^q \mathrm{d}x\right)^{1/q} \left(\int_A |P_\nu(x+\mathrm{i}y)|^p \mathrm{d}x\right)^{1/p}$$

$$\le C\left(\int_A |g(x)|^q \mathrm{d}x\right)^{1/q}.$$

Hence, the map $g \mapsto \int_A g(x)\mathrm{d}\nu(x)$ is a continuous linear functional on $L^q(A,\mathrm{d}x)$, and there is a function $\tilde{f} \in L^p(A,\mathrm{d}\mu)$ such that

$$\int_A g(x)\mathrm{d}\nu(x) = \int_A g(x)\tilde{f}(x)\mathrm{d}x.$$

This relation implies that $\nu \restriction A$ is absolutely continuous w.r.t. the Lebesgue measure and that $f(x) = \tilde{f}(x)$ for Lebesgue a.e. x. (1) and (2) follow. □

Theorem 3.8 has a partial converse which we will discuss in the problem set.

3.5 Local version of the Wiener theorem

In this subsection we prove Theorem 2.2 of [28].

Theorem 3.9. *Let ν be a signed measure and A_ν be the set of atoms of ν. Then for any finite interval $[a, b]$,*

$$\lim_{y\downarrow 0} y\int_a^b P_\nu(x+\mathrm{i}y)^2\mathrm{d}x = \frac{\pi}{2}\left(\frac{\nu(\{a\})^2}{2} + \frac{\nu(\{b\})^2}{2} + \sum_{x\in(a,b)\cap A_\nu} \nu(\{x\})^2\right).$$

Proof.

$$P_\nu(x+\mathrm{i}y)^2 = y^2\int_{\mathbb{R}^2}\frac{\mathrm{d}\nu(t)\mathrm{d}\nu(t')}{((x-t)^2+y^2)((x-t')^2+y^2)},$$

and

$$y\int_a^b P_\nu(x+\mathrm{i}y)^2\mathrm{d}x = \int_{\mathbb{R}^2} g_y(t,t')\mathrm{d}\nu(t)\mathrm{d}\nu(t'),$$

where

$$g_y(t,t') = \int_a^b \frac{y^3\mathrm{d}x}{((x-t)^2+y^2)((x-t')^2+y^2)}.$$

Notice now that:

(1) $0 \le g_y(t,t') \le \pi$.

(2) $\lim_{y\downarrow 0} g_y(t,t') = 0$ if $t \neq t'$, or $t \notin [a,b]$, or $t' \notin [a,b]$.

(3) If $t = t' \in (a,b)$, then

$$\lim_{y\downarrow 0} g_y(t,t) = \lim_{y\downarrow 0} y^3\int_{\mathbb{R}}\frac{\mathrm{d}x}{(x^2+y^2)^2} = \frac{\pi}{2},$$

(compute the integral using the residue calculus).

(4) If $t = t' = a$ or $t = t' = b$, then

$$\lim_{y\downarrow 0} g_y(t,t) = \lim_{y\downarrow 0} y^3\int_0^\infty \frac{\mathrm{d}x}{(x^2+y^2)^2} = \frac{\pi}{4}.$$

The result follows from these observations and the dominated convergence theorem. □

Corollary 3.10. *A signed measure ν has no atoms in $[a, b]$ iff*

$$\lim_{y\downarrow 0} y\int_a^b P_\nu(x+\mathrm{i}y)^2\mathrm{d}x = 0.$$

3.6 Poisson representation of harmonic functions

Theorem 3.11. *Let $V(z)$ be a positive harmonic function in $\mathbb{C}_+$. Then there is a constant $c \geq 0$ and a positive measure μ on $\mathbb{R}$ such that*

$$V(x + \mathrm{i}y) = cy + P_\mu(x + \mathrm{i}y).$$

The c and μ are uniquely determined by V.

Remark 1. The constant c is unique since $c = \lim_{y\to\infty} V(\mathrm{i}y)/y$. By Theorem 3.7, μ is also unique.
Remark 2. Theorems 3.5 and 3.11 yield that if V is a positive harmonic function in $\mathbb{C}_+$ and $\mathrm{d}\mu = f(x)\mathrm{d}x + \mu_{\mathrm{sing}}$ is the associated measure, then for Lebesgue a.e. x,

$$\lim_{y\downarrow 0} \pi^{-1} V(x + \mathrm{i}y) = f(x).$$

Let $D = \{z : |z| < 1\}$ and $\Gamma = \{z : |z| = 1\}$. For $z \in D$ and $w \in \Gamma$ let

$$p_z(w) = \mathrm{Re}\, \frac{w + z}{w - z} = \frac{1 - |z|^2}{|w - z|^2}.$$

We shall first prove:

Theorem 3.12. *Let U be a positive harmonic function in D. Then there exists a finite positive Borel measure ν on Γ such that for all $z \in D$,*

$$U(z) = \int_\Gamma p_z(w)\mathrm{d}\nu(w).$$

Proof. By the mean value property of harmonic functions, for any $0 < r < 1$,

$$U(0) = \frac{1}{2\pi} \int_{-\pi}^{\pi} U(r\mathrm{e}^{\mathrm{i}\theta})\mathrm{d}\theta.$$

In particular,

$$\sup_{0<r<1} \frac{1}{2\pi} \int_{-\pi}^{\pi} U(r\mathrm{e}^{\mathrm{i}\theta})\mathrm{d}\theta = U(0) < \infty.$$

For $f \in C(\Gamma)$ set

$$\Phi_r(f) = \frac{1}{2\pi} \int_{-\pi}^{\pi} U(r\mathrm{e}^{\mathrm{i}\theta}) f(\mathrm{e}^{\mathrm{i}\theta})\mathrm{d}\theta.$$

Each Φ_r is a continuous linear functional on the Banach space $C(\Gamma)$ and $\|\Phi_r\| = U(0)$. The standard diagonal argument yields that there is a sequence $r_j \to 1$ and a bounded linear functional Φ on $C(\Gamma)$ such that $\Phi_{r_j} \to \Phi$ weakly, that is, for all $f \in C(\Gamma)$, $\Phi_{r_j}(f) \to \Phi(f)$. Obviously, $\|\Phi\| = U(0)$. By the Riesz representation theorem there exists a complex measure ν on Γ such that $|\nu|(\Gamma) = U(0)$ and

$$\Phi(f)=\int_\Gamma f(w)\mathrm{d}\nu(w).$$

Since $\Phi_{r_j}(f)\geq 0$ if $f\geq 0$, the measure ν is positive. Finally, let $z\in D$. If $r_j>|z|$, then

$$U(zr_j)=\frac{1}{2\pi}\int_{-\pi}^{\pi}U(r_j\mathrm{e}^{\mathrm{i}\theta})p_z(\mathrm{e}^{\mathrm{i}\theta})\mathrm{d}\theta=\Phi_{r_j}(p_z), \tag{19}$$

(the proof of this relation is left for the problems—see Theorem 11.8 in [27]). Taking $j\to\infty$ we derive

$$U(z)=\Phi(p_z)=\int_\Gamma p_z(w)\mathrm{d}\nu(w).$$

□

Before proving Theorem 3.11, I would like to make a remark about change of variables in measure theory. Let $(M_1,\mathcal{F}_1)$ and $(M_2,\mathcal{F}_2)$ be measure spaces. A map $T:M_1\to M_2$ is called a measurable transformation if for all $F\in\mathcal{F}_2$, $T^{-1}(F)\in\mathcal{F}_1$. Let μ be a positive measure on $(M_1,\mathcal{F}_1)$, and let μ_T be a positive measure on $(M_2,\mathcal{F}_2)$ defined by $\mu_T(F)=\mu(T^{-1}(F))$. If f is a measurable function on $(M_2,\mathcal{F}_2)$, then $f_T=f\circ T$ is a measurable function on $(M_1,\mathcal{F}_1)$. Moreover, $f\in L^1(M_2,\mathrm{d}\mu_T)$ iff $f_T\in L^1(M_1,\mathrm{d}\mu)$, and in this case

$$\int_{M_2}f\mathrm{d}\mu_T=\int_{M_1}f_T\mathrm{d}\mu.$$

This relation is easy to check if f is a characteristic function. The general case follows by the usual approximation argument through simple functions.

If T is a bijection, then $g\in L^1(M_1,\mathrm{d}\mu)$ iff $g_{T^{-1}}\in L^1(M_2,\mathrm{d}\mu_T)$, and in this case

$$\int_{M_1}g\mathrm{d}\mu=\int_{M_2}g_{T^{-1}}\mathrm{d}\mu_T.$$

Proof of Theorem 3.11. We define a map $S:\mathbb{C}_+\to D$ by

$$S(z)=\frac{\mathrm{i}-z}{\mathrm{i}+z}. \tag{20}$$

This is the well-known conformal map between the upper half-plane and the unit disc. The map S extends to a homeomorphism $S:\overline{\mathbb{C}}_+\to\overline{D}\setminus\{-1\}$. Note that $S(\mathbb{R})=\Gamma\setminus\{-1\}$. If

$$K_z(t)=\frac{y}{(x-t)^2+y^2},\qquad z=x+\mathrm{i}y\in\mathbb{C}_+,$$

then

$$(1+t^2)K_z(t)=p_{S(z)}(S(t)).$$

Let $T=S^{-1}$. Explicitly,

$$T(\xi)=\frac{\xi-1}{\mathrm{i}\xi+\mathrm{i}}. \tag{21}$$

Let $U(\xi) = V(T(\xi))$. Then there exists a positive finite Borel measure ν on Γ such that

$$U(\xi) = \int_\Gamma p_\xi(w)\mathrm{d}\nu(w).$$

The map $T : \Gamma \setminus \{-1\} \to \mathbb{R}$ is a homeomorphism. Let ν_T be the induced Borel measure on $\mathbb{R}$. By the previous change of variables,

$$\int_{\Gamma\setminus\{-1\}} p_\xi(w)\mathrm{d}\nu(w) = \int (1+t^2)K_{T(\xi)}(t)\mathrm{d}\nu_T(t).$$

Hence, for $z \in \mathbb{C}_+$,

$$V(z) = p_{S(z)}(-1)\nu(\{-1\}) + \int_\mathbb{R} (1+t^2)K_z(t)\mathrm{d}\nu_T(t).$$

Since $p_{S(z)}(-1) = y$, setting $c = \nu(\{-1\})$ and $\mathrm{d}\mu(t) = (1+t^2)\mathrm{d}\nu_T(t)$, we derive the statement. □

3.7 The Hardy class $H^\infty(\mathbb{C}_+)$

The Hardy class $H^\infty(\mathbb{C}_+)$ is the vector space of all functions V analytic in $\mathbb{C}_+$ such that

$$\|V\| = \sup_{z\in\mathbb{C}_+} |V(z)| < \infty. \tag{22}$$

$H^\infty(\mathbb{C}_+)$ with norm (22) is a Banach space. In this subsection we will prove two basic properties of $H^\infty(\mathbb{C}_+)$.

Theorem 3.13. *Let $V \in H^\infty(\mathbb{C}_+)$. Then for Lebesgue a.e. $x \in \mathbb{R}$, the limit*

$$V(x) = \lim_{y\downarrow 0} V(x+\mathrm{i}y) \tag{23}$$

exists. Obviously, $V \in L^\infty(\mathbb{R}, \mathrm{d}x)$.

Theorem 3.14. *Let $V \in H^\infty(\mathbb{C}_+)$, $V \not\equiv 0$, and let $V(x)$ be given by (23). Then*

$$\int_\mathbb{R} \frac{|\log|V(x)||}{1+x^2}\mathrm{d}x < \infty.$$

In particular, if $\alpha \in \mathbb{C}$, then either $V(z) \equiv \alpha$ or the set $\{x \in \mathbb{R} : V(x) = \alpha\}$ has zero Lebesgue measure.

A simple and important consequence of Theorems 3.13 and 3.14 is:

Theorem 3.15. *Let F be an analytic function on $\mathbb{C}_+$ with positive imaginary part. Then:*
(1) *For Lebesgue a.e. $x \in \mathbb{R}$ the limit*

$$F(x) = \lim_{y \downarrow 0} F(x + \mathrm{i}y),$$

exists and is finite.
(2) *If $\alpha \in \mathbb{C}$, then either $F(z) \equiv \alpha$ or the set $\{x \in \mathbb{R} : F(x) = \alpha\}$ has zero Lebesgue measure.*

Proof. To prove (1), apply Theorem 3.13 to the function $(F(z)+\mathrm{i})^{-1}$. To prove (2), apply Theorem 3.14 to the function $(F(z)+\mathrm{i})^{-1} - (\alpha+\mathrm{i})^{-1}$. □

Proof of Theorem 3.13. Let $\mathrm{d}\nu_y(t) = V(t+\mathrm{i}y)\mathrm{d}t$. Then, for $f \in L^1(\mathbb{R}, \mathrm{d}t)$,

$$\left| \int_{\mathbb{R}} f(t)\mathrm{d}\nu_y(t) \right| \leq \|V\| \int_{\mathbb{R}} |f(t)|\mathrm{d}t.$$

The map $\Phi_y(f) = \int_{\mathbb{R}} f \mathrm{d}\nu_y$ is a linear functional on $L^1(\mathbb{R}, \mathrm{d}t)$ and $\|\Phi_y\| \leq \|V\|$. By the Banach-Alaoglu theorem, there a bounded linear functional Φ and a sequence $y_n \downarrow 0$ such that for all $f \in L^1(\mathbb{R}, \mathrm{d}t)$,

$$\lim_{n \to \infty} \Phi_{y_n}(f) = \Phi(f).$$

Let $V \in L^\infty(\mathbb{R}, \mathrm{d}t)$ be such that $\Phi(f) = \int_{\mathbb{R}} V(t) f(t)\mathrm{d}t$. Let

$$f(t) = \pi^{-1} y((x-t)^2 + y^2)^{-1}.$$

A simple residue calculation yields

$$\Phi_{y_n}(f) = \pi^{-1} y \int_{\mathbb{R}} \frac{V(t + \mathrm{i}y_n)\mathrm{d}t}{(x-t)^2 + y^2} = V(x + \mathrm{i}(y + y_n)).$$

Taking $n \to \infty$, we get

$$V(x + \mathrm{i}y) = \pi^{-1} y \int_{\mathbb{R}} \frac{V(t)\mathrm{d}t}{(x-t)^2 + y^2}, \tag{24}$$

and Theorem 3.5 yields the statement. □

Remark 1. Theorem 3.13 can be also proven using Theorem 3.11. The above argument has the advantage that it extends to any Hardy class $H^p(\mathbb{C}_+)$.

Remark 2. In the proof we have also established the Poisson representation of V (the relation (24)).

The next proposition is known as Jensen's formula.

Proposition 3.16. *Assume that $U(z)$ is analytic for $|z| < 1$ and that $U(0) \neq 0$. Let $r \in (0,1)$ and assume that U has no zeros on the circle $|z| = r$. Let $\alpha_1, \alpha_2, \ldots, \alpha_n$ be the zeros of $U(z)$ in the region $|z| < r$, listed with multiplicities. Then*

$$|U(0)| \prod_{j=1}^{n} \frac{r}{|\alpha_j|} = \exp\left\{ \frac{1}{2\pi} \int_{-\pi}^{\pi} \log |U(r\mathrm{e}^{\mathrm{i}t})|\mathrm{d}t \right\}. \tag{25}$$

Remark. The Jensen formula holds even if U has zeros on $|z| = r$. We will only need the above elementary version.

Proof. Set

$$V(z) = U(z) \prod_{j=1}^{n} \frac{r^2 - \overline{\alpha}_j z}{r(\alpha_j - z)}.$$

Then for some $\epsilon > 0$ $V(z)$ has no zeros in the disk $|z| < r + \epsilon$ and the function $\log |V(z)|$ is harmonic in the same disk (see Theorem 13.12 in [27]). By the mean value theorem for harmonic functions,

$$\log |V(0)| = \frac{1}{2\pi} \int_{-\pi}^{\pi} \log |V(r\mathrm{e}^{\mathrm{i}\theta})| \mathrm{d}\theta.$$

The substitution yields the statement. □

Proof of Theorem 3.14. Setting $U(\mathrm{e}^{\mathrm{i}t}) = V(\tan(t/2))$, we have that

$$\int_{\mathbb{R}} \frac{|\log |V(x)||}{1+x^2} \mathrm{d}x = \frac{1}{2} \int_{-\pi}^{\pi} |\log |U(\mathrm{e}^{\mathrm{i}t})|| \mathrm{d}t. \tag{26}$$

Hence, it suffices to show that the integral on the r.h.s. is finite.

In the rest of the proof we will use the same notation as in the proof of Theorem 3.11. Recall that S and T are defined by (20) and (21). Let $U(z) = V(T(z))$. Then, U is holomorphic in D and $\sup_{z \in D} |U(z)| < \infty$. Moreover, a change of variables and the formula (24) yield that

$$U(r\mathrm{e}^{\mathrm{i}\theta}) = \frac{1}{2\pi} \int_{-\pi}^{\pi} \frac{1 - r^2}{1 + r^2 - 2r\cos(\theta - t)} U(\mathrm{e}^{\mathrm{i}t}) \mathrm{d}t.$$

(The change of variables exercise is done in detail in [19], pages 106-107.) The analog of Theorem 3.5 for the circle yields that for Lebesgue a.e. θ

$$\lim_{r \to 1} U(r\mathrm{e}^{\mathrm{i}\theta}) = U(\mathrm{e}^{\mathrm{i}\theta}). \tag{27}$$

The proof is outlined in the problem set.

We will now make use of the Jensen formula. If $U(0) = 0$, let m be such $U_m(z) = z^{-m} U(z)$ satisfies $U_m(0) \neq 0$ (if $U(0) \neq 0$, then $m = 0$). Let $r_j \to 1$ be a sequence such that U has no zeros on $|z| = r_j$. Set

$$J_{r_j} = \frac{1}{2\pi} \int_{-\pi}^{\pi} \log |U_m(r_j \mathrm{e}^{\mathrm{i}t})| \mathrm{d}t = -m \log r_j + \frac{1}{2\pi} \int_{-\pi}^{\pi} \log |U(r_j \mathrm{e}^{\mathrm{i}t})| \mathrm{d}t.$$

The Jensen formula (applied to U_m) yields that $J_{r_i} \le J_{r_j}$ if $r_i \le r_j$. Write $\log^+ x = \max(\log x, 0)$, $\log^- x = -\min(\log x, 0)$. Note that

$$\sup_t \log^+ |U(\mathrm{e}^{\mathrm{i}t})| \le \sup_t |U(\mathrm{e}^{\mathrm{i}t})| < \infty.$$

Fatou's lemma, the dominated convergence theorem and (27) yield that

$$J_{r_1} + \frac{1}{2\pi}\int_{-\pi}^{\pi} \log^- |U(\mathrm{e}^{\mathrm{i}t})| \mathrm{d}t \leq \frac{1}{2\pi}\int_{-\pi}^{\pi} \log^+ |U(\mathrm{e}^{\mathrm{i}t})| \mathrm{d}t < \infty.$$

Hence,

$$\int_{-\pi}^{\pi} |\log |U(\mathrm{e}^{\mathrm{i}t})|| \mathrm{d}t < \infty,$$

and the identity (26) yields the statement. □

3.8 The Borel transform of measures

Recall that the Borel transform $F_\mu(z)$ is defined by (8).

Theorem 3.17. *Let μ be a complex or positive measure. Then:*
(1) *For Lebesgue a.e. x the limit*

$$F_\mu(x) = \lim_{y \downarrow 0} F_\mu(x + \mathrm{i}y),$$

exists and is finite.
(2) *If $F_\mu \not\equiv 0$, then*

$$\int_{\mathbb{R}} \frac{|\log |F_\mu(x)||}{1 + x^2} \mathrm{d}x < \infty. \tag{28}$$

(3) *If $F_\mu \not\equiv 0$, then for any complex number α the set*

$$\{x \in \mathbb{R} : F_\mu(x) = \alpha\}, \tag{29}$$

has zero Lebesgue measure.

Remark. It is possible that

$$\mu \neq 0 \qquad \text{and} \qquad F_\mu \equiv 0. \tag{30}$$

For example, this is the case if $\mathrm{d}\mu = (x - 2\mathrm{i})^{-1}(x - \mathrm{i})^{-1}\mathrm{d}x$. By the theorem of F. & M. Riesz (see, e.g., [19]), if (30) holds, then $\mathrm{d}\mu = h(x)\mathrm{d}x$, where $h(x) \neq 0$ for Lebesgue a.e. x. We will prove the F. & M. Riesz theorem in Section 5.

Proof. We will first show that

$$F_\mu(z) = \frac{R(z)}{G(z)}, \tag{31}$$

where $R, G \in H^\infty(\mathbb{C}_+)$ and G has no zeros in $\mathbb{C}_+$. If μ is positive, set

$$G(z) = \frac{1}{\mathrm{i} + F_\mu(z)}.$$

Then, $G(z)$ is holomorphic in $\mathbb{C}_+$, $|G(z)| \leq 1$ (since $\operatorname{Im} F_\mu(z) \geq 0$), and

$$F_\mu(z) = \frac{1 - \mathrm{i}G(z)}{G(z)}.$$

If μ is a complex measure, we first decompose $\mu = (\mu_1 - \mu_2) + \mathrm{i}(\mu_3 - \mu_4)$, where the μ_i's are positive measures, and then decompose

$$F_\mu(z) = (F_{\mu_1}(z) - F_{\mu_2}(z)) + \mathrm{i}(F_{\mu_3}(z) - F_{\mu_4}(z)).$$

Hence, (31) follows from the corresponding result for positive measures.

Proof of (1): By Theorems 3.13 and 3.14, the limits $R(x) = \lim_{y\downarrow 0} R(x + \mathrm{i}y)$ and $G(x) = \lim_{y\downarrow 0} G(x+\mathrm{i}y)$ exist and $G(x)$ is non-zero for Lebesgue a.e. x. Hence, for Lebesgue a.e. x,

$$F_\mu(x) = \lim_{y\downarrow 0} F_\mu(x + \mathrm{i}y) = \frac{R(x)}{G(x)}.$$

Proof of (2): $F_\mu(x)$ is zero on a set of positive measure iff $R(x)$ is, and if this is the case, $R \equiv 0$ and then $F_\mu \equiv 0$. Hence, if $F_\mu \not\equiv 0$, then $R \not\equiv 0$. Obviously,

$$|\log|F_\mu(x)|| \leq |\log|R(x)|| + |\log|G(x)||,$$

and (28) follows from Theorem 3.14.

Proof of (3): The sets $\{x : F_\mu(x) = \alpha\}$ and $\{x : R(x) - \alpha G(x) = 0\}$ have the same Lebesgue measure. If the second set has positive Lebesgue measure, then by Theorem 3.14, $R(z) = \alpha G(z)$ for all $z \in \mathbb{C}_+$, and $F_\mu(z) \equiv \alpha$. Since $\lim_{y\to\infty} |F_\mu(x+\mathrm{i}y)| = 0$, $\alpha = 0$, and so $F_\mu \equiv 0$. Hence, if the set $\{x : F_\mu(x) = \alpha\}$ has positive Lebesgue measure, then $\alpha = 0$ and $\mu = 0$. □

The final result we would like to mention is the theorem of Poltoratskii [21].

Theorem 3.18. *Let ν be a complex and μ a positive measure. Let $\nu = f\mu + \nu_{\mathrm{s}}$ be the Radon-Nikodym decomposition. Let μ_{sing} be the part of μ singular with respect to the Lebesgue measure. Then*

$$\lim_{y\downarrow 0} \frac{F_\nu(x + \mathrm{i}y)}{F_\mu(x + \mathrm{i}y)} = f(x), \quad \text{for } \mu_{\mathrm{sing}} - a.e.\ x. \tag{32}$$

This theorem has played an important role in the recent study of the spectral structure of Anderson type Hamiltonians [15, 16].

Poltoratskii's proof of Theorem 3.18 is somewhat complicated, partly since it is done in the framework of a theory that is also concerned with other questions. A relatively simple proof of Poltoratskii's theorem has recently been found in [14]. This new proof is based on the spectral theorem for self-adjoint operators and rank one perturbation theory, and will be discussed in Section 5.

3.9 Problems

[1] (1) *Prove Lemma 3.3.*
(2) *Assume that* μ *satisfies (9). Prove that the set of* x *for which (11) holds is* G_δ *(countable intersection of open sets) in* $\operatorname{supp}\mu$.

[2] (1) *Let* $C_0(\mathbb{R})$ *be the usual Banach space of continuous function on* $\mathbb{R}$ *vanishing at infinity with norm* $\|f\| = \sup|f(x)|$. *For* $f \in C_0(\mathbb{R})$ *let*

$$f_y(x) = \frac{y}{\pi}\int_{\mathbb{R}} \frac{f(t)}{(x-t)^2+y^2}\mathrm{d}t.$$

Prove that $\lim_{y\downarrow 0}\|f_y - f\| = 0$.
(2) *Prove that the linear span of the set of functions*

$$\{((x-a)^2+b^2)^{-1} : a \in \mathbb{R}, b > 0\},$$

is dense in $C_0(\mathbb{R})$.
(3) *Prove that the linear span of the set of functions* $\{(x-z)^{-1} : z \in \mathbb{C}\setminus\mathbb{R}\}$ *is dense in* $C_0(\mathbb{R})$.
Hint: To prove (1), you may argue as follows. Fix $\epsilon > 0$. *Let* $\delta > 0$ *be such that* $|t-x| < \delta \Rightarrow |f(t)-f(x)| < \epsilon$. *The estimates*

$$\begin{aligned}|f_y(x) - f(x)| &\le \frac{y}{\pi}\int_{\mathbb{R}} \frac{|f(t)-f(x)|}{(x-t)^2+y^2}\mathrm{d}t \\ &\le \epsilon + 2\|f\|\frac{y}{\pi}\int_{|t-x|>\delta}\frac{1}{(x-t)^2+y^2}\mathrm{d}t \\ &\le \epsilon + 4\pi^{-1}\|f\|y/\delta,\end{aligned}$$

yield that $\limsup_{y\downarrow 0}\|f_y - f\| \le \epsilon$. *Since* ϵ *is arbitrary, (1) follows. Approximating* f_y *by Riemann sums deduce that (1)* $\Rightarrow$ *(2). Obviously, (2)* $\Rightarrow$ *(3).*

[3] *Prove Part (3) of Theorem 3.8.*

[4] *Prove the following converse of Theorem 3.8: If (1) and (2) hold, then for* $[a,b] \subset A$,

$$\sup_{0<y<1}\int_a^b |P_\nu(x+\mathrm{i}y)|^p\mathrm{d}x < \infty.$$

[5] *The following extension of Theorem 3.9 holds. Let* ν *be a finite positive measure. Then for any* $p > 1$,

$$\lim_{y\downarrow 0} y^{p-1}\int_a^b P_\nu(x+\mathrm{i}y)^p\mathrm{d}x = C_p\left(\frac{\nu(\{a\})^p}{2} + \frac{\nu(\{b\})^p}{2} + \sum_{x\in(a,b)}\nu(\{x\})^p\right).$$

Prove this and compute C_p in terms of gamma functions. Hint: See Remark 1 after Theorem 2.2 in [28].

[6] *Prove the formula (19).*

[7] *Let μ be a complex measure. Prove that $F_\mu \equiv 0 \Rightarrow \mu = 0$ if either one of the following holds:*
(a) *μ is real-valued.*
(b) *$|\mu|(S) = 0$ for some open set S.*
(c) *$\int_{\mathbb{R}} \exp(p|x|)\mathrm{d}|\mu| < \infty$ for some $p > 0$.*

[8] *Let μ be a complex or positive measure on $\mathbb{R}$ and*

$$H_\mu(z) = \pi^{-1}(\mathrm{i}P_\mu(z) - F_\mu(z)) = \frac{1}{\pi}\int_{\mathbb{R}} \frac{(x-t)\mathrm{d}\mu(t)}{(x-t)^2+y^2}.$$

By Theorems 3.5 and 3.17, for Lebesgue a.e. x the limit

$$H_\mu(x) = \lim_{y\downarrow 0} H_\mu(x+\mathrm{i}y),$$

exists and is finite. If $\mathrm{d}\mu = f\mathrm{d}x$, we will denote $H_\mu(z)$ and $H_\mu(x)$ by $H_f(z)$ and $H_f(x)$. The function $H_\mu(x)$ is called the Hilbert transform of the measure μ (H_f is called the Hilbert transform of the function f).
(1) *Prove that for Lebesgue a.e. x the limit*

$$\lim_{\epsilon\to 0}\frac{1}{\pi}\int_{|t-x|>\epsilon}\frac{\mathrm{d}\mu(t)}{x-t},$$

exists and is equal to $H_\mu(x)$.
(2) *Assume that $f \in L^p(\mathbb{R}, \mathrm{d}x)$ for some $1 < p < \infty$. Prove that*

$$\sup_{y>0}\int_{\mathbb{R}} |H_f(x+\mathrm{i}y)|^p \mathrm{d}x < \infty,$$

and deduce that $H_f \in L^p(\mathbb{R}, \mathrm{d}x)$.
(3) *If $f \in L^2(\mathbb{R}, \mathrm{d}x)$, prove that $H_{H_f} = -f$ and deduce that*

$$\int_{\mathbb{R}} |H_f(x)|^2 \mathrm{d}x = \int_{\mathbb{R}} |f(x)|^2 \mathrm{d}x.$$

[9] *Let $1 \le p < \infty$. The Hardy class $H^p(\mathbb{C}_+)$ is the vector space of all analytic functions f on $\mathbb{C}_+$ such that*

$$\|f\|_p^p = \sup_{y>0}\int_{\mathbb{R}} |f(x+\mathrm{i}y)|^p \mathrm{d}x < \infty.$$

(1) *Prove that* $\|\cdot\|_p$ *is a norm and that* $H^p(\mathbb{C}_+)$ *is a Banach space.*
(2) *Let* $f \in H^p(\mathbb{C}_+)$. *Prove that the limit*

$$f(x) = \lim_{y \downarrow 0} f(x + \mathrm{i}y),$$

exists for Lebesgue a.e. x *and that* $f \in L^p(\mathbb{R}, \mathrm{d}x)$. *Prove that*

$$f(x + \mathrm{i}y) = \frac{y}{\pi} \int_{\mathbb{R}} \frac{f(t)}{(x-t)^2 + y^2} \mathrm{d}t.$$

(3) *Prove that* $H^2(\mathbb{C}_+)$ *is a Hilbert space and that*

$$\sup_{y>0} \int_{\mathbb{R}} |f(x + \mathrm{i}y)|^2 \mathrm{d}x = \int_{\mathbb{R}} |f(x)|^2 \mathrm{d}x.$$

Hence, $H^2(\mathbb{C}_+)$ *can be identified with a subspace of* $L^2(\mathbb{R}, \mathrm{d}x)$ *which we denote by the same letter. Let* $\overline{H}^2(\mathbb{C}_+) = \{f \in L^2(\mathbb{R}, \mathrm{d}x) : \overline{f} \in H^2(\mathbb{C}_+)\}$. *Prove that*

$$L^2(\mathbb{R}, \mathrm{d}x) = H^2(\mathbb{C}_+) \oplus \overline{H}^2(\mathbb{C}_+).$$

[10] *In this problem we will study the Poisson transform on the circle. Let* $\Gamma = \{z : |z| = 1\}$ *and let* μ *be a complex measure on* Γ. *The Poisson transform of the measure* μ *is*

$$P_\mu(z) = \int_\Gamma \frac{1 - |z|^2}{|z - w|^2} \mathrm{d}\mu(w).$$

If we parametrize Γ *by* $w = \mathrm{e}^{\mathrm{i}t}$, $t \in (-\pi, \pi]$ *and denote the induced complex measure by* $\mu(t)$, *then*

$$P_\mu(r\mathrm{e}^{\mathrm{i}\theta}) = \int_{-\pi}^{\pi} \frac{1 - r^2}{1 + r^2 - 2r\cos(\theta - t)} \mathrm{d}\mu(t).$$

Note also that if $\mathrm{d}\mu(t) = \mathrm{d}t$, *then* $P_\mu(z) = 2\pi$. *For* $w \in \Gamma$ *we denote by* $I(w, r)$ *the arc of length* $2r$ *centered at* w. *Let* ν *be a complex measure and* μ *a finite positive measure on* Γ. *The corresponding maximal function is defined by*

$$M_{\nu,\mu}(w) = \sup_{r>0} \frac{|\nu|(I(w,r))}{\mu(I(w,r))},$$

if $x \in \operatorname{supp} \mu$, *otherwise* $M_{\nu,\mu}(w) = \infty$.
(1) *Formulate and prove the Besicovitch covering lemma for the circle.*
(2) *Prove the following bound: For all* $r \in [0, 1)$ *and* $\theta \in (-\pi, \pi]$,

$$\frac{|P_\nu(r\mathrm{e}^{\mathrm{i}\theta})|}{P_\mu(r\mathrm{e}^{\mathrm{i}\theta})} \leq M_{\nu,\mu}(\mathrm{e}^{\mathrm{i}\theta}).$$

You may either mimic the proof of Lemma 3.2, or follow the proof of Theorem 11.20 in [27].

(3) *State and prove the analog of Theorem 3.5 for the circle.*

(4) *State and prove the analogs of Theorems 3.7 and 3.13 for the circle.*

[11] *In Part (4) of the previous problem you were asked to prove the relation (27). This relation could be also proved like follows. Let*

$$p_{r,\theta}(t) = \frac{1 - r^2}{1 + r^2 - 2r\cos(\theta - t)}.$$

Show first that

$$\begin{aligned} \limsup_{r \to 1} |U(r\mathrm{e}^{\mathrm{i}\theta}) - U(\mathrm{e}^{\mathrm{i}\theta})| &\le \limsup_{r \to 1} \frac{1}{2\pi} \int_{-\pi}^{\pi} p_{r,\theta}(t) |U(\mathrm{e}^{\mathrm{i}t}) - U(\mathrm{e}^{\mathrm{i}\theta})| \mathrm{d}t \\ &\le \limsup_{\epsilon \downarrow 0} \frac{1}{2\epsilon} \int_{I(\theta,\epsilon)} |U(\mathrm{e}^{\mathrm{i}t}) - U(\mathrm{e}^{\mathrm{i}\theta})| \mathrm{d}t, \end{aligned}$$

and then use Problem 9 of Section 2.

[12] *The goal of this problem is to extend all the results of this section to non-tangential limits. Our description of non-tangential limits follows* [21]. *Let again* $\Gamma = \{z : |z| = 1\}$ *and* $D = \{z : |z| < 1\}$. *Let* $w \in \Gamma$. *We say that* z *tends to* w *non-tangentially, and write*

$$z \underset{\angle}{\to} w$$

if z *tends to* w *inside the region*

$$\Delta_w^\varphi = \{z \in D : |\mathrm{Arg}(1 - z\overline{w})| < \varphi\},$$

for all $\varphi \in (0, \pi/2)$. $\mathrm{Arg}(z)$ *is the principal branch of the argument with values in* $(-\pi, \pi]$. *In the sector* Δ_w^φ *inscribe a circle centered at the origin (we denote it by* Γ_φ*). The two points on* $\Gamma_\varphi \cap \{z : \mathrm{Arg}(1 - z\overline{w}) = \pm\varphi\}$ *divide the circle into two arcs. The open region bounded by the shorter arc and the rays* $\mathrm{Arg}(1 - z\overline{w}) = \pm\varphi$ *is denoted* C_w^φ. *Let* ν *and* μ *be as in Problem 10.*

(1) *Let* $\varphi \in (0, \pi/2)$ *be given. Then there is a constant* C *such that*

$$\sup_{z \in \mathrm{C}_w^\varphi} \frac{|P_\nu(z)|}{P_\mu(z)} \le C M_{\nu,\mu}(w), \qquad \text{for } \mu - a.e.\, w. \tag{33}$$

This is the key result which extends the radial estimate of Part (2) of Problem 10. The passage from the radial estimate to (33) is similar to the proof of Harnack's lemma. Write the detailed proof following Lemma 1.2 of [21].

(2) *Let* $\nu = f\mu + \nu_{\mathrm{s}}$ *be the Radon-Nikodym decomposition. Prove that*

$$\lim_{z \underset{\angle}{\to} w} \frac{P_\nu(z)}{P_\mu(z)} = f(w), \qquad \text{for } \mu - a.e. w.$$

If ν is a positive measure, prove that

$$\lim_{\substack{z \to w \\ \angle}} \frac{P_\nu(z)}{P_\mu(z)} = \infty, \qquad \text{for } \nu_s - a.e.\, w.$$

(3) *Extend Parts (3) and (4) of Problem 10 to non-tangential limits.*
(4) *Consider now $\mathbb{C}_+$. We say that z tends to x non-tangentially if for all $\varphi \in (0, \pi/2)$ z tends to x inside the cone $\{z : |\mathrm{Arg}(z - x) - \pi/2| < \varphi\}$. Let T be the conformal mapping (21). Prove that $z \to w$ non-tangentially in D iff $T(z) \to T(w)$ non-tangentially in $\mathbb{C}_+$. Using this observation extend all the results of this section to non-tangential limits.*

4 Self-adjoint operators, spectral theory

4.1 Basic notions

Let $\mathcal{H}$ be a Hilbert space. We denote the inner product by $(\cdot|\cdot)$ (the inner product is linear w.r.t. the second variable).

A linear operator on $\mathcal{H}$ is a pair $(A, \mathrm{Dom}\,(A))$, where $\mathrm{Dom}\,(A) \subset \mathcal{H}$ is a vector subspace and $A : \mathrm{Dom}\,(A) \to \mathcal{H}$ is a linear map. We set

$$\mathrm{Ker}\, A = \{\psi \in \mathrm{Dom}\,(A) \,:\, A\psi = 0\}, \qquad \mathrm{Ran}\, A = \{A\psi \,:\, \psi \in \mathrm{Dom}\,(A)\}.$$

An operator A is densely defined if $\mathrm{Dom}\,(A)$ is dense in $\mathcal{H}$. If A and B are linear operators, then $A + B$ is defined on $\mathrm{Dom}\,(A + B) = \mathrm{Dom}\,(A) \cap \mathrm{Dom}\,(B)$ in the obvious way. For any $z \in \mathbb{C}$ we denote by $A + z$ the operator $A + z\mathbf{1}$, where $\mathbf{1}$ is the identity operator. Similarly,

$$\mathrm{Dom}\,(AB) = \{\psi : \psi \in \mathrm{Dom}\,(B), B\psi \in \mathrm{Dom}\,(A)\},$$

and $(AB)\psi = A(B\psi)$. $A = B$ if $\mathrm{Dom}\,(A) = \mathrm{Dom}\,(B)$ and $A\psi = B\psi$. The operator B is called an extension of A if $\mathrm{Dom}\,(A) \subset \mathrm{Dom}\,(B)$ and $A\psi = B\psi$ for $\psi \in \mathrm{Dom}\,(A)$. If B is an extension of A one writes $A \subset B$.

The operator A is called bounded if $\mathrm{Dom}\,(A) = \mathcal{H}$ and

$$\|A\| = \sup_{\|\psi\|=1} \|A\psi\| < \infty. \tag{34}$$

We denote by $\mathcal{B}(\mathcal{H})$ the vector space of all bounded operators on $\mathcal{H}$. $\mathcal{B}(\mathcal{H})$ with the norm (34) is a Banach space. If A is densely defined and there is a constant C such that for all $\psi \in \mathrm{Dom}\,(A)$, $\|A\psi\| \le C\|\psi\|$, then A has a unique extension to a bounded operator on $\mathcal{H}$. An operator $P \in \mathcal{B}(\mathcal{H})$ is called a projection if $P^2 = P$. An operator $U \in \mathcal{B}(\mathcal{H})$ is called unitary if U is onto and $(U\phi|U\psi) = (\phi|\psi)$ for all $\phi, \psi \in \mathcal{H}$.

The graph of a linear operator A is defined by

$$\Gamma(A) = \{(\psi, A\psi) \,:\, \psi \in \mathrm{Dom}\,(A)\} \subset \mathcal{H} \oplus \mathcal{H}.$$

Note that $A \subset B$ if $\Gamma(A) \subset \Gamma(B)$. A linear operator A is called *closed* if $\Gamma(A)$ is a closed subset of $\mathcal{H} \oplus \mathcal{H}$.

A is called closable if it has a closed extension. If A is closable, its smallest closed extension is called the closure of A and is denoted by $\overline{A}$. It is not difficult to show that A is closable iff $\overline{\Gamma(A)}$ is the graph of a linear operator and in this case $\Gamma(\overline{A}) = \overline{\Gamma(A)}$.

Let A be closed. A subset $D \subset \mathrm{Dom}\,(A)$ is called a *core* for A if $\overline{A \restriction D} = A$.

Let A be a densely defined linear operator. Its adjoint, A^*, is defined as follows. $\mathrm{Dom}\,(A^*)$ is the set of all $\phi \in \mathcal{H}$ for which there exists a $\psi \in \mathcal{H}$ such that

$$(A\varphi|\phi) = (\varphi|\psi), \qquad \text{for all } \varphi \in \mathrm{Dom}\,(A).$$

Obviously, such ψ is unique and $\mathrm{Dom}\,(A^*)$ is a vector subspace. We set $A^*\phi = \psi$. It may happen that $\mathrm{Dom}\,(A^*) = \{0\}$. If $\mathrm{Dom}\,(A^*)$ is dense, then $A^{**} = (A^*)^*$, etc.

Theorem 4.1. *Let A be a densely defined linear operator. Then:*
(1) *A^* is closed.*
(2) *A is closable iff* $\mathrm{Dom}\,(A^*)$ *is dense, and in this case* $\overline{A} = A^{**}$.
(3) *If A is closable, then* $\overline{A}^* = A^*$.

Let A be a closed densely defined operator. We denote by $\rho(A)$ the set of all $z \in \mathbb{C}$ such that

$$A - z : \mathrm{Dom}\,(A) \to \mathcal{H}$$

is a bijection. By the closed graph theorem, if $z \in \rho(A)$, then $(A - z)^{-1} \in \mathcal{B}(\mathcal{H})$. The set $\rho(A)$ is called the resolvent set of A. The spectrum of A, $\mathrm{sp}(A)$, is defined by

$$\mathrm{sp}(A) = \mathbb{C} \setminus \rho(A).$$

A point $z \in \mathbb{C}$ is called an eigenvalue of A if there is a $\psi \in \mathrm{Dom}\,(A)$, $\psi \neq 0$, such that $A\psi = z\psi$. The set of all eigenvalues is called the point spectrum of A and is denoted by $\mathrm{sp_p}(A)$. Obviously, $\mathrm{sp_p}(A) \subset \mathrm{sp}(A)$. It is possible that $\mathrm{sp}(A) = \mathrm{sp_p}(A) = \mathbb{C}$. It is also possible that $\mathrm{sp}(A) = \emptyset$. (For simple examples see [23], Example 5 in Chapter VIII).

Theorem 4.2. *Assume that $\rho(A)$ is non-empty. Then $\rho(A)$ is an open subset of $\mathbb{C}$ and the map*

$$\rho(A) \ni z \mapsto (A - z)^{-1} \in \mathcal{B}(\mathcal{H}),$$

is (norm) analytic. Moreover, if $z_1, z_2 \in \rho(A)$, then

$$(A - z_1)^{-1} - (A - z_2)^{-1} = (z_1 - z_2)(A - z_1)^{-1}(A - z_2)^{-1}.$$

The last relation is called the resolvent identity.

4.2 Digression: The notions of analyticity

Let $\Omega \subset \mathbb{C}$ be an open set and X a Banach space. A function $f : \Omega \to X$ is called norm analytic if for all $z \in \Omega$ the limit

$$\lim_{w \to z} \frac{f(w) - f(z)}{w - z},$$

exists in the norm of X. f is called weakly analytic if $x^* \circ f : \Omega \to \mathbb{C}$ is analytic for all $x^* \in X^*$. Obviously, if f is norm analytic, then f is weakly analytic. The converse also holds and we have:

Theorem 4.3. *f is norm analytic iff f is weakly analytic.*

For the proof, see [23].

The mathematical theory of Banach space valued analytic functions parallels the classical theory of analytic functions. For example, if γ is a closed path in a simply connected domain Ω, then

$$\oint_\gamma f(z)\mathrm{d}z = 0. \tag{35}$$

(The integral is defined in the usual way by the norm convergent Riemann sums.) To prove (35), note that for $x^* \in X^*$,

$$x^* \left(\oint_\gamma f(z)\mathrm{d}z \right) = \oint_\gamma x^*(f(z))\mathrm{d}z = 0.$$

Since X^* separates points in X, (35) holds. Starting with (35) one obtains in the usual way the Cauchy integral formula,

$$\frac{1}{2\pi \mathrm{i}} \oint_{|w-z|=r} \frac{f(w)}{w - z} \mathrm{d}w = f(z).$$

Starting with the Cauchy integral formula one proves that for $w \in \Omega$,

$$f(z) = \sum_{n=0}^{\infty} a_n (z - w)^n, \tag{36}$$

where $a_n \in X$. The power series converges and the representation (36) holds in the largest open disk centered at w and contained in Ω, etc.

4.3 Elementary properties of self-adjoint operators

Let A be a densely defined operator on a Hilbert space $\mathcal{H}$. A is called symmetric if $\forall \phi, \psi \in \mathrm{Dom}\,(A)$,

$$(A\phi|\psi) = (\phi|A\psi).$$

In other words, A is symmetric if $A \subset A^*$. Obviously, any symmetric operator is closable.

A densely defined operator A is called *self-adjoint* if $A = A^*$. A is self-adjoint iff A is symmetric and $\mathrm{Dom}\,(A) = \mathrm{Dom}\,(A^*)$.

Theorem 4.4. *Let A be a symmetric operator on $\mathcal{H}$. Then the following statements are equivalent:*
(1) *A is self-adjoint.*
(2) *A is closed and* $\operatorname{Ker}(A^* \pm \mathrm{i}) = \{0\}$.
(3) $\operatorname{Ran}(A \pm \mathrm{i}) = \mathcal{H}$.

A symmetric operator A is called essentially self-adjoint if $\overline{A}$ is self-adjoint.

Theorem 4.5. *Let A be a symmetric operator on $\mathcal{H}$. Then the following statements are equivalent:*
(1) *A is essentially self-adjoint.*
(2) $\operatorname{Ker}(A^* \pm \mathrm{i}) = \{0\}$.
(3) $\operatorname{Ran}(A \pm \mathrm{i})$ *are dense in $\mathcal{H}$.*

Remark. In Parts (2) and (3) of Theorems 4.4 and 4.5 $\pm\mathrm{i}$ can be replaced by $z, \overline{z}$, for any $z \in \mathbb{C} \setminus \mathbb{R}$.

Theorem 4.6. *Let A be self-adjoint. Then:*
(1) *If $z = x + \mathrm{i}y$, then for $\psi \in \operatorname{Dom}(A)$,*

$$\|(A - z)\psi\|^2 = \|(A - x)\psi\|^2 + y^2\|\psi\|^2.$$

(2) $\operatorname{sp}(A) \subset \mathbb{R}$ *and for* $z \in \mathbb{C} \setminus \mathbb{R}$, $\|(A - z)^{-1}\| \leq |\operatorname{Im} z|^{-1}$.
(3) *For any $x \in \mathbb{R}$ and $\psi \in \mathcal{H}$,*

$$\lim_{y\to\infty} \mathrm{i}y(A - x - \mathrm{i}y)^{-1}\psi = -\psi.$$

(4) *If $\lambda_1, \lambda_2 \in \operatorname{sp}_{\mathrm{p}}(A)$, $\lambda_1 \neq \lambda_2$, and ψ_1, ψ_2 are corresponding eigenvectors, then $\psi_1 \perp \psi_2$.*

Proof. (1) follows from a simple computation:

$$\begin{aligned}\|(A - x - \mathrm{i}y)\psi\|^2 &= ((A - x - \mathrm{i}y)\psi|(A - x - \mathrm{i}y)\psi)\\ &= \|(A - x)\psi\|^2 + y^2\|\psi\|^2\\ &\quad + \mathrm{i}y((A - x)\psi|\psi) - \mathrm{i}y((A - x)\psi|\psi)\\ &= \|(A - x)\psi\|^2 + y^2\|\psi\|^2.\end{aligned}$$

(2) Let $z \in \mathbb{C} \setminus \mathbb{R}$. By (1), if $(A - z)\psi = 0$, then $\psi = 0$, and so

$$A - z : \operatorname{Dom}(A) \to \mathcal{H},$$

is one-one. $\operatorname{Ran}(A - z) = \mathcal{H}$ by Theorem 4.4. Let us prove this fact directly. We will show first that $\operatorname{Ran}(A - z)$ is dense. Let $\psi \in \mathcal{H}$ such that $((A - z)\phi|\psi) = 0$ for all $\phi \in \operatorname{Dom}(A)$. Then $\psi \in \operatorname{Dom}(A)$ and $(\psi|A\psi) = \overline{z}\|\psi\|^2$. Since

$$(\psi|A\psi) \in \mathbb{R},$$

and $\mathrm{Im}\, z \neq 0$, $\psi = 0$. Hence, Ran $(A - z)$ is dense. Let $\psi_n = (A - z)\phi_n$ be a Cauchy sequence. Then, by (1), ϕ_n is also a Cauchy sequence, and since A is closed, Ran $(A-z)$ is closed. Hence, Ran $(A-z) = \mathcal{H}$ and $z \in \rho(A)$. Finally, the estimate $\|(A-z)^{-1}\| \leq |\mathrm{Im}\, z|^{-1}$ is an immediate consequence of (1).
(3) By replacing A with $A - x$, w.l.o.g. we may assume that $x = 0$. We consider first the case $\psi \in$ Dom (A). The identity

$$\mathrm{i}y(A - \mathrm{i}y)^{-1}\psi + \psi = (A - \mathrm{i}y)^{-1}A\psi,$$

and (2) yield that $\|\mathrm{i}y(A - \mathrm{i}y)^{-1}\psi + \psi\| \leq \|A\psi\|/y$, and so (3) holds. If

$$\psi \notin \mathrm{Dom}\,(A),$$

let $\psi_n \in$ Dom (A) be a sequence such that $\|\psi_n - \psi\| \leq 1/n$. We estimate

$$\begin{aligned}
\|\mathrm{i}y(A - \mathrm{i}y)^{-1}\psi + \psi\| &\leq \|\mathrm{i}y(A - \mathrm{i}y)^{-1}(\psi - \psi_n)\| + \|\psi - \psi_n\| \\
&\quad + \|(\mathrm{i}y(A - \mathrm{i}y)^{-1}\psi_n + \psi_n\| \\
&\leq 2\|\psi - \psi_n\| + \|(\mathrm{i}y(A - \mathrm{i}y)^{-1}\psi_n + \psi_n\| \\
&\leq 2/n + \|A\psi_n\|/y.
\end{aligned}$$

Hence,

$$\limsup_{y \to \infty} \|\mathrm{i}y(A - \mathrm{i}y)^{-1}\psi + \psi\| \leq 2/n.$$

Since n is arbitrary, (3) follows.
(4) Note that

$$\lambda_1(\psi_1|\psi_2) = (A\psi_1|\psi_2) = (\psi_1|A\psi_2) = \lambda_2(\psi_1|\psi_2).$$

Since $\lambda_1 \neq \lambda_2$, $(\psi_1|\psi_2) = 0$. □

A self-adjoint operator A is called positive if

$$(\psi|A\psi) \geq 0,$$

for all $\psi \in$ Dom (A). If A and B are bounded and self-adjoint, then obviously $A \pm B$ are also self-adjoint; we write $A \geq B$ if $A - B \geq 0$.

A self-adjoint projection P is called an orthogonal projection. In this case $\mathcal{H} = \mathrm{Ker}\, P \oplus \mathrm{Ran}\, P$. We write $\dim P = \dim \mathrm{Ran}\, P$.

Let A be a bounded operator on $\mathcal{H}$. The real and the imaginary part of A are defined by

$$\mathrm{Re}A = \frac{1}{2}(A + A^*), \qquad \mathrm{Im}A = \frac{1}{2\mathrm{i}}(A - A^*).$$

Clearly, ReA and ImA are self-adjoint operators and $A = \mathrm{Re}A + \mathrm{i}\mathrm{Im}A$.

4.4 Direct sums and invariant subspaces

Let $\mathcal{H}_1, \mathcal{H}_2$ be Hilbert spaces and A_1, A_2 self-adjoint operators on $\mathcal{H}_1, \mathcal{H}_2$. Then, the operator $A = A_1 \oplus A_2$ with the domain

$$\mathrm{Dom}\,(A) = \mathrm{Dom}\,(A_1) \oplus \mathrm{Dom}\,(A_2),$$

is self-adjoint. Obviously, $(A - z)^{-1} = (A_1 - z)^{-1} \oplus (A_2 - z)^{-1}$.

This elementary construction has a partial converse. Let A be a self-adjoint operator on a Hilbert space $\mathcal{H}$ and let $\mathcal{H}_1$ be a closed subspace of $\mathcal{H}$. The subspace $\mathcal{H}_1$ is *invariant* under A if for all $z \in \mathbb{C} \setminus \mathbb{R}$, $(A - z)^{-1}\mathcal{H}_1 \subset \mathcal{H}_1$. Obviously, if $\mathcal{H}_1$ is invariant under A, so is $\mathcal{H}_2 = \mathcal{H}_1^\perp$. Set

$$\mathrm{Dom}\,(A_i) = \mathrm{Dom}\,(A) \cap \mathcal{H}_i, \qquad A_i\psi = A\psi, \qquad i = 1, 2,$$

A_i is a self-adjoint operator on $\mathcal{H}_i$ and $A = A_1 \oplus A_2$. We will call A_1 the restriction of A to the invariant subspace $\mathcal{H}_1$ and write $A_1 = A \restriction \mathcal{H}_1$.

Let Γ be a countable set and $\mathcal{H}_n$, $n \in \Gamma$, a collection of Hilbert spaces. The direct sum of this collection,

$$\mathcal{H} = \bigoplus_n \mathcal{H}_n,$$

is the set of all sequences $\{\psi_n\}_{n\in\Gamma}$ such that $\psi_n \in \mathcal{H}_n$ and

$$\sum_{n\in\Gamma} \|\psi_n\|^2_{\mathcal{H}_n} < \infty.$$

$\mathcal{H}$ is a Hilbert space with the inner product

$$(\phi|\psi) = \sum_{n\in\Gamma} (\phi_n|\psi_n)_{\mathcal{H}_n}.$$

Let $B_n \in \mathcal{B}(\mathcal{H}_n)$ and assume that $\sup_n \|B_n\| < \infty$. Then

$$B\{\psi_n\}_{n\in\Gamma} = \{B_n\psi_n\}_{n\in\Gamma},$$

is a bounded operator on $\mathcal{H}$ and $\|B\| = \sup_n \|B_n\|$.

Proposition 4.7. *Let A_n be self-adjoint operators on $\mathcal{H}_n$. Set*

$$\mathrm{Dom}\,(A) = \{\psi = \{\psi_n\} \in \mathcal{H} : \psi_n \in \mathrm{Dom}\,(A_n),\ \sum_n \|A_n\psi_n\|^2_{\mathcal{H}_n} < \infty\},$$

$$A\psi = \{A_n\psi_n\}.$$

Then A is a self-adjoint operator on $\mathcal{H}$. We write

$$A = \bigoplus_{n\in\Gamma} A_n.$$

Moreover:
(1) *For $z \in \mathbb{C} \setminus \mathbb{R}$, $(A - z)^{-1} = \oplus_n (A_n - z)^{-1}$.*
(2) $\mathrm{sp}(A) = \overline{\cup_n \mathrm{sp}(A_n)}$.

The proof of Proposition 4.7 is easy and is left to the problems.

4.5 Cyclic spaces and the decomposition theorem

Let $\mathcal{H}$ be a separable Hilbert space and A a self-adjoint operator on $\mathcal{H}$. A collection of vectors $\mathcal{C} = \{\psi_n\}_{n\in\Gamma}$, where Γ is a countable set, is called *cyclic* for A if the closure of the linear span of the set of vectors

$$\{(A-z)^{-1}\psi_n \,:\, n \in \Gamma, z \in \mathbb{C}\setminus\mathbb{R}\},$$

is equal to $\mathcal{H}$. A cyclic set for A always exists (take an orthonormal basis for $\mathcal{H}$). If $\mathcal{C} = \{\psi\}$, then ψ is called a cyclic vector for A.

Theorem 4.8. (The decomposition theorem) *Let $\mathcal{H}$ be a separable Hilbert space and A a self-adjoint operator on $\mathcal{H}$. Then there exists a countable set Γ, a collection of mutually orthogonal closed subspaces $\{\mathcal{H}_n\}_{n\in\Gamma}$ of $\mathcal{H}$ ($\mathcal{H}_n \perp \mathcal{H}_m$ if $n \neq m$), and self-adjoint operators A_n on $\mathcal{H}_n$ such that:*
(1) *For all $n \in \Gamma$ there is a $\psi_n \in \mathcal{H}_n$ cyclic for A_n.*
(2) *$\mathcal{H} = \oplus_n \mathcal{H}_n$ and $A = \oplus_n A_n$.*

Proof. Let $\{\phi_n : n = 1, 2, \cdots\}$ be a given cyclic set for A. Set $\psi_1 = \phi_1$ and let $\mathcal{H}_1$ be the cyclic space generated by A and ψ_1 ($\mathcal{H}_1$ is the closure of the linear span of the set of vectors $\{(A-z)^{-1}\psi_1 : z \in \mathbb{C}\setminus\mathbb{R}\}$). By Theorem 4.6, $\psi_1 \in \mathcal{H}_1$. Obviously, $\mathcal{H}_1$ is invariant under A and we set $A_1 = A \restriction \mathcal{H}_1$.

We define ψ_n, $\mathcal{H}_n$ and A_n inductively as follows. If $\mathcal{H}_1 \neq \mathcal{H}$, let ϕ_{n_2} be the first element of the sequence $\{\phi_2, \phi_3, \cdots\}$ which is not in $\mathcal{H}_1$. Decompose $\phi_{n_2} = \phi'_{n_2} + \phi''_{n_2}$, where $\phi'_{n_2} \in \mathcal{H}_1$ and $\phi''_{n_2} \in \mathcal{H}_1^{\perp}$. Set $\psi_2 = \phi''_{n_2}$ and let $\mathcal{H}_2$ be the cyclic space generated by A and ψ_2. It follows from the resolvent identity that $\mathcal{H}_1 \perp \mathcal{H}_2$. Set $A_2 = A \restriction \mathcal{H}_2$. In this way we inductively define $\psi_n, \mathcal{H}_n, A_n$, $n \in \Gamma$, where Γ is a finite set $\{1, \ldots, N\}$ or $\Gamma = \mathbb{N}$. By the construction, $\{\phi_n\}_{n\in\Gamma} \subset \cup_{n\in\Gamma}\mathcal{H}_n$. Hence, (1) holds and $\mathcal{H} = \oplus_n \mathcal{H}_n$.

To prove the second part of (2), note first that by the construction of A_n,

$$(A-z)^{-1} = \bigoplus_n (A_n - z)^{-1}.$$

If $\tilde{A} = \oplus A_n$, then by Proposition 4.7, $\tilde{A}$ is self-adjoint and

$$(\tilde{A}-z)^{-1} = \bigoplus_n (A_n - z)^{-1}.$$

Hence $A = \tilde{A}$. □

4.6 The spectral theorem

We start with:

Theorem 4.9. *Let* $(M, \mathcal{F})$ *be a measure space and* μ *a finite positive measure on* $(M, \mathcal{F})$. *Let* $f : M \to \mathbb{R}$ *be a measurable function and let* A_f *be a linear operator on* $L^2(M, \mathrm{d}\mu)$ *defined by*

$$\mathrm{Dom}\,(A_f) = \{\psi \in L^2(M, \mathrm{d}\mu) \,:\, f\psi \in L^2(M, \mathrm{d}\mu)\}, \qquad A_f\psi = f\psi.$$

Then:
(1) A_f *is self-adjoint.*
(2) A_f *is bounded iff* $f \in L^\infty(M, \mathrm{d}\mu)$, *and in this case* $\|A_f\| = \|f\|_\infty$.
(3) $\mathrm{sp}(A_f)$ *is equal to the essential range of* f:

$$\mathrm{sp}(A_f) = \{\lambda \in \mathbb{R} \,:\, \mu(f^{-1}(\lambda - \epsilon, \lambda + \epsilon)) > 0 \textit{ for all } \epsilon > 0\}.$$

The proof of this theorem is left to the problems.

The content of the spectral theorem for self-adjoint operators is that *any* self-adjoint operator is unitarily equivalent to A_f for some f.

Let $\mathcal{H}_1$ and $\mathcal{H}_2$ be two Hilbert spaces. A linear bijection $U : \mathcal{H}_1 \to \mathcal{H}_2$ is called unitary if for all $\phi, \psi \in \mathcal{H}_1$, $(U\phi|U\psi)_{\mathcal{H}_2} = (\phi|\psi)_{\mathcal{H}_1}$. Let A_1, A_2 be linear operators on $\mathcal{H}_1, \mathcal{H}_2$. The operators A_1, A_2 are unitarily equivalent if there exists a unitary $U : \mathcal{H}_1 \to \mathcal{H}_2$ such that $U\mathrm{Dom}\,(A_1) = \mathrm{Dom}\,(A_2)$ and $UA_1U^{-1} = A_2$.

Theorem 4.10. (Spectral theorem for self-adjoint operators) *Let* A *be a self-adjoint operator on a Hilbert space* $\mathcal{H}$. *Then there is a measure space* $(M, \mathcal{F})$, *a finite positive measure* μ *and measurable function* $f : M \to \mathbb{R}$ *such that* A *is unitarily equivalent to the operator* A_f *on* $L^2(M, \mathrm{d}\mu)$.

We will prove the spectral theorem only for separable Hilbert spaces.

In the literature one can find many different proofs of Theorem 4.10. The proof below is constructive and allows to explicitly identify M and f while the measure μ is directly related to $(A - z)^{-1}$.

4.7 Proof of the spectral theorem—the cyclic case

Let A be a self-adjoint operator on a Hilbert space $\mathcal{H}$ and $\psi \in \mathcal{H}$.

Theorem 4.11. *There exists a unique finite positive Borel measure* μ_ψ *on* $\mathbb{R}$ *such that* $\mu_\psi(\mathbb{R}) = \|\psi\|^2$ *and*

$$(\psi|(A - z)^{-1}\psi) = \int_{\mathbb{R}} \frac{\mathrm{d}\mu_\psi(t)}{t - z}, \qquad z \in \mathbb{C} \setminus \mathbb{R}. \tag{37}$$

The measure μ_ψ *is called the spectral measure for* A *and* ψ.

Proof. Since $(A - \overline{z})^{-1} = (A - z)^{-1*}$, we need only to consider $z \in \mathbb{C}_+$. Set $U(z) = (\psi|(A-z)^{-1}\psi)$ and $V(z) = \mathrm{Im}\, U(z)$, $z \in \mathbb{C}_+$. It follows from the resolvent identity that

$$V(x+\mathrm{i}y) = y\|(A-x-\mathrm{i}y)^{-1}\psi\|^2, \tag{38}$$

and so V is harmonic and strictly positive in $\mathbb{C}_+$. Theorem 3.11 yields that there is a constant $c \geq 0$ and a unique positive Borel measure μ_ψ on $\mathbb{R}$ such that for $y > 0$,

$$V(x+\mathrm{i}y) = cy + P_{\mu_\psi}(x+\mathrm{i}y) = cy + y\int_{\mathbb{R}} \frac{\mathrm{d}\mu_\psi(t)}{(x-t)^2+y^2}. \tag{39}$$

By Theorem 4.6,

$$V(x+\mathrm{i}y) \leq \|\psi\|^2/y \qquad \text{and} \qquad \lim_{y\to\infty} yV(x+\mathrm{i}y) = \|\psi\|^2.$$

The first relation yields that $c = 0$. The second relation and the dominated convergence theorem yield that $\mu_\psi(\mathbb{R}) = \|\psi\|^2$.

The functions

$$F_{\mu_\psi}(z) = \int_{\mathbb{R}} \frac{\mathrm{d}\mu_\psi(t)}{t-z},$$

and $U(z)$ are analytic in $\mathbb{C}_+$ and have equal imaginary parts. The Cauchy-Riemann equations imply that $F_{\mu_\psi}(z) - U(z)$ is a constant. Since $F_{\mu_\psi}(z)$ and $U(z)$ vanish as $\operatorname{Im} z \to \infty$, $F_{\mu_\psi}(z) = U(z)$ for $z \in \mathbb{C}_+$ and (37) holds. □

Corollary 4.12. *Let $\varphi, \psi \in \mathcal{H}$. Then there exists a unique complex measure $\mu_{\varphi,\psi}$ on $\mathbb{R}$ such that*

$$(\varphi|(A-z)^{-1}\psi) = \int_{\mathbb{R}} \frac{\mathrm{d}\mu_{\varphi,\psi}(t)}{t-z}, \qquad z \in \mathbb{C}\setminus\mathbb{R}. \tag{40}$$

Proof. The uniqueness is obvious (the set of functions $\{(x-z)^{-1} : z \in \mathbb{C}\setminus\mathbb{R}\}$ is dense in $C_0(\mathbb{R})$). The existence follows from the polarization identity:

$$\begin{aligned}4(\varphi|(A-z)^{-1}\psi) &= (\varphi+\psi|(A-z)^{-1}(\varphi+\psi)) - (\varphi-\psi|(A-z)^{-1}(\varphi-\psi)) \\ &\quad + \mathrm{i}(\varphi-\mathrm{i}\psi|(A-z)^{-1}(\varphi-\mathrm{i}\psi)) - \mathrm{i}(\varphi+\mathrm{i}\psi|(A-z)^{-1}(\varphi+\mathrm{i}\psi)).\end{aligned}$$

In particular,

$$\mu_{\varphi,\psi} = \frac{1}{4}\left(\mu_{\varphi+\psi} - \mu_{\varphi-\psi} + \mathrm{i}(\mu_{\varphi-\mathrm{i}\psi} - \mu_{\varphi+\mathrm{i}\psi})\right). \tag{41}$$

□

The main result of this subsection is:

Theorem 4.13. *Assume that ψ is a cyclic vector for A. Then A is unitarily equivalent to the operator of multiplication by x on $L^2(\mathbb{R}, \mathrm{d}\mu_\psi)$. In particular,* $\mathrm{sp}(A) = \operatorname{supp} \mu_\psi$.

Proof. Clearly, we may assume that $\psi \neq 0$. Note that

$$(A-z)^{-1}\psi = (A-w)^{-1}\psi,$$

iff $z = w$.

For $z \in \mathbb{C} \setminus \mathbb{R}$ we set $r_z(x) = (x-z)^{-1}$. $r_z \in L^2(\mathbb{R}, \mathrm{d}\mu_\psi)$ and the linear span of $\{r_z\}_{z\in\mathbb{C}\setminus\mathbb{R}}$ is dense in $L^2(\mathbb{R}, \mathrm{d}\mu_\psi)$. Set

$$U(A-z)^{-1}\psi = r_z. \tag{42}$$

If $\overline{z} \neq w$, then

$$\begin{aligned}(r_z|r_w)_{L^2(\mathbb{R},\mathrm{d}\mu_\psi)} &= \int_{\mathbb{R}} r_{\overline{z}} r_w \mathrm{d}\mu_\psi = \frac{1}{\overline{z}-w}\int_{\mathbb{R}}(r_{\overline{z}} - r_w)\mathrm{d}\mu_\psi \\ &= \frac{1}{\overline{z}-w}\left[(\psi|(A-\overline{z})^{-1}\psi) - (\psi|(A-w)^{-1}\psi)\right] \\ &= ((A-z)^{-1}\psi|(A-w)^{-1}\psi).\end{aligned}$$

By a limiting argument, the relation

$$(r_z|r_w)_{L^2(\mathbb{R},\mathrm{d}\mu_\psi)} = ((A-z)^{-1}\psi|(A-w)^{-1}\psi),$$

holds for all $z, w \in \mathbb{C} \setminus \mathbb{R}$. Hence, the map (42) extends to a unitary

$$U : \mathcal{H} \to L^2(\mathbb{R}, \mathrm{d}\mu_\psi).$$

Since

$$U(A-z)^{-1}(A-w)^{-1}\psi = r_z(x)r_w(x) = r_z(x)U(A-w)^{-1}\psi,$$

$(A-z)^{-1}$ is unitarily equivalent to the operator of multiplication by $(x-z)^{-1}$ on $L^2(\mathbb{R}, \mathrm{d}\mu_\psi)$. For any $\phi \in \mathcal{H}$,

$$\begin{aligned}UA(A-z)^{-1}\phi &= U\phi + zU(A-z)^{-1}\phi = (1 + z(x-z)^{-1})U\phi \\ &= x(x-z)^{-1}U\phi = xU(A-z)^{-1}\phi,\end{aligned}$$

and so A is unitarily equivalent to the operator of multiplication by x. □

We finish this subsection with the following remark. Assume that ψ is a cyclic vector for A and let A_x be the operator of multiplication by x on $L^2(\mathbb{R}, \mathrm{d}\mu_\psi)$. Then, by Theorem 4.13, there exists a unitary $U : \mathcal{H} \to L^2(\mathbb{R}, \mathrm{d}\mu_\psi)$ such that

$$UAU^{-1} = A_x. \tag{43}$$

However, a unitary satisfying (43) is not unique. If U is such a unitary, then $U\psi$ is a cyclic vector for A_x. On the other hand, if $f \in L^2(\mathbb{R}, \mathrm{d}\mu_\psi)$ is a cyclic vector for A_x, then there is a unique unitary $U : \mathcal{H} \to L^2(\mathbb{R}, \mathrm{d}\mu_\psi)$ such that (43) holds and $U\psi = f\|\psi\|/\|f\|$. The unitary constructed in the proof of Theorem 4.13 satisfies $U\psi = \mathbb{1}$.

4.8 Proof of the spectral theorem—the general case

Let A be a self-adjoint operator on a separable Hilbert space $\mathcal{H}$. Let $\mathcal{H}_n$, A_n, ψ_n, $n \in \Gamma$ be as in the decomposition theorem (Theorem 4.8). Let

$$U_n : \mathcal{H}_n \to L^2(\mathbb{R}, \mathrm{d}\mu_{\psi_n}),$$

be unitary such that A_n is unitarily equivalent to the operator of multiplication by x. We denote this last operator by $\tilde{A}_n$. Let $U = \oplus_n U_n$. An immediate consequence of the decomposition theorem and Theorem 4.13 is

Theorem 4.14. *The map* $U : \mathcal{H} \to \bigoplus_{n\in\Gamma} L^2(\mathbb{R}, \mathrm{d}\mu_{\psi_n})$ *is unitary and* A *is unitarily equivalent to the operator* $\bigoplus_{n\in\Gamma} \tilde{A}_n$. *In particular,*

$$\mathrm{sp}(A) = \overline{\bigcup_{n\in\Gamma} \mathrm{supp}\ \mu_{\psi_n}}.$$

Note that if $\phi \in \mathcal{H}$ *and* $U\phi = \{\phi_n\}_{n\in\Gamma}$, *then* $\mu_\phi = \sum_{n\in\Gamma} \mu_{\phi_n}$.

Theorem 4.10 is a reformulation of Theorem 4.14. To see that, choose cyclic vectors ψ_n so that $\sum_{n\in\Gamma} \|\psi_n\|^2 < \infty$. For each $n \in \Gamma$, let $\mathbb{R}_n$ be a copy of $\mathbb{R}$ and

$$M = \bigcup_{n\in\Gamma} \mathbb{R}_n.$$

You may visualize M as follows: enumerate Γ so that $\Gamma = \{1, \ldots, N\}$ or $\Gamma = \mathbb{N}$ and set $\mathbb{R}_n = \{(n, x) : x \in \mathbb{R}\} \subset \mathbb{R}^2$. Hence, M is just a collection of lines in $\mathbb{R}^2$ parallel to the y-axis and going through the points $(n, 0)$, $n \in \Gamma$. Let $\mathcal{F}$ be the collection of all sets $F \subset M$ such that $F \cap \mathbb{R}_n$ is Borel for all n. Then $\mathcal{F}$ is a σ-algebra and

$$\mu(F) = \sum_{n\in\Gamma} \mu_{\psi_n}(F \cap \mathbb{R}_n),$$

is a finite measure on M ($\mu(M) = \sum_{n\in\Gamma} \|\psi_n\|^2 < \infty$). Let $f : M \to \mathbb{R}$ be the identity function ($f(n, x) = x$). Then

$$L^2(M, \mathrm{d}\mu) = \bigoplus_{n\in\Gamma} L^2(\mathbb{R}, \mathrm{d}\mu_{\psi_n}), \qquad A_f = \bigoplus_{n\in\Gamma} \tilde{A}_n,$$

and Theorem 4.10 follows.

Set

$$\mu_{\mathrm{ac}}(F) = \sum_{n\in\Gamma} \mu_{\psi_n,\mathrm{ac}}(F \cap \mathbb{R}_n),$$

$$\mu_{\mathrm{sc}}(F) = \sum_{n\in\Gamma} \mu_{\psi_n,\mathrm{sc}}(F \cap \mathbb{R}_n),$$

$$\mu_{\mathrm{pp}}(F) = \sum_{n\in\Gamma} \mu_{\psi_n,\mathrm{pp}}(F \cap \mathbb{R}_n).$$

Then $L^2(M, \mathrm{d}\mu_{\mathrm{ac}})$, $L^2(M, \mathrm{d}\mu_{\mathrm{sc}})$, and $L^2(M, \mathrm{d}\mu_{\mathrm{pp}})$ are closed subspaces of the space $L^2(M, \mathrm{d}\mu)$ invariant under A_f and

$$L^2(M, \mathrm{d}\mu) = L^2(M, \mathrm{d}\mu_{\mathrm{ac}}) \oplus L^2(M, \mathrm{d}\mu_{\mathrm{sc}}) \oplus L^2(M, \mathrm{d}\mu_{\mathrm{pp}}).$$

Set

$$\mathcal{H}_{\mathrm{ac}} = U^{-1}L^2(M, \mathrm{d}\mu_{\mathrm{ac}}), \;\; \mathcal{H}_{\mathrm{sc}} = U^{-1}L^2(M, \mathrm{d}\mu_{\mathrm{sc}}), \;\; \mathcal{H}_{\mathrm{pp}} = U^{-1}L^2(M, \mathrm{d}\mu_{\mathrm{pp}}).$$

These subspaces are invariant under A. Moreover, $\psi \in \mathcal{H}_{\mathrm{ac}}$ iff the spectral measure μ_ψ is absolutely continuous w.r.t. the Lebesgue measure, $\psi \in \mathcal{H}_{\mathrm{sc}}$ iff μ_ψ is singular continuous and $\psi \in \mathcal{H}_{\mathrm{pp}}$ iff μ_ψ is pure point. Obviously,

$$\mathcal{H} = \mathcal{H}_{\mathrm{ac}} \oplus \mathcal{H}_{\mathrm{sc}} \oplus \mathcal{H}_{\mathrm{pp}}.$$

The spectra

$$\mathrm{sp}_{\mathrm{ac}}(A) = \mathrm{sp}(A \restriction \mathcal{H}_{\mathrm{ac}}) = \overline{\bigcup_{n\in\Gamma} \mathrm{supp}\, \mu_{\psi_n,\mathrm{ac}}},$$

$$\mathrm{sp}_{\mathrm{sc}}(A) = \mathrm{sp}(A \restriction \mathcal{H}_{\mathrm{sc}}) = \overline{\bigcup_{n\in\Gamma} \mathrm{supp}\, \mu_{\psi_n,\mathrm{sc}}},$$

$$\mathrm{sp}_{\mathrm{pp}}(A) = \mathrm{sp}(A \restriction \mathcal{H}_{\mathrm{pp}}) = \overline{\bigcup_{n\in\Gamma} \mathrm{supp}\, \mu_{\psi_n,\mathrm{pp}}}$$

are called, respectively, the absolutely continuous, the singular continuous, and the pure point spectrum of A.

Note that

$$\mathrm{sp}(A) = \mathrm{sp}_{\mathrm{ac}}(A) \cup \mathrm{sp}_{\mathrm{sc}}(A) \cup \mathrm{sp}_{\mathrm{pp}}(A),$$

and $\overline{\mathrm{sp}_{\mathrm{p}}(A)} = \mathrm{sp}_{\mathrm{pp}}(A)$. The singular and the continuous spectrum of A are defined by

$$\mathrm{sp}_{\mathrm{sing}}(A) = \mathrm{sp}_{\mathrm{sc}}(A) \cup \mathrm{sp}_{\mathrm{pp}}(A), \qquad \mathrm{sp}_{\mathrm{cont}}(A) = \mathrm{sp}_{\mathrm{ac}}(A) \cup \mathrm{sp}_{\mathrm{sc}}(A).$$

The subspaces $\mathcal{H}_{\mathrm{ac}}$, $\mathcal{H}_{\mathrm{sc}}$, $\mathcal{H}_{\mathrm{pp}}$ are called the spectral subspaces associated, respectively, to the absolutely continuous, singular continuous, and pure point spectrum. The projections on these subspaces are denoted by $\mathbf{1}_{\mathrm{ac}}(A)$, $\mathbf{1}_{\mathrm{sc}}(A)$, $\mathbf{1}_{\mathrm{pp}}(A)$. The spectral subspaces associated to the singular and the continuous spectrum are $\mathcal{H}_{\mathrm{sing}} = \mathcal{H}_{\mathrm{sc}} \oplus \mathcal{H}_{\mathrm{pp}}$ and $\mathcal{H}_{\mathrm{cont}} = \mathcal{H}_{\mathrm{ac}} \oplus \mathcal{H}_{\mathrm{sc}}$. The corresponding projections are $\mathbf{1}_{\mathrm{sing}}(A) = \mathbf{1}_{\mathrm{sc}}(A) + \mathbf{1}_{\mathrm{pp}}(A)$ and $\mathbf{1}_{\mathrm{cont}}(A) = \mathbf{1}_{\mathrm{ac}}(A) + \mathbf{1}_{\mathrm{sc}}(A)$.

When we wish to indicate the dependence of the spectral subspaces on the operator A, we will write $\mathcal{H}_{\mathrm{ac}}(A)$, etc.

4.9 Harmonic analysis and spectral theory

Let A be a self-adjoint operator on a Hilbert space $\mathcal{H}$, $\psi \in \mathcal{H}$, and μ_ψ the spectral measure for A and ψ. Let F_{μ_ψ} and P_{μ_ψ} be the Borel and the Poisson transform of μ_ψ. The formulas

$$(\psi|(A-z)^{-1}\psi) = F_{\mu_\psi}(z),$$

$$\mathrm{Im}\,(\psi|(A-z)^{-1}\psi) = P_{\mu_\psi}(z),$$

provide the key link between the harmonic analysis (the results of Section 3) and the spectral theory. Recall that $\mu_{\psi,\mathrm{sing}} = \mu_{\psi,\mathrm{sc}} + \mu_{\psi,\mathrm{pp}}$.

Theorem 4.15. (1) *For Lebesgue a.e. $x \in \mathbb{R}$ the limit*

$$(\psi|(A-x-\mathrm{i}0)^{-1}\psi) = \lim_{y\downarrow 0}(\psi|(A-x-\mathrm{i}y)^{-1}\psi),$$

exists and is finite and non-zero.
(2) $\mathrm{d}\mu_{\psi,\mathrm{ac}} = \pi^{-1}\mathrm{Im}(\psi|(A-x-\mathrm{i}0)^{-1}\psi)\mathrm{d}x$.
(3) $\mu_{\psi,\mathrm{sing}}$ *is concentrated on the set*

$$\{x : \lim_{y\downarrow 0}\mathrm{Im}\,(\psi|(A-x-\mathrm{i}y)^{-1}\psi) = \infty\}.$$

Theorem 4.15 is an immediate consequence of Theorems 3.5 and 3.17. Similarly, Theorems 3.6, 3.8 and Corollary 3.10 yield:

Theorem 4.16. *Let $[a,b]$ be a finite interval.*
(1) $\mu_{\psi,\mathrm{ac}}([a,b]) = 0$ *iff for some $p \in (0,1)$*

$$\lim_{y\downarrow 0}\int_a^b \left[\mathrm{Im}\,(\psi|(A-x-\mathrm{i}y)^{-1}\psi)\right]^p \mathrm{d}x = 0.$$

(2) *Assume that for some $p > 1$*

$$\sup_{0<y<1}\int_a^b \left[\mathrm{Im}\,(\psi|(A-x-\mathrm{i}y)^{-1}\psi)\right]^p \mathrm{d}x < \infty.$$

Then $\mu_{\psi,\mathrm{sing}}([a,b]) = 0$.
(3) $\mu_{\psi,\mathrm{pp}}([a,b]) = 0$ *iff*

$$\lim_{y\downarrow 0} y\int_a^b \left[\mathrm{Im}\,(\psi|(A-x-\mathrm{i}y)^{-1}\psi)\right]^2 \mathrm{d}x = 0.$$

Let $\mathcal{H}_\psi$ be the cyclic subspace spanned by A and ψ. W.l.o.g. we may assume that $\|\psi\| = 1$. By Theorem 4.13 there exists a (unique) unitary $U_\psi : \mathcal{H}_\psi \to L^2(\mathbb{R}, \mathrm{d}\mu_\psi)$

such that $U_\psi \psi = \mathbb{1}$ and $U_\psi A U_\psi^{-1}$ is the operator of multiplication by x on $L^2(\mathbb{R}, \mathrm{d}\mu_\psi)$. We extend U_ψ to $\mathcal{H}$ by setting $U_\psi \phi = 0$ for $\phi \in \mathcal{H}_\psi^\perp$. Recall that

$$\mathrm{Im}\,(A - z)^{-1} = \frac{1}{2\mathrm{i}}((A - z)^{-1} - (A - \overline{z})^{-1}).$$

The interplay between spectral theory and harmonic analysis is particularly clearly captured in the following result.

Theorem 4.17. *Let $\phi \in \mathcal{H}$. Then:*
(1)

$$(U_\psi \mathbf{1}_{\mathrm{ac}}\phi)(x) = \lim_{y \downarrow 0} \frac{(\psi | \mathrm{Im}\,(A - x - \mathrm{i}y)^{-1}\phi)}{\mathrm{Im}\,(\psi | (A - x - \mathrm{i}y)^{-1}\psi)}, \qquad \text{for } \mu_{\psi,\mathrm{ac}} - a.e.\ x.$$

(2)

$$(U_\psi \mathbf{1}_{\mathrm{sing}}\phi)(x) = \lim_{y \downarrow 0} \frac{(\psi | (A - x - \mathrm{i}y)^{-1}\phi)}{(\psi | (A - x - \mathrm{i}y)^{-1}\psi)}, \qquad \text{for } \mu_{\psi,\mathrm{sing}} - a.e.\ x.$$

Proof. Since

$$\frac{(\psi | \mathrm{Im}\,(A - x - \mathrm{i}y)^{-1}\phi)}{\mathrm{Im}\,(\psi | (A - x - \mathrm{i}y)^{-1}\psi)} = \frac{P_{(U_\psi \phi)\mu_\psi}(x + \mathrm{i}y)}{P_{\mu_\psi}(x + \mathrm{i}y)},$$

(1) follows from Theorem 3.5. Similarly, since

$$\frac{(\psi | (A - x - \mathrm{i}y)^{-1}\phi)}{(\psi | (A - x - \mathrm{i}y)^{-1}\psi)} = \frac{F_{(U_\psi \phi)\mu_\psi}(x + \mathrm{i}y)}{F_{\mu_\psi}(x + \mathrm{i}y)},$$

(2) follows from the Poltoratskii theorem (Theorem 3.18). □

4.10 Spectral measure for A

Let A be a self-adjoint operator on a separable Hilbert space $\mathcal{H}$ and let $\{\phi_n\}_{n\in\Gamma}$ be a cyclic set for A. Let $\{a_n\}_{n\in\Gamma}$ be a sequence such that $a_n > 0$ and

$$\sum_{n\in\Gamma} a_n \|\phi_n\|^2 < \infty.$$

The spectral measure for A, μ_A, is a Borel measure on $\mathbb{R}$ defined by

$$\mu_A = \sum_{n\in\Gamma} a_n \mu_{\phi_n}.$$

Obviously, μ_A depends on the choice of $\{\phi_n\}$ and a_n. Two positive Borel measures ν_1 and ν_2 on $\mathbb{R}$ are called equivalent (we write $\nu_1 \sim \nu_2$) iff ν_1 and ν_2 have the same sets of measure zero.

Theorem 4.18. *Let μ_A and ν_A be two spectral measures for A. Then $\mu_A \sim \nu_A$. Moreover, $\mu_{A,\mathrm{ac}} \sim \nu_{A,\mathrm{ac}}$, $\mu_{A,\mathrm{sc}} \sim \nu_{A,\mathrm{sc}}$, and $\mu_{A,\mathrm{pp}} \sim \nu_{A,\mathrm{pp}}$.*

Theorem 4.19. *Let μ_A be a spectral measure for A. Then*

$$\mathrm{supp}\ \mu_{A,\mathrm{ac}} = \mathrm{sp}_{\mathrm{ac}}(A), \qquad \mathrm{supp}\ \mu_{A,\mathrm{sc}} = \mathrm{sp}_{\mathrm{sc}}(A), \qquad \mathrm{supp}\ \mu_{A,\mathrm{pp}} = \mathrm{sp}_{\mathrm{pp}}(A).$$

The proofs of these two theorems are left to the problems.

4.11 The essential support of the ac spectrum

Let B_1 and B_2 be two Borel sets in $\mathbb{R}$. Let $B_1 \sim B_2$ iff the Lebesgue measure of the symmetric difference $(B_1 \setminus B_2) \cup (B_2 \setminus B_1)$ is zero. $\sim$ is an equivalence relation. Let μ_A be a spectral measure of a self-adjoint operator A and $f(x)$ its Radon-Nikodym derivative w.r.t. the Lebesgue measure ($\mathrm{d}\mu_{A,\mathrm{ac}} = f(x)\mathrm{d}x$). The equivalence class associated to $\{x : f(x) > 0\}$ is called the essential support of the absolutely continuous spectrum and is denoted by $\Sigma_{\mathrm{ac}}^{\mathrm{ess}}(A)$. With a slight abuse of terminology we will also refer to a particular element of $\Sigma_{\mathrm{ac}}^{\mathrm{ess}}(A)$ as an essential support of the ac spectrum (and denote it by the same symbol $\Sigma_{\mathrm{ac}}^{\mathrm{ess}}(A)$). For example, the set

$$\left\{x : 0 < \lim_{r \downarrow 0} (2r)^{-1} \mu_A(I(x,r)) < \infty \right\},$$

is an essential support of the absolutely continuous spectrum.

Note that the essential support of the ac spectrum is independent on the choice of μ_A.

Theorem 4.20. *Let $\Sigma_{\mathrm{ac}}^{\mathrm{ess}}(A)$ be an essential support of the absolutely continuous spectrum. Then* $\mathrm{cl}(\Sigma_{\mathrm{ac}}^{\mathrm{ess}}(A) \cap \mathrm{sp}_{\mathrm{ac}}(A)) = \mathrm{sp}_{\mathrm{ac}}(A)$.

The proof is left to the problems.

Although the closure of an essential support $\Sigma_{\mathrm{ac}}^{\mathrm{ess}}(A) \subset \mathrm{sp}_{\mathrm{ac}}(A)$ equals $\mathrm{sp}_{\mathrm{ac}}(A)$, $\Sigma_{\mathrm{ac}}^{\mathrm{ess}}(A)$ could be substantially "smaller" than $\mathrm{sp}_{\mathrm{ac}}(A)$; see Problem 6.

4.12 The functional calculus

Let A be a self-adjoint operator on a separable Hilbert space $\mathcal{H}$. Let

$$U : \mathcal{H} \to L^2(M, \mathrm{d}\mu),$$

f, and A_f be as in the spectral theorem. Let $B_{\mathrm{b}}(\mathbb{R})$ be the vector space of all bounded Borel functions on $\mathbb{R}$. For $h \in B_{\mathrm{b}}(\mathbb{R})$, consider the operator $A_{h \circ f}$. This operator is bounded and $\|A_{h \circ f}\| \le \sup h(x)$. Set

$$h(A) = U^{-1} A_{h \circ f} U. \tag{44}$$

Let $\Phi : B_{\mathrm{b}}(\mathbb{R}) \to \mathcal{B}(\mathcal{H})$ be given by $\Phi(h) = h(A)$. Recall that $r_z(x) = (x - z)^{-1}$.

Theorem 4.21. (1) *The map Φ is an algebraic $*$-homomorphism.*
(2) $\|\Phi(h)\| \leq \max |h(x)|$.
(3) $\Phi(r_z) = (A - z)^{-1}$ *for all* $z \in \mathbb{C} \setminus \mathbb{R}$.
(4) *If* $h_n(x) \to h(x)$ *for all* x, *and* $\sup_{n,x} |h_n(x)| < \infty$, *then* $h_n(A)\psi \to h(A)\psi$ *for all* ψ.
The map Φ is uniquely specified by (1)-(4). Moreover, it has the following additional properties:
(5) *If* $A\psi = \lambda\psi$, *then* $\Phi(h)\psi = h(\lambda)\psi$.
(6) *If* $h \geq 0$, *then* $\Phi(h) \geq 0$.

We remark that the uniqueness of the functional calculus is an immediate consequence of Problem 11 in Section 2.

Let $\mathcal{K} \subset \mathcal{H}$ be a closed subspace. If $\mathcal{K}$ is invariant under A, then for all $h \in B_{\mathrm{b}}(\mathbb{R})$, $h(A)\mathcal{K} \subset \mathcal{K}$.

For any Borel function $h : \mathbb{R} \to \mathbb{C}$ we define $h(A)$ by (44). Of course, $h(A)$ could be an unbounded operator. Note that $h_1(A)h_2(A) \subset h_1 \circ h_2(A)$, $h_1(A) + h_2(A) \subset (h_1 + h_2)(A)$. Also, $h(A)^* = \overline{h}(A)$ and $h(A)$ is self-adjoint iff $h(x) \in \mathbb{R}$ for μ_A-a.e. $x \in M$.

In fact, to define $h(A)$, we only need that Ran $f \subset$ Dom h. Hence, if $A \geq 0$, we can define $\sqrt{A}$, if $A > 0$ we can define $\ln A$, etc.

The two classes of functions, characteristic functions and exponentials, play a particularly important role.

Let F be a Borel set in $\mathbb{R}$ and χ_F its characteristic function. The operator $\chi_F(A)$ is an orthogonal projection, called the spectral projection on the set F. In these notes we will use the notation $\mathbf{1}_F(A) = \chi_F(A)$ and $\mathbf{1}_{\{e\}}(A) = \mathbf{1}_e(A)$. Note that $\mathbf{1}_e(A) \neq 0$ iff $e \in \mathrm{sp_p}(A)$. By definition, the multiplicity of the eigenvalue e is $\dim \mathbf{1}_e(A)$.

The subspace Ran $\mathbf{1}_F(A)$ is invariant under A and

$$\mathrm{cl}(\mathrm{int}(F) \cap \mathrm{sp}(A)) \subset \mathrm{sp}(A \restriction \mathrm{Ran}\, \mathbf{1}_F(A)) \subset \mathrm{sp}(A) \cap \mathrm{cl}(F). \tag{45}$$

Note in particular that $e \in \mathrm{sp}(A)$ iff for all $\epsilon > 0$ Ran $\mathbf{1}_{(e-\epsilon,e+\epsilon)}(A) \neq \{0\}$. The proof of (45) is left to the problems.

Theorem 4.22. (Stone's formula) *For* $\psi \in \mathcal{H}$,

$$\lim_{y \downarrow 0} \frac{y}{\pi} \int_a^b \mathrm{Im}(A - x - \mathrm{i}y)^{-1}\psi \mathrm{d}x = \frac{1}{2}\left[\mathbf{1}_{[a,b]}(A)\psi + \mathbf{1}_{(a,b)}(A)\psi\right].$$

Proof. Since

$$\lim_{y \downarrow 0} \frac{y}{\pi} \int_a^b \frac{1}{(t - x)^2 + y^2} \mathrm{d}x = \begin{cases} 0 & \text{if } t \notin [a,b], \\ 1/2 & \text{if } t = a \text{ or } t = b, \\ 1 & \text{if } t \in (a,b), \end{cases}$$

the Stone formula follows from Theorem 4.21. □

Another important class of functions are exponentials. For $t \in \mathbb{R}$, set $U(t) = \exp(\mathrm{i}tA)$. Then $U(t)$ is a group of unitary operators on $\mathcal{H}$. The group $U(t)$ is strongly continuous, i.e. for all $\psi \in \mathcal{H}$,

$$\lim_{s \to t} U(s)\psi = U(t)\psi.$$

For $\psi \in \mathrm{Dom}\,(A)$ the function $\mathbb{R} \ni t \mapsto U(t)\psi$ is strongly differentiable and

$$\lim_{t \to 0} \frac{U(t)\psi - \psi}{t} = \mathrm{i}A\psi. \tag{46}$$

On the other hand, if the limit on the l.h.s. exists for some ψ, then

$$\psi \in \mathrm{Dom}\,(A),$$

and (46) holds.

The above results have a converse:

Theorem 4.23. (Stone's theorem) *Let $U(t)$ be a strongly continuous group on a separable Hilbert space $\mathcal{H}$. Then there is a self-adjoint operator A such that $U(t) = \exp(\mathrm{i}tA)$.*

4.13 The Weyl criteria and the RAGE theorem

Let A be a self-adjoint operator on a separable Hilbert space $\mathcal{H}$.

Theorem 4.24. (Weyl criterion 1) $e \in \mathrm{sp}(A)$ *iff there exists a sequence of unit vectors $\psi_n \in \mathrm{Dom}\,(A)$ such that*

$$\lim_{n \to \infty} \|(A - e)\psi_n\| = 0. \tag{47}$$

Remark. A sequence of unit vectors for which (47) holds is called a Weyl sequence.

Proof. Recall that $e \in \mathrm{sp}(A)$ iff $\mathbf{1}_{(e-\epsilon, e+\epsilon)}(A) \neq 0$ for all $\epsilon > 0$.

Assume that $e \in \mathrm{sp}(A)$. Let $\psi_n \in \mathrm{Ran}\,\mathbf{1}_{(e-1/n, e+1/n)}(A)$ be unit vectors. Then, by the functional calculus,

$$\|(A - e)\psi_n\| \leq \sup_{x \in (e-1/n, e+1/n)} |x - e| \leq 1/n.$$

On the other hand, assume that there is a sequence ψ_n such that (47) holds and that $e \notin \mathrm{sp}(A)$. Then

$$\|\psi_n\| = \|(A - e)^{-1}(A - e)\psi_n\| \leq C\|(A - e)\psi_n\|,$$

and so $1 = \|\psi_n\| \to 0$, contradiction. □

The discrete spectrum of A, denoted $\mathrm{sp}_{\mathrm{disc}}(A)$, is the set of all isolated eigenvalues of finite multiplicity. Hence $e \in \mathrm{sp}_{\mathrm{disc}}(A)$ iff

$$1 \leq \dim \mathbf{1}_{(e-\epsilon,e+\epsilon)}(A) < \infty,$$

for all ϵ small enough. The essential spectrum of A is defined by

$$\mathrm{sp}_{\mathrm{ess}}(A) = \mathrm{sp}(A) \setminus \mathrm{sp}_{\mathrm{disc}}(A).$$

Hence, $e \in \mathrm{sp}_{\mathrm{ess}}(A)$ iff for all $\epsilon > 0$ $\dim \mathbf{1}_{(e-\epsilon,e+\epsilon)}(A) = \infty$. Obviously, $\mathrm{sp}_{\mathrm{ess}}(A)$ is a closed subset of $\mathbb{R}$.

Theorem 4.25. (Weyl criterion 2) $e \in \mathrm{sp}_{\mathrm{ess}}(A)$ *iff there exists an orthonormal sequence* $\psi_n \in \mathrm{Dom}\,(A)$ *(*$\|\psi_n\| = 1$*,* $(\psi_n|\psi_m) = 0$ *if* $n \neq m$*), such that*

$$\lim_{n\to\infty} \|(A-e)\psi_n\| = 0. \tag{48}$$

Proof. Assume that $e \in \mathrm{sp}_{\mathrm{ess}}(A)$. Then $\dim \mathbf{1}_{(e-1/n,e+1/n)}(A) = \infty$ for all n, and we can choose an orthonormal sequence ψ_n such that

$$\psi_n \in \mathrm{Ran}\, \mathbf{1}_{(e-1/n,e+1/n)}(A).$$

Clearly,

$$\|(A-e)\psi_n\| \leq 1/n,$$

and (48) holds.

On the other hand, assume that there exists an orthonormal sequence ψ_n such that (48) holds and that $e \in \mathrm{sp}_{\mathrm{disc}}(A)$. Choose $\epsilon > 0$ such that $\dim \mathbf{1}_{(e-\epsilon,e+\epsilon)}(A) < \infty$. Then, $\lim_{n\to\infty} \mathbf{1}_{(e-\epsilon,e+\epsilon)}(A)\psi_n = 0$ and

$$\lim_{n\to\infty} \|(A-e)\mathbf{1}_{\mathbb{R}\setminus(e-\epsilon,e+\epsilon)}(A)\psi_n\| = 0.$$

Since $(A-e) \restriction \mathrm{Ran}\, \mathbf{1}_{\mathbb{R}\setminus(e-\epsilon,e+\epsilon)}(A)$ is invertible and the norm of its inverse is $\leq 1/\epsilon$, we have that

$$\|\mathbf{1}_{\mathbb{R}\setminus(e-\epsilon,e+\epsilon)}(A)\psi_n\| \leq \epsilon^{-1}\|(A-e)\mathbf{1}_{\mathbb{R}\setminus(e-\epsilon,e+\epsilon)}(A)\psi_n\|,$$

and so $\lim_{n\to\infty} \mathbf{1}_{\mathbb{R}\setminus(e-\epsilon,e+\epsilon)}(A)\psi_n = 0$. Hence $1 = \|\psi_n\| \to 0$, contradiction. □

Theorem 4.26. (RAGE) (1) *Let* K *be a compact operator. Then for all* $\psi \in \mathcal{H}_{\mathrm{cont}}$,

$$\lim_{T\to\infty} \frac{1}{T}\int_0^T \|K\mathrm{e}^{-\mathrm{i}tA}\psi\|^2 \mathrm{d}t = 0. \tag{49}$$

(2) *The same result holds if* K *is bounded and* $K(A+\mathrm{i})^{-1}$ *is compact.*

Proof. (1) First, recall that any compact operator is a norm limit of finite rank operators. In other words, there exist vectors $\phi_n, \varphi_n \in \mathcal{H}$ such that $K_n = \sum_{j=1}^n (\phi_j|\cdot)\varphi_j$ satisfies $\lim_{n\to\infty} \|K - K_n\| = 0$. Hence, it suffices to prove the statement for the finite rank operators K_n. By induction and the triangle inequality, it suffices to prove the statement for the rank one operator $K = (\phi|\cdot)\varphi$. Thus, it suffices to show that for $\phi \in \mathcal{H}$ and $\psi \in \mathcal{H}_{\rm cont}$,

$$\lim_{T\to\infty} \frac{1}{T}\int_0^T |(\phi|\mathrm{e}^{-\mathrm{i}tA}\psi)|^2 \mathrm{d}t = 0.$$

Moreover, since

$$(\phi|\mathrm{e}^{-\mathrm{i}tA}\psi) = (\phi|\mathrm{e}^{-\mathrm{i}tA}\mathbf{1}_{\rm cont}(A)\psi) = (\mathbf{1}_{\rm cont}(A)\phi|\mathrm{e}^{-\mathrm{i}tA}\psi),$$

w.l.o.g we may assume that $\phi \in \mathcal{H}_{\rm cont}$. Finally, by the polarization identity, we may assume that $\varphi = \psi$. Since for $\psi \in \mathcal{H}_{\rm cont}$ the spectral measure μ_ψ has no atoms, the result follows from the Wiener theorem (Theorem 2.6 in Section 2).

(2) Since $\mathrm{Dom}\,(A) \cap \mathcal{H}_{\rm cont}$ is dense in $\mathcal{H}_{\rm cont}$, it suffices to prove the statement for $\psi \in \mathrm{Dom}\,(A) \cap \mathcal{H}_{\rm cont}$. Write

$$\|K\mathrm{e}^{-\mathrm{i}tA}\psi\| = \|K(A+\mathrm{i})^{-1}\mathrm{e}^{-\mathrm{i}tA}(A+\mathrm{i})\psi\|,$$

and use (1). □

4.14 Stability

We will first discuss stability of self-adjointness—if A and B are self-adjoint operators, we wish to discuss conditions under which $A + B$ is self-adjoint on $\mathrm{Dom}\,(A) \cap \mathrm{Dom}\,(B)$. One obvious sufficient condition is that either A or B is bounded. A more refined result involves the notion of relative boundedness.

Let A and B be densely defined linear operators on a separable Hilbert space $\mathcal{H}$. The operator B is called *A-bounded* if $\mathrm{Dom}\,(A) \subset \mathrm{Dom}\,(B)$ and for some positive constants a and b and all $\psi \in \mathrm{Dom}\,(A)$,

$$\|B\psi\| \leq a\|A\psi\| + b\|\psi\|. \tag{50}$$

The number a is called a relative bound of B w.r.t. A.

Theorem 4.27. (Kato-Rellich) *Suppose that A is self-adjoint, B is symmetric, and B is A-bounded with a relative bound $a < 1$. Then:*
(1) *$A + B$ is self-adjoint on* $\mathrm{Dom}\,(A)$.
(2) *$A + B$ is essentially self-adjoint on any core of A.*
(3) *If A is bounded from below, then $A + B$ is also bounded from below.*

Proof. We will prove (1) and (2); (3) is left to the problems. In the proof we will use the following elementary fact: if V is a bounded operator and $\|V\| < 1$, then $0 \notin \mathrm{sp}(\mathbf{1} + V)$. This is easily proven by checking that the inverse of $\mathbf{1} + V$ is given by $\mathbf{1} + \sum_{k=1}^{\infty}(-1)^k V^k$.

By Theorem 4.4 (and the Remark after Theorem 4.5), to prove (1) it suffices to show that there exists $y > 0$ such that $\mathrm{Ran}(A + B \pm \mathrm{i}y) = \mathcal{H}$. In what follows $y = (1 + b)/(1 - a)$. The relation (50) yields

$$\|B(A \pm \mathrm{i}y)^{-1}\| \leq a\|A(A \pm \mathrm{i}y)^{-1}\| + b\|(A \pm \mathrm{i}y)^{-1}\| \leq a + by^{-1} < 1,$$

and so $\mathbf{1} + B(A \pm \mathrm{i}y)^{-1} : \mathcal{H} \to \mathcal{H}$ are bijections. Since $A \pm \mathrm{i}y : \mathrm{Dom}\,(A) \to \mathcal{H}$ are also bijections, the identity

$$A + B \pm \mathrm{i}y = (\mathbf{1} + B(A \pm \mathrm{i}y)^{-1})(A \pm \mathrm{i}y)$$

yields $\mathrm{Ran}(A + B \pm \mathrm{i}y) = \mathcal{H}$.

The proof of (2) is based on Theorem 4.5. Let D be a core for A. Then the sets $(A \pm \mathrm{i}y)(D) = \{(A \pm \mathrm{i}y)\psi : \psi \in D\}$ are dense in $\mathcal{H}$, and since $\mathbf{1} + B(A \pm \mathrm{i}y)^{-1}$ are bijections,

$$(A + B \pm \mathrm{i}y)(D) = (\mathbf{1} + B(A \pm \mathrm{i}y)^{-1})(A \pm \mathrm{i}y)(D)$$

are also dense in $\mathcal{H}$. □

We now turn to stability of the essential spectrum. The simplest result in this direction is:

Theorem 4.28. (Weyl) *Let A and B be self-adjoint and B compact. Then* $\mathrm{sp}_{\mathrm{ess}}(A) = \mathrm{sp}_{\mathrm{ess}}(A + B)$.

Proof. By symmetry, it suffices to prove that $\mathrm{sp}_{\mathrm{ess}}(A + B) \subset \mathrm{sp}_{\mathrm{ess}}(A)$. Let $e \in \mathrm{sp}_{\mathrm{ess}}(A + B)$ and let ψ_n be an orthonormal sequence such that

$$\lim_{n\to\infty} \|(A + B - e)\psi_n\| = 0.$$

Since ψ_n converges weakly to zero and B is compact, $B\psi_n \to 0$. Hence, $\|(A - e)\psi_n\| \to 0$ and $e \in \mathrm{sp}_{\mathrm{ess}}(A)$. □

Section XIII.4 of [RS4] deals with various extensions and generalizations of Theorem 4.28.

4.15 Scattering theory and stability of ac spectra

Let A and B be self-adjoint operators on a Hilbert space $\mathcal{H}$. Assume that for all $\psi \in \mathrm{Ran}\,\mathbf{1}_{\mathrm{ac}}(A)$ the limits

$$\Omega^{\pm}(A, B)\psi = \lim_{t\to\pm\infty} \mathrm{e}^{\mathrm{i}tA}\mathrm{e}^{-\mathrm{i}tB}\psi, \tag{51}$$

exist. The operators $\Omega^{\pm}(A, B) : \mathrm{Ran}\,\mathbf{1}_{\mathrm{ac}}(A) \to \mathcal{H}$ are called *wave operators*.

Proposition 4.29. *Assume that the wave operators exist. Then*
(1) $(\Omega^{\pm}(A, B)\phi|\Omega^{\pm}(A, B)\psi) = (\phi|\psi)$.
(2) *For any* $f \in B_{\mathrm{b}}(\mathbb{R})$, $\Omega^{\pm}(A, B)f(A) = f(B)\Omega^{\pm}(A, B)$.
(3) $\mathrm{Ran}\,\Omega^{\pm}(A, B) \subset \mathrm{Ran}\,\mathbf{1}_{\mathrm{ac}}(B)$.
The wave operators $\Omega^{\pm}(A, B)$ are called complete if

$$\mathrm{Ran}\,\Omega^{\pm}(A, B) = \mathrm{Ran}\,\mathbf{1}_{\mathrm{ac}}(B);$$

(4) *Wave operators $\Omega^{\pm}(A, B)$ are complete iff $\Omega^{\pm}(B, A)$ exist.*

The proof of this proposition is simple and is left to the problems (see also [25]).

Let $\mathcal{H}$ be a separable Hilbert space and $\{\psi_n\}$ an orthonormal basis. A bounded positive self-adjoint operator A is called *trace class* if

$$\mathrm{Tr}(A) = \sum_n (\psi_n | A\psi_n) < \infty.$$

The number $\mathrm{Tr}(A)$ does not depend on the choice of orthonormal basis. More generally, a bounded self-adjoint operator A is called trace class if $\mathrm{Tr}(|A|) < \infty$. A trace class operator is compact.

Concerning stability of the ac spectrum, the basic result is:

Theorem 4.30. (Kato-Rosenblum) *Let A and B be self-adjoint and B trace class. Then the wave operators $\Omega^{\pm}(A+B, A)$ exist and are complete. In particular,* $\mathrm{sp}_{\mathrm{ac}}(A+B) = \mathrm{sp}_{\mathrm{ac}}(A)$ *and* $\Sigma_{\mathrm{ac}}^{\mathrm{ess}}(A+B) = \Sigma_{\mathrm{ac}}^{\mathrm{ess}}(A)$.

For the proof of Kato-Rosenblum theorem see [25], Theorem XI.7.

The singular and the pure point spectra are in general unstable under perturbations — they may appear or disappear under the influence of a rank one perturbation. We will discuss in Section 5 criteria for "generic" stability of the singular and the pure point spectra.

4.16 Notions of measurability

In mathematical physics one often encounters self-adjoint operators indexed by elements of some measure space $(M, \mathcal{F})$, namely one deals with functions $M \ni \omega \mapsto A_\omega$, where A_ω is a self-adjoint operator on some fixed separable Hilbert space $\mathcal{H}$. In this subsection we address some issues concerning measurability of such functions.

Let $(M, \mathcal{F})$ be a measure space and X a topological space. A function $f : M \to X$ is called measurable if the inverse image of every open set is in $\mathcal{F}$.

Let $\mathcal{H}$ be a separable Hilbert space and $\mathcal{B}(\mathcal{H})$ the vector space of all bounded operators on $\mathcal{H}$. We distinguish three topologies in $\mathcal{B}(\mathcal{H})$, the uniform topology, the strong topology, and the weak topology. The uniform topology is induced by the operator norm on $\mathcal{B}(\mathcal{H})$. The strong topology is the minimal topology w.r.t. which the functions $\mathcal{B}(\mathcal{H}) \ni A \mapsto A\psi \in \mathcal{H}$ are continuous for all $\psi \in \mathcal{H}$. The weak topology is the minimal topology w.r.t. which the functions $\mathcal{B}(\mathcal{H}) \ni A \mapsto (\phi|A\psi) \in \mathbb{C}$ are continuous for all $\phi, \psi \in \mathcal{H}$. The weak topology is weaker than the strong topology, and the strong topology is weaker than the uniform topology.

A function $f : M \to \mathcal{B}(\mathcal{H})$ is called *measurable* if it is measurable with respect to the weak topology. In other words, f is measurable iff the function $M \ni \omega \mapsto (\phi|f(\omega)\psi) \in \mathbb{C}$ is measurable for all $\phi, \psi \in \mathcal{H}$.

Let $\omega \mapsto A_\omega$ be a function with values in (possibly unbounded) self-adjoint operators on $\mathcal{H}$. We say that A_ω is *measurable* if for all $z \in \mathbb{C} \setminus \mathbb{R}$ the function

$$\omega \mapsto (A_\omega - z)^{-1} \in \mathcal{B}(\mathcal{H}),$$

is measurable.

Until the end of this subsection $\omega \mapsto A_\omega$ is a given measurable function with values in self-adjoint operators.

Theorem 4.31. *The function $\omega \mapsto h(A_\omega)$ is measurable for all $h \in B_{\rm b}(\mathbb{R})$.*

Proof. Let $\mathcal{T} \subset B_{\rm b}(\mathbb{R})$ be the class of functions such that $\omega \mapsto h(A_\omega)$ is measurable By definition, $(x-z)^{-1} \in \mathcal{T}$ for all $z \in \mathbb{C}\setminus\mathbb{R}$. Since the linear span of $\{(x-z)^{-1} : z \in \mathbb{C}\setminus\mathbb{R}\}$ is dense in the Banach space $C_0(\mathbb{R})$, $C_0(\mathbb{R}) \subset \mathcal{T}$. Note also that if $h_n \in \mathcal{T}$, $\sup_{n,x}|h_n(x)| < \infty$, and $h_n(x) \to h(x)$ for *all* x, then $h \in \mathcal{T}$. Hence, by Problem 11 in Section 2, $\mathcal{T} = B_{\rm b}(\mathbb{R})$. □

From this theorem it follows that the functions $\omega \mapsto \mathbf{1}_B(A_\omega)$ (B Borel) and $\omega \mapsto \exp(\mathrm{i}tA_\omega)$ are measurable. One can also easily show that if $h : \mathbb{R} \mapsto \mathbb{R}$ is an arbitrary Borel measurable real valued function, then $\omega \mapsto h(A_\omega)$ is measurable.

We now turn to the measurability of projections and spectral measures.

Proposition 4.32. *The function $\omega \mapsto \mathbf{1}_{\rm cont}(A_\omega)$ is measurable.*

Proof. Let $\{\phi_n\}_{n\in\mathbb{N}}$ be an orthonormal basis of $\mathcal{H}$ and let P_n be the orthogonal projection on the subspace spanned by $\{\phi_k\}_{k\geq n}$. The RAGE theorem yields that for $\varphi, \psi \in \mathcal{H}$,

$$(\varphi|\mathbf{1}_{\rm cont}(A_\omega)\psi) = \lim_{n\to\infty}\lim_{T\to\infty}\frac{1}{T}\int_0^T(\varphi|\mathrm{e}^{\mathrm{i}tA_\omega}P_n\mathrm{e}^{-\mathrm{i}tA_\omega}\psi)\mathrm{d}t, \quad (52)$$

(the proof of (52) is left to the problems), and the statement follows. □

Proposition 4.33. *The function $\omega \mapsto \mathbf{1}_{\rm ac}(A_\omega)$ is measurable.*

Proof. By Theorem 3.6, for all $\psi \in \mathcal{H}$,

$$(\psi|\mathbf{1}_{\rm ac}(A_\omega)\psi) = \lim_{M\to\infty}\lim_{p\uparrow 1}\lim_{\epsilon\downarrow 0}\frac{1}{\pi^p}\int_{-M}^{M}\left[\mathrm{Im}\,(\psi|(A_\omega - x - \mathrm{i}\epsilon)^{-1}\psi)\right]^p \mathrm{d}x,$$

and so $\omega \mapsto (\psi|\mathbf{1}_{\rm ac}(A_\omega)\psi)$ is measurable. The polarization identity yields the statement. □

Proposition 4.34. *The functions $\omega \mapsto \mathbf{1}_{\rm sc}(A_\omega)$ and $\omega \mapsto \mathbf{1}_{\rm pp}(A_\omega)$ are measurable.*

Proof. $\mathbf{1}_{\rm sc}(A_\omega) = \mathbf{1}_{\rm cont}(A_\omega) - \mathbf{1}_{\rm ac}(A_\omega)$ and $\mathbf{1}_{\rm pp}(A_\omega) = \mathbf{1} - \mathbf{1}_{\rm cont}(A_\omega)$. □

Let $M(\mathbb{R})$ be the Banach space of all complex Borel measures on $\mathbb{R}$ (the dual of $C_0(\mathbb{R})$). A map $\omega \mapsto \mu^\omega \in M(\mathbb{R})$ is called measurable iff for all $f \in B_{\rm b}(\mathbb{R})$ the map $\omega \mapsto \mu^\omega(f)$ is measurable.

We denote by μ_ψ^ω the spectral measure for A_ω and ψ.

Proposition 4.35. *The functions $\omega \mapsto \mu^\omega_{\psi,\rm ac}$, $\omega \mapsto \mu^\omega_{\psi,\rm sc}$, $\omega \mapsto \mu^\omega_{\psi,\rm pp}$ are measurable.*

Proof. Since for any Borel set B,

$$(\mathbf{1}_B(A_\omega)\psi|\mathbf{1}_{\rm ac}(A_\omega)\psi) = \mu^\omega_{\psi,\rm ac}(B),$$

$$(\mathbf{1}_B(A_\omega)\psi|\mathbf{1}_{\rm sc}(A_\omega)\psi) = \mu^\omega_{\psi,\rm sc}(B),$$

$$(\mathbf{1}_B(A_\omega)\psi|\mathbf{1}_{\rm pp}(A_\omega)\psi) = \mu^\omega_{\psi,\rm pp}(B),$$

the statement follows from Propositions 4.33 and 4.34. □

Let $\{\psi_n\}$ be a cyclic set for A_ω and let $a_n > 0$ be such that

$$\sum_n a_n\|\psi_n\|^2 < \infty.$$

We denote by

$$\mu^\omega = \sum_n a_n\mu^\omega_{\psi_n},$$

the corresponding spectral measure for A_ω. Proposition 4.35 yields

Proposition 4.36. *The functions* $\omega \mapsto \mu^\omega_{\rm ac}$, $\omega \mapsto \mu^\omega_{\rm sc}$, $\omega \mapsto \mu^\omega_{\rm pp}$ *are measurable.*

Let $\Sigma^{{\rm ess},\omega}_{\rm ac}$ be the essential support of the ac spectrum of A_ω. The map

$$\omega \mapsto (1+x^2)^{-1}\chi_{\Sigma^{{\rm ess},\omega}_{\rm ac}}(x) \in L^1(\mathbb{R}, {\rm d}x),$$

does not depend on the choice of representative in $\Sigma^{{\rm ess},\omega}_{\rm ac}$.

Proposition 4.37. *The function*

$$\omega \mapsto \int_{\mathbb{R}} h(x)(1+x^2)^{-1}\chi_{\Sigma^{{\rm ess},\omega}_{\rm ac}}(x){\rm d}x,$$

is measurable for all $h \in L^\infty(\mathbb{R}, {\rm d}x)$.

Proof. Clearly, it suffices to consider

$$h(x) = (1+x)^2\chi_B(x),$$

where B is a bounded interval. Let μ^ω be a spectral measure for A_ω. By the dominated convergence theorem,

$$\int_B \chi_{\Sigma^{{\rm ess},\omega}_{\rm ac}}(x){\rm d}x = 2\lim_{\epsilon\downarrow 0}\lim_{\delta\downarrow 0}\int_B \frac{P_{\mu^\omega_{\rm ac}}(x+{\rm i}\delta)}{P_{\mu^\omega_{\rm ac}}(x+{\rm i}\epsilon)+P_{\mu^\omega_{\rm ac}}(x+{\rm i}\delta)}{\rm d}x, \tag{53}$$

and the statement follows. □

4.17 Non-relativistic quantum mechanics

According to the usual axiomatization of quantum mechanics, a physical system is described by a Hilbert space $\mathcal{H}$. Its observables are described by bounded self-adjoint operators on $\mathcal{H}$. Its states are described by density matrices on $\mathcal{H}$, i.e. positive trace class operators with trace 1. If the system is in a state ρ, then the expected value of the measurement of an observable A is

$$\langle A\rangle_\rho = \mathrm{Tr}(\rho A).$$

The variance of the measurement is

$$D_\rho(A) = \langle (A - \langle A\rangle_\rho)^2\rangle_\rho = \langle A^2\rangle_\rho - \langle A\rangle_\rho^2.$$

The Cauchy-Schwarz inequality yields the *Heisenberg principle*: For self-adjoint $A, B \in \mathcal{B}(\mathcal{H})$,

$$|\mathrm{Tr}(\rho \mathrm{i}[A,B])| \leq D_\rho(A)^{1/2} D_\rho(B)^{1/2}.$$

Of particular importance are the so called pure states $\rho = (\varphi|\cdot)\varphi$. In this case, for a self-adjoint A,

$$\langle A\rangle_\rho = \mathrm{Tr}(\rho A) = (\varphi|A\varphi) = \int_{\mathbb{R}} x \mathrm{d}\mu_\varphi(x),$$

$$D_\rho(A) = \int_{\mathbb{R}} x^2 \mathrm{d}\mu_\varphi - \left(\int_{\mathbb{R}} x \mathrm{d}\mu_\varphi\right)^2,$$

where μ_φ is the spectral measure for A and φ. If the system is in a pure state described by a vector φ, the possible results R of the measurement of A are numbers in $\mathrm{sp}(A)$ randomly distributed according to

$$\mathrm{Prob}(R \in [a,b]) = \int_{[a,b]} \mathrm{d}\mu_\varphi,$$

(recall that μ_φ is supported on $\mathrm{sp}(A)$). Obviously, in this case $\langle A\rangle_\rho$ and $D_\rho(A)$ are the usual expectation and variance of the random variable R.

The dynamics is described by a strongly continuous unitary group $U(t)$ on $\mathcal{H}$. In the Heisenberg picture, one evolves observables and keeps states fixed. Hence, if the system is initially in a state ρ, then the expected value of A at time t is

$$\mathrm{Tr}(\rho[U(t)AU(t)^*]).$$

In the Schrödinger picture, one keeps observables fixed and evolves states—the expected value of A at time t is $\mathrm{Tr}([U(t)^*\rho U(t)]A)$. Note that if $\rho = |\varphi)(\varphi|$, then

$$\mathrm{Tr}([U(t)^*\rho U(t)]A) = \|AU(t)\varphi\|^2.$$

The generator of $U(t)$, H, is called the Hamiltonian of the system. The spectrum of H describes the possible energies of the system. The discrete spectrum of H describes energy levels of bound states (the eigenvectors of H are often called bound states). Note that if φ is an eigenvector of H, then $\|AU(t)\varphi\|^2 = \|A\varphi\|^2$ is independent of t.

By the RAGE theorem, if $\varphi \in \mathcal{H}_{\text{cont}}(H)$ and A is compact, then

$$\lim_{T\to\infty} \frac{1}{T} \int_0^T \|AU(t)\varphi\|^2 \mathrm{d}t = 0. \tag{54}$$

Compact operators describe what one might call sharply localized observables. The states associated to $\mathcal{H}_{\text{cont}}(H)$ move to infinity in the sense that after a sufficiently long time the sharply localized observables are not seen by these states. On the other hand, if $\varphi \in \mathcal{H}_{\text{pp}}(H)$, then for any bounded A,

$$\lim_{T\to\infty} \frac{1}{T} \int_0^T \|AU(t)\varphi\|^2 \mathrm{d}t = \sum_{e \in \text{sp}_{\text{pp}}(H)} \|\mathbf{1}_e(H)A\mathbf{1}_e(H)\varphi\|^2.$$

The mathematical formalism sketched above is commonly used for a description of non-relativistic quantum systems with finitely many degrees of freedom. It can be used, for example, to describe non-relativistic matter—a finite assembly of interacting non-relativistic atoms and molecules. In this case H is the usual N-body Schrödinger operator. This formalism, however, is not suitable for a description of quantum systems with infinitely many degrees of freedom like non-relativistic QED, an infinite electron gas, quantum spin-systems, etc.

4.18 Problems

[1] *Prove Proposition 4.7*

[2] *Prove Theorem 4.9.*

[3] *Prove Theorem 4.18.*

[4] *Prove Theorem 4.19.*

[5] *Prove Theorem 4.20.*

[6] *Let* $0 < \epsilon < 1$. *Construct an example of a self-adjoint operator* A *such that* $\text{sp}_{\text{ac}}(A) = [0,1]$ *and that the Lebesgue measure of* $\Sigma_{\text{ac}}^{\text{ess}}(A)$ *is equal to* ϵ.

[7] *Prove that* $\psi \in \mathcal{H}_{\text{cont}}$ *iff (49) holds for all compact* K.

[8] *Prove that* $A \geq 0$ *iff* $\text{sp}(A) \subset [0,\infty)$.

[9] *Prove Relation (45).*

[10] *Prove Part (3) of Theorem 4.27.*

[11] *Prove Proposition 4.29.*

[12] *Let* $M \ni \omega \mapsto A_\omega$ *be a function with values in self-adjoint operators on* $\mathcal{H}$. *Prove that the following statements are equivalent:*
(1) $\omega \mapsto (A_\omega - z)^{-1}$ *is measurable for all* $z \in \mathbb{C} \setminus \mathbb{R}$.
(2) $\omega \mapsto \exp(\mathrm{i}tA_\omega)$ *is measurable for all* $t \in \mathbb{R}$.
(3) $\omega \mapsto \mathbf{1}_B(A_\omega)$ *is measurable for all Borel sets* B.

[13] *Prove Relation (52).*

[14] *Let* $\omega \mapsto A_\omega$ *be a measurable function with values in self-adjoint operators.*
(1) *In the literature, the proof of the measurability of the function* $\omega \mapsto \mathbf{1}_{\mathrm{sc}}(A_\omega)$ *is usually based on Carmona's lemma. Let* μ *be a finite, positive Borel measure on* $\mathbb{R}$, *and let* $\mathcal{I}$ *be the set of finite unions of open intervals, each of which has rational endpoints. Then, for any Borel set* B,

$$\mu_{\mathrm{sing}}(B) = \lim_{n\to\infty} \sup_{I\in\mathcal{I}, |I|<1/n} \mu(B \cap I).$$

Prove Carmona's lemma and using this result show that $\omega \mapsto \mathbf{1}_{\mathrm{sc}}(A_\omega)$ *is a measurable function. Hint: See* [3] *or Section 9.1 in* [2].
(2) *Prove that* $\omega \mapsto \mathbf{1}_{\mathrm{pp}}(A_\omega)$ *is a measurable function by using Simon's local Wiener theorem (Theorem 3.9).*

The next set of problems deals with spectral theory of a closed operator A *on a Hilbert space* $\mathcal{H}$.

[15] *Let* $F \subset \mathrm{sp}(A)$ *be an isolated, bounded subset of* $\mathrm{sp}(A)$. *Let* γ *be a closed simple path in the complex plane that separates* F *from* $\mathrm{sp}(A) \setminus F$. *Set*

$$\mathbf{1}_F(A) = \frac{1}{2\pi \mathrm{i}} \oint_\gamma (z - A)^{-1} \mathrm{d}z.$$

(1) *Prove that* $\mathbf{1}_F(A)$ *is a (not necessarily orthogonal) projection.*
(2) *Prove that* $\mathrm{Ran}\,\mathbf{1}_F(A)$ *and* $\mathrm{Ker}\,\mathbf{1}_F(A)$ *are complementary (not necessarily orthogonal) subspaces:* $\mathrm{Ran}\,\mathbf{1}_F(A) + \mathrm{Ker}\,\mathbf{1}_F(A) = \mathcal{H}$ *and*

$$\mathrm{Ran}\,\mathbf{1}_F(A) \cap \mathrm{Ker}\,\mathbf{1}_F(A) = \{0\}.$$

(3) *Prove that* $\mathrm{Ran}\,\mathbf{1}_F(A) \subset \mathrm{Dom}\,(A)$ *and that* $A : \mathrm{Ran}\,\mathbf{1}_F(A) \to \mathrm{Ran}\,\mathbf{1}_F(A)$. *Prove that* $A \restriction \mathrm{Ran}\,\mathbf{1}_F(A)$ *is a bounded operator and that its spectrum is* F.
(4) *Prove that* $\mathrm{Ker}\,\mathbf{1}_F(A) \cap \mathrm{Dom}\,(A)$ *is dense and that*

$$A \restriction (\mathrm{Ker}\,\mathbf{1}_F(A) \cap \mathrm{Dom}\,(A)) \to \mathrm{Ker}\,\mathbf{1}_F(A). \tag{55}$$

Prove that the operator (55) is closed and that its spectrum is $\mathrm{sp}(A) \setminus F$.
(5) *Assume that* $F = \{z_0\}$ *and that* $\mathrm{Ran}\, \mathbf{1}_{z_0}(A)$ *is finite dimensional. Prove that if* $\psi \in \mathrm{Ran}\, \mathbf{1}_{z_0}(A)$, *then* $(A - z_0)^n \psi = 0$ *for some* n.
Hint: Consult Theorem XII.5 in [26].
Remark. *The set of* $z_0 \in \mathrm{sp}(A)$ *which satisfy (5) is called the discrete spectrum of the closed operator operator* A *and is denoted* $\mathrm{sp}_{\mathrm{disc}}(A)$. *The essential spectrum is defined by* $\mathrm{sp}_{\mathrm{ess}}(A) = \mathrm{sp}(A) \setminus \mathrm{sp}_{\mathrm{disc}}(A)$

[16] *Prove that* $\mathrm{sp}_{\mathrm{ess}}(A)$ *is a closed subset of* $\mathbb{C}$.

[17] *Prove that* $z \mapsto (A - z)^{-1}$ *is a meromorphic function on* $\mathbb{C} \setminus \mathrm{sp}_{\mathrm{ess}}(A)$ *with singularities at points* $z_0 \in \mathrm{sp}_{\mathrm{disc}}(A)$. *Prove that the negative coefficents of the Laurent expansion at* $z_0 \in \mathrm{sp}_{\mathrm{disc}}(A)$ *are finite rank operators. Hint: See Lemma 1 in* [26], *Section XIII.4.*

[18] *The numerical range of* A *is defined by* $N(A) = \{(\psi|A\psi) : \psi \in \mathrm{Dom}\,(A)\}$. *In general,* $N(A)$ *is neither open nor closed subset of* $\mathbb{C}$. *Prove that if* $\mathrm{Dom}\,(A) = \mathrm{Dom}\,(A^*)$, *then* $\mathrm{sp}(A) \subset \overline{N(A)}$. *For additional information about numerical range, the reader may consult* [11].

[19] *Let* $z \in \mathrm{sp}(A)$. *A sequence* $\psi_n \in \mathrm{Dom}\,(A)$ *is called a Weyl sequence if* $\|\psi_n\| = 1$ *and* $\|(A - z)\psi_n\| \to 0$. *If* A *is not self-adjoint, then a Weyl sequence may not exist for some* $z \in \mathrm{sp}(A)$. *Prove that a Weyl sequence exists for every* z *on the topological boundary of* $\mathrm{sp}(A)$. *Hint: See Section XIII.4 of* [26] *or* [32].

[20] *Let* A *and* B *be densely defined linear operators. Assume that* B *is* A*-bounded with a relative bound* $a < 1$. *Prove that* $A + B$ *is closable iff* A *is closable, and that in this case the closures of* A *and* $A + B$ *have the same domain. Deduce that* $A + B$ *is closed iff* A *is closed.*

[21] *Let* A *and* B *be densely defined linear operators. Assume that* A *is closed and that* B *is* A*-bounded with constants* a *and* b. *If* A *is invertible (that is,* $0 \notin \mathrm{sp}(A)$*), and if* a *and* b *satisfy*

$$a + b\|A^{-1}\| < 1,$$

prove that $A + B$ *is closed, invertible, and that*

$$\|(A + B)^{-1}\| \leq \frac{\|A^{-1}\|}{1 - a - b\|A^{-1}\|},$$

$$\|(A + B)^{-1} - A^{-1}\| \leq \frac{\|A^{-1}\|(a + b\|A^{-1}\|)}{1 - a - b\|A^{-1}\|}.$$

Hint: See Theorem IV.1.16 in [17].

[22] *In this problem we will discuss the regular perturbation theory for closed operators. Let* A *be a closed operator and let* B *be* A*-bounded with constants* a *and*

b. For $\lambda \in \mathbb{C}$ we set $A_\lambda = A + \lambda B$. If $|\lambda| a < 1$, then A_λ is a closed operator and $\mathrm{Dom}\,(A_\lambda) = \mathrm{Dom}\,(A)$. *Let F be an isolated, bounded subset of A and γ a simple closed path that separates F and* $\mathrm{sp}(H) \setminus F$.
(1) *Prove that for $z \in \gamma$,*

$$\|B(A-z)^{-1}\| \le a + (a|z| + b)\|(A-z)^{-1}\|.$$

(2) *Prove that if A is self-adjoint and $z \notin \mathrm{sp}(A)$, then*

$$\|(A-z)^{-1}\| = 1/\mathrm{dist}\{z, \mathrm{sp}(A)\}.$$

(3) *Assume that* $\mathrm{Dom}\,(A) = \mathrm{Dom}\,(A^*)$ *and let $N(A)$ be the numerical range of A. Prove that for all $z \notin \overline{N(A)}$,*

$$\|(A-z)^{-1}\| \le 1/\mathrm{dist}\{z, \overline{N(A)}\}.$$

(4) *Let $\Lambda = \left[a + \sup_{z\in\gamma}(a|z|) + b)\|(A-z)^{-1}\|\right]^{-1}$ and assume that $|\lambda| < \Lambda$. Prove that* $\mathrm{sp}(A_\lambda) \cap \gamma = \emptyset$ *and that for $z \in \gamma$,*

$$(z - A_\lambda)^{-1} = \sum_{n=0}^{\infty} \lambda^n (z-A)^{-1}\left[B(z-A)^{-1}\right]^n.$$

Hint: Start with $z - A_\lambda = (1 - \lambda B(z-A)^{-1})(z-A)$.
(5) *Let F_λ be the spectrum of A_λ inside γ (so $F_0 = F$). For $|\lambda| < \Lambda$ the projection of A_λ onto F_λ is given by*

$$P_\lambda = \mathbf{1}_{F_\lambda}(A_\lambda) = \frac{1}{2\pi\mathrm{i}}\oint_\gamma (z - A_\lambda)^{-1}\mathrm{d}z.$$

Prove that the projection-valued function $\lambda \mapsto P_\lambda$ is analytic for $|\lambda| < \Lambda$.
(6) *Prove that the differential equation $U'_\lambda = [P'_\lambda, P_\lambda]U_\lambda$, $U_0 = \mathbf{1}$, (the derivatives are w.r.t. λ and $[A, B] = AB - BA$) has a unique solution for $|\lambda| < \Lambda$, and that U_λ is an analytic family of bounded invertible operators such that $U_\lambda P_0 U_\lambda^{-1} = P_\lambda$. Hint: See* [26], *Section XII.2.*
(7) *Set $\tilde{A}_\lambda = U_\lambda^{-1} A_\lambda U_\lambda$ and $\Sigma_\lambda = P_0 \tilde{A}_\lambda P_0$. Σ_λ is a bounded operator on the Hilbert space* $\mathrm{Ran}\, P_0$. *Prove that $\mathrm{sp}(\Sigma_\lambda) = F_\lambda$ and that the operator-valued function $\lambda \mapsto \Sigma_\lambda$ is analytic for $|\lambda| < \Lambda$. Compute the first three terms in the expansion*

$$\Sigma_\lambda = \sum_{n=0}^{\infty} \lambda^n \Sigma_n. \tag{56}$$

The term Σ_1 is sometimes called the Feynman-Hellman term. The term Σ_2, often called the Level Shift Operator (LSO), plays an important role in quantum mechanics and quantum field theory. For example, the formal computations in physics involving Fermi's Golden Rule are often best understood and most easily proved with the help of LSO.
(8) *Assume that* $\dim P_0 = \dim \mathrm{Ran}\, P_0 < \infty$. *Prove that* $\dim P_\lambda = \dim P_0$ *for*

$|\lambda| < \Lambda$ and conclude that the spectrum of A_λ inside γ is discrete and consists of at most $\dim P_0$ *distinct eigenvalues. Prove that the eigenvalues of A_λ inside γ are all the branches of one or more multi-valued analytic functions with at worst algebraic singularities.*
(9) *Assume that $F_0 = \{z_0\}$ and* $\dim P_0 = 1$ *(namely that the spectrum of A inside γ is a semisimple eigenvalue z_0). In this case $\Sigma_\lambda = z(\lambda)$ is an analytic function for $|\lambda| < \Lambda$. Compute the first five terms in the expansion*

$$z(\lambda) = \sum_{n=0}^{\infty} \lambda^n z_n.$$

5 Spectral theory of rank one perturbations

The Hamiltonians which arise in non-relativistic quantum mechanics typically have the form

$$H_V = H_0 + V, \tag{57}$$

where H_0 and V are two self-adjoint operators on a Hilbert space $\mathcal{H}$. H_0 is the "free" or "reference" Hamiltonian and V is the "perturbation". For example, the Hamiltonian of a free non-relativistic quantum particle of mass m moving in $\mathbb{R}^3$ is $-\frac{1}{2m}\Delta$, where Δ is the Laplacian in $L^2(\mathbb{R}^3)$. If the particle is subject to an external potential field $V(x)$, then the Hamiltonian describing the motion of the particle is

$$H_V = -\frac{1}{2m}\Delta + V, \tag{58}$$

where V denotes the operator of multiplication by the function $V(x)$. Operators of this form are called Schrödinger operators.

We will not study in this section the spectral theory of Schrödinger operators. Instead, we will keep H_0 general and focus on simplest non-trivial perturbations V. More precisely, let $\mathcal{H}$ be a Hilbert space, H_0 a self-adjoint operator on $\mathcal{H}$ and $\psi \in \mathcal{H}$ a given unit vector. We will study spectral theory of the family of operators

$$H_\lambda = H_0 + \lambda(\psi|\cdot)\psi, \qquad \lambda \in \mathbb{R}. \tag{59}$$

This simple model is of profound importance in mathematical physics. The classical reference for the spectral theory of rank one perturbations is [29].

The cyclic subspace generated by H_λ and ψ does not depend on λ and is equal to the cyclic subspace generated by H_0 and ψ which we denote $\mathcal{H}_\psi$ (this fact is an immediate consequence of the formulas (61) below). Let μ^λ be the spectral measure for H_λ and ψ. This measure encodes the spectral properties of $H_\lambda \restriction \mathcal{H}_\psi$. Note that $H_\lambda \restriction \mathcal{H}_\psi^\perp = H_0 \restriction \mathcal{H}_\psi^\perp$. In this section we will always assume that $\mathcal{H} = \mathcal{H}_\psi$, namely that ψ is a cyclic vector for H_0.

The identities

$$
\begin{aligned}
(H_\lambda - z)^{-1} - (H_0 - z)^{-1} &= (H_\lambda - z)^{-1}(H_0 - H_\lambda)(H_0 - z)^{-1} \\
&= (H_0 - z)^{-1}(H_0 - H_\lambda)(H_\lambda - z)^{-1},
\end{aligned}
\tag{60}
$$

yield

$$
\begin{aligned}
(H_\lambda - z)^{-1}\psi &= (H_0 - z)^{-1}\psi - \lambda(\psi|(H_0 - z)^{-1}\psi)(H_\lambda - z)^{-1}\psi, \\
(H_0 - z)^{-1}\psi &= (H_\lambda - z)^{-1}\psi + \lambda(\psi|(H_\lambda - z)^{-1}\psi)(H_0 - z)^{-1}\psi.
\end{aligned}
\tag{61}
$$

Let

$$
F_\lambda(z) = (\psi|(H_\lambda - z)^{-1}\psi) = \int_{\mathbb{R}} \frac{\mathrm{d}\mu^\lambda(t)}{t - z}.
$$

Note that if $z \in \mathbb{C}_+$, then $F_\lambda(z)$ is the Borel transform and $\mathrm{Im}\, F_\lambda(z)$ is the Poisson transform of μ^λ.

The second identity in (61) yields

$$
F_0(z) = F_\lambda(z)(1 + \lambda F_0(z)),
$$

and so

$$
F_\lambda(z) = \frac{F_0(z)}{1 + \lambda F_0(z)}, \tag{62}
$$

$$
\mathrm{Im}\, F_\lambda(z) = \frac{\mathrm{Im}\, F_0(z)}{|1 + \lambda F_0(z)|^2}. \tag{63}
$$

These elementary identities will play a key role in our study. The function

$$
G(x) = \int_{\mathbb{R}} \frac{\mathrm{d}\mu^0(t)}{(x - t)^2},
$$

will also play an important role. Recall that $G(x) = \infty$ for μ^0-a.e. x (Lemma 3.3).

In this section we will occasionally denote by $|B|$ the Lebesgue measure of a Borel set B.

5.1 Aronszajn-Donoghue theorem

Recall that the limit

$$
F_\lambda(x) = \lim_{y \downarrow 0} F_\lambda(x + \mathrm{i}y),
$$

exists and is finite and non-zero for Lebesgue a.e. x.

For $\lambda \neq 0$ define

$$
\begin{aligned}
S_\lambda &= \{x \in \mathbb{R} : F_0(x) = -\lambda^{-1},\ G(x) = \infty\}, \\
T_\lambda &= \{x \in \mathbb{R} : F_0(x) = -\lambda^{-1},\ G(x) < \infty\}, \\
L &= \{x \in \mathbb{R} : \mathrm{Im}\, F_0(x) > 0\}.
\end{aligned}
$$

In words, S_λ is the set of all $x \in \mathbb{R}$ such that $\lim_{y\downarrow 0} F_0(x+iy)$ exists and is equal to $-\lambda^{-1}$, etc. Any two sets in the collection $\{S_\lambda, T_\lambda, L\}_{\lambda\neq 0}$ are disjoint. By Theorem 4.11, $|S_\lambda| = |T_\lambda| = 0$.

As usual, $\delta(y)$ denotes the delta-measure of $y \in \mathbb{R}$; $\delta(y)(f) = f(y)$.

Theorem 5.1. (1) *T_λ is the set of eigenvalues of H_λ. Moreover,*

$$\mu_{\rm pp}^\lambda = \sum_{x_n\in T_\lambda} \frac{1}{\lambda^2 G(x_n)}\delta(x_n).$$

(2) *$\mu_{\rm sc}^\lambda$ is concentrated on S_λ.*
(3) *For all λ, L is the essential support of the ac spectrum of H_λ and $\mathrm{sp}_{\rm ac}(H_\lambda) = \mathrm{sp}_{\rm ac}(H_0)$.*
(4) *The measures $\{\mu_{\rm sing}^\lambda\}_{\lambda\in\mathbb{R}}$ are mutually singular. In other words, if $\lambda_1 \neq \lambda_2$, then the measures $\mu_{\rm sing}^{\lambda_1}$ and $\mu_{\rm sing}^{\lambda_2}$ are concentrated on disjoint sets.*

Proof. (1) The eigenvalues of H_λ are precisely the atoms of μ^λ. Let

$$\tilde{T}_\lambda = \{x \in \mathbb{R} : \mu^\lambda(\{x\}) \neq 0\}.$$

Since

$$\mu^\lambda(\{x\}) = \lim_{y\downarrow 0} y\mathrm{Im}\, F_\lambda(x+iy) = \lim_{y\downarrow 0} \frac{y\mathrm{Im}\, F_0(x+iy)}{|1+\lambda F_0(x+iy)|^2}, \tag{64}$$

$\tilde{T}_\lambda \subset \{x : F_0(x) = -\lambda^{-1}\}$. The relation (64) yields

$$\mu^\lambda(\{x\}) \le \lambda^{-2} \lim_{y\downarrow 0} \frac{y}{\mathrm{Im}\, F_0(x+iy)} = \frac{1}{\lambda^2 G(x)},$$

and so $\tilde{T}_\lambda \subset \{x : F_0(x) = -\lambda^{-1}, G(x) < \infty\} = T_\lambda$. On the other hand, if $F_0(x) = -\lambda^{-1}$ and $G(x) < \infty$, then

$$\lim_{y\downarrow 0} \frac{F_0(x+iy) - F_0(x)}{iy} = G(x), \tag{65}$$

(the proof of this relation is left to the problems). Hence, if $x \in T_\lambda$, then

$$F_0(x+iy) = iyG(x) - \lambda^{-1} + o(y),$$

and

$$\mu^\lambda(\{x\}) = \lim_{y\downarrow 0} \frac{y\mathrm{Im}\, F_0(x+iy)}{|1+\lambda F_0(x+iy)|^2} = \frac{1}{\lambda^2 G(x)} > 0.$$

Hence $T_\lambda = \tilde{T}_\lambda$, and for $x \in \tilde{T}_\lambda$, $\mu^\lambda(\{x\}) = 1/\lambda^2 G(x)$. This yields (1).
(2) By Theorem 3.5, $\mu_{\rm sing}^\lambda$ is concentrated on the set

$$\{x : \lim_{y\downarrow 0} \mathrm{Im}\, F_\lambda(x+iy) = \infty\}.$$

The formula (63) yields that $\mu_{\rm sing}^\lambda$ is concentrated on the set

$$\{x : F_0(x) = -\lambda^{-1}\}. \tag{66}$$

If $F_0(x) = -\lambda^{-1}$ and $G(x) < \infty$, then by (1) x is an atom of μ^λ. Hence, $\mu^\lambda_{\rm sc}$ is concentrated on the set $\{x : F_0(x) = -\lambda^{-1}, G(x) = \infty\} = S_\lambda$.
(3) By Theorem 3.5,

$$\mathrm{d}\mu^\lambda_{\rm ac}(x) = \pi^{-1}\mathrm{Im}\, F_\lambda(x)\mathrm{d}x.$$

On the other hand, by the formula (63), the sets $\{x : \mathrm{Im}\, F_0(x) > 0\}$ and $\{x : \mathrm{Im}\, F_\lambda(x) > 0\}$ coincide up to a set of Lebesgue measure zero. Hence, L is the essential support of the ac spectrum of H_λ for all λ. Since $\mu^0_{\rm ac}$ and $\mu^\lambda_{\rm ac}$ are equivalent measures, $\mathrm{sp}_{\rm ac}(H_0) = \mathrm{sp}_{\rm ac}(H_\lambda)$.
(4) For $\lambda \neq 0$, $\mu^\lambda_{\rm sing}$ is concentrated on the set (66). By Theorem 3.5, $\mu^0_{\rm sing}$ is concentrated on $\{x : \mathrm{Im}\, F_0(x) = \infty\}$. This yields the statement. □

5.2 The spectral theorem

By Theorem 4.13, for all λ there exists a unique unitary $U_\lambda : \mathcal{H}_\psi \to L^2(\mathbb{R}, \mathrm{d}\mu_\lambda)$ such that $U_\lambda \psi = \mathbb{1}$ and $U_\lambda H_\lambda U_\lambda^{-1}$ is the operator of multiplication by x on $L^2(\mathbb{R}, \mathrm{d}\mu_\lambda)$. In this subsection we describe U_λ.

For $\phi \in \mathcal{H}$ and $z \in \mathbb{C} \setminus \mathbb{R}$ let

$$M_\phi(z) = (\psi|(H_0 - z)^{-1}\phi),$$

and

$$M_\phi(x \pm \mathrm{i}0) = \lim_{y \downarrow 0}(\psi|(H_0 - x \mp \mathrm{i}y)^{-1}\phi),$$

whenever the limits exist. By Theorem 3.17 the limits exist and are finite for Lebesgue a.e. x.

For consistency, in this subsection we write $F_0(x + \mathrm{i}0) = \lim_{y\downarrow 0} F_0(x + \mathrm{i}y)$.

Theorem 5.2. *Let* $\phi \in \mathcal{H}$.
(1) *For all* λ *and for* $\mu_{\lambda,\rm ac}$*-a.e.* x,

$$(U_\lambda \mathbf{1}_{\rm ac}\phi)(x) = \frac{1}{2\mathrm{i}}\frac{M_\phi(x + \mathrm{i}0) - M_\phi(x - \mathrm{i}0)}{\mathrm{Im}\, F_0(x + \mathrm{i}0)} - \lambda M_\phi(x + \mathrm{i}0)$$
$$+ \frac{\lambda}{2\mathrm{i}}\frac{(M_\phi(x + \mathrm{i}0) - M_\phi(x - \mathrm{i}0))F_0(x + \mathrm{i}0)}{\mathrm{Im}\, F_0(x + \mathrm{i}0)}.$$

(2) *Let* $\lambda \neq 0$. *Then for* $\mu_{\lambda,\rm sing}$*-a.e.* x *the limit* $M_\phi(x + \mathrm{i}0)$ *exists and*

$$(U_\lambda \mathbf{1}_{\rm sing}\phi)(x) = -\lambda M_\phi(x + \mathrm{i}0).$$

Proof. The identities (60) yield

$$(\psi|(H_\lambda - z)^{-1}\phi) = \frac{M_\phi(z)}{1+\lambda F_0(z)}.$$

Combining this relation with (62) and (63) we derive

$$\begin{aligned}\frac{(\psi|\mathrm{Im}\,(H_\lambda - z)^{-1}\phi)}{\mathrm{Im}\,(\psi|(H_\lambda - z)^{-1}\psi)} &= \frac{1}{2\mathrm{i}}\frac{M_\phi(z) - M_\phi(\overline{z})}{\mathrm{Im}\,F_0(z)} \\ &\quad + \frac{\lambda}{2\mathrm{i}}\frac{F_0(\overline{z})M_\phi(z) - F_0(z)M_\phi(\overline{z})}{\mathrm{Im}\,F_0(z)}.\end{aligned} \tag{67}$$

Similarly,

$$\frac{(\psi|(H_\lambda - z)^{-1}\phi)}{(\psi|(H_\lambda - z)^{-1}\psi)} = \frac{M_\phi(z)}{F_0(z)}. \tag{68}$$

(1) follows from the identity (67) and Part 1 of Theorem 4.17. Since $\mu_{\lambda,\mathrm{sing}}$ is concentrated on the set $\{x : \lim_{y\downarrow 0} F_0(x + \mathrm{i}0) = -\lambda^{-1}\}$, the identity (68) and Part 2 of Theorem 4.17 yield (2). □

Note that Part 2 of Theorem 5.2 yields that for every eigenvalue x of H_λ (i.e. for all $x \in T_\lambda$),

$$(U_\lambda \mathbf{1}_{\mathrm{pp}}\phi)(x) = -\lambda M_\phi(x + \mathrm{i}0). \tag{69}$$

This special case of Theorem 5.2 (which can be easily proven directly) has been used in the proofs of dynamical localization in the Anderson model; see [1, 6]. The extension of (69) to singular continuous spectrum depends critically on the full strength of the Poltoratskii theorem. For some applications of this result see [16].

5.3 Spectral averaging

In the sequel we will freely use the measurability results established in Subsection 4.16.

Let

$$\overline{\mu}(B) = \int_{\mathbb{R}} \mu^\lambda(B)\mathrm{d}\lambda,$$

where $B \subset \mathbb{R}$ is a Borel set. Obviously, $\overline{\mu}$ is a Borel measure on $\mathbb{R}$. The following (somewhat surprising) result is often called *spectral averaging*:

Theorem 5.3. *The measure $\overline{\mu}$ is equal to the Lebesgue measure and for all $f \in L^1(\mathbb{R}, \mathrm{d}x)$,*

$$\int_{\mathbb{R}} f(x)\mathrm{d}x = \int_{\mathbb{R}} \left[\int_{\mathbb{R}} f(x)\mathrm{d}\mu^\lambda(x)\right] \mathrm{d}\lambda.$$

Proof. For any positive Borel function f,

$$\int_{\mathbb{R}} f(t)\mathrm{d}\overline{\mu}(t) = \int_{\mathbb{R}}\left[\int_{\mathbb{R}} f(t)\mathrm{d}\mu^{\lambda}(t)\right]\mathrm{d}\lambda,$$

(both sides are allowed to be infinity). Let

$$f(t) = \frac{y}{(t-x)^2+y^2},$$

where $y > 0$. Then

$$\int_{\mathbb{R}} f(t)\mathrm{d}\mu^{\lambda}(t) = \mathrm{Im}\, F_{\lambda}(x+\mathrm{i}y) = \frac{\mathrm{Im}\, F_0(x+\mathrm{i}y)}{|1+\lambda F_0(x+\mathrm{i}y)|^2}.$$

By the residue calculus,

$$\int_{\mathbb{R}} \frac{\mathrm{Im}\, F_0(x+\mathrm{i}y)}{|1+\lambda F_0(x+\mathrm{i}y)|^2}\mathrm{d}\lambda = \pi, \tag{70}$$

and so the Poisson transform of $\overline{\mu}$ exists and is identically equal to π, the Poisson transform of the Lebesgue measure. By Theorem 3.7, $\overline{\mu}$ is equal to the Lebesgue measure. □

Spectral averaging is a mathematical gem which has been rediscovered by many authors. A detailed list of references can be found in [30].

5.4 Simon-Wolff theorems

Theorem 5.4. *Let $B \subset \mathbb{R}$ be a Borel set. Then the following statements are equivalent:*
(1) $G(x) < \infty$ *for Lebesgue a.e.* $x \in B$.
(2) $\mu^{\lambda}_{\mathrm{cont}}(B) = 0$ *for Lebesgue a.e.* λ.

Proof. (1) ⇒ (2). If $G(x) < \infty$ for Lebesgue a.e. $x \in B$, then $\mathrm{Im}\, F_0(x) = 0$ for Lebesgue a.e. $x \in B$. Hence, for all λ, $\mathrm{Im}\, F_{\lambda}(x) = 0$ for Lebesgue a.e. $x \in B$, and

$$\mu^{\lambda}_{\mathrm{ac}}(B) = \pi^{-1}\int_B \mathrm{Im}\, F_{\lambda}(x)\mathrm{d}x = 0.$$

By Theorem 5.1, the measure $\mu^{\lambda}_{\mathrm{sc}} \restriction B$ is concentrated on the set $A = \{x \in B : G(x) = \infty\}$. Since A has Lebesgue measure zero, by spectral averaging,

$$\int_{\mathbb{R}} \mu^{\lambda}_{\mathrm{sc}}(A)\mathrm{d}\lambda \le \int_{\mathbb{R}} \mu^{\lambda}(A)\mathrm{d}\lambda = |A| = 0.$$

Hence, $\mu^{\lambda}_{\mathrm{sc}}(A) = 0$ for Lebesgue a.e. $\lambda \in \mathbb{R}$, and so $\mu^{\lambda}_{\mathrm{sc}}(B) = 0$ for Lebesgue a.e. λ.

(2) ⇒ (1). Assume that the set $A = \{x \in B : G(x) = \infty\}$ has positive Lebesgue measure. By Theorem 5.1, $\mu_{\text{pp}}^{\lambda}(A) = 0$ for all $\lambda \neq 0$. By spectral averaging,

$$\int_{\mathbb{R}} \mu_{\text{cont}}^{\lambda}(A)\mathrm{d}\lambda = \int_{\mathbb{R}} \mu^{\lambda}(A)\mathrm{d}\lambda = |A| > 0.$$

Hence, for a set of λ of positive Lebesgue measure, $\mu_{\text{cont}}^{\lambda}(B) > 0$. □

Theorem 5.5. *Let B be a Borel set. Then the following statements are equivalent:*
(1) $\operatorname{Im} F_0(x) > 0$ *for Lebesgue a.e.* $x \in B$.
(2) $\mu_{\text{sing}}^{\lambda}(B) = 0$ *for Lebesgue a.e.* λ.

Proof. (1) ⇒ (2). By Theorem 5.1, for $\lambda \neq 0$ the measure $\mu_{\text{sing}}^{\lambda} \restriction B$ is concentrated on the set $A = \{x \in B : \operatorname{Im} F_0(x) = 0\}$. Since A has Lebesgue measure zero, by spectral averaging,

$$\int_{\mathbb{R}} \mu_{\text{sing}}^{\lambda}(A)\mathrm{d}\lambda \leq \int_{\mathbb{R}} \mu^{\lambda}(A)\mathrm{d}\lambda = 0.$$

Hence, for Lebesgue a.e. λ, $\mu_{\text{sing}}^{\lambda}(B) = 0$.
(2) ⇒ (1). Assume that the set $A = \{x \in B : \operatorname{Im} F_0(x) = 0\}$ has positive Lebesgue measure. Clearly, $\mu_{\text{ac}}^{\lambda}(A) = 0$ for all λ, and by spectral averaging,

$$\int_{\mathbb{R}} \mu_{\text{sing}}^{\lambda}(A)\mathrm{d}\lambda = \int_{\mathbb{R}} \mu^{\lambda}(A)\mathrm{d}\lambda = |A| > 0.$$

Hence, for a set of λ of positive Lebesgue measure, $\mu_{\text{sing}}^{\lambda}(B) > 0$. □

Theorem 5.6. *Let B be a Borel set. Then the following statements are equivalent:*
(1) $\operatorname{Im} F_0(x) = 0$ *and* $G(x) = \infty$ *for Lebesgue a.e.* $x \in B$.
(2) $\mu_{\text{ac}}^{\lambda}(B) + \mu_{\text{pp}}^{\lambda}(B) = 0$ *for Lebesgue a.e.* λ.

The proof of Theorem 5.6 is left to the problems.

Theorem 5.4 is the celebrated result of Simon-Wolff [31]. Although Theorems 5.5 and 5.6 are well known to the workers in the field, I am not aware of a convenient reference.

5.5 Some remarks on spectral instability

By the Kato-Rosenblum theorem, the absolutely continuous spectrum is stable under trace class perturbations, and in particular under rank one perturbations. In the rank one case this result is also an immediate consequence of Theorem 5.1.

The situation is more complicated in the case of the singular continuous spectrum. There are examples where sc spectrum is stable, namely when H_λ has purely singular continuous spectrum in (a, b) for all $\lambda \in \mathbb{R}$. There are also examples where H_0 has purely sc spectrum in (a, b), but H_λ has pure point spectrum for all $\lambda \neq 0$.

A. Gordon [10] and del Rio-Makarov-Simon [7] have proven that pp spectrum is *always* unstable for generic λ.

Theorem 5.7. *The set*

$$\{\lambda : H_\lambda \text{ has no eigenvalues in } \mathrm{sp}(H_0)\},$$

is dense G_δ *in* $\mathbb{R}$.

Assume that

$$(a, b) \subset \mathrm{sp}(H_0),$$

and that $G(x) < \infty$ for Lebesgue a.e. $x \in (a, b)$. Then the spectrum of H_λ in (a, b) is pure point for Lebesgue a.e. λ. However, by Theorem 5.7, there is a dense G_δ set of λ's such that H_λ has purely singular continuous spectrum in (a, b) (of course, H_λ has no ac spectrum in (a, b) for all λ).

5.6 Boole's equality

So far we have used the rank one perturbation theory and harmonic analysis to study spectral theory. In the last three subsections we will turn things around and use rank one perturbation theory and spectral theory to reprove some well known results in harmonic analysis. This subsection deals with Boole's equality and is based on [6] and [22].

Let ν be a finite positive Borel measure on $\mathbb{R}$ and $F_\nu(z)$ its Borel transform. As usual, we denote

$$F_\nu(x) = \lim_{y \downarrow 0} F_\nu(x + \mathrm{i}y).$$

The following result is known as Boole's equality:

Proposition 5.8. *Assume that* ν *is a pure point measure with finitely many atoms. Then for all* $t > 0$

$$|\{x : F_\nu(x) > t\}| = |\{x : F_\nu(x) < -t\}| = \frac{\nu(\mathbb{R})}{t}.$$

Proof. We will prove that $|\{x : F_\nu(x) > t\}| = \nu(\mathbb{R})/t$. Let

$$\{x_j\}_{1 \le j \le n}, \qquad x_1 < \cdots < x_n,$$

be the support of ν and $\alpha_j = \nu(\{x_j\})$ the atoms of ν. W.l.o.g. we may assume that $\nu(\mathbb{R}) = \sum_j \alpha_j = 1$. Clearly,

$$F_\nu(x) = \sum_{j=1}^{n} \frac{\alpha_j}{x_j - x}.$$

Set $x_0 = -\infty$, $x_{n+1} = \infty$. Since $F'_\nu(x) > 0$ for $x \neq x_j$, the function $F_\nu(x)$ is strictly increasing on (x_j, x_{j+1}), with vertical asymptotes at $x_j, 1 \le j \le n$. Let $r_1 < \cdots < r_n$ be the solutions of the equation $F_\nu(x) = t$. Then

$$|\{x : F_\nu(x) > t\}| = \sum_{j=1}^{n}(x_j - r_j).$$

On the other hand, the equation $F_\nu(x) = t$ is equivalent to

$$\sum_{k=1}^{n} \alpha_k \prod_{j\neq k}(x_j - x) = t\prod_{j=1}^{n}(x_j - x),$$

or

$$\prod_{j=1}^{n}(x_j - x) - t^{-1}\sum_{k=1}^{n} \alpha_k \prod_{j\neq k}(x_j - x) = 0.$$

Since $\{r_j\}$ are all the roots of the polynomial on the l.h.s.,

$$\sum_{j=1}^{n} r_j = -t^{-1} + \sum_{j=1}^{n} x_j$$

and this yields the statement. □

Proposition 5.8 was first proven by G. Boole in 1867. The Boole equality is another gem that has been rediscovered by many authors; see [22] for the references.

The rank one perturbation theory allows for a simple proof of the optimal version of the Boole equality.

Theorem 5.9. *Assume that ν is a purely singular measure. Then for all $t > 0$*

$$|\{x : F_\nu(x) > t\}| = |\{x : F_\nu(x) < -t\}| = \frac{\nu(\mathbb{R})}{t}.$$

Proof. W.l.o.g. we may assume that $\nu(\mathbb{R}) = 1$. Let H_0 be the operator of multiplication by x on $L^2(\mathbb{R}, \mathrm{d}\nu)$ and $\psi \equiv 1$. Let $H_\lambda = H_0 + \lambda(\psi|\cdot)\psi$ and let μ^λ be the spectral measure for H_λ and ψ. Obviously, $\mu^0 = \nu$ and $F_0 = F_\nu$. Since ν is a singular measure, μ^λ is singular for all $\lambda \in \mathbb{R}$.

By Theorem 5.1, for $\lambda \neq 0$, the measure μ^λ is concentrated on the set $\{x : F_0(x) = -\lambda^{-1}\}$. Let

$$\Gamma_t = \{x : F_0(x) > t\}.$$

Then for $\lambda \neq 0$,

$$\mu^\lambda(\Gamma_t) = \begin{cases} 1 & \text{if } -t^{-1} < \lambda < 0, \\ 0 & \text{if } \lambda \leq -t^{-1} \text{ or } \lambda > 0. \end{cases}$$

By the spectral averaging,

$$|\Gamma_t| = \int_{\mathbb{R}} \mu^\lambda(\Gamma_t)\mathrm{d}\lambda = t^{-1}.$$

A similar argument yields that $|\{x : F_\nu(x) < -t\}| = t^{-1}$. □

The Boole equality fails if ν is not a singular measure. However, in general we have

Theorem 5.10. *Let ν be a finite positive Borel measure on $\mathbb{R}$. Then*

$$\lim_{t\to\infty} t\,|\{x : |F_\nu(x)| > t\}| = 2\nu_{\rm sing}(\mathbb{R}).$$

Theorem 5.10 is due to Vinogradov-Hruschev. Its proof (and much additional information) can be found in the paper of Poltoratskii [22].

5.7 Poltoratskii's theorem

This subsection is devoted to the proof of Theorem 3.18. We follow [14].

We first consider the case $\nu_{\rm s} = 0$, μ compactly supported, $f \in L^2(\mathbb{R}, \mathrm{d}\mu)$ real valued. W.l.o.g. we may assume that $\mu(\mathbb{R}) = 1$.

Consider the Hilbert space $L^2(\mathbb{R}, \mathrm{d}\mu)$ and let H_0 be the operator of multiplication by x. Note that

$$F_\mu(z) = (\,\mathbb{1}|(H_0 - z)^{-1}\mathbb{1}\,), \qquad F_{f\mu}(z) = (\,\mathbb{1}|(H_0 - z)^{-1}f\,).$$

For $\lambda \in \mathbb{R}$, let

$$H_\lambda = H_0 + \lambda(\,\mathbb{1}|\cdot\,)\mathbb{1},$$

and let μ^λ be the spectral measure for H_λ and $\mathbb{1}$. To simplify the notation, we write

$$F_\lambda(z) = (\,\mathbb{1}|(H_\lambda - z)^{-1}\mathbb{1}\,) = F_{\mu^\lambda}(z).$$

Note that with this notation, $F_0 = F_\mu$!

By Theorem 5.1, the measures $\{\mu^\lambda_{\rm sing}\}_{\lambda\in\mathbb{R}}$ are mutually singular. By Theorem 3.5, the measure $\mu_{\rm sing} = \mu^0_{\rm sing}$ is concentrated on the set

$$\{x \in \mathbb{R} : \lim_{y\downarrow 0} \mathrm{Im}\, F_0(x + \mathrm{i}y) = \infty\}.$$

We also recall the identity

$$F_\lambda(z) = \frac{F_0(z)}{1 + \lambda F_0(z)}. \tag{71}$$

By the spectral theorem, there exists a unitary

$$U_\lambda : L^2(\mathbb{R}, \mathrm{d}\mu) \to L^2(\mathbb{R}, \mathrm{d}\mu^\lambda),$$

such that $U_\lambda \mathbb{1} = \mathbb{1}$ and $U_\lambda H_\lambda U_\lambda^{-1}$ is the operator of multiplication by x on $L^2(\mathbb{R}, \mathrm{d}\mu^\lambda)$. Hence

$$(\,\mathbb{1}|(H_\lambda - z)^{-1}f) = \int_{\mathbb{R}} \frac{(U_\lambda f)(x)}{x - z}\mathrm{d}\mu^\lambda(x) = F_{(U_\lambda f)\mu^\lambda}(z).$$

In what follows we set $\lambda = 1$ and write $U = U_1$.

For $a \in \mathbb{R}$ and $b > 0$ let $h_{ab}(x) = 2b((x-a)^2 + b^2)^{-1}$, $w = a + ib$, and $r_w(x) = (x - w)^{-1}$ (hence $h_{ab} = \mathrm{i}^{-1}(r_w - r_{\overline{w}})$). The relation

$$Uh_{ab} = h_{ab} + \lambda \mathrm{i}^{-1}(F_0(w)r_w - F_0(\overline{w})r_{\overline{w}}), \tag{72}$$

yields that Uh_{ab} is a real-valued function. The proof of (72) is simple and is left to the problems. Since the linear span of $\{h_{ab} : a \in \mathbb{R}, b > 0\}$ is dense in $C_0(\mathbb{R})$, U takes real-valued functions to real-valued functions. In particular, Uf is a real-valued function.

The identity

$$(\mathbb{1}|(H_0 - z)^{-1}f) = (1 + (\mathbb{1}|(H_0 - z)^{-1}\mathbb{1}))(\mathbb{1}|(H_1 - z)^{-1}f),$$

can be rewritten as

$$(\mathbb{1}|(H_0 - z)^{-1}f) = (1 + F_0(z))F_{(Uf)\mu^1}(z). \tag{73}$$

It follows that

$$\frac{\mathrm{Im}\,(\mathbb{1}|(H_0 - z)^{-1}f)}{\mathrm{Im}\,F_0(z)} = \mathrm{Re}\,F_{(Uf)\mu^1}(z) + L(z), \tag{74}$$

where

$$L(z) = \frac{\mathrm{Re}\,(1 + F_0(z))}{\mathrm{Im}\,F_0(z)}\mathrm{Im}\,F_{(Uf)\mu^1}(z).$$

We proceed to prove that

$$\lim_{y\downarrow 0} \mathrm{Im}\,F_{(Uf)\mu^1}(x + \mathrm{i}y) = 0, \qquad \text{for } \mu_{\text{sing}} - a.e.\ x, \tag{75}$$

$$\lim_{y\downarrow 0} L(x + \mathrm{i}y) = 0, \qquad \text{for } \mu_{\text{sing}} - a.e.\ x. \tag{76}$$

We start with (75). Using first that Uf is real-valued and then the Cauchy-Schwarz inequality, we derive

$$\mathrm{Im}\,F_{(Uf)\mu^1}(x + \mathrm{i}y) = P_{(Uf)\mu^1}(x + \mathrm{i}y) \le \sqrt{P_{\mu^1}(x + \mathrm{i}y)}\sqrt{P_{(Uf)^2\mu^1}(x + \mathrm{i}y)}.$$

Since the measures $(Uf)^2\mu^1_{\text{sing}}$ and μ_{sing} are mutually singular,

$$\lim_{y\downarrow 0} \frac{P_{(Uf)^2\mu^1}(x + \mathrm{i}y)}{P_\mu(x + \mathrm{i}y)} = 0, \qquad \text{for } \mu_{\text{sing}} - a.e.\ x,$$

(see Problem 4). Hence,

$$\lim_{y\downarrow 0} \frac{\mathrm{Im}\,F_{(Uf)\mu^1}(x + \mathrm{i}y)}{\sqrt{P_{\mu^1}(x + \mathrm{i}y)}\sqrt{P_\mu(x + \mathrm{i}y)}} = 0, \qquad \text{for } \mu_{\text{sing}} - a.e.\ x. \tag{77}$$

Since

$$P_{\mu^1}(x+\mathrm{i}y)P_\mu(x+\mathrm{i}y) = \mathrm{Im}\, F_1(x+\mathrm{i}y)\mathrm{Im}\, F_0(x+\mathrm{i}y) = \frac{(\mathrm{Im}\, F_0(x+\mathrm{i}y))^2}{|1+F_0(x+\mathrm{i}y)|^2} \leq 1$$

for all $x \in \mathbb{R}$, (77) yields (75).

To prove (76), note that

$$|L(x+\mathrm{i}y)| = \frac{\mathrm{Im}\, F_{(Uf)\mu^1}(x+\mathrm{i}y)}{\sqrt{P_{\mu^1}(x+\mathrm{i}y)}\sqrt{P_\mu(x+\mathrm{i}y)}} \frac{|\mathrm{Re}\,(1+F_0(x+\mathrm{i}y))|}{\mathrm{Im}\, F_0(x+\mathrm{i}y)} \frac{\mathrm{Im}\, F_0(x+\mathrm{i}y)}{|1+F_0(x+\mathrm{i}y)|}$$

$$\leq \frac{\mathrm{Im}\, F_{(Uf)\mu^1}(x+\mathrm{i}y)}{\sqrt{P_{\mu^1}(x+\mathrm{i}y)}\sqrt{P_\mu(x+\mathrm{i}y)}}.$$

Hence, (77) yields (76).

Rewrite (74) as

$$F_{(Uf)\mu^1}(z) = \frac{\mathrm{Im}\,(\mathbb{1}|(H_0-z)^{-1}f)}{\mathrm{Im}\, F_0(z)} + \mathrm{Im}\, F_{(Uf)\mu^1}(z) - L(z). \tag{78}$$

By Theorem 3.5,

$$\lim_{y\downarrow 0} \frac{\mathrm{Im}\,(\mathbb{1}|(H_0-x-\mathrm{i}y)^{-1}f)}{\mathrm{Im}\, F_0(x+\mathrm{i}y)} = \lim_{y\downarrow 0} \frac{P_{f\mu}(x+\mathrm{i}y)}{P_\mu(x+\mathrm{i}y)} = f(x), \qquad \text{for } \mu - a.e.\ x.$$

Hence, (78), (75), and (76) yield that

$$\lim_{y\downarrow 0} F_{(Uf)\mu^1}(x+\mathrm{i}y) = f(x), \qquad \text{for } \mu_{\mathrm{sing}} - a.e.\ x. \tag{79}$$

Rewrite (73) as

$$\frac{F_{f\mu}(x+\mathrm{i}y)}{F_\mu(x+\mathrm{i}y)} = \left(\frac{1}{F_0(x+\mathrm{i}y)} + 1\right) F_{(Uf)\mu^1}(x+\mathrm{i}y). \tag{80}$$

Since $|F_0(x+\mathrm{i}y)| \to \infty$ as $y \downarrow 0$ for μ_{sing}-a.e. x, (79) and (80) yield

$$\lim_{y\downarrow 0} \frac{F_{f\mu}(x+\mathrm{i}y)}{F_\mu(x+\mathrm{i}y)} = f(x), \qquad \text{for } \mu_{\mathrm{sing}} - a.e.\ x.$$

This proves the Poltoratskii theorem in the special case where $\nu_{\mathrm{s}} = 0$, μ is compactly supported, and $f \in L^2(\mathbb{R}, \mathrm{d}\mu)$ is real-valued.

We now remove the assumptions $f \in L^2(\mathbb{R}, \mathrm{d}\mu)$ and that f is real valued (we still assume that μ is compactly supported and that $\nu_{\mathrm{s}} = 0$). Assume that $f \in L^1(\mathbb{R}, \mathrm{d}\mu)$ and that f is *positive*. Set $g = 1/(1+f)$ and $\rho = (1+f)\mu$. Then

$$\lim_{y\downarrow 0} \frac{F_\mu(x+\mathrm{i}y)}{F_{(1+f)\mu}(x+\mathrm{i}y)} = \lim_{y\downarrow 0} \frac{F_{g\rho}(x+\mathrm{i}y)}{F_\rho(x+\mathrm{i}y)} = \frac{1}{1+f(x)},$$

for μ_{sing}-a.e. x. By the linearity of the Borel transform,

$$\lim_{y\downarrow 0}\frac{F_{f\mu}(x+\mathrm{i}y)}{F_\mu(x+\mathrm{i}y)}=\lim_{y\downarrow 0}\frac{F_{(1+f)\mu}(x+\mathrm{i}y)}{F_\mu(x+\mathrm{i}y)}-1=f(x),$$

for μ_{sing}-a.e. x. Since every $f\in L^1(\mathbb{R},\mathrm{d}\mu)$ is a linear combination of four positive functions in $L^1(\mathbb{R},\mathrm{d}\mu)$, the linearity of the Borel transform implies the statement for all $f\in L^1(\mathbb{R},\mathrm{d}\mu)$.

Assume that μ is not compactly supported (we still assume $\nu_{\text{s}}=0$) and let $[a,b]$ be a finite interval. Decompose $\mu=\mu_1+\mu_2$, where $\mu_1=\mu\restriction[a,b]$, $\mu_2=\mu\restriction\mathbb{R}\setminus[a,b]$. Since

$$\frac{F_{f\mu}(z)}{F_\mu(z)}=\frac{F_{f\mu_1}(z)+F_{f\mu_2}(z)}{F_{\mu_1}(z)(1+F_{\mu_2}(z)/F_{\mu_1}(z))},$$

and $\lim_{y\downarrow 0}|F_{\mu_1}(x+\mathrm{i}y)|\to\infty$ for $\mu_{1,\text{sing}}$-a.e. $x\in[a,b]$,

$$\lim_{y\downarrow 0}\frac{F_{f\mu}(x+\mathrm{i}y)}{F_\mu(x+\mathrm{i}y)}=f(x),\qquad \text{for } \mu_{\text{sing}}\text{-a.e. } x\in(a,b).$$

Since $[a,b]$ is arbitrary, we have removed the assumption that μ is compactly supported.

Finally, to finish the proof we need to show that if $\nu\perp\mu$, then

$$\lim_{y\downarrow 0}\frac{F_\nu(x+\mathrm{i}y)}{F_\mu(x+\mathrm{i}y)}=0, \tag{81}$$

for μ_{sing}-a.e. x. Since ν can be written as a linear combination of four positive measures each of which is singular w.r.t. μ, w.l.o.g. me may assume that ν is positive. Let S be a Borel set such that $\mu(S)=0$ and that ν is concentrated on S. Then

$$\lim_{y\downarrow 0}\frac{F_{\chi_S(\mu+\nu)}(x+\mathrm{i}y)}{F_{\mu+\nu}(x+\mathrm{i}y)}=\chi_S(x),$$

for $\mu_{\text{sing}}+\nu_{\text{sing}}$-a.e. x. Hence,

$$\lim_{y\downarrow 0}\frac{F_\nu(x+\mathrm{i}y)}{F_\mu(x+\mathrm{i}y)+F_\nu(x+\mathrm{i}y)}=0,$$

for μ_{sing}- a.e. x, and this yields (81). The proof of the Poltoratskii theorem is complete.

The Poltoratskii theorem also holds for complex measures μ:

Theorem 5.11. *Let ν and μ be complex Borel measures and $\nu=f\mu+\nu_{\text{s}}$ be the Radon-Nikodym decomposition. Let $|\mu|_{\text{sing}}$ be the part of $|\mu|$ singular with respect to the Lebesgue measure. Then*

$$\lim_{y\downarrow 0}\frac{F_\nu(x+\mathrm{i}y)}{F_\mu(x+\mathrm{i}y)}=f(x),\qquad \text{for } |\mu|_{\text{sing}}-a.e.\ x.$$

Theorem 5.11 follows easily from Theorem 3.18.

5.8 F. & M. Riesz theorem

The celebrated theorem of F. & M. Riesz states:

Theorem 5.12. *Let $\mu \neq 0$ be a complex measure and $F_\mu(z)$ its Borel transform. If $F_\mu(z) = 0$ for all $z \in \mathbb{C}_+$, then $|\mu|$ is equivalent to the Lebesgue measure.*

In the literature one can find many different proofs of this theorem (for example, three different proofs are given in [19]). However, it has been only recently noticed that F. & M. Riesz theorem is an easy consequence of the Poltoratskii theorem. The proof below follows [16].

Proof. For $z \in \mathbb{C} \setminus \mathbb{R}$ we set

$$F_\mu(z) = \int_{\mathbb{R}} \frac{\mathrm{d}\mu(t)}{t - z},$$

and write

$$F_\mu(x \pm \mathrm{i}0) = \lim_{y \downarrow 0} F_\mu(x \pm \mathrm{i}y).$$

By Theorem 3.17 (and its obvious analog for the lower half-plane), $F_\mu(x \pm \mathrm{i}0)$ exists and is finite for Lebesgue a.e. x.

Write $\mu = h|\mu|$, where $|h(x)| = 1$ for all x. By the Poltoratskii theorem,

$$\lim_{y \downarrow 0} \frac{|F_\mu(x + \mathrm{i}y)|}{|F_{|\mu|}(x + \mathrm{i}y)|} = |h(x)| = 1,$$

for $|\mu|_{\mathrm{sing}}$-a.e. x. Since by Theorem 3.5, $\lim_{y\downarrow 0} |F_{|\mu|}(x + \mathrm{i}y)| = \infty$ for $|\mu|_{\mathrm{sing}}$-a.e. x, we must have $\lim_{y\downarrow 0} |F_\mu(x + \mathrm{i}y)| = \infty$ for $|\mu|_{\mathrm{sing}}$-a.e. x. Hence, if $|\mu|_{\mathrm{sing}} \neq 0$, then $F_\mu(z)$ cannot vanish on $\mathbb{C}_+$.

It remains to prove that $|\mu|$ is equivalent to the Lebesgue measure. By Theorem 3.5, $\mathrm{d}|\mu| = \pi^{-1}\mathrm{Im}\, F_{|\mu|}(x + \mathrm{i}0)\mathrm{d}x$, so we need to show that

$$\mathrm{Im}\, F_{|\mu|}(x + \mathrm{i}0) > 0,$$

for Lebesgue a.e. x. Assume that $\mathrm{Im}\, F_{|\mu|}(x + \mathrm{i}0) = 0$ for $x \in S$, where S has positive Lebesgue measure. The formula

$$F_\mu(x + \mathrm{i}y) = \int_{\mathbb{R}} \frac{(t - x)\mathrm{d}\mu(t)}{(t - x)^2 + y^2} + \mathrm{i} \int_{\mathbb{R}} \frac{y\mathrm{d}\mu(t)}{(t - x)^2 + y^2},$$

and the bound

$$\left| \int_{\mathbb{R}} \frac{y\mathrm{d}\mu(t)}{(t - x)^2 + y^2} \right| \leq \mathrm{Im}\, F_{|\mu|}(x + \mathrm{i}y),$$

yield that for $x \in S$,

$$\lim_{y \to 0} F_\mu(x + \mathrm{i}y) = \lim_{y \to 0} \int_{\mathbb{R}} \frac{(t - x)\mathrm{d}\mu(t)}{(t - x)^2 + y^2}.$$

Hence,

$$F_\mu(x - \mathrm{i}0) = F_\mu(x + \mathrm{i}0) = 0, \qquad \text{for Lebesgue a.e. } x \in S. \tag{82}$$

Since F_μ vanishes on $\mathbb{C}_+$, F_μ does not vanish on $\mathbb{C}_-$ (otherwise, since the linear span of the set of functions $\{(x-z)^{-1} : z \in \mathbb{C} \setminus \mathbb{R}\}$ is dense in $C_0(\mathbb{R})$, we would have $\mu = 0$). Then, by Theorem 3.17 (i.e., its obvious analog for the lower half-plane), $F_\mu(x - \mathrm{i}0) \neq 0$ for Lebesgue a.e $x \in \mathbb{R}$. This contradicts (82). □

5.9 Problems and comments

[1] *Prove Relation (65). Hint: See Theorem I.2 in* [29].

[2] *Prove Theorem 5.6.*

[3] *Prove Relation (72).*

[4] *Let ν and μ be positive measures such that $\nu_{\rm sing} \perp \mu_{\rm sing}$. Prove that for $\mu_{\rm sing}$-a.e. x*

$$\lim_{y \downarrow 0} \frac{P_\nu(x + \mathrm{i}y)}{P_\mu(x + \mathrm{i}y)} = 0.$$

Hint: Write

$$\frac{P_\nu(z)}{P_\mu(z)} = \frac{\dfrac{P_{\nu_{\rm ac}}(z)}{P_{\mu_{\rm sing}}(z)} + \dfrac{P_{\nu_{\rm sing}}(z)}{P_{\mu_{\rm sing}}(z)}}{\dfrac{P_{\mu_{\rm ac}}(z)}{P_{\mu_{\rm sing}}(z)} + 1},$$

and use Theorem 3.5.

[5] *Prove the Poltoratskii theorem in the case where ν and μ are positive pure point measures.*

[6] *In the Poltoratskii theorem one cannot replace $\mu_{\rm sing}$ by μ. Find an example justifying this claim.*

The next set of problems deals with various examples involving rank one perturbations. Note that the model (59) is completely determined by a choice of a Borel probability measure μ^0 on $\mathbb{R}$. Setting $\mathcal{H} = L^2(\mathbb{R}, \mathrm{d}\mu^0)$, H_0 = operator of multiplication by x, $\psi \equiv 1$, we obtain a class of Hamiltonians $H_\lambda = H_0 + \lambda(\psi|\cdot)\psi$ of the form (59). On the other hand, by the spectral theorem, any family Hamiltonians (59), when restricted to the cyclic subspace $\mathcal{H}_\psi$, is unitarily equivalent to such a class.

[7] *Let μ_C be the standard Cantor measure (see Example 3 in Section I.4 of* [23]*) and $\mathrm{d}\mu^0 = (\mathrm{d}x \restriction [0,1] + \mathrm{d}\mu_C)/2$. The ac spectrum of H_0 is $[0,1]$. The singular continuous part of μ^0 is concentrated on the Cantor set C. Since C is closed,*

$\mathrm{sp_{sing}}(H_0) = C$. *Prove that for* $\lambda \neq 0$ *the spectrum of* H_λ *in* $[0,1]$ *is purely absolutely continuous. Hint: See the last example in Section XIII.7 of* [26].

[8] *Assume that* $\mu^0 = \mu_C$. *Prove that for all* $\lambda \neq 0$, H_λ *has only pure point spectrum. Compute the spectrum of* H_λ. *Hint: This is Example 1 in* [31]. *See also Example 3 in Section II.5 of* [29].

[9] *Let*

$$\mu_n = 2^{-n} \sum_{j=1}^{2^n} \delta(j/2^n),$$

and $\mu = \sum_n a_n \mu_n$, *where* $a_n > 0$, $\sum_n a_n = 1$, $\sum_n 2^n a_n = \infty$. *The spectrum of* H_0 *is pure point and equal to* $[0,1]$. *Prove that the spectrum of* H_λ *in* $[0,1]$ *is purely singular continuous for all* $\lambda \neq 0$. *Hint: This is Example 2 in* [31]. *See also Example 4 in Section II.5 of* [29].

[10] *Let* $\nu_{j,n}(A) = \mu_C(A + j/2^n)$ *and*

$$\mu^0 = c\chi_{[0,1]} \sum_{n=1}^{\infty} n^{-2} \sum_{j=1}^{2^n} \nu_{j,n},$$

where c *is the normalization constant. Prove that the spectrum of* H_λ *on* $[0,1]$ *is purely singular continuous for all* λ. *Hint: This is Example 5 in Section II.5 of* [29].

[11] *Find* μ^0 *such that:*
(1) *The spectrum of* H_0 *is purely absolutely continuous and equal to* $[0,1]$.
(2) *For a set of* λ*'s of positive Lebesgue measure,* H_λ *has embedded point spectrum in* $[0,1]$.
Hint: See [8] *and Example 7 in Section II.5 of* [29].

[12] *Find* μ^0 *such that:*
(1) *The spectrum of* H_0 *is purely absolutely continuous and equal to* $[0,1]$.
(2) *For a set of* λ*'s of positive Lebesgue measure,* H_λ *has embedded singular continuous spectrum in* $[0,1]$.
Hint: See [8] *and Example 8 in Section II.5 of* [29].

[13] *del Rio and Simon* [8] *have shown that there exists* μ^0 *such that:*
(1) *For all* λ, $\mathrm{sp_{ac}}(H_\lambda) = [0,1]$.
(2) *For a set of* λ*'s of positive Lebesgue measure,* H_λ *has embedded point spectrum in* $[0,1]$.
(3) *For a set of* λ*'s of positive Lebesgue measure,* H_λ *has embedded singular continuous spectrum in* $[0,1]$.

[14] *del Rio-Fuentes-Poltoratskii* [5] *have shown that there exists* μ^0 *such that:*
(1) *For all* λ $\mathrm{sp_{ac}}(H_\lambda) = [0,1]$. *Moreover, for all* $\lambda \in [0,1]$, *the spectrum of* H_λ *is*

purely absolutely continuous.
(2) *For all* $\lambda \notin [0,1]$, $[0,1] \subset \mathrm{sp}_{\mathrm{sing}}(H_\lambda)$.

[15] *Let* μ^0 *be a pure point measure with atoms* $\mu^0(\{x_n\}) = a_n$, $n \in \mathbb{N}$, *where* $x_n \in [0,1]$. *Clearly,*

$$G(x) = \sum_{n=1}^{\infty} \frac{a_n}{(x - x_n)^2}.$$

(1) *Prove that if* $\sum_n \sqrt{a_n} < \infty$, *then* $G(x) < \infty$ *for Lebesgue a.e.* $x \in [0,1]$.
(2) *Assume that* $x_n = x_n(\omega)$ *are independent random variables uniformly distributed on* $[0,1]$ *(we keep* a_n *deterministic). Assume that* $\sum_n \sqrt{a}_n = \infty$. *Prove that for a.e.* ω, $G(x) = \infty$ *for Lebesgue a.e.* $x \in [0,1]$.
(3) *What can you say about the spectrum of* H_λ *in the cases (1) and (2)?*
Hint: (1) and (2) are proven in [12].

References

1. Aizenman M.: Localization at weak disorder: Some elementary bounds. Rev. Math. Phys. **6** (1994), 1163.
2. Cycon H.L., Froese R.G., Kirsch W., Simon B.: *Schrödinger Operators*, Springer-Verlag, Berlin, 1987.
3. Carmona R., Lacroix J.: *Spectral Theory of Random Schrödinger Operators*, Birkhauser, Boston, 1990.
4. Davies E.B.: *Spectral Theory and Differential Operators*, Cambridge University Press, Cambridge, 1995.
5. del Rio R., Fuentes S., Poltoratskii A.G.: Coexistence of spectra in rank-one perturbation problems. Bol. Soc. Mat. Mexicana **8** (2002), 49.
6. del Rio R., Jitomirskaya S., Last Y., Simon B.: Operators with singular continuous spectrum, IV. Hausdorff dimensions, rank one perturbations, and localization. J. d'Analyse Math. **69** (1996), 153.
7. del Rio R., Makarov N., Simon B.: Operators with singular continuous spectrum: II. Rank one operators. Commun. Math. Phys. **165** (1994), 59.
8. del Rio R., Simon B.: Point spectrum and mixed spectral types for rank one perturbations. Proc. Amer. Math. Soc. **125** (1997), 3593.
9. Evans L., Gariepy G.: *Measure Theory and Fine Properties of Functions*, CRC Press, 1992.
10. Gordon A.: Pure point spectrum under 1-parameter perturbations and instability of Anderson localization. Commun. Math. Phys. **164** (1994), 489.
11. Gustafson K., Rao D.K.M.: *Numerical Range*, Springer-Verlag, Berlin, 1996.
12. Howland J.: Perturbation theory of dense point spectra. J. Func. Anal. **74** (1987), 52.
13. Jakšić V., Kritchevski E., Pillet C.-A.: Spectral theory of the Wigner-Weisskopf atom. Nordfjordeid Lecture Notes.
14. Jakšić V., Last Y.: A new proof of Poltoratskii's theorem. J. Func. Anal. **215** (2004), 103.
15. Jakšić V., Last Y.: Spectral structure of Anderson type Hamiltonians. Invent. Math. **141** (2000), 561.

16. Jakšić V., Last Y.: Simplicity of singular spectrum in Anderson type Hamiltonians. Submitted.
17. Kato T.: *Perturbation Theory for Linear Operators*, second edition, Springer-Verlag, Berlin, 1976.
18. Katznelson A.: *An Introduction to Harmonic Analysis*, Dover, New York, 1976.
19. Koosis P.: *Introduction to H_p Spaces*, second edition, Cambridge University Press, New York, 1998.
20. Lyons R.: Seventy years of Rajchman measures. J. Fourier Anal. Appl. Publ. Res., Kahane Special Issue (1995), 363.
21. Poltoratskii A.G.: The boundary behavior of pseudocontinuable functions. St. Petersburgh Math. J. **5** (1994), 389.
22. Poltoratskii A.G.: On the distribution of the boundary values of Cauchy integrals. Proc. Amer. Math. Soc. **124** (1996), 2455.
23. Reed M., Simon B.: *Methods of Modern Mathematical Physics, I. Functional Analysis*, Academic Press, London, 1980.
24. Reed M., Simon B.: *Methods of Modern Mathematical Physics, II. Fourier Analysis, Self-Adjointness*, Academic Press, London, 1975.
25. Reed M., Simon B.: *Methods of Modern Mathematical Physics, III. Scattering Theory*, Academic Press, London, 1978.
26. Reed M., Simon B.: *Methods of Modern Mathematical Physics, IV. Analysis of Operators*, Academic Press, London, 1978.
27. Rudin W.: *Real and Complex Analysis*, 3rd edition, McGraw-Hill, Boston, 1987.
28. Simon B.: L^p norms of the Borel transforms and the decomposition of measures. Proc. Amer. Math. Soc. **123** (1995), 3749
29. Simon B.: Spectral analysis of rank one perturbations and applications. CRM Lecture Notes Vol. **8**, pp. 109-149, AMS, Providence, RI, 1995.
30. Simon B.: Spectral averaging and the Krein spectral shift. Proc. Amer. Math. Soc. **126** (1998), 1409.
31. Simon B., Wolff T.: Singular continuous spectrum under rank one perturbations and localization for random Hamiltonians. Communications in Pure and Applied Mathematics **49** (1986), 75.
32. Vock E., Hunziker W.: Stability of Schrödinger eigenvalue problems. Commun. Math. Phys. **83** (1982), 281.

Index of Volume I

Information about the other two volumes

Contents of Volume II

Quantum Noises

Complete Positivity and the Markov structure of Open Quantum Systems

Quantum Stochastic Differential Equations and Dilation of Completely Positive Semigroups

Information about the other two volumes

Index of Volume II

Contents of Volume III

Continual Measurements in Quantum Mechanics and Quantum Stochastic Calculus

Index of Volume III

Lecture Notes in Mathematics

For information about earlier volumes please contact your bookseller or Springer LNM Online archive: springerlink.com

Vol. 1681: G. J. Wirsching, The Dynamical System Generated by the 3n+1 Function (1998)
Vol. 1682: H.-D. Alber, Materials with Memory (1998)
Vol. 1683: A. Pomp, The Boundary-Domain Integral Method for Elliptic Systems (1998)
Vol. 1684: C. A. Berenstein, P. F. Ebenfelt, S. G. Gindikin, S. Helgason, A. E. Tumanov, Integral Geometry, Radon Transforms and Complex Analysis. Firenze, 1996. Editors: E. Casadio Tarabusi, M. A. Picardello, G. Zampieri (1998)
Vol. 1685: S. König, A. Zimmermann, Derived Equivalences for Group Rings (1998)
Vol. 1686: J. Azéma, M. Émery, M. Ledoux, M. Yor (Eds.), Séminaire de Probabilités XXXII (1998)
Vol. 1687: F. Bornemann, Homogenization in Time of Singularly Perturbed Mechanical Systems (1998)
Vol. 1688: S. Assing, W. Schmidt, Continuous Strong Markov Processes in Dimension One (1998)
Vol. 1689: W. Fulton, P. Pragacz, Schubert Varieties and Degeneracy Loci (1998)
Vol. 1690: M. T. Barlow, D. Nualart, Lectures on Probability Theory and Statistics. Editor: P. Bernard (1998)
Vol. 1691: R. Bezrukavnikov, M. Finkelberg, V. Schechtman, Factorizable Sheaves and Quantum Groups (1998)
Vol. 1692: T. M. W. Eyre, Quantum Stochastic Calculus and Representations of Lie Superalgebras (1998)
Vol. 1694: A. Braides, Approximation of Free-Discontinuity Problems (1998)
Vol. 1695: D. J. Hartfiel, Markov Set-Chains (1998)
Vol. 1696: E. Bouscaren (Ed.): Model Theory and Algebraic Geometry (1998)
Vol. 1697: B. Cockburn, C. Johnson, C.-W. Shu, E. Tadmor, Advanced Numerical Approximation of Nonlinear Hyperbolic Equations. Cetraro, Italy, 1997. Editor: A. Quarteroni (1998)
Vol. 1698: M. Bhattacharjee, D. Macpherson, R. G. Möller, P. Neumann, Notes on Infinite Permutation Groups (1998)
Vol. 1699: A. Inoue,Tomita-Takesaki Theory in Algebras of Unbounded Operators (1998)
Vol. 1700: W. A. Woyczyński, Burgers-KPZ Turbulence (1998)
Vol. 1701: Ti-Jun Xiao, J. Liang, The Cauchy Problem of Higher Order Abstract Differential Equations (1998)
Vol. 1702: J. Ma, J. Yong, Forward-Backward Stochastic Differential Equations and Their Applications (1999)
Vol. 1703: R. M. Dudley, R. Norvaiša, Differentiability of Six Operators on Nonsmooth Functions and p-Variation (1999)
Vol. 1704: H. Tamanoi, Elliptic Genera and Vertex Operator Super-Algebras (1999)
Vol. 1705: I. Nikolaev, E. Zhuzhoma, Flows in 2-dimensional Manifolds (1999)
Vol. 1706: S. Yu. Pilyugin, Shadowing in Dynamical Systems (1999)
Vol. 1707: R. Pytlak, Numerical Methods for Optimal Control Problems with State Constraints (1999)
Vol. 1708: K. Zuo, Representations of Fundamental Groups of Algebraic Varieties (1999)
Vol. 1709: J. Azéma, M. Émery, M. Ledoux, M. Yor (Eds.), Séminaire de Probabilités XXXIII (1999)
Vol. 1710: M. Koecher, The Minnesota Notes on Jordan Algebras and Their Applications (1999)
Vol. 1711: W. Ricker, Operator Algebras Generated by Commuting Projećtions: A Vector Measure Approach (1999)
Vol. 1712: N. Schwartz, J. J. Madden, Semi-algebraic Function Rings and Reflectors of Partially Ordered Rings (1999)
Vol. 1713: F. Bethuel, G. Huisken, S. Müller, K. Steffen, Calculus of Variations and Geometric Evolution Problems. Cetraro, 1996. Editors: S. Hildebrandt, M. Struwe (1999)
Vol. 1714: O. Diekmann, R. Durrett, K. P. Hadeler, P. K. Maini, H. L. Smith, Mathematics Inspired by Biology. Martina Franca, 1997. Editors: V. Capasso, O. Diekmann (1999)
Vol. 1715: N. V. Krylov, M. Röckner, J. Zabczyk, Stochastic PDE's and Kolmogorov Equations in Infinite Dimensions. Cetraro, 1998. Editor: G. Da Prato (1999)
Vol. 1716: J. Coates, R. Greenberg, K. A. Ribet, K. Rubin, Arithmetic Theory of Elliptic Curves. Cetraro, 1997. Editor: C. Viola (1999)
Vol. 1717: J. Bertoin, F. Martinelli, Y. Peres, Lectures on Probability Theory and Statistics. Saint-Flour, 1997. Editor: P. Bernard (1999)
Vol. 1718: A. Eberle, Uniqueness and Non-Uniqueness of Semigroups Generated by Singular Diffusion Operators (1999)
Vol. 1719: K. R. Meyer, Periodic Solutions of the N-Body Problem (1999)
Vol. 1720: D. Elworthy, Y. Le Jan, X-M. Li, On the Geometry of Diffusion Operators and Stochastic Flows (1999)
Vol. 1721: A. Iarrobino, V. Kanev, Power Sums, Gorenstein Algebras, and Determinantal Loci (1999)
Vol. 1722: R. McCutcheon, Elemental Methods in Ergodic Ramsey Theory (1999)
Vol. 1723: J. P. Croisille, C. Lebeau, Diffraction by an Immersed Elastic Wedge (1999)
Vol. 1724: V. N. Kolokoltsov, Semiclassical Analysis for Diffusions and Stochastic Processes (2000)
Vol. 1725: D. A. Wolf-Gladrow, Lattice-Gas Cellular Automata and Lattice Boltzmann Models (2000)
Vol. 1726: V. Marić, Regular Variation and Differential Equations (2000)
Vol. 1727: P. Kravanja M. Van Barel, Computing the Zeros of Analytic Functions (2000)
Vol. 1728: K. Gatermann Computer Algebra Methods for Equivariant Dynamical Systems (2000)
Vol. 1729: J. Azéma, M. Émery, M. Ledoux, M. Yor (Eds.)

Séminaire de Probabilités XXXIV (2000)
Vol. 1730: S. Graf, H. Luschgy, Foundations of Quantization for Probability Distributions (2000)
Vol. 1731: T. Hsu, Quilts: Central Extensions, Braid Actions, and Finite Groups (2000)
Vol. 1732: K. Keller, Invariant Factors, Julia Equivalences and the (Abstract) Mandelbrot Set (2000)
Vol. 1733: K. Ritter, Average-Case Analysis of Numerical Problems (2000)
Vol. 1734: M. Espedal, A. Fasano, A. Mikelić, Filtration in Porous Media and Industrial Applications. Cetraro 1998. Editor: A. Fasano. 2000.
Vol. 1735: D. Yafaev, Scattering Theory: Some Old and New Problems (2000)
Vol. 1736: B. O. Turesson, Nonlinear Potential Theory and Weighted Sobolev Spaces (2000)
Vol. 1737: S. Wakabayashi, Classical Microlocal Analysis in the Space of Hyperfunctions (2000)
Vol. 1738: M. Émery, A. Nemirovski, D. Voiculescu, Lectures on Probability Theory and Statistics (2000)
Vol. 1739: R. Burkard, P. Deuflhard, A. Jameson, J.-L. Lions, G. Strang, Computational Mathematics Driven by Industrial Problems. Martina Franca, 1999. Editors: V. Capasso, H. Engl, J. Periaux (2000)
Vol. 1740: B. Kawohl, O. Pironneau, L. Tartar, J.-P. Zolesio, Optimal Shape Design. Tróia, Portugal 1999. Editors: A. Cellina, A. Ornelas (2000)
Vol. 1741: E. Lombardi, Oscillatory Integrals and Phenomena Beyond all Algebraic Orders (2000)
Vol. 1742: A. Unterberger, Quantization and Non-holomorphic Modular Forms (2000)
Vol. 1743: L. Habermann, Riemannian Metrics of Constant Mass and Moduli Spaces of Conformal Structures (2000)
Vol. 1744: M. Kunze, Non-Smooth Dynamical Systems (2000)
Vol. 1745: V. D. Milman, G. Schechtman (Eds.), Geometric Aspects of Functional Analysis. Israel Seminar 1999-2000 (2000)
Vol. 1746: A. Degtyarev, I. Itenberg, V. Kharlamov, Real Enriques Surfaces (2000)
Vol. 1747: L. W. Christensen, Gorenstein Dimensions (2000)
Vol. 1748: M. Ruzicka, Electrorheological Fluids: Modeling and Mathematical Theory (2001)
Vol. 1749: M. Fuchs, G. Seregin, Variational Methods for Problems from Plasticity Theory and for Generalized Newtonian Fluids (2001)
Vol. 1750: B. Conrad, Grothendieck Duality and Base Change (2001)
Vol. 1751: N. J. Cutland, Loeb Measures in Practice: Recent Advances (2001)
Vol. 1752: Y. V. Nesterenko, P. Philippon, Introduction to Algebraic Independence Theory (2001)
Vol. 1753: A. I. Bobenko, U. Eitner, Painlevé Equations in the Differential Geometry of Surfaces (2001)
Vol. 1754: W. Bertram, The Geometry of Jordan and Lie Structures (2001)
Vol. 1755: J. Azéma, M. Émery, M. Ledoux, M. Yor (Eds.), Séminaire de Probabilités XXXV (2001)
Vol. 1756: P. E. Zhidkov, Korteweg de Vries and Nonlinear Schrödinger Equations: Qualitative Theory (2001)
Vol. 1757: R. R. Phelps, Lectures on Choquet's Theorem (2001)
Vol. 1758: N. Monod, Continuous Bounded Cohomology of Locally Compact Groups (2001)
Vol. 1759: Y. Abe, K. Kopfermann, Toroidal Groups (2001)
Vol. 1760: D. Filipović, Consistency Problems for Heath-Jarrow-Morton Interest Rate Models (2001)
Vol. 1761: C. Adelmann, The Decomposition of Primes in Torsion Point Fields (2001)
Vol. 1762: S. Cerrai, Second Order PDE's in Finite and Infinite Dimension (2001)
Vol. 1763: J.-L. Loday, A. Frabetti, F. Chapoton, F. Goichot, Dialgebras and Related Operads (2001)
Vol. 1764: A. Cannas da Silva, Lectures on Symplectic Geometry (2001)
Vol. 1765: T. Kerler, V. V. Lyubashenko, Non-Semisimple Topological Quantum Field Theories for 3-Manifolds with Corners (2001)
Vol. 1766: H. Hennion, L. Hervé, Limit Theorems for Markov Chains and Stochastic Properties of Dynamical Systems by Quasi-Compactness (2001)
Vol. 1767: J. Xiao, Holomorphic Q Classes (2001)
Vol. 1768: M.J. Pflaum, Analytic and Geometric Study of Stratified Spaces (2001)
Vol. 1769: M. Alberich-Carramiñana, Geometry of the Plane Cremona Maps (2002)
Vol. 1770: H. Gluesing-Luerssen, Linear Delay-Differential Systems with Commensurate Delays: An Algebraic Approach (2002)
Vol. 1771: M. Émery, M. Yor (Eds.), Séminaire de Probabilités 1967-1980. A Selection in Martingale Theory (2002)
Vol. 1772: F. Burstall, D. Ferus, K. Leschke, F. Pedit, U. Pinkall, Conformal Geometry of Surfaces in S^4 (2002)
Vol. 1773: Z. Arad, M. Muzychuk, Standard Integral Table Algebras Generated by a Non-real Element of Small Degree (2002)
Vol. 1774: V. Runde, Lectures on Amenability (2002)
Vol. 1775: W. H. Meeks, A. Ros, H. Rosenberg, The Global Theory of Minimal Surfaces in Flat Spaces. Martina Franca 1999. Editor: G. P. Pirola (2002)
Vol. 1776: K. Behrend, C. Gomez, V. Tarasov, G. Tian, Quantum Comohology. Cetraro 1997. Editors: P. de Bartolomeis, B. Dubrovin, C. Reina (2002)
Vol. 1777: E. García-Río, D. N. Kupeli, R. Vázquez-Lorenzo, Osserman Manifolds in Semi-Riemannian Geometry (2002)
Vol. 1778: H. Kiechle, Theory of K-Loops (2002)
Vol. 1779: I. Chueshov, Monotone Random Systems (2002)
Vol. 1780: J. H. Bruinier, Borcherds Products on O(2,1) and Chern Classes of Heegner Divisors (2002)
Vol. 1781: E. Bolthausen, E. Perkins, A. van der Vaart, Lectures on Probability Theory and Statistics. Ecole d' Eté de Probabilités de Saint-Flour XXIX-1999. Editor: P. Bernard (2002)
Vol. 1782: C.-H. Chu, A. T.-M. Lau, Harmonic Functions on Groups and Fourier Algebras (2002)
Vol. 1783: L. Grüne, Asymptotic Behavior of Dynamical and Control Systems under Perturbation and Discretization (2002)
Vol. 1784: L.H. Eliasson, S. B. Kuksin, S. Marmi, J.-C. Yoccoz, Dynamical Systems and Small Divisors. Cetraro, Italy 1998. Editors: S. Marmi, J.-C. Yoccoz (2002)
Vol. 1785: J. Arias de Reyna, Pointwise Convergence of Fourier Series (2002)
Vol. 1786: S. D. Cutkosky, Monomialization of Morphisms from 3-Folds to Surfaces (2002)
Vol. 1787: S. Caenepeel, G. Militaru, S. Zhu, Frobenius and Separable Functors for Generalized Module Categories and Nonlinear Equations (2002)

Vol. 1788: A. Vasil'ev, Moduli of Families of Curves for Conformal and Quasiconformal Mappings (2002)
Vol. 1789: Y. Sommerhäuser, Yetter-Drinfel'd Hopf algebras over groups of prime order (2002)
Vol. 1790: X. Zhan, Matrix Inequalities (2002)
Vol. 1791: M. Knebusch, D. Zhang, Manis Valuations and Prüfer Extensions I: A new Chapter in Commutative Algebra (2002)
Vol. 1792: D. D. Ang, R. Gorenflo, V. K. Le, D. D. Trong, Moment Theory and Some Inverse Problems in Potential Theory and Heat Conduction (2002)
Vol. 1793: J. Cortés Monforte, Geometric, Control and Numerical Aspects of Nonholonomic Systems (2002)
Vol. 1794: N. Pytheas Fogg, Substitution in Dynamics, Arithmetics and Combinatorics. Editors: V. Berthé, S. Ferenczi, C. Mauduit, A. Siegel (2002)
Vol. 1795: H. Li, Filtered-Graded Transfer in Using Noncommutative Gröbner Bases (2002)
Vol. 1796: J.M. Melenk, hp-Finite Element Methods for Singular Perturbations (2002)
Vol. 1797: B. Schmidt, Characters and Cyclotomic Fields in Finite Geometry (2002)
Vol. 1798: W.M. Oliva, Geometric Mechanics (2002)
Vol. 1799: H. Pajot, Analytic Capacity, Rectifiability, Menger Curvature and the Cauchy Integral (2002)
Vol. 1800: O. Gabber, L. Ramero, Almost Ring Theory (2003)
Vol. 1801: J. Azéma, M. Émery, M. Ledoux, M. Yor (Eds.), Séminaire de Probabilités XXXVI (2003)
Vol. 1802: V. Capasso, E. Merzbach, B.G. Ivanoff, M. Dozzi, R. Dalang, T. Mountford, Topics in Spatial Stochastic Processes. Martina Franca, Italy 2001. Editor: E. Merzbach (2003)
Vol. 1803: G. Dolzmann, Variational Methods for Crystalline Microstructure – Analysis and Computation (2003)
Vol. 1804: I. Cherednik, Ya. Markov, R. Howe, G. Lusztig, Iwahori-Hecke Algebras and their Representation Theory. Martina Franca, Italy 1999. Editors: V. Baldoni, D. Barbasch (2003)
Vol. 1805: F. Caó, Geometric Curve Evolution and Image Processing (2003)
Vol. 1806: H. Broer, I. Hoveijn. G. Lunther, G. Vegter, Bifurcations in Hamiltonian Systems. Computing Singularities by Gröbner Bases (2003)
Vol. 1807: V. D. Milman, G. Schechtman (Eds.), Geometric Aspects of Functional Analysis. Israel Seminar 2000-2002 (2003)
Vol. 1808: W. Schindler, Measures with Symmetry Properties (2003)
Vol. 1809: O. Steinbach, Stability Estimates for Hybrid Coupled Domain Decomposition Methods (2003)
Vol. 1810: J. Wengenroth, Derived Functors in Functional Analysis (2003)
Vol. 1811: J. Stevens, Deformations of Singularities (2003)
Vol. 1812: L. Ambrosio, K. Deckelnick, G. Dziuk, M. Mimura, V. A. Solonnikov, H. M. Soner, Mathematical Aspects of Evolving Interfaces. Madeira, Funchal, Portugal 2000. Editors: P. Colli, J. F. Rodrigues (2003)
Vol. 1813: L. Ambrosio, L. A. Caffarelli, Y. Brenier, G. Buttazzo, C. Villani, Optimal Transportation and its Applications. Martina Franca, Italy 2001. Editors: L. A. Caffarelli, S. Salsa (2003)
Vol. 1814: P. Bank, F. Baudoin, H. Föllmer, L.C.G. Rogers, M. Soner, N. Touzi, Paris-Princeton Lectures on Mathematical Finance 2002 (2003)
Vol. 1815: A. M. Vershik (Ed.), Asymptotic Combinatorics with Applications to Mathematical Physics. St. Petersburg, Russia 2001 (2003)
Vol. 1816: S. Albeverio, W. Schachermayer, M. Talagrand, Lectures on Probability Theory and Statistics. Ecole d'Eté de Probabilités de Saint-Flour XXX-2000. Editor: P. Bernard (2003)
Vol. 1817: E. Koelink, W. Van Assche(Eds.), Orthogonal Polynomials and Special Functions. Leuven 2002 (2003)
Vol. 1818: M. Bildhauer, Convex Variational Problems with Linear, nearly Linear and/or Anisotropic Growth Conditions (2003)
Vol. 1819: D. Masser, Yu. V. Nesterenko, H. P. Schlickewei, W. M. Schmidt, M. Waldschmidt, Diophantine Approximation. Cetraro, Italy 2000. Editors: F. Amoroso, U. Zannier (2003)
Vol. 1820: F. Hiai, H. Kosaki, Means of Hilbert Space Operators (2003)
Vol. 1821: S. Teufel, Adiabatic Perturbation Theory in Quantum Dynamics (2003)
Vol. 1822: S.-N. Chow, R. Conti, R. Johnson, J. Mallet-Paret, R. Nussbaum, Dynamical Systems. Cetraro, Italy 2000. Editors: J. W. Macki, P. Zecca (2003)
Vol. 1823: A. M. Anile, W. Allegretto, C. Ringhofer, Mathematical Problems in Semiconductor Physics. Cetraro, Italy 1998. Editor: A. M. Anile (2003)
Vol. 1824: J. A. Navarro González, J. B. Sancho de Salas, $\mathscr{C}^\infty$ – Differentiable Spaces (2003)
Vol. 1825: J. H. Bramble, A. Cohen, W. Dahmen, Multiscale Problems and Methods in Numerical Simulations, Martina Franca, Italy 2001. Editor: C. Canuto (2003)
Vol. 1826: K. Dohmen, Improved Bonferroni Inequalities via Abstract Tubes. Inequalities and Identities of Inclusion-Exclusion Type. VIII, 113 p, 2003.
Vol. 1827: K. M. Pilgrim, Combinations of Complex Dynamical Systems. IX, 118 p, 2003.
Vol. 1828: D. J. Green, Gröbner Bases and the Computation of Group Cohomology. XII, 138 p, 2003.
Vol. 1829: E. Altman, B. Gaujal, A. Hordijk, Discrete-Event Control of Stochastic Networks: Multimodularity and Regularity. XIV, 313 p, 2003.
Vol. 1830: M. I. Gil', Operator Functions and Localization of Spectra. XIV, 256 p, 2003.
Vol. 1831: A. Connes, J. Cuntz, E. Guentner, N. Higson, J. E. Kaminker, Noncommutative Geometry, Martina Franca, Italy 2002. Editors: S. Doplicher, L. Longo (2004)
Vol. 1832: J. Azéma, M. Émery, M. Ledoux, M. Yor (Eds.), Séminaire de Probabilités XXXVII (2003)
Vol. 1833: D.-Q. Jiang, M. Qian, M.-P. Qian, Mathematical Theory of Nonequilibrium Steady States. On the Frontier of Probability and Dynamical Systems. IX, 280 p, 2004.
Vol. 1834: Yo. Yomdin, G. Comte, Tame Geometry with Application in Smooth Analysis. VIII, 186 p, 2004.
Vol. 1835: O.T. Izhboldin, B. Kahn, N.A. Karpenko, A. Vishik, Geometric Methods in the Algebraic Theory of Quadratic Forms. Summer School, Lens, 2000. Editor: J.-P. Tignol (2004)
Vol. 1836: C. Năstăsescu, F. Van Oystaeyen, Methods of Graded Rings. XIII, 304 p, 2004.
Vol. 1837: S. Tavaré, O. Zeitouni, Lectures on Probability Theory and Statistics. Ecole d'Eté de Probabilités de Saint-Flour XXXI-2001. Editor: J. Picard (2004)
Vol. 1838: A.J. Ganesh, N.W. O'Connell, D.J. Wischik, Big Queues. XII, 254 p, 2004.
Vol. 1839: R. Gohm, Noncommutative Stationary Processes. VIII, 170 p, 2004.

Vol. 1840: B. Tsirelson, W. Werner, Lectures on Probability Theory and Statistics. Ecole d'Eté de Probabilités de Saint-Flour XXXII-2002. Editor: J. Picard (2004)
Vol. 1841: W. Reichel, Uniqueness Theorems for Variational Problems by the Method of Transformation Groups (2004)
Vol. 1842: T. Johnsen, A.L. Knutsen, K3 Projective Models in Scrolls (2004)
Vol. 1843: B. Jefferies, Spectral Properties of Noncommuting Operators (2004)
Vol. 1844: K.F. Siburg, The Principle of Least Action in Geometry and Dynamics (2004)
Vol. 1845: Min Ho Lee, Mixed Automorphic Forms, Torus Bundles, and Jacobi Forms (2004)
Vol. 1846: H. Ammari, H. Kang, Reconstruction of Small Inhomogeneities from Boundary Measurements (2004)
Vol. 1847: T.R. Bielecki, T. Björk, M. Jeanblanc, M. Rutkowski, J.A. Scheinkman, W. Xiong, Paris-Princeton Lectures on Mathematical Finance 2003 (2004)
Vol. 1848: M. Abate, J. E. Fornaess, X. Huang, J. P. Rosay, A. Tumanov, Real Methods in Complex and CR Geometry, Martina Franca, Italy 2002. Editors: D. Zaitsev, G. Zampieri (2004)
Vol. 1849: Martin L. Brown, Heegner Modules and Elliptic Curves (2004)
Vol. 1850: V. D. Milman, G. Schechtman (Eds.), Geometric Aspects of Functional Analysis. Israel Seminar 2002-2003 (2004)
Vol. 1851: O. Catoni, Statistical Learning Theory and Stochastic Optimization (2004)
Vol. 1852: A.S. Kechris, B.D. Miller, Topics in Orbit Equivalence (2004)
Vol. 1853: Ch. Favre, M. Jonsson, The Valuative Tree (2004)
Vol. 1854: O. Saeki, Topology of Singular Fibers of Differential Maps (2004)
Vol. 1855: G. Da Prato, P.C. Kunstmann, I. Lasiecka, A. Lunardi, R. Schnaubelt, L. Weis, Functional Analytic Methods for Evolution Equations. Editors: M. Iannelli, R. Nagel, S. Piazzera (2004)
Vol. 1856: K. Back, T.R. Bielecki, C. Hipp, S. Peng, W. Schachermayer, Stochastic Methods in Finance, Bressanone/Brixen, Italy, 2003. Editors: M. Fritelli, W. Runggaldier (2004)
Vol. 1857: M. Émery, M. Ledoux, M. Yor (Eds.), Séminaire de Probabilités XXXVIII (2005)
Vol. 1858: A.S. Cherny, H.-J. Engelbert, Singular Stochastic Differential Equations (2005)
Vol. 1859: E. Letellier, Fourier Transforms of Invariant Functions on Finite Reductive Lie Algebras (2005)
Vol. 1860: A. Borisyuk, G.B. Ermentrout, A. Friedman, D. Terman, Tutorials in Mathematical Biosciences I. Mathematical Neurosciences (2005)
Vol. 1861: G. Benettin, J. Henrard, S. Kuksin, Hamiltonian Dynamics – Theory and Applications, Cetraro, Italy, 1999. Editor: A. Giorgilli (2005)
Vol. 1862: B. Helffer, F. Nier, Hypoelliptic Estimates and Spectral Theory for Fokker-Planck Operators and Witten Laplacians (2005)
Vol. 1863: H. Fürh, Abstract Harmonic Analysis of Continuous Wavelet Transforms (2005)
Vol. 1864: K. Efstathiou, Metamorphoses of Hamiltonian Systems with Symmetries (2005)
Vol. 1865: D. Applebaum, B.V. R. Bhat, J. Kustermans, J. M. Lindsay, Quantum Independent Increment Processes I. From Classical Probability to Quantum Stochastic Calculus. Editors: M. Schürmann, U. Franz (2005)
Vol. 1866: O.E. Barndorff-Nielsen, U. Franz, R. Gohm, B. Kümmerer, S. Thorbjønsen, Quantum Independent Increment Processes II. Structure of Quantum Lévy Processes, Classical Probability, and Physics. Editors: M. Schürmann, U. Franz, (2005)
Vol. 1867: J. Sneyd (Ed.), Tutorials in Mathematical Biosciences II. Mathematical Modeling of Calcium Dynamics and Signal Transduction. (2005)
Vol. 1868: J. Jorgenson, S. Lang, $Pos_n(R)$ and Eisenstein Sereies. (2005)
Vol. 1869: A. Dembo, T. Funaki, Lectures on Probability Theory and Statistics. Ecole d'Eté de Probabilités de Saint-Flour XXXIII-2003. Editor: J. Picard (2005)
Vol. 1870: V.I. Gurariy, W. Lusky, Geometry of Müntz Spaces and Related Questions. (2005)
Vol. 1871: P. Constantin, G. Gallavotti, A.V. Kazhikhov, Y. Meyer, S. Ukai, Mathematical Foundation of Turbulent Viscous Flows, Martina Franca, Italy, 2003. Editors: M. Cannone, T. Miyakawa (2006)
Vol. 1872: A. Friedman (Ed.), Tutorials in Mathematical Biosciences III. Cell Cycle, Proliferation, and Cancer (2006)
Vol. 1873: R. Mansuy, M. Yor, Random Times and Enlargements of Filtrations in a Brownian Setting (2006)
Vol. 1874: M. Yor, M. Émery (Eds.), In Memoriam Paul-André Meyer - Séminaire de Probabilités XXXIX (2006)
Vol. 1875: J. Pitman, Combinatorial Stochastic Processes. Ecole d'Eté de Probabilités de Saint-Flour XXXII-2002. Editor: J. Picard (2006)
Vol. 1876: H. Herrlich, Axiom of Choice (2006)
Vol. 1877: J. Steuding, Value Distributions of *L*-Functions (2006)
Vol. 1878: R. Cerf, The Wulff Crystal in Ising and Percolation Models, Ecole d'Eté de Probabilités de Saint-Flour XXXIV-2004. Editor: Jean Picard (2006)
Vol. 1879: G. Slade, The Lace Expansion and its Applications, Ecole d'Eté de Probabilités de Saint-Flour XXXIV-2004. Editor: Jean Picard (2006)
Vol. 1880: S. Attal, A. Joye, C.-A. Pillet, Open Quantum Systems I, The Hamiltonian Approach (2006)
Vol. 1881: S. Attal, A. Joye, C.-A. Pillet, Open Quantum Systems II, The Markovian Approach (2006)
Vol. 1882: S. Attal, A. Joye, C.-A. Pillet, Open Quantum Systems III, Recent Developments (2006)
Vol. 1883: W. Van Assche, F. Marcellàn (Eds.), Orthogonal Polynomials and Special Functions, Computation and Application (2006)
Vol. 1884: N. Hayashi, E.I. Kaikina, P.I. Naumkin, I.A. Shishmarev, Asymptotics for Dissipative Nonlinear Equations (2006)
Vol. 1885: A. Telcs, The Art of Random Walks (2006)

Recent Reprints and New Editions

Vol. 1471: M. Courtieu, A.A. Panchishkin, Non-Archimedean L-Functions and Arithmetical Siegel Modular Forms. – Second Edition (2003)
Vol. 1618: G. Pisier, Similarity Problems and Completely Bounded Maps. 1995 – Second, Expanded Edition (2001)
Vol. 1629: J.D. Moore, Lectures on Seiberg-Witten Invariants. 1997 – Second Edition (2001)
Vol. 1638: P. Vanhaecke, Integrable Systems in the realm of Algebraic Geometry. 1996 – Second Edition (2001)
Vol. 1702: J. Ma, J. Yong, Forward-Backward Stochastic Differential Equations and their Applications. 1999. – Corrected 3rd printing (2005)